CAMBRIDGE LIBRARY COLLECTION

Books of enduring scholarly value

Technology

The focus of this series is engineering, broadly construed. It covers technological innovation from a range of periods and cultures, but centres on the technological achievements of the industrial era in the West, particularly in the nineteenth century, as understood by their contemporaries. Infrastructure is one major focus, covering the building of railways and canals, bridges and tunnels, land drainage, the laying of submarine cables, and the construction of docks and lighthouses. Other key topics include developments in industrial and manufacturing fields such as mining technology, the production of iron and steel, the use of steam power, and chemical processes such as photography and textile dyes.

Motor Vehicles and Motors

'Cheap or rapid or convenient road transport for man and goods is one of the most important of all contributions to national comfort and prosperity.' An early evangelist for the automobile, William Worby Beaumont (1848–1929) drew on his engineering background to produce the first volume of this work in 1900, when motor vehicles were still relatively new to British roads. Rapid developments in the automotive industry prompted the publication of a second volume in 1906. Replete with technical drawings and photographs, the work describes in great detail the design, construction and operation of the earliest motor vehicles, including those powered by steam, electricity and fuels derived from oil. Volume 2 describes the advances made both in the technological development of automobiles and in the volume produced. Detailed descriptions and illustrations are provided for the leading examples of the time from manufacturers such as Renault, Cadillac, Daimler and Wolseley.

Motor Vehicles
and Motors

*Their Design, Construction
and Working by Steam, Oil and Electricity*

VOLUME 2

W. WORBY BEAUMONT

CAMBRIDGE
UNIVERSITY PRESS

CAMBRIDGE
UNIVERSITY PRESS

University Printing House, Cambridge, CB2 8BS, United Kingdom

Cambridge University Press is part of the University of Cambridge.

It furthers the University's mission by disseminating knowledge in the pursuit of
education, learning and research at the highest international levels of excellence.

www.cambridge.org
Information on this title: www.cambridge.org/9781108070614

© in this compilation Cambridge University Press 2014

This edition first published 1906
This digitally printed version 2014

ISBN 978-1-108-07061-4 Paperback

This book reproduces the text of the original edition. The content and language reflect
the beliefs, practices and terminology of their time, and have not been updated.

Cambridge University Press wishes to make clear that the book, unless originally published
by Cambridge, is not being republished by, in association or collaboration with,
or with the endorsement or approval of, the original publisher or its successors in title.

MOTOR VEHICLES AND MOTORS

MOTOR VEHICLES
AND MOTORS

THEIR DESIGN CONSTRUCTION AND WORKING BY STEAM OIL AND ELECTRICITY

BY W. WORBY BEAUMONT

Member of the Institution of Civil Engineers, Member of the Institution of Mechanical Engineers,
and Member of the Institution of Electrical Engineers ; Hon. Consulting Engineer
to the Automobile Club of Great Britain and Ireland ; Consulting
Technical Adviser to the Commissioner of Metropolitan
Police, etc.

VOLUME II

LONDON
ARCHIBALD CONSTABLE & COMPANY Ltd
16 JAMES STREET
1906

Butler & Tanner,
The Selwood Printing Works,
Frome, and London.

PREFACE TO VOLUME II

WHEN the Preface to the first volume of this work was written in 1900 it was thought that if its reception proved a second edition to be required, some supplementary chapters and engravings would enable me to present to the reader a sufficient account of the advance likely to be made in design and construction of motor vehicles. The favour, however, and, I might add, indulgence with which the book was received by the public and the Press made a second edition necessary more than two years ago. It was then found that the space which the supplementary matter and illustrations, which even a general survey of the improvements made in two years would require, would occupy, not the space of a supplement, but a new volume.

The progress made in the past four years in motor vehicle construction has been phenomenal, and more remarkable than that in any other mechanical engineering industry ever established. The value of the output in Great Britain alone exceeded two millions sterling in the past year, and our imports amounted to about the same sum. The British output in the year—September, 1905, to September, 1906—will probably be little, if at all, short of £4,000,000. My anticipations, as expressed in lectures in 1895-6, have been more than realized, though then received with polite incredulity. When it is remembered that of the 18,500,000 of the population actively engaged in various occupations in the United Kingdom, 1 in every 12·3 is engaged in conveyance services of one kind or another, it will be readily understood that an immense field is open for new mechanical means of road transport.

All the essential elements of design, described and illustrated in the first volume, and the principles of construction, remain the same now in 1905 as in March, 1900 ; but the details of design and utilization of the results of experience have made such great strides towards the fulfilment of general motor vehicle requirements, that this second volume is necessary ; partly from the illustrative and descriptive point of view, and partly in order that the results of experience in many matters may be placed before the reader in a way that may be useful in practice. There are also numerous questions of a practical-theoretical character with which it has been found necessary to deal. Some of these arise out of experience in the new directions of development, and some of them are required in order that some principles of design and construction may

v

PREFACE

be presented in a simple practical manner, as applied in motor vehicle considerations.

Of this kind are the chapters on turning effort and sequence of cycles in motors, design and strength of crankshafts and axles, on radiators and water-cooling requirements ; power speed and tractive effort, carburettors and carburation ; engine dimensions, overturning and skidding efforts and stresses ; transmission efficiency, and on the cost of motor transport.

The drawings for this volume have been more than three years in preparation, and cover the period 1900–1905. It is impossible to include the most recent changes in general design, or detail ; but those herein presented may, it is hoped, prove sufficiently illustrative or indicative of the change and the trend of design to make this volume as acceptable to the constructor and to the automobilist as the first volume proved to be.

It has not been found necessary in this volume to deal with several of the subjects taken up in the first ; others are again dealt with where experience and experiment have either modified opinions, or have introduced new practical applications of principles previously well understood.

Those interested in motor vehicles have now an illustrated press with contributors, probably a hundred to one that existed at the time of the preparation of my first volume. These illustrated journals show the character of the changes from week to week, and show how great is the industry which only ten years ago was struggling to be allowed to exist. To two of these journals—the *Autocar* and the *Automotor Journal* —frequent reference is herein made.

As it is impossible or inaccurate to treat some kinds of vehicles as of more importance than others, or to make any rigid classification of types, except as dictated by the employment of either internal combustion engines, steam engines, or electric motors, no particular order of presentment of different cars is herein attempted. It has been common to classify vehicles in the trials of the Automobile Club by their prices and seating capacity. This classification cannot be observed here, as it is inconvenient to separate the different sizes of similar designs or of the same make. Any apparent want of sequence in this respect must be assumed to be rendered unimportant by reference to the index.

I have to crave the indulgence of reader and critic, as nothing but the fact that I am committed to the realization of the promise made in the Preface to the second edition of Volume I. would have induced me to add the labours of this one to occupations already sufficiently engrossing.

I have to acknowledge the assistance of several of those whose vehicles are herein described, and more particularly the assistance of my son, Mr. E. G. E. Beaumont, Assoc.M.Inst.C.E.

W. W. B.

Contents

CONTENTS

viii

List of Illustrations

ix

LIST OF ILLUSTRATIONS

LIST OF ILLUSTRATIONS

LIST OF ILLUSTRATIONS

LIST OF ILLUSTRATIONS

xiii

LIST OF ILLUSTRATIONS

LIST OF ILLUSTRATIONS

XV

LIST OF ILLUSTRATIONS

Introduction

THE past two years have been marked by a greater activity in motor vehicle manufacture and use than most people expected, and perhaps by nearly as much as could be hoped for by those who properly realized the inevitable results of the triumphs of the years 1895-6.

Industrial activity

The great spread of the area of strenuous emulation to the United States, Belgium, and over larger parts of France and of Germany and Italy, has given rise to a rapidity of development that could not take place with a costly thing unless really wanted.

The pleasure vehicle has occupied by far the most attention, and the petrol motor car has received, for various reasons, by far the greatest support—the great races and the great feats of long journey endurance have been performed by petrol cars.

Petrol

The steam car has also received unstinted inventive attention, and great feats have been performed by some of them, while at least two types of car have renewed confidence in the possible practical value of this form of motor. For the propulsion of the heavier cars, steam has had but one serious rival, and at the present time the continuance of that rivalry is watched with the interest which is proportional to the importance of the result. In the opinion, however, of some of the best informed, and of those capable of estimating the value of apparent success of one type and of partial failure of another, the internal combustion motor is the more likely to reach commercial success in the end—for the largest field of employment, though not for all purposes.

Steam

Many things have been done with electrical vehicles in the past two years, and much practical success has been achieved. Commercial success in some respects has been attained, though it be in a field in which the user has been in a position to esteem convenience, ease, and comfort as of higher importance than lowness of cost per mile run, or distance covered.

Electrical

The distinguishing feature of the development of the automobile since the appearance of the first volume of this book is the enormous increase in the power with which it is now possible to furnish a vehicle weighing not more than a ton. This has been the outcome of the construction of the voiturette and the racing car. The most powerful cars at the beginning of 1900 were the Mors 12 HP. racing car, Panhard & Levassor 16 HP. racing cars, Peugeot 18 HP., and Cannstadt 24 HP., the greatest average speed

Increase of power

INTRODUCTION

in March, 1900, over the Nice racecourse of 150 miles being then 36·6 miles by a Panhard & Levassor, but 37 miles average was reached by a Perfecta tricycle, carrying no less than 4½ declared HP. In the same races a Perfecta quadricycle carried 6 HP.

The Renault light car, with a De Dion-Bouton motor of only 2¼ nominal HP., had already appeared, and had reached an average speed of 23·75 miles per hour in a run of 64·62 miles from Paris to Rambouillet. Other light cars then designed with the object of meeting the demand for a car to carry two at moderate cost were the Decauville, the Mors, the Peugeot, the Panhard (Krebs) light car, and the De Dion-Bouton voiturette, the last with a 3 HP. motor. There were also the beginnings in this direction made in England with the Daimler light belt-driven car with a 6 HP. motor, the same power as that which only a little more than a year before was employed in the Panhard & Levassor racing cars.

Horse power These light cars were soon made to carry larger and larger engines, modifications being of course made in the change-speed gear. Thus did the racing tricycle, the quadricycle, and the voiturette open the eyes of designers to the extraordinary capabilities of light but well made frames and gearing, and the possibilities of the quadruple cylinder engine for driving them. The racing car of to-day, with a body carrying only two seats, has from 60 to 120 HP. engines, and has reached on the road a speed of over 80 miles per hour.

There had, of course, all this time been the Benz light cars, but their design did not lend itself to the enormous increase of engine power possible with the others.

As I write, the Automobile Club of Great Britain and Ireland is busy in the preparations for the race of the year, namely, that for the Gordon-Bennet cup, which was won in 1902 by a Napier car, driven by Mr. S. F. Edge. The entries for this race include cars from France, Germany, Italy, Belgium, Austria, Switzerland, the United States, and England, some of them having engines declared by the makers to be able to give off 100 B.HP., and all of them being only 1,000 kilogrammes, or under one ton in weight.

These facts serve to show how great has been the development of the car for pleasure and racing purposes, and how great the possibilities in the application of the experience gained in the construction of vehicles for other purposes.

These other purposes include the passenger car, cab, brougham, victoria, omnibus, and the like, for everyday purposes, vehicles which must be trustworthy for all occasions.

Next, and even more numerous, are the vehicles, large and small, for hundreds of different business purposes, the leading qualification of which must be trustworthiness, almost every other necessary qualification being either secondary to this, or contributing to it.

Commercial vehicles Further, there are the heavy vehicles for loads of over a ton and up to, say, four or five tons. Even for many of these the experience gained

2

INTRODUCTION

in the development of the racing car has been of great value ; but for the very heavy work done by the steam wagon or lorry, the light steam vehicle experience affords less help, although even here the modern light high-speed engine, high quality gear construction, and boiler pipe and cooler construction are pointers which may, and even have been regarded as of the greatest importance by those who are not incapable of seeing that traction engine and railway locomotive experience may be most usefully supplemented by that afforded by the modern types of motor wagon.

It is not, however, probable that any very long continued commercial success will attend the use of the very heavy motor wagon for conducting a heavy regular goods transport service between any two or more places. The heavy or medium heavy load vehicle will be of great and increasing value, but the heavy motor wagon, with or without its trailer, running anything like regular services dealing with daily consignments or large total weights, cannot be a commerical success on common roads. Where such conditions of large transport trade along definite and constantly traversed routes obtain, then the conditions are those requiring railways, tramways, or tram-plate ways. From some great trading centres to places scattered all round the centre and within such distances as are short relatively to the collection and delivery distances, the motor wagon will do useful and paying work. Even for this, however, the tram-plate way is necessary for that part of the road which is constantly traversed, and in many cases this might, and should be, on a new road, not the common highway.

Heavy vehicles and frequency of service

The motor wagon for loads of up to, say, three tons will, however, have a very wide field of usefulness in connexion with numerous trades and manufactures, the goods or products of which require distribution in all directions and to consumers' premises.

The vehicle for moderate weight and moderate or even low speed will be the commercially successful vehicle, and the traction engine, with its slow-speed engine, heavy-type boiler, enormous and heavy gear, and its train of wagons, is not the precursor of this type of vehicle.

The things which have led to the developments at first mentioned, the developments themselves as shown by structural detail, and some indications as to the trend of improvements and invention, will form the subjects of this volume, in which descriptions will be illustrated as far as possible by engravings, which, through various circumstances which will be readily understood, must be a few months behind the maker of the things illustrated.

Attention will be directed to some of those essential elements in the organism of the automobile which have undergone most improvement or have been the subject of most investigation or experiment.

3

Chapter I

LIGHT PETROL MOTOR VEHICLES

IN chapter xiv. of vol. i. several different types of the light petrol vehicle were described, the first being the little Decauville car. This was a clever car, and was fully illustrated. Its remarkable handiness was well displayed in numerous trials and gymkhanas, but it was of a type which has not survived, though the influence of its design is still traceable. It was fitted with an air-cooled motor and with a form of change-speed gear which was defective as to the size of its smaller wheels and pinions, as to the mode **Decauville** of shifting them (pp. 218 to 223), which involved passing one pinion through the teeth of two wheels to make one change, and as to the absence of all protection of motor and gear from dust and mud. It was consequently very noisy, and the small gear wore quickly. The car was, however, one of the first driven by a bevel wheel on a live axle, was light and easy running and not costly. Modern practice covers all such gears, and uses water cooling for the motor instead of air cooling.

The Mors light car, illustrated in the same chapter, was of a type which has not survived, except with considerable modification, the greatest change being the use of a vertical engine in place of the horizontal engine **Mors** then used, and of much higher power. The 4 HP. is no longer made, and all the Mors cars now belong to another class, which will be dealt with hereafter.

The Daimler light car, in which the gear was belt driven and carried by an independent short bogie frame, on which the car body was supported by inverted leaf springs, has not survived, and no sufficient efforts **Daimler** appear to have been made to develop the leading ideas embodied in it. It was the first car in which the belt was given a good length, and a comparatively small modification would have converted it into a useful single belt car.

Belts have, however, fallen into disuse, although a few cars are still driven by them, including some of the Benz, the Pick light car, some of the Delahaye, and some other simple light cars. The belt was seldom used properly, generally too short and too narrow, but it is a simple, flexible and quiet drive, and is very likely to be revived for some kinds of light vehicles.

4

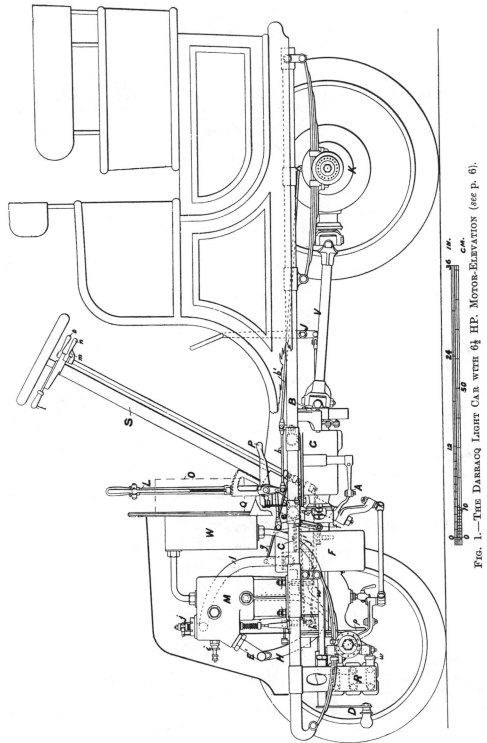

Fig. 1.—The Darracq Light Car with 6½ HP. Motor-Elevation (*see* p. 6).

MOTOR VEHICLES AND MOTORS

The Darracq-Bollee (p. 233) was another of the light cars exhibited in Paris in 1899–1900. It was fitted with a pair of five-step coned pulleys, with an ingeniously arranged belt shifter for changing speeds. The gear driven by the driven belt pulley was enclosed, and was only for a single and final speed reduction. The belt arrangement might have survived for the purposes of some light vehicles, but the car was fitted with a single cylinder air cooled motor, and the exigencies of design put the fly-wheel outside the frame. The car soon ceased to be made, and was replaced by a light car entirely different and of novel design, gear driven throughout, and of remarkably light construction. Illustrations of this car will be found on pages 6 to 13.

Renault The Renault light car (p. 237) was the forerunner of a type which has been and is followed on a very large scale. It is still in general principles of design of transmission gear the same now as in 1900, a fact which speaks highly for the designer.

The Accles-Turrell light belt-driven car (p. 239) has ceased to be made, but as a heavier chain-driven car it is now made from new designs.

The Peugeot light car (p. 246), with horizontal engine with valve gear driven by a tongue in a snitch groove in the fly-wheel, has also, as might be expected, given place to a different design with vertical engine of the De Dion-Bouton high-speed type placed in the front of the car.

Another type of light car, an elegant little car, was the Panhard & Levassor, Krebs' 4 HP. car (p. 248), in which a Krebs motor and gear were placed under the seat and the steering wheels were on a through axle, pivoted by a spring across it at its centre. A car similar to this was offered by one firm in England during 1902.

De Dion The De Dion-Bouton voiturette (p. 255), like all the others, has grown in power from 3 to 6 or 8 HP., and the all-prevailing front position of the motor has been adopted.

Such were the light cars dealt with in the book which was completed before March, 1900, or five years ago.

The changes in these will be shown as far as possible, and some of the reasons for them by the engravings now to be described.

THE DARRACQ LIGHT CAR

Although the Darracq light car which appeared in the latter part of 1900 was a complete reversal of the policy of the Darracq-Bollee of the year before, and was in many respects a great advance on most of the preceding designs, it nevertheless showed that the little Renault car had proved a pathmaker. The path was adopted, and the Darracq design paved the way for very much that has followed, not only in the chainless type of light and heavier car, but in careful attention to the minutiæ of a design made complete before the drawings were put into the works.

The Darracq car shown by Figs. 1 to 9 was the embodiment of a consistent attempt to combine satisfactory arrangement of essential organs

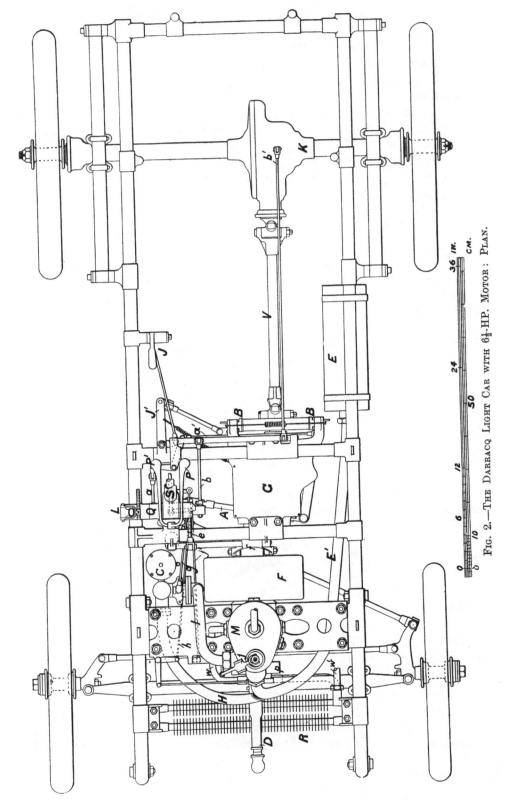

Fig. 2.—The Darracq Light Car with 6½-HP. Motor : Plan.

with extreme lightness of parts and workshop facility of construction. It was a great attempt toward a standardization which must be studied by all manufacturers who aim at low cost of production of a machine of high qualifications.

It in its turn has become a pointer, and although the first cars showed defects in several respects, they were mostly those which naturally resulted from a bold attempt at lightness, defects which the designer of the original could most easily remedy. Some of the defects were due to the use of materials which would not stand the test of the very severe work to which almost every part of a motor carriage must submit.

Design material wear

Beautifully made malleable iron castings were employed, but in the bevel wheel, driving wheel and case, and elsewhere, it was not, on the one hand, stiff enough, and on the other, even where case-hardened, not the material for the work, which requires good steel hardened. The differential gear was too light and too small; the bearing which carried the spindle of the driving bevel pinion was of insufficient length, and would have been better if a plain bearing; the pins of the universal joints of the coupling or tail-driving rod were too small, although mere calculation of the torque effort they had to transmit would have shown them to be of ample size. They wanted larger diameter merely to give larger surface, and this same reason for more ample dimensions was found in several details.

The drawings herein given of this car are interesting for these very reasons, and the success of the most recent of the Darracq designs has shown this conspicuously.

The car illustrated by Figs. 1 to 9 is that of the type purchased by Mr. Claude Johnson, and used by him in all weathers in travelling over the course of the Glasgow trials when he was organizing the arrangements and obtaining the road guide particulars for them. The arrangement of mechanism in this car is shown by Figs. 1 and 2, and details of the engine and transmission gear by Figs. 3 to 9. From Figs. 1 to 9 it will be seen that the chief items of the mechanism are attached to a rectangular underframe constructed of steel tubing, while the final transmission to the driving wheels is effected by means of a jointed tail shaft and bevel gearing on the live rear axle.

The motor M is mounted towards the front of the frame on a light pressed-steel base (see Figs. 2 and 4), and, together with the water tank, W and carburettor C, is enclosed within a light sheet-metal bonnet. As will be gathered from Figs. 3 and 4, the motor is of the single cylinder vertical

Engine

type, water-cooled and fitted with electric ignition. The cylinder is 3·94 in diam. (100mm.), the stroke the same, and the engine is rated at 6·5 B.HP. at 1,400 revs. per minute. The induction and exhaust valves A and B are superimposed, the former being automatic, and the latter mechanically operated by means of a cam C on a short spindle driven at half the speed of the crankshaft; this spindle at its exterior end also carries the make and break circuit switch G for the electric ignition.

The cooling water for the cylinder jacket is circulated by the gear-

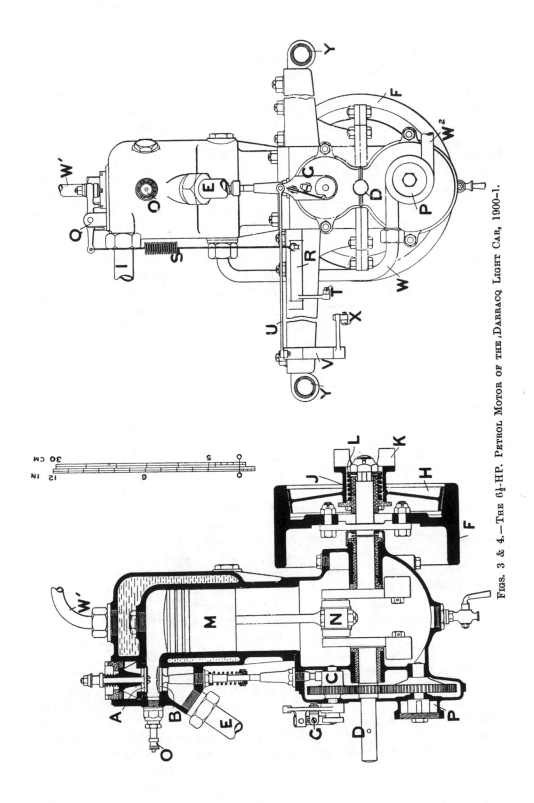

FIGS. 3 & 4.—THE 6½-HP. PETROL MOTOR OF THE ,DARRACQ LIGHT CAR, 1900-1.

9

driven centrifugal pump P, which sucks from the tank w, Fig. 1, by way of the radiator R, and pumps through the jacket and thence into the tank by pipe w¹. The speed and power of the motor may be adjusted by the

Engine control

driver by controlling the lift of the admission valve and by adjusting the point of ignition, these operations being effected by means of the hand-levers O and N on the steering pillar. The lift of the admission valve is decreased by increasing the tension of the spring s, Fig. 4. The mixture is taken from the carburettor C, which is of the float feed spraying type, by way of pipe I, Figs. 1 and 2, and the exhaust passes by pipe E, Fig. 4, and E′, Fig. 2, into the silencer E, which is attached to one of the main tubes of the underframe. The petrol supply is carried in the tank O, Fig. 1, mounted on the dashboard. The power is transmitted from motor to change-speed gear through the cone clutch H in the fly-wheel F, the cones being engaged by the spring L. The claw coupling K M, Figs. 3 and 6, transmits from the clutch to the first motion shaft U, Fig. 6, which may be engaged with the second motion shaft I by one of three sets of spur-wheels and pinions for forward driving, or a third pinion K, Fig. 6 may be engaged with the wheel and pinion of the slowest forward speed when

Transmission gearing

these are out of gear with each other, in order to impart a reverse motion to the car. As shown in Fig. 6, the slow-speed wheel and pinion are in gear for forward motion. By moving P a little way towards I this wheel and pinion are out of gear with each other but are connected with the wide pinion K, so that a reverse motion is imparted to P when K is forced into gear with them. The power is conveyed from the shaft I to the bevel gear B C, Fig. 8, by means of the tailshaft V, Figs. 1 and 2, which is provided with Hook joints at either end. The wheel C of the bevel gear is mounted on the box D containing the differential gear on the live axle F, and the axle runs in ball bearings carried in the tubes E, to which the spring plates H are attached, and are joined at the centre by the gear enclosing case K. The shafts of the change-speed gear and the bevel pinion B run in ball bearings.

The three forward speeds are engaged from the hand lever 1, Fig. 1, below the steering wheel, motion being conveyed through the link A (Figs. 1, 2, 5 and 6) and bell crank N to the sliding gear sleeve P, Fig. 6. The reverse pinion is brought into gear by means of the lever J, Figs. 1 and 2, and rocking arm J, Fig. 6. The vehicle is controlled by double acting brakes, which were at the time of the design among the best then made. It has since been flattered in the sincerest way. It is seen at B B on the end of the second motion shaft of the change gear (Figs. 1, 5 and 7). Its action is obvious. A coil band-brake I, Fig. 8, acting on the exterior of the differential gear box D, is put into action by the pedal P. The brake B B, which is shown in detail in Figs. 5 and 7, is operated from the hand-lever L,

Brakes

(Figs. 1 and 2). It consists of a drum, C, and two blocks, B B, pivoted on levers F F′, which are in turn pivoted at G G on extensions from the gear box. The levers carrying the blocks B are coupled by the link passing across from E to F, Fig. 5, and the lever E, which pivots on an extension of the pin of block B, is coupled to the bell crank D (Figs. 5 and 7), and thence to the lever L. The first action on pulling back the lever L is to disengage

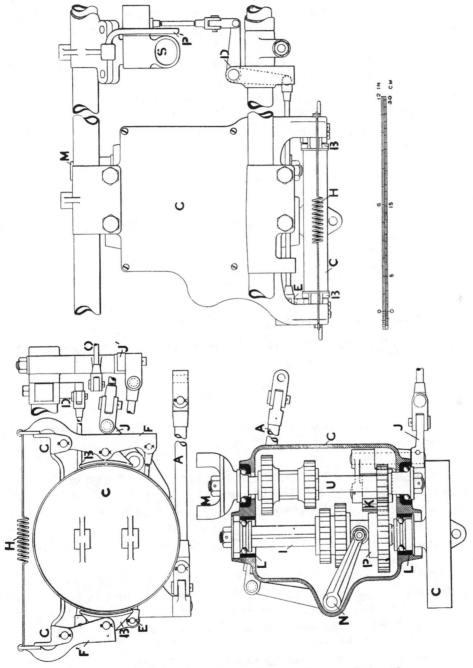

FIGS. 5, 6, 7.—CHANGE-SPEED GEAR OF DARRACQ LIGHT CAR.

the cone clutch H. Then tension is applied to the link D E, and the lever E pivoting on its pin pulls on the block B on lever F′, then the lever F pivots about G, and the other block B is applied. The blocks are normally held off the drum C by the spring H, Figs. 5 and 7.

The band-brake I on the differential is operated by the pedal P, which also disengages the clutch H. The clutch alone is controlled by pedal P'. The steering of the car is controlled direct from a hand-wheel without gearing.

Experience soon proved that the very short distance between the ball races of the bearing between A and B, Fig. 8, permitted the side thrust of the bevel gearing to throw a very heavy stress on the two ball races, and some of the cars which were used in this country were improved by taking out the ball bearings and putting in as long a solid bearing as could be got between pinion B and the joint piece A on a slightly lengthened spindle, the tail rod V being shortened for the purpose.

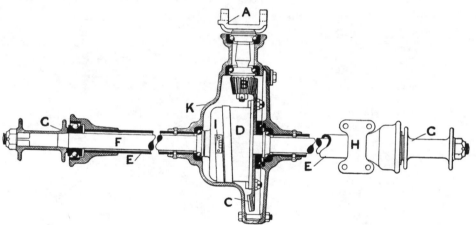

FIG. 8.—DRIVING GEAR, AXLE AND DIFFERENTIAL GEAR, AND BRAKE OF DARRACQ LIGHT CAR, 1900–1.

In the succeeding year M. Darracq improved the design of the whole of this part of the car by the arrangements shown by Fig. 9. The coil brake on the differential case was discarded, and a brake drum, D D, placed on each wheel. This brake arrangement was not only safer in the ordinary sense, but much better as against side slipping, which may be aggravated by differential gear action.

Improved design

The improvement of the bearing of the spindle carrying the pinion B is also seen to be very great, the spindle receiving support on each side of the pinion. Severe stresses on the axle and the whole of the driving gear case, in which the live axle carrying tubes are held, contributed to the rapid wear of the parts shown in Fig. 8. These stresses were avoided by the employment of the strong tubular stay K, Fig. 9. The remarkable lightness secured by the use of wonderful specimens of the malleable cast iron founders' work may be gathered from the drawing of the casing carrying the differential gear J in Fig. 9.

12

Many hundreds of these carriages have been made, but most of them have been the cars with the larger powers, the 9-HP. in particular. The motor for this power has a cylinder 3·54 in (90 mm.) diameter 3·94 in (100 mm.) stroke, and runs normally at 1,400 revs. per minute.

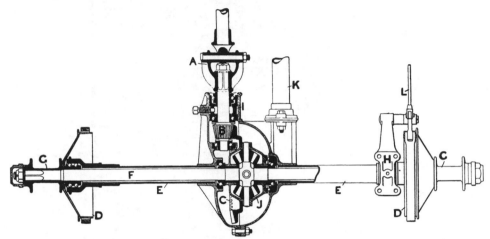

Fig. 9.—Driving Gear, Differential Gear, Axle and Brakes of Darracq Light Car, 1902.

The motors are known as the "Perfecta," and were certainly of remarkable power for their size when they first appeared five years ago.

During the past few years M. Darracq has made some very remarkable developments of his voiturettes and light cars, and like some other constructors has turned his attention to the racing machine. Two of his 20-HP. racers were shown at the Bexhill meet of the Automobile Club in May, 1902, and made some very high speeds.

They weighed only about 12 cwt., and one of them covered the measured kilometre with a flying start at the rate of 52 miles per hour. These were racing machines, but the Darracq Co. have made since then very great advances in every detail of design of frame, gear and motor, which place them in the front ranks of touring car makers.

The Renault Light Car.

The small Renault cars have undergone fewer changes than any of the others mentioned in the light or voiturette class in vol. i., a fact which speaks highly for the original design. The transmission gear, p. 237, was, I believe, the first to secure a through drive from clutch to driving axle when running at the highest speed. The motors employed by Renault are of the De Dion-Bouton make or type, such as that which will be found illustrated on page 30.

The general design of the 9-HP. car as now constructed will be understood from the engravings, Figs. 10 to 14, while a special form of carriage fitted with a motor of about 5-HP. is sufficiently shown by Fig. 15.

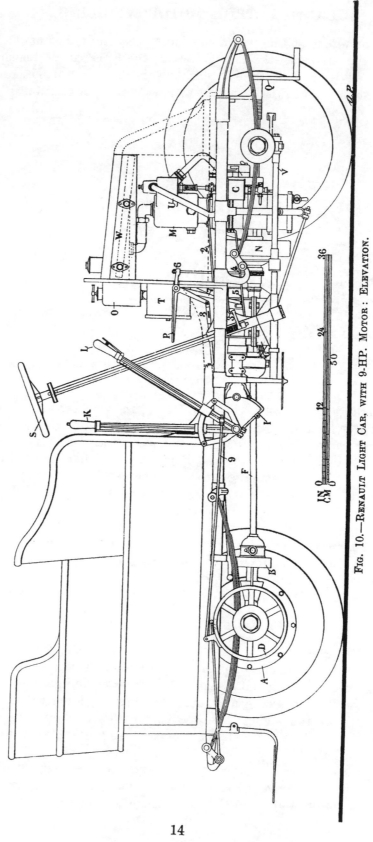

Fig. 10.—Renault Light Car, with 9-HP. Motor: Elevation.

LIGHT PETROL MOTOR VEHICLES

From Figs. 10 and 11 it will be seen that with the exception of the length of the frame, the use of straight frame side members, the longer wheel base, and the use of wheels of equal size, the general arrangement and main features are those illustrated at p. 237, vol. i. There are, however, numerous changes in the details. For instance, the car of 1900 was fitted with a 2¼ HP. De Dion motor; the smallest cars are now fitted with motors of double that power, and the car illustrated, though a light car, is again double that. The method of transmission from the engine to the driving axle is much the same as three years ago, but the cone clutch is, as shown in Fig. 13, so made that the thrust of the spring which keeps the male portion of the clutch in position for driving does not throw any end thrust on either crankshaft or tailshaft To accomplish this the clutch cone is turned in the opposite direction to that formerly used, so that for declutching it is forced into the fly-wheel, instead of being withdrawn from it. In Fig. 13 the fly-wheel F on the end of the crankshaft M is fitted with an internal ring which is put into place, and when the clutch cone C is in position, held there by bolts, one of which is shown. Between the surface C and the ring in the fly-wheel a leather cover on the former is employed as is usual. The spring by means of which the surfaces of the cones are forced together is seen at S. When the cone surfaces are engaged the spring is resisted by the two parts of the clutch only, but when declutching by means of the nut T and screw Q, when operated by the pedal P and the pitch chain at the end of the rod B, a ball thrust bearing as shown receives the thrust of the spring.

Change in details

Balanced clutch

This bearing is of course only in operation when the engine is running declutched, the thrust upon it being received by the nut T on the screw Q which is fixed to the gear box G.

The arrangement of the pedal P is seen in Figs. 10 and 11, the arm marked 6 on the pedal spindle being clearly seen in both cases.

A change has been made also in the method of manipulating the speed gear in the box G, this being done by a quadrant or sector in the box I called the mandolin, actuating a pinion and bevel gear, in the small chamber by the side of the gear box, the bevel pinions giving motion to a spindle, which operates the change-speed gear as already described at p. 237. vol. i.

This will be clearly understood by reference to Fig 14, in which the screw on the clutchshaft is seen at Q, the gear box G being shown in horizontal section. As in Figs 10 and 11, F is the jointed tail shaft which is in line with the clutchshaft S, on which is the sleeve A, on the end of which is a claw clutch B engaging with the wheel C. The shaft S is in two parts, the part within the wheel C and the bearing next it being tubular and the squared part being turned down to work therein. When the clutch B is engaged with C, the two parts of the shaft are locked together and run as one. None of the gearing is then in mesh, and there is a straight drive through from clutch to live axle.

Change-speed gear

To change speed to the first speed, for instance, the lever K is moved

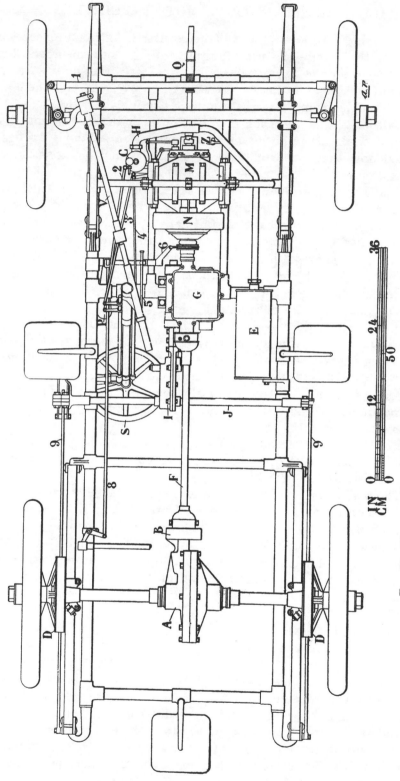

Fig. 11.—Renault Light Motor with 9-H.P. Motor: Plan of Underside.

16

operating the sector in the box I and thereby through the bevel wheel and pinion therein, rotating the spindle T upon which is the quick thread shown

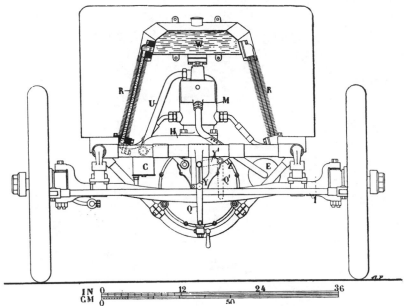

FIG. 12.—RENAULT LIGHT CAR: FRONT VIEW.

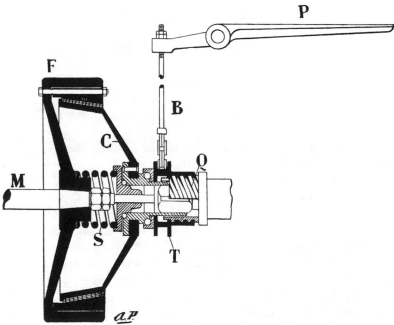

FIG. 13.—CLUTCH OF RENAULT CAR.

working in the nut U, carrying the fork V which shifts the sleeve A over towards the clutch so that the pinion D is opposite the wheel E. The gear-wheels E G H are on a sleeve loose upon a spindle, fixed upon the ends of which are excentrics N N. On one end of the same spindle is a pinion P, operated by the rack R shown below it, which is actuated by the disc cam X on the spindle T. The pinion D being then opposite E, the further movement of the sector at I gives motion to the excentric spindle N N and after the manner of lathe back gear throws the spindle towards the shaft S, and

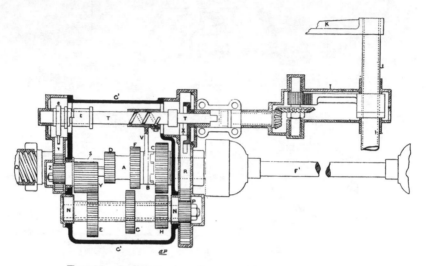

FIG. 14.—SPEED-CHANGE GEAR OF RENAULT 8-HP. CAR.

thereby puts wheel E and pinion D into mesh. The intermediate speed obtained by means of the wheels F and G are obtained in the same way. Reversal is effected by means of the pinion Y which is upon an excentric pin partially rotated by the pinion Z and rack r and an excentric at e, all the movements being effected, as will be readily understood, by the successive parts of a rotation of the spindle T, upon which is the clutch t which only comes into action at the later part of the movement of the sector and the divided spindle operated by it.

System of control   The pedal P also operates an arm on the pedal spindle, to which the rod 8 is connected, and thereby a brake at B. Thus, when the clutch is put out of engagement, the internal brake at B is brought by further depression of the pedal into action. At the same time through the lever 7 (see Fig. 10) the rod 5 is moved and it actuates through a pivoted lever the rod 4 (see Fig. 11) connected with the bell crank towards the front of the motor, which operates an exhaust lifter to be described hereafter in connexion with the De Dion light car. The bottom end of the arm 7 on the pedal spindle is simply an eye which moves freely on the rod 5 until the collar or head at the end of it is reached. The operation of declutching is thus followed by control of the exhaust and by putting on a brake at B.

There is another pedal near the steering pillar, but only partly seen in Fig. 11, by which the rod 4 for controlling the exhaust lifter can be operated independently of the clutch pedal, for controlling the car in traffic. Attached to the steering pillar are two rods with handles and short levers,

18

the one for operating the rod 2 which varies the air admission to the carburettor, and the other for operating the rod 3, which by means of a bell crank and connecting rod operates the contact maker disc and thereby the period of ignition.

The lever for operating the speed gear seen at κ is fixed upon the tube J, and a lever L for operating the brakes D D through the rods 9, is fixed upon one end of the solid spindle which passes through J. The steering is done by means of a hand-wheel s instead of by means of the bicycle handle as in the earlier car, and the direct steering by short arm and con-

Steering gear

Fig. 15.—Renault 5-HP. Motor Hansom.

necting rods is replaced by a skew pinion on the bottom end of the steering shaft, which gears into a rack sliding in the tube carried at the bottom of the steering pillar bracket and operating the rod v and thereby the connecting rod 1.

The carburettor c is of the De Dion float feed spray-making type. The induction pipe from carburettor to admission valve is seen at v. The end of the pipe H by which air is conveyed to the carburettor is splayed so as to form a wide embrace of the exhaust pipe for heating the air.

Carburettor

The jacket water is cooled by means of the radiator R R, consisting of gilled pipes connected at their upper ends to a water tank w, and at their lower ends to trunk pipes, resting on the frame, these trunks being connected to the bottom of the cylinder jacket as shown in Fig. 12. The upper part of the cylinder jacket is connected with the bottom of the water tank, and the hottest water rising there passes off in the two directions down

Circulation of jacket water

the radiator pipes, a natural circulation thus being obtained which is sufficient for the purpose.

The starting handle is seen at Q (Figs. 10, 11 and 12), where it engages directly with the end of the crankshaft, but is so arranged with separate links (Y and Y') as to be capable of taking the position Q' when not in use, and in which position it cannot rotate.

In the elevation (Fig. 10) a lubricator for the motor is shown at O, and the induction coil is shown at T.

It should be mentioned that in preparing the photograph for engravers' purposes the jointed tail rod F, shown in Figs. 10 and 11, has been omitted in Fig. 15, and that in the same figure the change speed lever is in the centre of the footboard as in the earlier cars.

Power and weight

Some remarkable developments of the car illustrated have been made by Renault Frères, in the production of small racing cars of great power for their weight. For instance, those cars which ran in the Paris Berlin and Paris Vienna Races, were very little or no heavier than that shown by Figs. 10 and 11, but the bodies consisted merely of two seats. They were really only racing machines—with two seats on them. The motors had cylinders 3·54 in. dia. (90 mm.) and 3·94 in. stroke (100 mm.) and they ran at anything between 800 and 1,500 revolutions per minute. The complete vehicle only weighed 900 lbs. Their performance as to mere speed was astonishing.

Speed

The machine that won the first prize in the voiturette class of the Paris-Berlin race in 1901 covered the 744 miles in 19 hours 16 minutes, or a mean speed of 38·6 miles per hour. Another of the 16 HP. racing machines was a winner in the Paris-Vienna Race of 870 miles which it covered in 25 hours 51 minutes, in competition with cars of very high powers, thus showing that beyond a certain maximum power no great gain need follow higher powers in a car carrying only two men.

The Renault Racing Cars, have, however, grown in power, and this year the car for the Auvergne race is of 72 HP., and the frame is suspended under the axles.

The Roadway Autocar Company of London have exhibited Renault cars of various sizes up to 14 HP. with four cylinders and tonneau body.

THE DE DION-BOUTON LIGHT CAR.

One of the most successful light cars referred to in vol. i. was the 3 HP. De Dion-Bouton voiturette (page 255), in the sense that it was the parent of the several thousands of light cars which have since been made on its general principles though with numerous and important modifications.

Universal joints

All of them are, however, made with the articulated axle-driving rods with rigid hollow axle as adopted many years ago in the De Dion-Bouton heavy steam vehicles (vol. i. p. 486).

The driving gear in the later cars has been similar to that described in vol. .i, but the front position of motor has been adopted, and three speeds and modified arrangement of reversing gear have become necessary. With these modifications and with a variation upon the Longuemare carburettor,

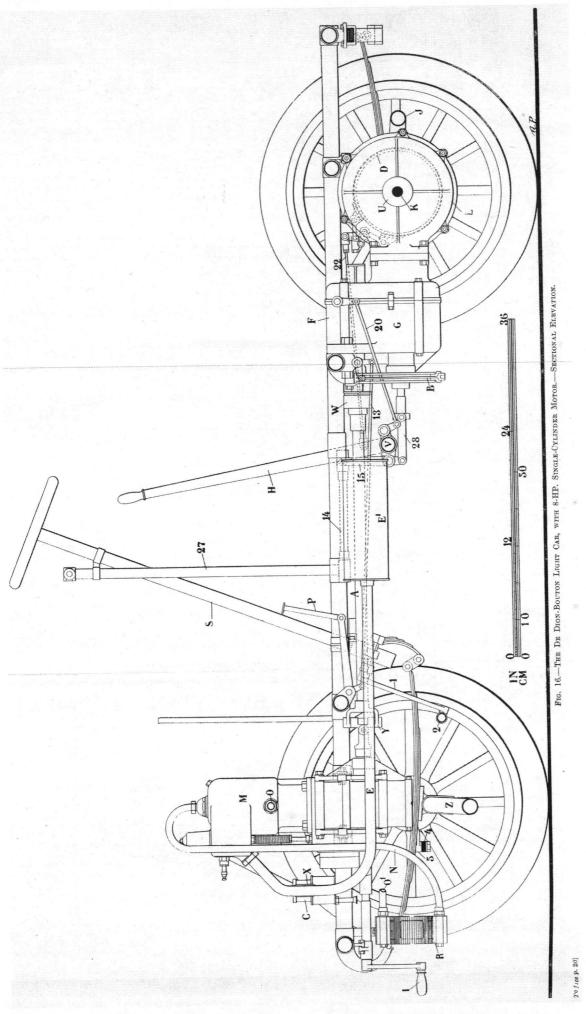

Fig. 16.—The De Dion-Bouton Light Car, with 8-HP. Single-Cylinder Motor.—Sectional Elevation.

To face p. 20]

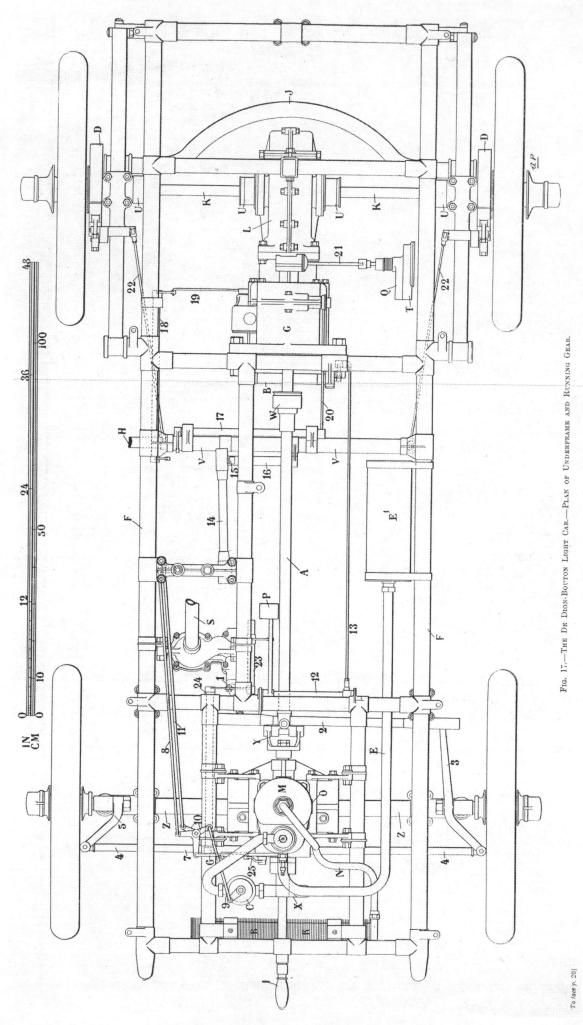

FIG. 17.—THE DE DION-BOUTON LIGHT CAR.—PLAN OF UNDERFRAME AND RUNNING GEAR.

To face p. 20]

the light car with an 8 HP. single cylinder motor as illustrated by Figs. 16 to 25, has become a standard form by these justly celebrated makers.

One great feature in the arrangement of this car is the complete flexibility or freedom of movement of every part that is likely to receive a twisting or racking stress due to the varying and contrary movements of the axles with relation to the frame, and to elastic flexure of the frame. **Freedom from straining**

The frame is made of stiff longitudinal straight tubes F, transversely connected by equally stiff tubes, all fixed together by sockets of different forms brazed to the tubes. The necessary fittings for spring hangers and lugs for different parts of the machinery and lugs for carrying the car body are similarly attached, every necessary in this respect being completely set out in the drawings. **Tubular frame**

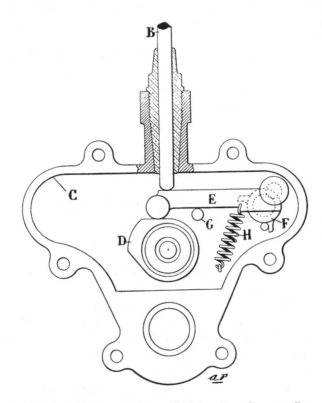

FIG. 18.—DE DION-BOUTON EXHAUST LIFT CONTROL DEVICE.

The motor M is of the type and design now so commonly known as the De Dion type, and made by so many manufacturers that it is unnecessary to introduce here an engraving to show its construction, this being in most respects identical with that shown by Figs. 27–30, p. 30, made by the Motor Manufacturing Company.

As compared with these, however, the motor like that already referred to, with reference to the Renault light car has this difference—that it is fitted with the exhaust lift and control device shown in diagram form by Fig. 18. In this D is the exhaust valve lifting cam on the half speed spindle, the valve spindle or pusher B being lifted by the intermediation of the rod E pivoted above F on a crank or excentric pin which may be caused to move from the position shown towards the spindle B so that the cam, in its rotation, will raise the valve later and drop it earlier than in the position illustrated. The piece F, with the pin on which E is pivoted, is operated by a lever under the control of the driver, its extreme positions being fixed by the two lugs shown, one of which is resting against a stop pin with E in its normal position. By means of this apparatus the speed of the motor may be varied through a sufficient range to meet most of the requirements of control in traffic. It is operated by hand lever on the steering column through a connexion at I, Fig. 17, and by the pedal P which bears on an arm on spindle 23 and connexion at I to spindle 24 and thence to lever 25. Continued depression of pedal P puts the brake B, Figs. 16 and 17, into action by spindle 12 and rod 13.

(margin note: Variable exhaust lift)

The motor of the 8 HP. size has a cylinder 3·94 in diameter (100 mm.) and a stroke of 4·3 in. (110 mm.), and it runs normally for 8 HP. at from 1,000 to 1,500 revolutions per minute.

It is placed directly over the front axle z, which for the greater part of its length is tubular and curved downwardly to clear the bottom of the crank chamber.

To the crankshaft is attached a universal joint Y, transmitting motion to the rod A, and by another joint W to the upper spindle D of the speed-change gear, Figs. 19, 20.

The gear box is shown at G in Figs. 16 and 17, and at R in Figs. 19 and 20. The gear is similar to that illustrated by Fig. 181A, page 256 of vol. i., but has been rearranged, so as to effect a reverse motion without the reversing gear formerly used. In Fig. 19, D is the spindle driven by the motor and connecting rod and upon it are the three pinions 1, 5 and 2 gearing into the wheels 3, 6 and 4.

(margin note: Change-speed gear)

The pinion 2 and wheel 4 and the gear wheels 1 and 3 are always in mesh and the one or the other pair, for low speed or high speed respectively, is caused to transmit motion to the spindle H, and thereby to the bevel-wheel E on the driving axle, by the friction clutches within the drums c c'. When the clutch in c' is in operation, i.e. fast in its drum, the speed of the second motion shaft H is the slow speed due to pinion 2 and wheel 4.

When c' clutch is out of contact with its drum, the wheel 4 and drum run loose upon the hollow shaft without rotating the clutch parts fixed upon the spindle H.

When the high speed is required the pinion 5, which slides upon a feather, is by means of the arm and finger J, passed into the hollow back of wheel 1, and with it forms a clutch, fixing wheel 1 upon spindle D. By now using the friction clutch in c the high speed is obtained.

Reverse motion is obtained by putting pinion 5 in the position shown and then by putting the pinion 7 into gear with it and with wheel 6, wheel 1 now being free. The pinion 7 is carried by a pivoted arm G, Fig. 20, which is actuated by the crank K and by the rod 19, Fig. 17, and the spindle 18 and connexion thereto. The reverse motion is thus obtained through the drum of the high speed gear wheels 1 and 3, but the speed is that due to the diameter relation of the pinion 5 and wheel 6.

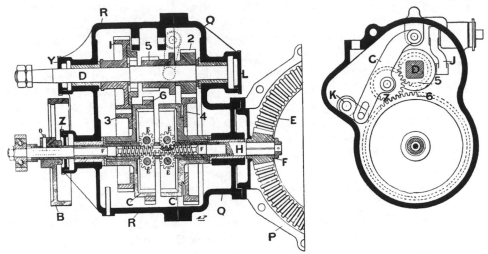

FIGS. 19, 20.—THE DE DION-BOUTON SPEED-CHANGE AND DRIVING GEAR.

It will, of course, be seen that the reverse motion does not occur until clutch c is put into action by the longitudinal movement of the rod F F.

Reverse motion

It is a very useful feature of this combination that when the reverse motion gear is in mesh the car may be worked backward and forward at the slow speed by simply operating the clutches and without touching the gear lever. The reversals should of course only be made to take place when the car has come to rest from motion in either direction.

The friction clutches in c c' are of the nearly semi-circular form described in vol. i. and are forced into contact with the interior of the drums by right and left handed screws on which are the small pinions E E. These are operated by the large right and left handed screw on the internal spindle F F, which takes the place of the rack formerly employed. It acts as a rack and not as a screw, the screw form being adopted for reasons now to be described. It is a modification which secures great facility of adjustment of the pressure on the friction clutches. By giving the screw spindle F a partial rotation with relation to the hollow spindle H containing it, the wear on the friction clutch blocks may be easily taken up without taking the gear box to pieces. A fourth of a turn, for instance, may be sufficient at any one time to move the little pinions E on the clutch operating screws enough for this readjustment. By moving the pin o the spindle F, otherwise only free to move longitudinally, can be rotated a fourth of a turn, and it is again prevented from rotating by replacing the pin in the next of the several grooves indicated by dotted lines. The spindle F is moved backwards or forwards longitudinally by a bridle on the ball-bearing collar seen at the left hand of the engraving. This motion by rotating the little pinions

Gear clutches

23

E and their screws alternatively separates or draws together the internal clutch pieces of either clutch.

The collar referred to is connected to the links 28, Figs. 16 and 17, and by the arm 15, spindle 16 and rod 14 to a handle on the top of the pillar 27, the spindle 16 works in lug bearings fixed on the cross tube within which is the brake operating shaft v actuated by the lever H.

This gear, it may be here remarked, has been arranged so as to give three speeds and reverse, but although many cars have been fitted with it, it will not in future be used. Another form is now in use on the larger two cylinder engine cars, and will be referred to hereafter in connexion with them. The later form of sliding gear, with cone clutch arrangement, as now employed on the cars sent to this country, will also be described.

Transmission of power

The transmission of the power from the change-speed gear to the driving wheels is by bevel pinion F and wheel E, Fig. 19, the wheel E being fixed to, and driving the differential gear, the whole being enclosed in the gear case L. The ends of the differential gear spindles are fitted with one part of a pair of universal joints U U, the counter parts of the joints being on the ends of the driving rods K K, the outer ends of which are fitted with similar joints on the inner ends of driving spindles, through the centre of the fixed axles as illustrated in vol. i. p. 486, but having a squared end in the wheel nave instead of the spring driving arms there shown. The fixed axle J is curved round to the back of the gear box instead of under it as in the heavy vehicles.

The jacket water is circulated by a pump Q, Fig 17, a connexion being made at T with the bottom of the water jacket at O, the hot water passing from the top of the cylinder by the pipe N to the cooler R, thence to the tank; and from the tank it returns to the pump.

The carburettor is placed in front of the engine at c. Its supply of air is taken in through the pipe, the end of which is splayed open and partly embraces the hot exhaust pipe E so that the air taken in is heated. Throttling of the charge to the cylinder is effected by a handle on the steering pillar connected to the rod 11, bell crank 10 and lever 9. The contact breaker is moved for advance or retardation of ignition by rod 8, bell crank lever and rod 6.

Carburettor

The construction of the carburettor is illustrated by the engravings Figs. 21 to 24. It is of the spraying type, arranged with float vessel and regulating parts superposed on the same axis. The float vessel H contains an annular float F surrounding the air inlet passage within which is a central downwardly projecting tube surrounding the petrol jet D. The tube immediately surrounding the jet is formed as part of a chambered piece B, the position of which can be altered by means of the arm Y so that the air entering at N is readily altered through a considerable range, though the throttling of the passage to the pipe o leading to the engine remains unrestricted, and the area of the passage round the jet is unchanged. The petrol enters at P, passes through a strainer at s and is admitted by a small valve within the cylindrical piece T, the top of the valve being an adjustable head, which is lifted more or less by the descent of the float, which rests upon a forked lever

24

G, resting in pivots seen behind the valve head in Fig 21, one end of the lever seen in section at G, Fig 21, lifting the valve. It will be noticed that this carburettor differs from the Longuemare carburettor not only in the

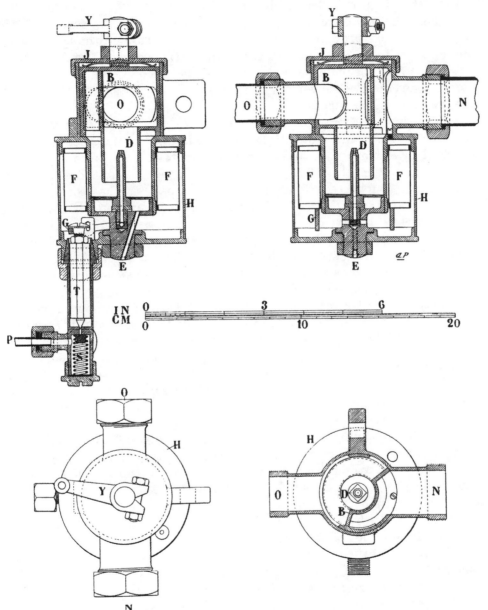

FIGS. 21–24.—THE DE DION-BOUTON CARBURETTOR.

superposition of the parts, but in that only one adjustment instead of two is required, or at all events only one is used. Any petrol which is not used, in case of the flooding of the carburettor, escapes through the little pipe E.

The ignition of the charges by the high tension system is described at pp. 357 and 364, vol. i.

MOTOR VEHICLES AND MOTORS

MOTOR MANUFACTURING COMPANY'S 8 HP. LIGHT CAR

Another light car of which a considerable number has been made is the 5 HP. single cylinder car illustrated by Figs. 25–30. It is constructed very much on what are known as Panhard lines, but with several details which are departures from the designs of that firm. Although brought out as a 5 HP. car, it is now fitted with an 8 HP. motor of the kind shown by Figs. 27–30, or with a 7 HP. double cylinder engine of Daimler type, throttle governed. The general arrangement of mechanism in this vehicle is shown in Figs. 25 and 26.

The frame of the car is constructed of steel tubes, and is of very substantial design. The motor 1 is mounted towards its front end, and enclosed by a bonnet. The power is transmitted from a cone clutch 2 to the first motion shaft 3 of the change-speed gear, and thence by one of four sets of spur-wheels which are always in mesh to the second motion shaft 23. At the back end of this shaft is a bevel pinion gearing with a bevel wheel on an extension of the differential gear box 5 on the cross countershaft 4. From the ends of this shaft the power is transmitted by chains 6, 7 to the rear road wheels.

The motor, which is of the single cylinder vertical type, is modelled closely on the well known De Dion lines, and is shown by Figs. 27–30. The body and head of the cylinder form a single casting. A point of novelty consists in the method of controlling the speed by varying the spring pressure on the inlet valve. This is effected by raising the rod 19 by which the

Control of engine subsidiary valve spring 17 is compressed, so that a greater suction is required to lift the valve and admit the charge. The quantity admitted is thus reduced according to the pressure exerted in lifting rod 19, and the power of the explosion lessened accordingly. This affords a ready means of governing the speed.

The carburettor, which is seen at 29, Fig. 26, is of the float feed spraying Longuemare type. High tension electric ignition is employed, the form of make and break switch being identical with that on the De Dion engines. The cooling water circulation for the cylinder jacket is maintained by a centrifugal pump 33 (Fig. 26), driven by a friction wheel 34, which presses against the engine fly-wheel. The water is cooled in a radiator 37, mounted at the extreme front of the frame.

By means of the change gear three forward speeds and one reverse are obtained. This gear is controlled by the hand levers 21 and 24. The lever 8 a controls the rear wheel band brakes 8 8, and the clutch is operated by the pedal 41.

Two levers mounted at 19, on the steering pillar, control the timing of the electric ignition and the throttle gear on the inlet valve respectively.

Iden's change-speed gear As already mentioned, the change gear is always in mesh, and is constructed from Mr. G. Idens' designs (Patent No. 5,139 of 1900). The gear wheels have large bores, and are formed with four strong internal teeth cut out of the solid. These teeth may be looked upon as forming an internal

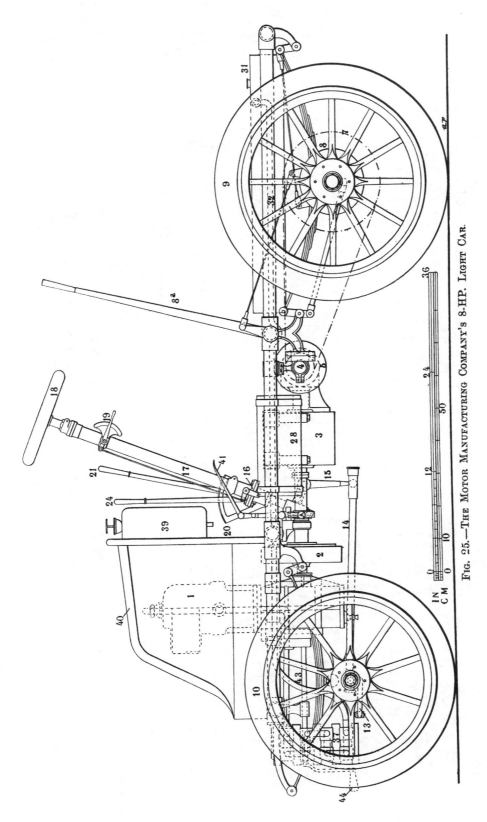

Fig. 25.—The Motor Manufacturing Company's 8-H.P. Light Car.

27

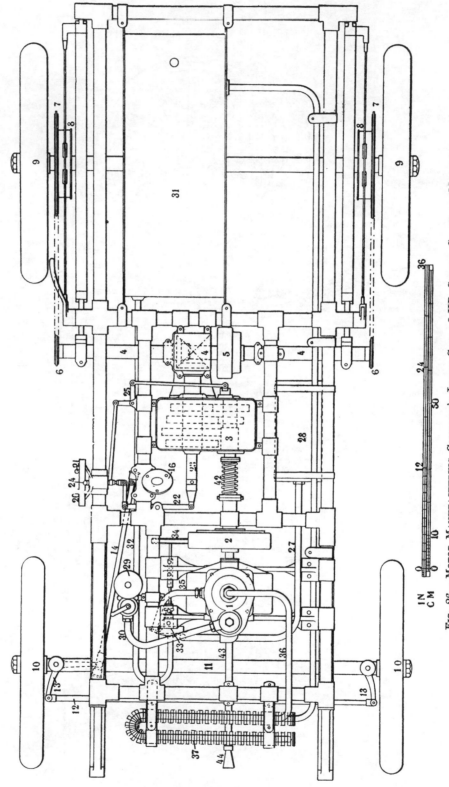

Fig. 26.—Motor Manufacturing Company's Light Car: 8-H.P. Single Cylinder Motor.

28

cylindrical claw clutch as contrasted with the claw clutch formed on the end of a cylinder. On the spindle 23 (see plan, p. 28), corresponding teeth are formed from the solid. By moving the spindle 23, the teeth upon it can be caused to engage with either of the two rear wheels upon it, or with the two front wheels, one of which gears with the long pinion 3. The gear wheel nearest the front of the gear box is only in gear when a pinion between it and pinion 3 is interposed for reversing by means of the spindle and fork operated by the bell crank lever 25 and the hand lever 24.

One of these cars, with the single cylinder motor as illustrated by Figs. 27–30, but with stroke of 130 mm., was run throughout the Automobile Club reliability trials last September with such satisfactory results that it was awarded the gold medal in its class. The motor, of 8 HP., **Engine** has a cylinder 3·94 in. diam. (100 mm.) and a stroke of 5·11 in. (130 mm.). **dimensions** The 5 HP. motors had cylinders of the same diameter, but a stroke of only 3·94 inches.

Although numbers relating to the various parts of the motor have been placed on the drawings given on page 30, it is not necessary here to refer to them or to describe the motor minutely, as most readers will readily gather all they convey without description.

It will be seen that for the size of the engine its power is large, due to the high speed at which it runs, though of course the antecedent requirements for the high speed are satisfied, including free admission, large exhaust, and large or sufficient water jacket spaces, cooler surface, and quick circulation.

THE ORLEANS LIGHT CAR.

A light car of simple design, and of which a good many have been made, is the Orleans gear-driven, two-cylinder 9 HP. car. Of this car some illustrations are given in Figs. 31–38.

This car is the outcome of the Vivinus car of early 1900, by whom and the Orleans Co. a small voiturette was made with a 3 HP. single air-cooled motor, the cooling being assisted by a small fan. The car was driven by a belt on a pulley about 5 in. diameter on an extension of the crank- **Belt-driven** shaft, which drove a very simple form of two speed gear on the differential **cars** gear and live axle. The car was sold at about £140, but as few ordinary users have either the ability or the will to attend properly to a belt, and as the performance of a 3 HP. car was very limited as compared with other somewhat more expensive cars, the construction of the cars has been discontinued. The same may be said of the similar Orleans car with two 3 HP. air-cooled motors side by side and a fan between them. The driving pulley for the belt was also between the two engines, and the general arrangement is worth recording, as even now the belt and a somewhat similar construction of gear is used by more than one firm who find users for a belt-driven car. Moreover it is not improbable that the belt may be restored to favour for some light cars of inexpensive construction suitable for the requirements of many.

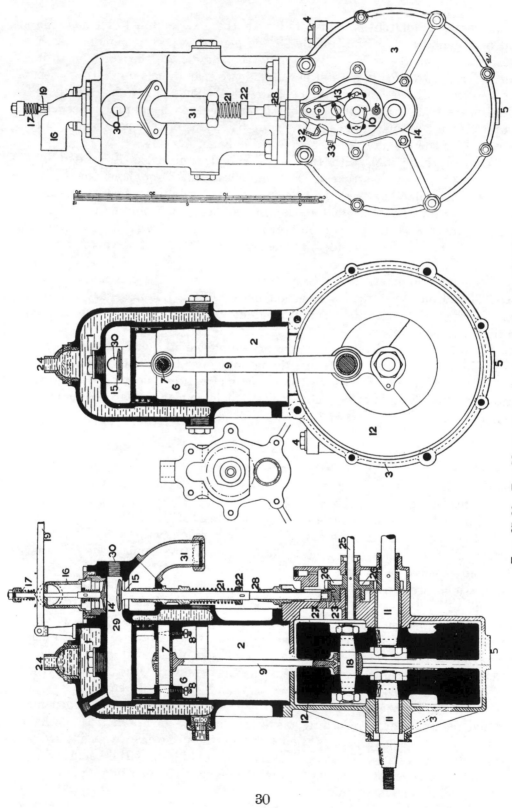

FIGS. 27-30.—THE MOTOR MANUFACTURING COMPANY'S 5-HP. MOTOR.

30

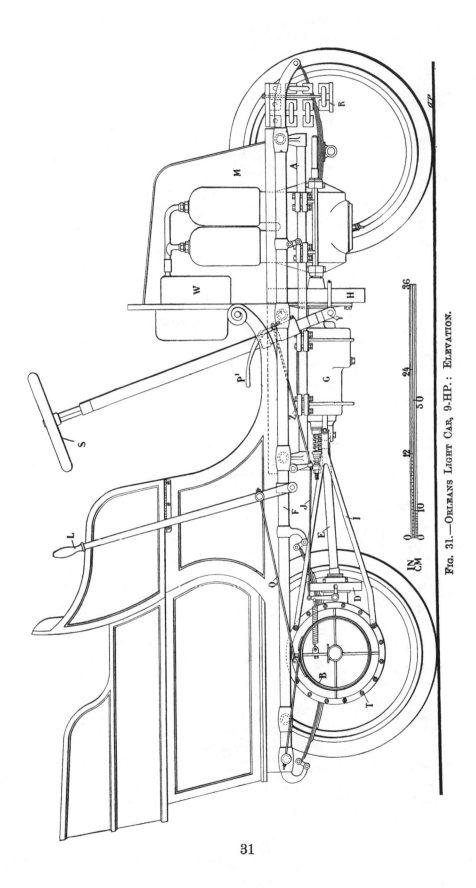

FIG. 31.—ORLEANS LIGHT CAR, 9-HP. : ELEVATION.

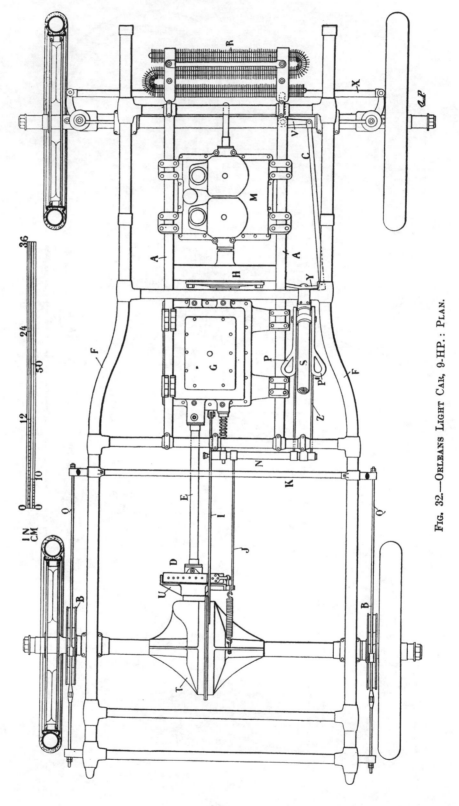

FIG. 32.—ORLEANS LIGHT CAR, 9-HP. : PLAN.

32

LIGHT PETROL MOTOR VEHICLES

Fig. 33 sufficiently shows the arrangement of the car with the two-cylinder motor of the Vivinus make. This motor presented several points of useful novelty. The cylinder ends, for instance, simply sat in a groove in the upper part of the crank chamber, and the whole of the cylinder was held in place with a bridle which passed over the cylinder, the bridle at the centre of the top having one set screw. To remove the cylinder it was only necessary to loosen this one screw. The valve box attachments were also very simple, and the design has points worth remembering now. The **Separate valve seats** central part of the valve chamber only was cast with the cylinder. The seats of both inlet and exhaust valves were separately made and removable and held only by a couple of bolts. When first these little cars came here, the carburettor was of the surface evaporation type and consisted of a piece of felt cut like a comb, the teeth of the comb being of an inch or so **Surface carburettor** in width and suspended in a cylindrical chamber through which the air passed on its way to the engine. The back of the comb was kept soaked with petrol and it acted as a wick for the fingers which offered large evaporative surface. The system was, however, a failure in this country, although in Belgium it was used long after it was discarded here in consequence of the gradual cessation of evaporation due to the deposition of moisture from the air drawn in by the engine. The two cylinder 6 HP. car was sold at £240 and it weighed only about 7½ cwt.

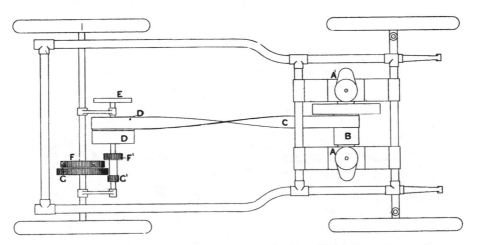

FIG. 33.—ORLEANS 6-HP. LIGHT CAR, 1900.

From the plan of the 9 HP. car (Fig. 32) made in 1902, it will be seen that the frame is of the tubular kind and that the arrangement of the engine, gear and gear-driven live axle is very simple, the attachment of the engine crank base and the speed gear box G being well devised, a separate pair of tubes for carrying the machinery being carried by two of the frame cross tubes.

The main features of the engine are shown by Figs. 34 and 35, which show the 7 HP. engine with cylinders 95 mm. × 100 mm. or 3·74 in. ×

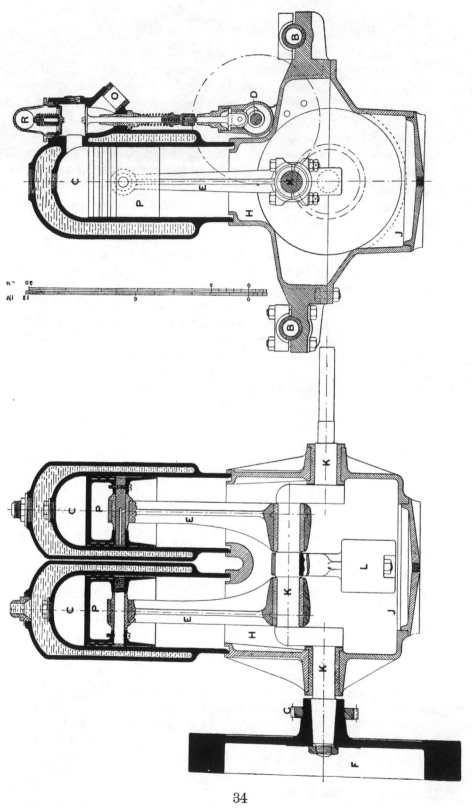

FIGS. 34, 35.—ORLEANS LIGHT CAR: SECTIONS OF 7-HP. MOTOR.

34

Engine dimensions 3·94 in. The engine used on this car was precisely as shown, but in order to give 9 HP. the cylinders were made 3·94 ins. diam. (100 mm.), and the stroke 4·33 ins. (110 mm.), the normal speed of the engine being 900 revolutions per minute. The cranks K are in line, the inlet valves R being above the exhaust valves which are actuated by a cam D and well-guided lifting rod. The connecting rods E are of cast steel stiffened to compensate for the overhang of one side of the large ends, The cranks and part of the weight of the connecting rods are balanced by the weight L attached to the centre of the cranks. Access to the connecting rod ends is by a large door J. The cylinder and head are cast in one piece, with large jacket spaces, large valves and clear way for induction and exhaust.

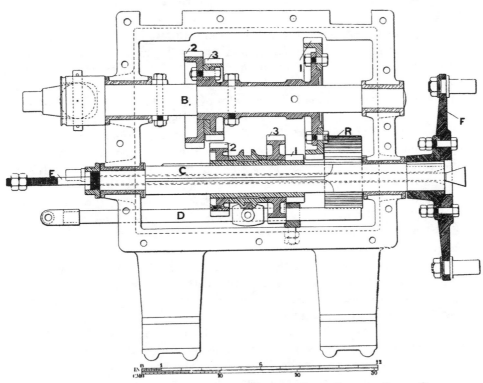

Fig. 36.—Speed Change and Reverse Gear, Orleans Light Car.

The engine gives motion to the first spindle C in the change-speed gear (Fig. 36) by means of an expanding clutch within the fly-wheel, the parts of the expanding clutch being pivoted on the two pins in the disc F, their release being effected by the pedal P, rod and connexions to spindle N **Expanding clutch** (Fig. 32) and thence by pull on the rod E (Fig. 36), the clutch end of which is wedge-shaped. The pressure on the pedal pulls this wedge backward and forces the clutch parts to approach each other in opposition to the springs by which the parts are normally forced to expand into the fly-wheel.

The gear gives three speeds and reverse motion and is shown in its

Change-speed gear

box in Fig. 36. The wheels and pinions 1, 1, 2, 2, and 3, 3 respectively give the lowest second and third speeds by the movement of the rod D acting by a fork on the sleeve carrying the first motion gear on spindle C. Reversal is effected by the movement of a lever not shown, by which the wide pinion R is put into mesh with wheel and pinion 1, 1, while in their positions as shown. The rod D is connected to a handle on the steering pillar and that for moving the reversing pinion was in the centre of the foot-board, but is now combined with levers at the side of the car for changing speed and holding in out of gear position.

The carburettor employed is of the spray maker float feed type and the governor acts by throttling the induction pipe.

This car has gained several of the Automobile Club prizes and certificates, and in last year's (1903) durability trial won a gold medal.

The construction of the driving axle, differential gear and bearings is shown by Fig. 37, in which the axle A runs in bearings at the ends of tubes firmly fixed in long bosses on the sides of the gear case C, this case also

Rear axle

carries a bearing for the short spindle J on the inner end of which is the pinion E gearing with the bevel-wheel F on the outer part of the differential gear box H which rotates freely on the bosses of the two pinions on the inner ends of the divided axle A A. The brake pulleys B are made of malleable cast iron, and are screwed on the axle, the wheel naves being also screwed on at the same part and the ends keyed, right and left hand screws being employed.

The tendency of the application of the brakes in the backward direction is to lock the pulley B against the wheel nave, but for forward direction they tend to turn from the nave and to hold only by jamming on the inner end of the thread.

Experience has shown that it is undesirable to have a screw thread at this part of an axle and has confirmed the wisdom of warning makers against sharp corners in shoulders in any part of an axle or such part where journals or changes of diameter occur.

For their larger cars, 15 H.P. with four cylinders of 90 × 110 mm., the Orleans Company is now using an improved form of axle in which the axle proper ceases to be a live axle, the tube from casing to spring support at D being carried out at each end and forming a fixed axle on which the driving wheels run, the present axle A being relieved of all load and only used as the torsional medium. This method will probably be adopted for the 9 H.P. cars in future, and is illustrated here, although at present only used with the heavier cars.

Improved design of rear axle

Fig. 37 shows the new axle A on which is fixed the short sleeve and spring attachment B, forming an oil drip at C, the outer end of the hollow axle carrying the wheel nave D, the nose of which is squared inside to receive the end of the driving shaft E. The axle which is $2\frac{7}{16}$ in. diameter and $\frac{3}{8}''$ thick will afford a good stiff axle and the large size inside the wheel will be no detriment. The driving shaft E is 33 mm or nearly $1\frac{5}{16}$ diameter and carries at its end a nut and washer by which the wheel is kept on. The

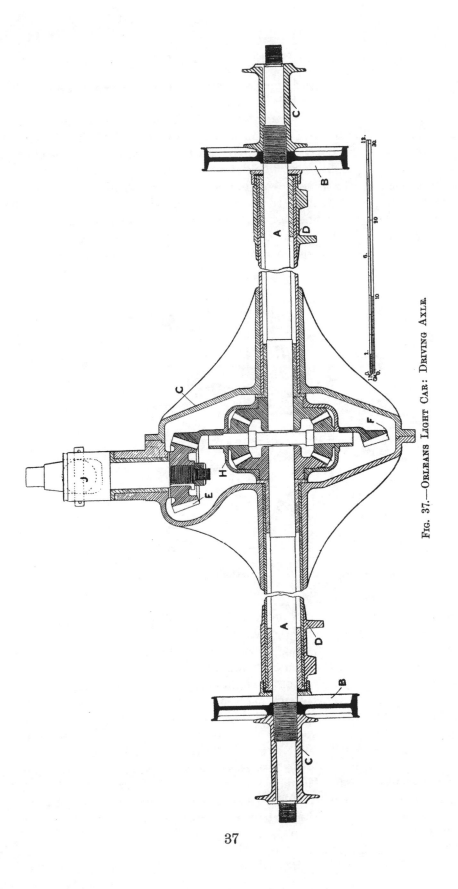

FIG. 37.—ORLEANS LIGHT CAR: DRIVING AXLE.

37

nose of the hub is reduced in diameter where the cap is screwed on, but it will probably be found desirable to enlarge this so that a larger washer under the nut can be used. The annular space between the axle and driving shaft provides an oil space and easy means of lubricating the wheel from a supply entering at F, but holes at the top of the axle instead of, or as well as, the holes shown at the bottom should be provided. The washer at the inner side of the nave in the oil drip ring is of vulcanized fibre, the others are of steel. The brake drum G forms part of the wheel nave, and an oil drain from the axle is provided at H. The length of the divided driving shaft between the nose of the wheel nave and the tube A is not great, so probably the driving axle of strong steel will be quite sufficient in

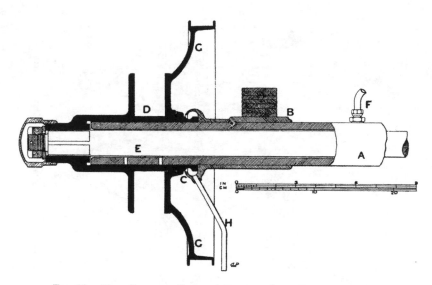

FIG. 38.—THE ORLEANS FIXED AXLE FOR GEAR-DRIVEN CARS.

diameter, especially as it is only subject to torsional stress, although this stress is, it must be remembered, reversed in direction sometimes.

The reference to live driving axles recalls a feature of weakness in many of these, apart from the weakness inherent to a divided axle, however well connected by fixing in the bosses of the differential gear-case. This is the loss of strength at screw threads made upon the axle for adjusting ball bearings, and in the sudden changes of diameter from a larger to a smaller size, especially when the change is made at a square, or nearly square, shoulder. This is especially to be avoided in axles in which the reversal of stress is not only constant and at high speed, but is frequently aggravated by heavy percussive stress. All sharp corners at shoulders should be avoided by using rounded corners with as large a radius as possible, and the change of diameter at any one place should be small, especially if the larger part is more rigidly held than the smaller part.

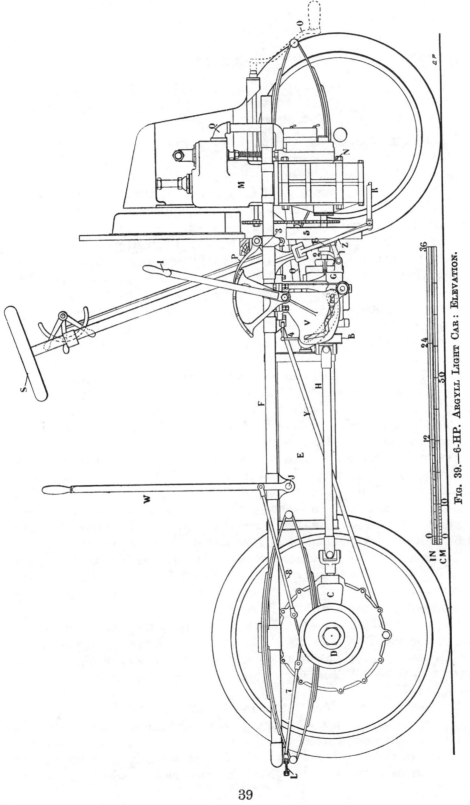

Fig. 39.—6-HP. Argyll Light Car : Elevation.

39

Chapter II

LIGHT PETROL MOTOR VEHICLES (*Continued*).

THE ARGYLL LIGHT CARS

THE Hozier Engineering Company of Glasgow have now for some time been making a light car known as the Argyll. It was at first fitted with a 5-HP. single cylinder motor of the De Dion type as made by the Motor Manufacturing Company of Coventry. A motor of larger power, namely that of the Simms Manufacturing Company, with the Simms-Bosch magneto electric ignition apparatus, was subsequently fitted, and the Company has also fitted the Argyll cars with the Clement motor.

Since the preparation of the drawings illustrating the light cars here described, larger and more powerful types have been introduced. They are fitted with two, three, or four cylinder engines of 13 to 24 B.HP. of the Argyll or Argyll-Aster type. Various detail changes have been made in the design of these cars, notably in the use of pressed steel frames, but it has nevertheless been found unnecessary to depart from much that is here shown.

The car illustrated herewith, as made in 1902, is fitted with the 5 HP. M.M.C. type of motor, and presents several features of interest. The general arrangement is shown by Figs. 39 and 40, which are respectively a side elevation and a plan of the car without the body, and various details are shown by Figs. 41 to 55.

Arrangement The underframe F is constructed of steel tube and carries the motor M and change-speed gear G towards its front end. The power is transmitted to the live rear axle by a flexible tailshaft H and bevel gear C on the exterior of the differential gear. The rods Y, which are connected to the top and bottom of the case enclosing the bevel driving and differential gear, and which pivot on a cross member of the underframe above the change-gear box, preserve the radial distance between axle and gear box as the former rises and falls, due to the inequalities of the road surface. The motor is enclosed by a bonnet which constitutes both water tank and radiator. No pump is employed for circulating the water.

The petrol supply is carried in the tank mounted on the front of the dashboard, the feed being by gravity to a float feed spray-making carburettor A, Fig. 40. The exhaust from the motor passes by way of pipe Q to the silencer E, which is of decidedly more liberal dimensions than is usual on these light vehicles.

Simms Engine The motor shown only in outline in Figs. 39 and 40 is illustrated in the slightly larger size by Figs. 27 to 30, p. 30. The Simms motor, fitted in the more recently-constructed vehicles, is shown by Figs 47, 48, and the details are shown by Figs. 49 to 53. This is a single cylinder vertical engine, with cylinder body and head in one casting, which is bolted

40

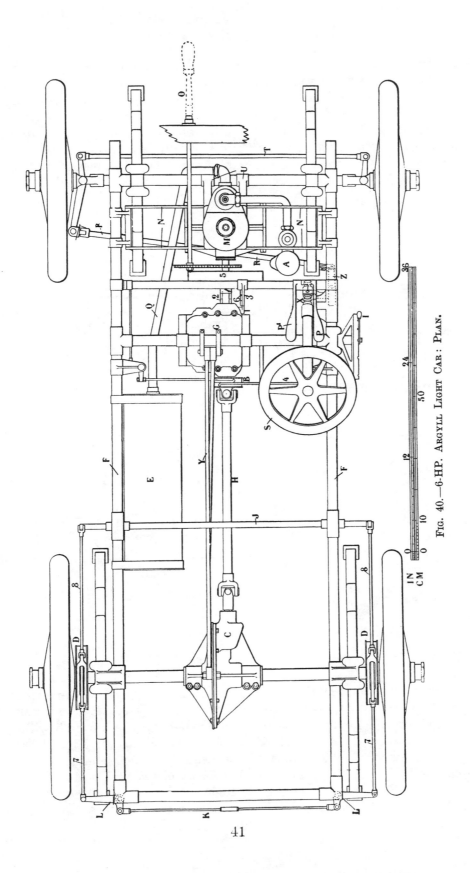

FIG. 40.—6-HP. ARGYLL LIGHT CAR: PLAN.

41

to an aluminium crank enclosing chamber. The cylinder dimensions are Bore 4·33 in. × 3·94 stroke and 6 B.HP. is developed at 1,000 revs. per minute.

The inlet valve A is automatic and controlled by an external spring. The exhaust valve E, which is directly below the admission valve, is operated by a cam on the half-speed shaft I, the lift of the cam being transmitted through the guided lifting rod L. The crankshaft is a single forging with balance weights W W attached to the crank webs. The fly-wheel is mounted

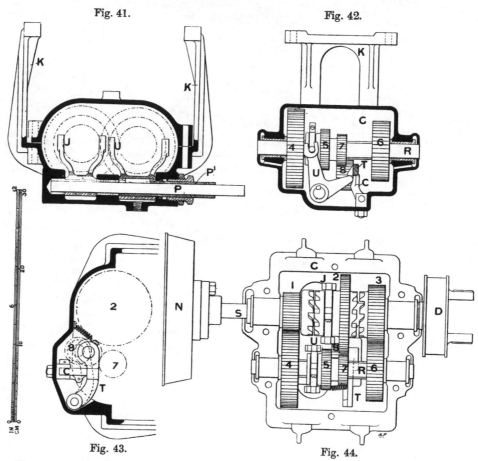

Fig. 41.

Fig. 42.

Fig. 43.

Fig. 44.

FIGS. 41 TO 44.—TRANSMISSION AND SPEED-CHANGE GEAR. ARGYLL LIGHT CAR, 1902.

on one end of the crankshaft external to the crank chamber, and constitutes the female portion X of the cone clutch. This engine is fitted with the Simms-Bosch ignition, the method of operating the make and break switch within the combustion chamber being shown by Figs. 49 and 50. In Fig. 49, M is the cam for operating the striker H, by means of which contact at the switch U is broken at the pallet seen between the two valves in Fig. 48, within the combustion chamber of the engine. The cam M is mounted on an extension of the half speed spindle which carries the exhaust cam T.

Low tension magneto ignition.

This smaller part of the half-speed spindle may be partially rotated relatively to the part carrying the exhaust cam spindle by means of the sliding collar Q, Figs. 49 and 53, a pin in which enters a helical groove on the half-speed spindle. The collar Q carries a feather, which is fixed in cam M. Longitudinal movement on the divided spindle causes the pin in the collar Q to follow the helical groove, so that it receives a partial rotation. In this rotative movement it carries with it the cam M, so that the striker rod H is **Ignition** lifted earlier or later and the circuit is thus broken sooner or later by the **timing** pallet on the spindle U in the combustion chamber. The movement along the cam spindle is given by the fork on the spindle O (Fig. 53), and the lever

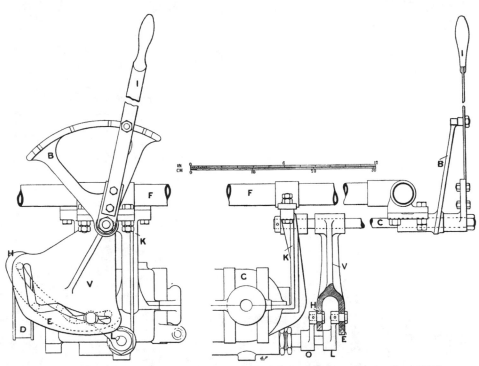

FIGS. 45 AND 46.—SPEED CHANGE-GEAR LEVERS OF ARGYLL LIGHT CAR, 1902.

O (Fig. 50). The advance or delay of ignition is thus variable from the driver's seat by moving the connexion to the lever O'.

On the end of the cam-spindle is a small crank-pin seen in section at F (Fig. 49) for operating the oscillating armature or shield of the magneto electric machine. A connecting link from this crank-pin transmits motion to a corresponding crank-pin on the spindle of the magneto.

The change-speed gear employed on the earlier type of the Argyll car is shown by Figs. 41 to 44, and a modified construction subsequently employed by Figs. 54, 55. The operation of the transmission gear may be described with reference to the later type of gear. The power is trans- **Change** mitted from the cone clutch and claws C to a short spindle A, which carries **speed gear** the spur pinion I. In line with A is a free spindle A, which at its outer

43

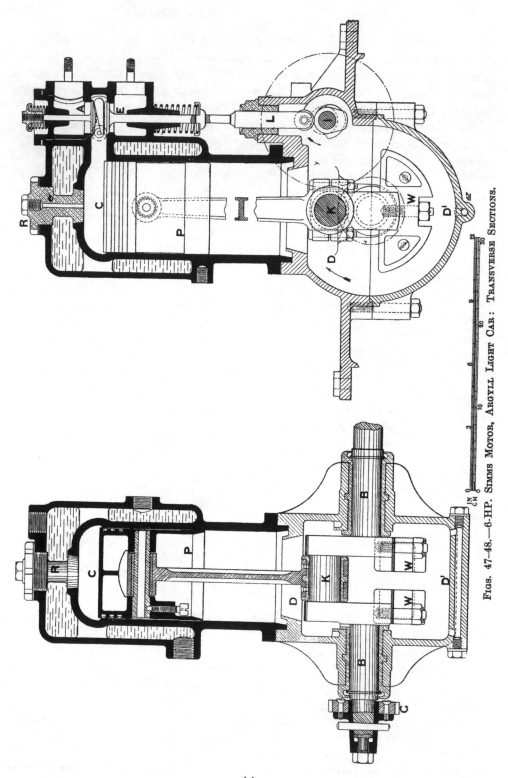

FIGS. 47-48.—6-HP. SIMMS MOTOR, ARGYLL LIGHT CAR: TRANSVERSE SECTIONS.

44

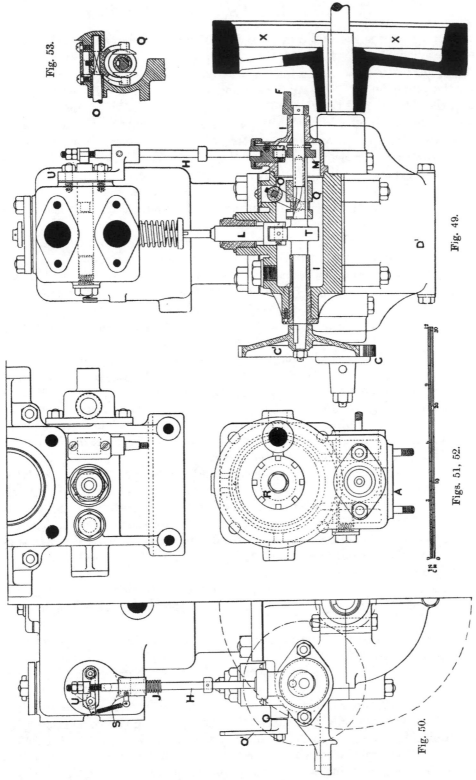

FIGS. 49 TO 53.—6-HP. SIMMS MOTOR, ARGYLL LIGHT CAR.

45

extremity carries the brake drum D, to which is connected one of the joints of the tailshaft H, Figs. 39, 40. Loose on spindle A is a spur pinion 3, and sliding on feathers on it, is a clutch sleeve operated by the fork U and carrying the spur-wheel 2. The pinions 1 and 3 run in mesh with wheels 4 and 6 on the shaft B, and the pinion 5, which slides on feathers on this shaft, may be engaged with wheel 2. A wide pinion on an eccentric spindle R may be engaged with wheel 2 and pinion 5 for the reverse speed. For the slowest forward speed the sleeve moved by the fork U stands with its claw faces out of engagement with those on 1 and 3, as shown, and pinion 5, by means of fork J is brought into mesh with 2 ; then pinion 1 drives wheel 4 and pinion 5 drives wheel 2, motion being transmitted through A and the joint on the brake drum D to the tailshaft H, Fig. 40.

For the intermediate speed the sleeve carrying wheel 2 is moved to the right hand Fig. 54, so that one of its claw faces and that on 3 are engaged, then pinion 1 drives wheel 4, and 6 drives 3, which drives A. For the highest speed the fork U is moved to the left and the left-hand claw faces of the sliding sleeve engage with pinion 1. The power is then transmitted direct through to the tail shaft, and shaft B with its gear wheels runs idle.

Engagement of gears The operations controlled by the forks U and J are effected by the hand lever I, Figs. 39, 40, 45 and 46, which is mounted on a spindle C, Fig. 46, carrying a double sector V, having a double set of cam slots E H, which control the movement of the levers O, L, Fig. 46 and Fig. 55, on the spindle P and sleeve P', Fig. 55, to which are connected the forks J and U, operating the pinion 5 and sliding sleeve of wheel 2 respectively. The lever I moves over a notched sector B, Figs. 45, 46, which determines its position for reverse, neutral, and the three forward speeds respectively.

Interconnection of brake and clutch The tailshaft band brake B, Figs. 39 and 40, is operated by a pedal P, which at the same time disengages the cone clutch 5. The pedal P' controls the clutch only. The rear wheel band brakes D are operated by the lever W at one end of the cross spindle J, the ends of which are coupled to one end of each band. The other ends of the bands are interconnected by means of the bell cranks L L and links J and K, Fig. 40, so that the brakes may act equally on both wheels.

The steering of the car is controlled by a hand-wheel S, operating the front wheels through a screw and nut X at the lower end of the steering pillar.

It will be seen that the principle of the form of transmission gear described with reference to Figs. 54 and 55, is the same as that of the arrangement shown in Figs. 41–44 of the earlier cars, a through drive being obtained direct from motor to tailshaft at full speed. The wide reversing pinion 8, seen in Figs. 42 and 44, has been placed in the upper part of the gear box instead of the lower part, and the method of operating the reversing pinion is modified accordingly. In the arrangement shown in Figs. 41, 44, the fork U (*see* Fig. 42) was provided with an arm, which with U in the proper position raises the arm T carrying the reversing pinion pivot and permits a trigger support C to pass under it and hold it in proper position for the gear to

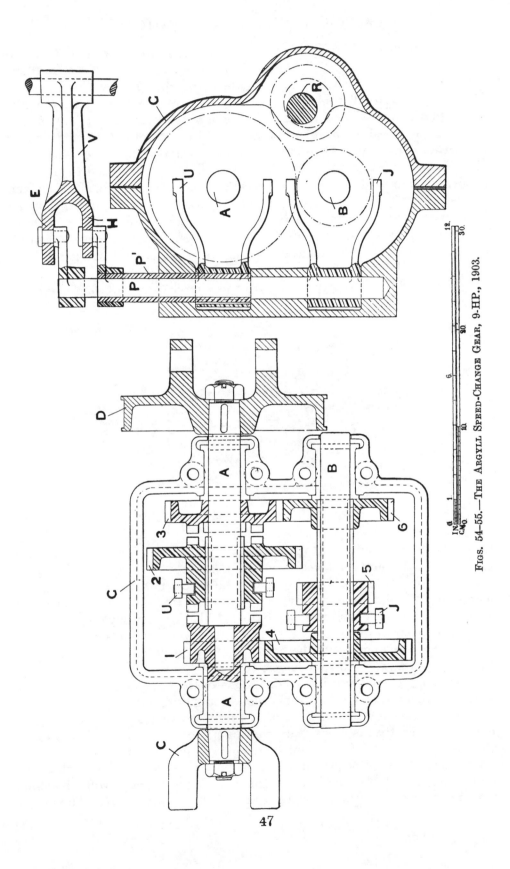

FIGS. 54-55.—THE ARGYLL SPEED-CHANGE GEAR, 9-HP., 1903.

remain in pitch while doing work. The first result of moving u from the position shown in Fig. 42 is to displace the trigger support c, which allows the arm t to drop under the action of the spiral spring shown in Fig. 43.

The larger Argyll car has attracted even more attention than its predecessor just described, and it has done some noteworthy long distance running. It is illustrated by Figs. 56 to 66. In this car the tubular frame of the earlier car is replaced by an ash frame A strengthened by a thin steel plate along the inside, the transverse members being connected to the main longitudinals by thin well-fastened steel gusset-plates. As against racking

Frame stresses the frame receives a good deal of support from these, and the engine supports B , and gear box supports D   The frame is stiffened vertically by the tension rods A′ connected to the spring hangers K′ and the plate near s′, the tension rod passing under and resting upon the central bracket for the speed-change lever.

The wood frame is carried forward to a simple form of spring hanger in

FIG. 56.—THE HOZIER CO.'S ARGYLL LIGHT CAR, 10-HP., 1903.

front, the pump-handle front spring hanger, which has often been the cause of trouble, not being employed.

The front springs are directly under the frame, but the rear springs are carried outside the frame, see Fig. 58.   At their front ends they are carried by links on the ends of a cross bar, but at the back ends they are carried on

Springs cantilever pins, which throw a torsional stress upon the ends of the frame. A bar across this end would be better for this purpose, as it would avoid this stress, and it would supplement the cross-frame member.

The engine employed in this car is one of the Clement motors running at a normal speed of 1,100 revs. per min., and having pistons of 3·35 ins. or 85 mm. in diameter and 4·32 ins. or 110 mm. stroke.   The Clement engine has in it much that is common to the engines of a number of French and Belgian makers, and these things having now settled down to something approaching fixed principles of design, it may be expected that with a few larger machine tools than they ordinarily employ, the sewing machine makers will soon be turning them out at a fixed price per gross.   The motor carriage builders may in some cases make their own engines, but in many cases the engine question will resolve itself into selecting that one which will best suit any particular car arrangement and is best in other matters, such as economy, accessibility, durability, which are less onerous points in a difficult matter of selection and adaptation. The Clement engine is a good one and represents the survival of that which has proved the fittest in the course of the experience of many of the purchasers of inexperience or experiment. See p. 186. The jacket in these engines is of very ample capacity ; a good feature, which costs little more than a jacket in which ingenuity is wasted in showing how thin a core can be made.

Clement engine.

The water circulation in this jacket is what is known as natural circulation, that is to say, no pump is employed to maintain it, the common arrangement of the kitchen boiler and overhead cistern being found sufficient. The top of the bonnet N is a cistern, half the Clarkson cooler tubes connected to it (*see* Fig. 56), are down flow tubes and half upflow, these being respectively fitted to the lower and upper parts of the cylinder jackets, as in the smaller car already described. The power of the engine is transmitted to the gearing by means of the conical clutch shown by Fig. 59, in which E is the engine fly-wheel fixed at D to the flange on the end A of the engine crank-shaft. Within the fly-wheel is an adjustable coned ring F, into which the leather-lined cone G fits. The direction of the taper of the cone is outward from the fly-wheel, so that when pressed home by the spring L the pressure upon the clutch is balanced or exerts no end thrust on any spindle or bearing. The male cone G is carried by a centrepiece, in the centre of which is the ball-bearing K, which receives the pressure of the spring L and comes into play as a bearing when the clutch is released by forcing G inward towards the fly-wheel by a fork in the groove near the letter K.

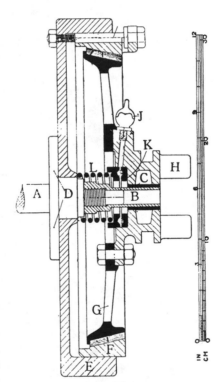

FIG. 59.—CONE CLUTCH OF THE ARGYLL 10-HP. CAR, 1903.

Balanced clutch

Screwed on the end of the crankshaft is a nut formed with a long sleeve B, into which the end of the spindle connecting it with the gear box enters, and on it is a bush shown black, in which the part B rotates when the clutch is not driving. Motion is imparted to the gear-driving spindle by a simple form of universal joint of which the lugs are shown at H.

The transmission gear is practically the same as that shown by Figs.

4

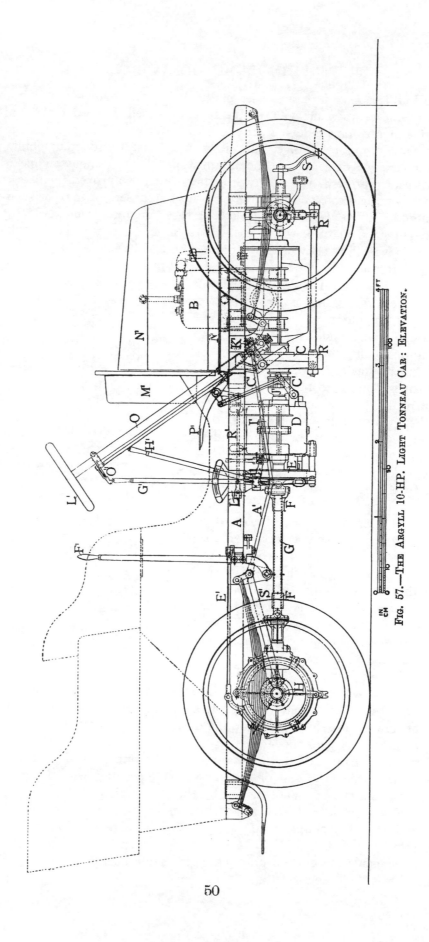

FIG. 57.—THE ARGYLL 10-HP. LIGHT TONNEAU CAR: ELEVATION.

50

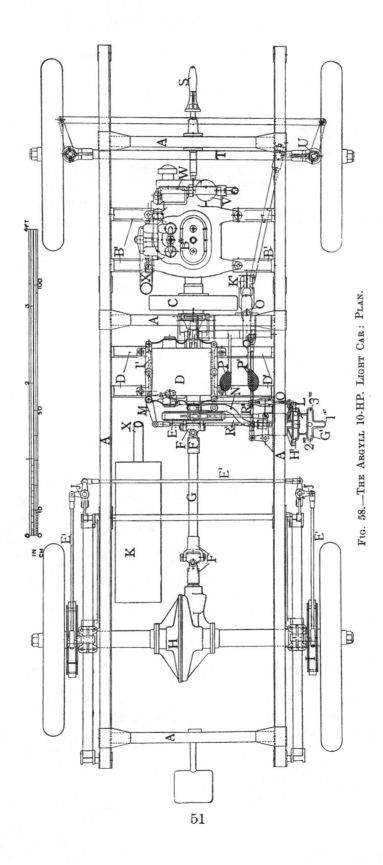

FIG. 58.—THE ARGYLL 10-HP. LIGHT CAR: PLAN.

51

54, 55, but the method of operating it is quite different, as will be seen in Figs. 60, 61 and 62.

In these Figs. it will be seen that there are two levers, x and v, for effecting the several operations. The lever x has three movements, two longitudinally, or in the direction of the length of the car, and one transversely in the T-shaped guide quadrant, seen in plan in Fig. 61. When moved into the position marked 1st, the rod N moves the arm M of a bell crank which by arm o′ commands the spindle R′ on which is the fork L′ by which the gears already described are put into position for the first speed. When the lever is moved into the rear part of the longitudinal slot it rotates the spindle T, on the inner end of which is the lever P, which moves the lever

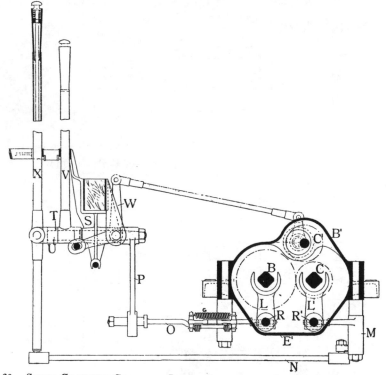

FIG. 60.—SPEED-CHANGING GEAR AND LEVER CONNEXIONS, ARGYLL 10-HP. CAR, 1903.

o, which slides the spindle R, on which is a fork L, which moves the gears into mesh for the second speed. By pushing the lever x forward, the arm P pushes the lever o backward, and by the collar R the third speed gear wheels are put into mesh, i.e. the motor drives direct through from crankshaft to tailshaft. When the lever x is in the central position shown in Figs. 57 and 61, the whole of the gear wheels are disengaged and the engine runs free. An advantage of the arrangement of the lever x and its connexions with the gearing is that it is difficult to make a mistake as to the position of the lever even in the dark.

Operation of change-speed gear.

The second hand lever H′, Fig. 57 and v, Figs. 60, 61, has at its lower end a bell crank arm connected to a bell crank w, by means of which the eccentric spindle carrying the broad pinion c is put into gear with the low

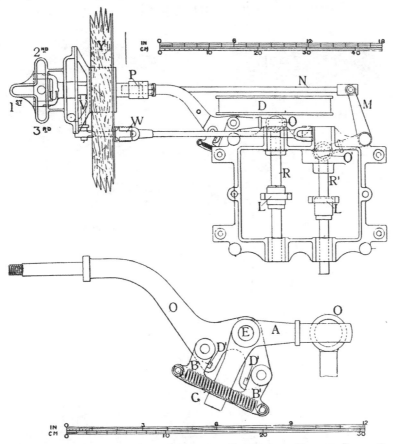

FIGS. 61, 62.—SPEED-CHANGING GEAR AND LEVER CONNEXIONS, ARGYLL LIGHT CAR. 1903.

FIG. 63.—TRANSMISSION, SPEED-CHANGE AND REVERSING GEARING, ARGYLL CAR, 1903.

speed wheel and pinion, and the reverse motions thus obtained. This gear is seen in perspective in Fig. 63.

Turning now to the novel feature of the lever gear of the 1903 car, as

FIG. 64.—DRIVING AXLE, BEARINGS AND PARTS, ARGYLL 10-HP. CAR, 1903.

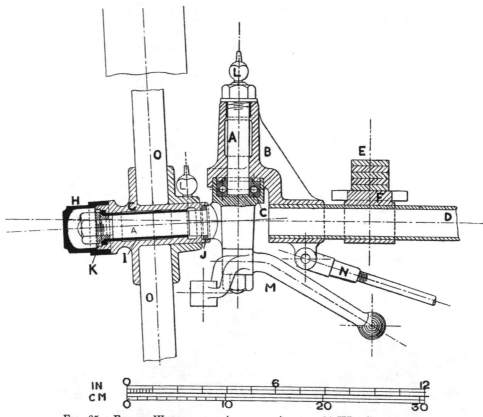

FIG. 65.—FRONT WHEEL AND AXLE OF ARGYLL 10-HP. CAR, 1903.

shown particularly in Fig. 62, namely a spring permissive system by which the speed-change lever may be put to the position required for a given speed, and the gear clutches left to find their places under the pressure of the springs. In Fig. 62 it will be seen that the lever o is pivoted at E, and that it takes a forked shape, there being on the ends of the fork the two stops or detents D D'. Pivoted on the arms of the fork are the short ⊤-shaped levers B and B' pulled towards each other by the spring G. The heads of the levers B B' bear against the bent lever A also pivoted at E, and carrying at its outer end the finger of the pivot Q, by which the spindle R and fork L are moved.

Spring assisted gear engagement

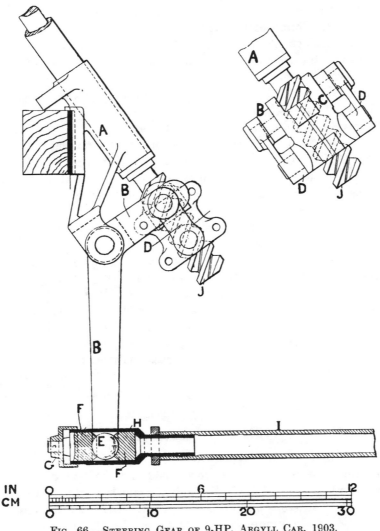

FIG. 66.—STEERING GEAR OF 9-HP. ARGYLL CAR, 1903.

When the lever o is pushed over by the lever x into the forward position—for example, for the third speed—the arm P pushes the lever o backward. The lever B' then pushes against the arm of the bent lever A, Fig. 62, not by the force put into the lever x, but by the pull of the spring G. If the claw clutch on the gear wheels happens to be opposite the corresponding sliding clutch, it goes in at once, but if not opposite, the wheel turns until the claw and claw tooth space are opposite, then the spring acts and pulls A over, the spring pull upon the lever B being resisted by the stop D′ Thus a driver has only to change the position of the lever X at any time to make the speed change for second or third speeds, without any consideration of relative speed of gear and engine, except that the clutch must be used. This of course could not be done with the usual sliding change gear wheels.

The driving axle is in two parts, connected at the centre to the differential gear on which is the large bevel driving wheel in the aluminium case H, Figs. 57 and 58. The axle runs in roller bearings, as seen in Fig. 64, the outer shells of the bearings, shown separately, being fixed in the ends of the tubes which are fixed by their flanges to the sides of the aluminium central case.

Axles The front axle, steering arms and wheel are shown in section by Fig. 65. In this the hollow fixed axle D is seen fixed at C in the boss of the head socket B for the vertical part of the pivoted axle A A. On the fixed tubular part of the axle D is brazed the pad F for the spring E. The wheel G is fitted with a lining bush I and a ball bearing is fitted at C to the vertical pivot of the steering axle. The steering arm M is provided with a ball end for connexion to the rod R, Fig. 57, between M and the arm actuated by the screw on the steering wheel spindle. Both the front and rear axles are trussed by means of tension rods as seen at N, Fig. 65.

Steering gear The steering screw and parts are shown by Fig. 66, in which A is a bracket attached to a cross member of the main frame, a nut on the double-threaded screw J actuating a forked bell crank B B by means of the links D D. The lower single end of the bell crank B is a ball E fitting within adjustable blocks F F in the socket H screwed into the hollow connecting rod I, the end of the socket being closed by a nut G screwing over the end of the socket.

Brakes The pedal brake is actuated by the pedal P′, Fig. 58, and bell crank lever connexions. The main brakes I I are actuated by the lever F′, Fig. 57, on a cross shaft in the brackets near S′, the cross shaft carrying short arm levers on which are pivoted the bell crank levers J J, Fig. 58, connected by a cross rod E′ and by side rods E′ to the brake bands, the arrangement securing a compensated equal pull on the two brakes.

THE GOBRON-BRILLIÉ CAR

In vol. i. p. 188 a full description, with numerous illustrations, of the Gobron-Brillié car was given. Since 1900 the makers of this car have adopted several modifications, although for most of these cars they have adhered to the form of engine peculiar to themselves, namely that with two pistons in each cylinder connected to cranks at 180 deg.

The mechanism generally has been rearranged, and the frame design somewhat modified. The wheel base is considerably longer, and the height from ground to frame has been reduced. One of these cars is now illustrated by Figs. 67 to 70. The engine M is centrally placed between the axles and side frame trusses, and the gear-box G is immediately behind it The rear wheels are attached to a live axle driven from the second motion shaft of the change gear by a roller chain on the sprocket wheels K K′.

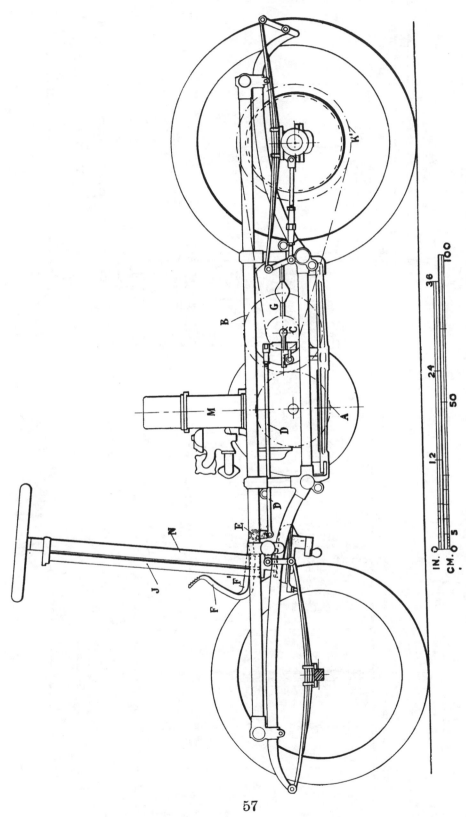

FIG. 67.—THE GOBRON-BRILLIÉ 14-HP. CAR: ELEVATION OF FRAME AND GEAR.

57

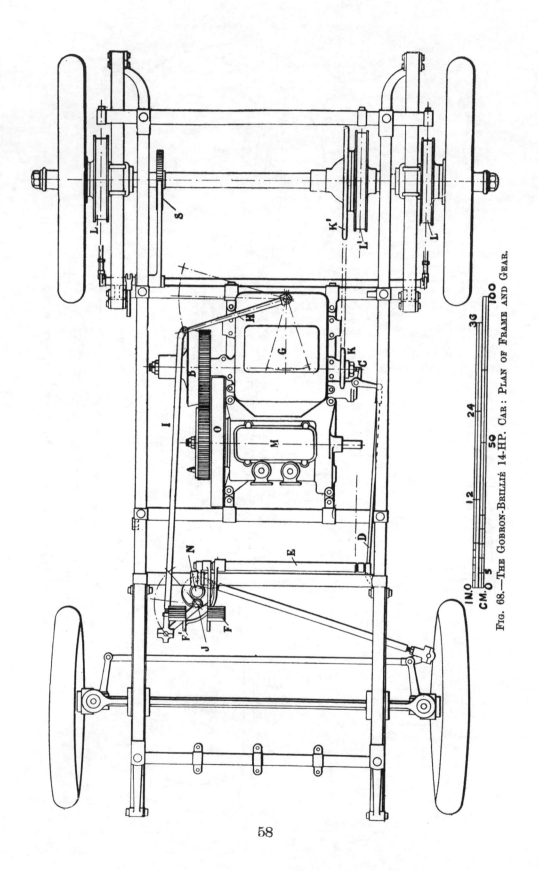

Fig. 68.—The Gobron-Brillié 14-H.P. Car: Plan of Frame and Gear.

58

The design of the motor, with two vertical cylinders in which move four pistons, is practically identical with that previously described. The power is transmitted from a raw hide pinion A, Figs. 67 and 68, mounted by the side of the fly-wheel O, and gearing with a spur-wheel and clutch B (Figs. 67, 68, 70) on a spindle carrying a pinion H, on the side of which is a claw clutch. The spindle upon which the male part of the cone clutch is fixed by a feather, is turned down to a small size within the gear box, and runs within a tubular part E' of the spindle. The gear wheel next the cone clutch is loose on the spindle, the pinion H being fixed upon it. The tubular part of the spindle carries the wheels A' and D loose upon it, but driven by a feather. The pinion H is always in gear with and drives the wheel H. For the slowest speed the wheel A' is put into gear with the pinion A, the speed of the hollow spindle E' and the sprocket wheel K being

Arrangement of gearing

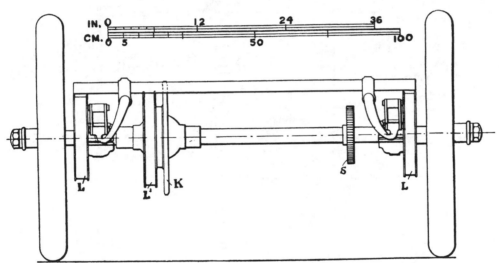

FIG. 69.—GOBRON-BRILLIÈ CAR: BACK ELEVATION.

then less than that of the cone B in proportion to the relations between the pinion H with the wheel H multiplied by the relation of pinion A with wheel A'.

For the second speed the wheels D D are put into gear, and for the highest speed the wheels A' D are slid along the spindle until the two claw clutches on wheel D and pinion H engage. This connects the tubular spindle E' with the spindle E; all the wheels and pinions except H and H are out of mesh, and the sprocket wheel K is driven directly by the cone clutch at the speed of the engine, the gear wheels exterior to the gear box being of the same size.

For reversing the wheel A' is put into gear with the pinion R' driven by the pinion R on the spindle F.

The driving clutch B is operated by a pedal F, Fig. 67, which by a small arm on the end of the lever carrying it gives motion to the rod D and the bell-crank which pushes a rod, the end of which is seen at c. This rod c passes through the spindles E E' and pushes the inner part of the cone clutch out of the exterior part against the resistance of the spring contained within the boss of the cone.

The speed-changing wheels A'D, are moved by a fork which fits the groove on the sleeve between them, the fork being moved by a hand-lever on the spindle J parallel with the steering column. On the lower part of this spindle is an arm to which the rod I is connected, this rod being also connected to the lever H, Fig. 68. This lever H actuates the fork referred to.

Brakes There are three brakes, one L', Figs. 68, 69, on the differential gear box, and one on each of the driving wheels at L L. That on the differential gear box is operated by the pedal F' on the tubular spindle E, on the near

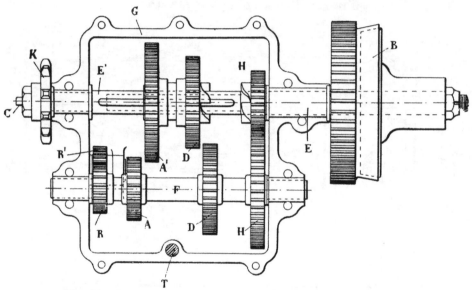

FIG. 70.—GOBRON-BRILLIÉ CHANGE-SPEED GEAR AND CLUTCH.

end of which is an arm to which a rod is connected to the band on L' in-**Ratchet** dicated in the drawing only by a central dotted line. The side brakes are **sprag** operated by a hand-lever not shown. At s, Figs. 68, 69, is a ratchet wheel, and a catch, which may be dropped on to it, to be used in the place of the ordinary sprag, to prevent the car running backward downhill.

It will be noticed that the three speeds and the reversing gear are operated by one lever, and that the same system could be applied to a four-speed gear. It is simple, and the single chain drive to the live axle is also a simple one, and should be mechanically efficient.

Since the date of the car shown by these engravings, 1901-2, MM. Gobron-Brillié have followed the example set them by nearly all makers, **110-HP.** of placing the engine longitudinally in the front of the car, under the usual **Racing Car** bonnet. The arrangement they have adopted is in fact very similar to that which they chose for their racing cars, one of which is shown by the engraving Fig. 70A, as made for the 1903 Gordon-Bennett race—a design which has been closely adhered to by them in their racers of 1904 and 1905. Of the car illustrated, it may be mentioned that the engine is of nominally 110-HP., and it has cylinders of 140 mm. or 5·51 inches bore.   The combined

FIG. 70A.—GOBRON-BRILLIÉ 110-H.P. RACING CAR. 1903 TYPE.

strokes of the two pistons in each cylinder is 200 mm., or 7·87 inches, the stroke of the upper pistons, with the long connecting rods, being shorter than that of the lower pistons, and as arranged in the smaller engines.   The stroke of the lower pistons is about 1·27 times greater than that of the upper

61

pistons. During 1904 Rigolly drove the racing car of that year's construction at a recorded speed of nearly 95 miles per hour.

Since 1904, pressed steel frames have been adopted instead of the trussed tubular frames previously used, and still retained for the racing cars. The 20-HP. two-cylinder car has been followed by more recent types of 25- and 35-HP., with four-cylinder engines of the Gobron-Brillié type. The combined stroke of these engines is 190 mm., or 7·48 inches, and the diameter of the cylinders of the smaller and larger types 90 and 100 mm. respectively, or 3·54 and 3·93 inches. A carburettor of the float feed spray type is used, designed to provide a more nearly constant richness of the **Automatic Carburettor** carburetted air mixture. To obtain this result a movable cone, capable of receiving a sliding movement on the spraying nipple, obstructs or throttles the area of the annular air passage to the mixing chamber at about the position of the upper end of the nipple. The sliding cone is interconnected with the throttle valve through a series of levers and a cam, in order that what is deemed a suitable rate of flow of air past the spraying nipple may be obtained for various degrees of opening of the throttle valve. The method of obtaining differential movements of the throttle valve and sliding cone is ingenious, but it is questionable whether, with this arrangement of parts, any other than a simple throttle valve effect is obtained, except through a limited range or at a critical position of the sliding cone, when, by direction of the air stream upon the tip of the nipple, additional inductive effect is obtained, with perhaps more complete atomisation of the petrol spray. That part of the carburettor which may be described as the mixing chamber, and between the spraying nipple and the throttle valve, is water jacketed, branches being taken from the cylinder jacket pipes.

The distinctive form of measured feed carburettor described in vol. i., at page 191, is no longer used, but with it satisfactory use could be made of alcohol and other spirits heavier than petrol.

Double Cone Clutch A particular form of double cone clutch is used with these cars, in which a centre cone engages first, the friction surfaces being metal, and the driving effort having been partly taken up, the leather-lined cone clutch of greater diameter then engages. This arrangement permits the use of a clutch of ample dimensions, whilst the provision of the smaller clutch, with metal friction surfaces, allows some slip to occur at the moment of engagement.

The arrangement of change-speed gear box now adopted provides for four speeds forward, and the reverse speed, with a direct drive through dog clutches, on the faces of adjacent gear wheels for the highest speed. The bevel driving gear wheels and the differential gear are in the gear box, and Oldham universal joints are used between the differential shaft ends and the side chain pinion spindles

THE WOLSELEY LIGHT CAR

Another small power car which has attracted a great deal of attention is the 5-HP. Wolseley car, made by the Wolseley Tool and Motor Company, Birmingham, from the designs of Mr. H. Austin. This car, like the 10-HP.

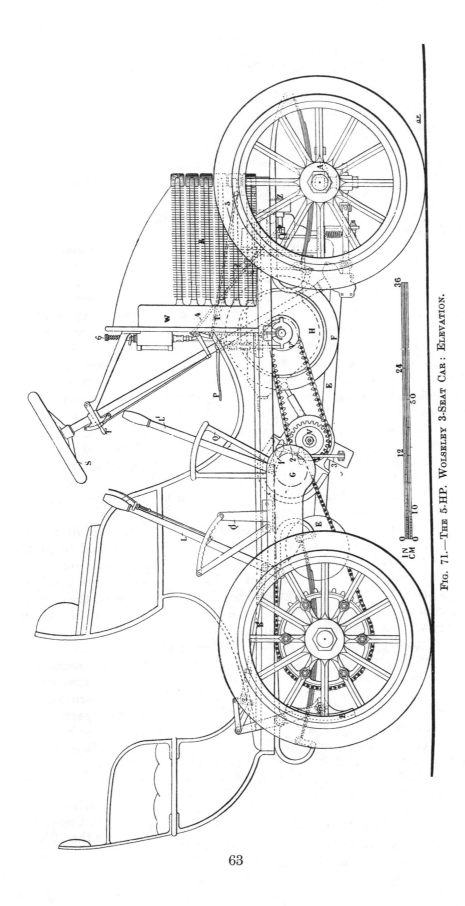

Fig. 71.—The 5-H.P. Wolseley 3-Seat Car: Elevation.

which will be hereafter illustrated, and the still larger cars by the same manufacturers, is of English design and make throughout, there being no part of it which is due to Continental design. It is illustrated by Figs. 71 to 80.

It will be seen that the chief parts of the mechanism and the body of the carriage are attached to a channel steel under-frame, which is supported on the front and rear axles by semi-elliptic plate springs. The engine M, which is of the single cylinder horizontal type, is located towards the front end of the frame, its longitudinal centre line being somewhat below the level of the frame, but on centre line of the frame in plan. The crankshaft lies across the frame, and at either end outside the crank chamber are mounted the fly-wheels F′ F″, Fig. 72; see also Fig. 76. On the off fly-wheel is the cone friction clutch, by means of which the power is transmitted to the chain pinion v′, whence it is conveyed by the chain drive v′ v to the speed-changing gear enclosed in the case G, and finally by chain drives to the rear road wheels of the vehicle. The petrol supply is carried in a tank placed under the bonnet at the front of the underframe, and is supplied to the carburettor by the pipe o and float vessel C.

General Arrangement

The exhaust from the engine passes through pipe E′ into the silencer E and escapes thence into the atmosphere. The cooling water is contained in the vertical tanks w and in the pipes of the radiator R and other parts of the water circuit w w. The throttle valve in the inlet pipe is controlled from the lever T′ on the steering pillar, the connexion being effected through a bell-crank and rod T. A similar lever controls the timing of the electric ignition, the connexion being by rod 5 to the contact maker z on the extremity of the camshaft of the engine.

To facilitate starting the engine the exhaust valve may be held open during part of the suction stroke; this is effected from the dashboard by the connexions 4, J′, Figs. 71 and 72, which slides a small cam tongue into position opposite the main exhaust cam, and thus raises the exhaust-lifting roller a second time. The tongue can be seen in its relation to the main exhaust cam in the plan, Fig. 72, at J. All the vital points on the engine are oiled from the sight feed lubricator 6 on the dashboard.

The change-speed gear is operated from the hand-lever L′, which moves in the sector Q, and the connexions to which will be hereafter described. Steering is controlled from the hand-wheel s on the pillar S, the connexion to the road wheel steering links A being effected through a bevel and worm gearing, to be hereafter described. The vehicle is controlled by three brakes, a countershaft band brake at B″ operated by the pedal P′, which works on the same spindle as the clutch pedal P, and two internal shoe brakes B B acting on drums attached to the rear road wheels. These brakes are controlled by the hand-lever L in the sector Q′, the motion being transmitted to the brake block spindles by the link B′ and the equalizing links connected to it.

Brakes

Referring now to the details of the 5 HP. engine, Figs. 74, 75, and 76, the central portion of the crank chamber B and the head and body of the

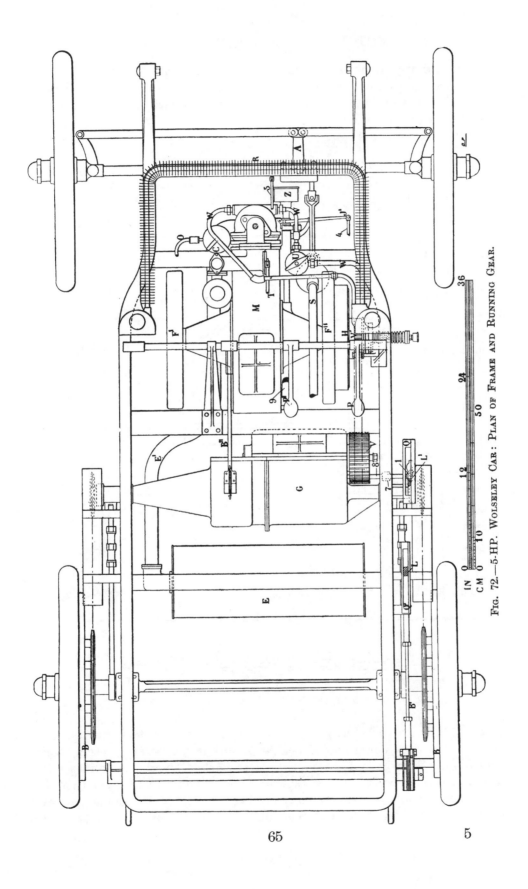

FIG. 72.—5-HP. WOLSELEY CAR: PLAN OF FRAME AND RUNNING GEAR.

cylinder H C are of cast iron, while the long bearings of the crankshaft E are carried in aluminium end pieces, N N. The outer jacket shell is now **Engine** also made of aluminium. The induction valve A, which is automatically operated, is placed directly over the mechanically operated exhaust valve E. The induction valve is controlled by the double spring P, outside the valve

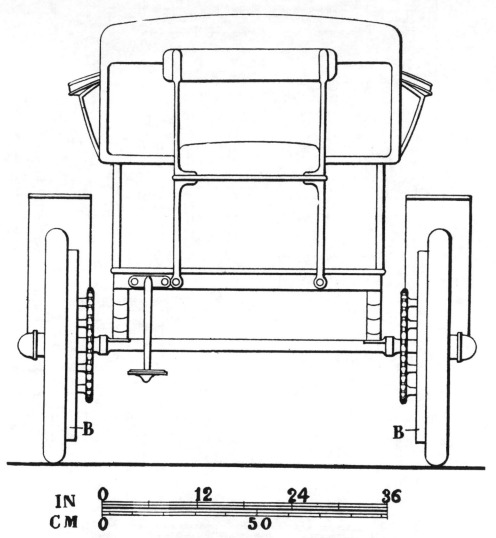

FIG. 73.—WOLSELEY 5-HP. 3-SEAT CAR: BACK ELEVATION.

box J. The exhaust valve is of hard cast iron, with a steel stem screwed and riveted in. The valve is heated to admit the screwed stem, which would not otherwise enter, and the latter is then riveted over in a slight taper. It is held on its seat by a spring S, the tension of which is adjustable. The exhaust valve cam K is seen in Fig. 75, the valve being operated through the bell crank L. The sparking plug is inserted at I between the

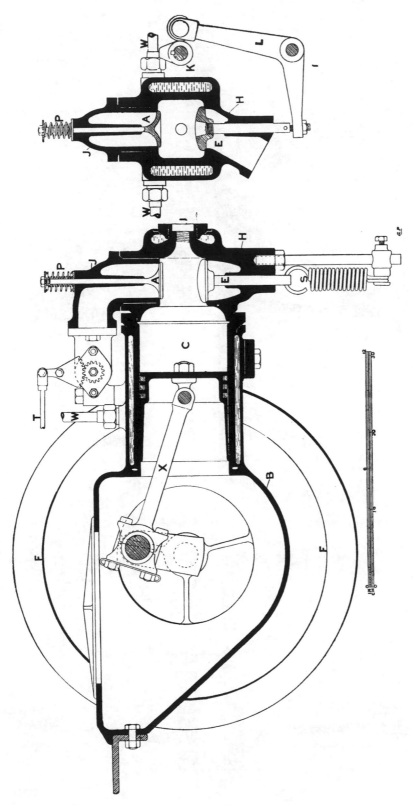

FIGS. 74 AND 75.—ENGINE OF WOLSELEY 5-HP. CAR.

67

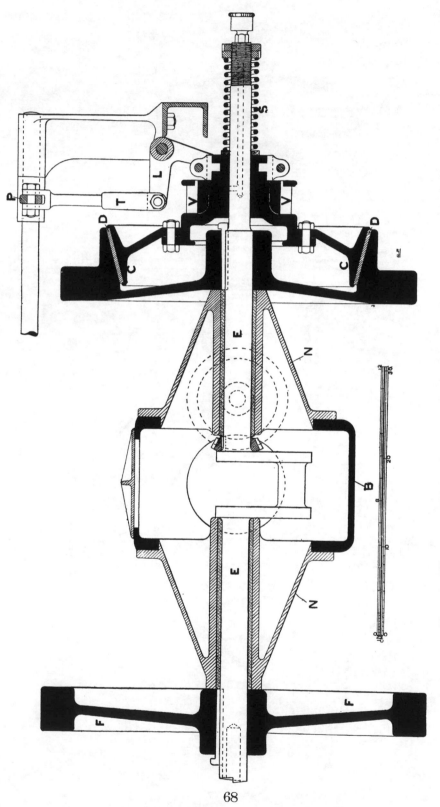

FIG. 76.—WOLSELEY 5-HP. CAR: SECTION THROUGH CRANK CHAMBER AND MAIN CLUTCH.

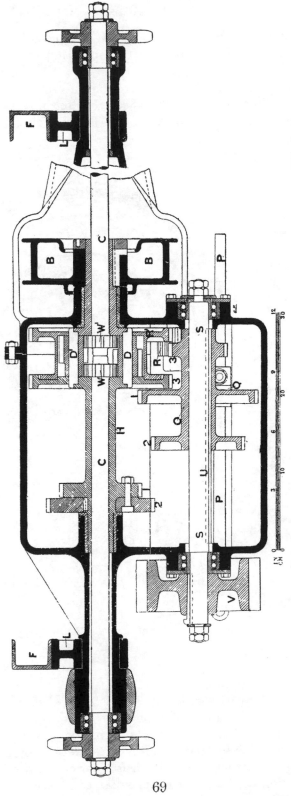

FIG. 77.—Transmission Gear, Differential and Speed-Change Gear: Wolseley 5-HP. Car.

69

inlet and exhaust valves. W w, Fig. 74, are the inlet and outlet pipes for the jacket water. The throttle valve operating mechanism is seen at T, Fig. 74.

Fig. 76, which is a vertical section in the plane of the crankshaft, shows the position of the two fly-wheels, one forming part of the cone clutch C D, together with the pinion v′ of the Renolds chain drive. The clutch

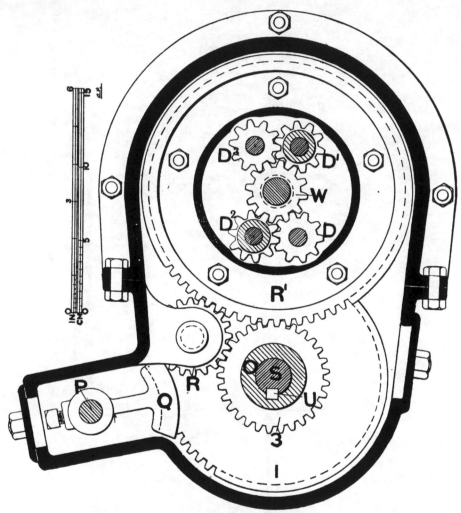

FIG. 78.—GEAR AND DIFFERENTIAL GEAR WOLSELEY 5-HP. CAR.

is held engaged by the spring s, and is controlled from the pedal lever through the links P T and bell crank I.

Engine dimensions The cylinder of this engine is 4·5 bore × 5″ stroke, and 5 BHP. is developed at 700 revs. per minute. The details of construction of the speed-changing gear and the differential gear are shown by Figs. 77 and 78. The first motion shaft s of the change gear carries the wheel v of the Renold silent chain drive, and on it is mounted the sleeve O, carrying the three spur pinions 1, 2 and 3. This sleeve is driven by the feather U. By means of a sliding rod P, seen at 8, Fig. 72, and arm Q, Fig. 78, the sleeve O may be moved along shaft S, so as to alternatively engage the pinions 1, 2, 3 with their corresponding wheels, 1, 2, 3, two of which, 1 and 2, are fixed on the sleeve H of the differential gear box, while 3 is mounted direct on the differential gear box. A reverse motion is obtained by causing pinion 3 to engage, as shown in Fig. 78, with the broad pinion R, which is always in mesh with wheel R′ on the differential gear box. The differential is of the spur-wheel type. In it the pinions D, D_1, D_2, D_3 run on pins fixed to

Change speed gear.

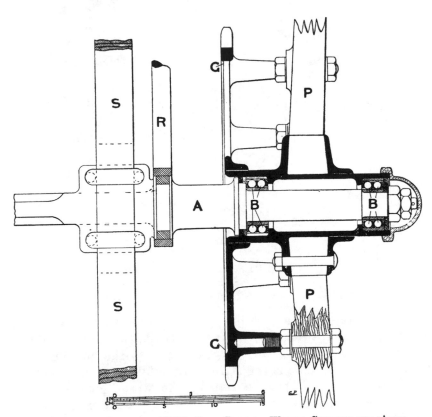

FIG. 79.—WOLSELEY 5-HP. CAR: DRIVING-WHEEL CENTRE AND AXLE.

the differential box, and engage, two with wheel W on one half of the differential cross shaft C, and two with W′ on the other half of this shaft.

The extremities of the change gear box, in which are located the ball bearings of the cross shaft, are carried in brackets L, attached to the underframe F of the vehicle. The cross shaft band brake acts on a drum B, Fig. 77, which is keyed to an extension of the differential box sleeve. The arrangement of this pedal band brake, and that of the brakes on the rear wheels, are similar to those of the 10-HP. car, which will be described in connexion with that car. All the four road wheels in this car run on ball

71

Ball bearings to road wheels

bearings. Fig. 79 shows a section of one of the rear wheels. The method of attaching the chain-wheel rings G to the spokes of the wheels is also shown. At B B are double ball bearings. At R is seen one of the radius rods, by which the fixed but adjustable distance is maintained between the rear axle and the sprocket pinion shaft C of Fig. 77.

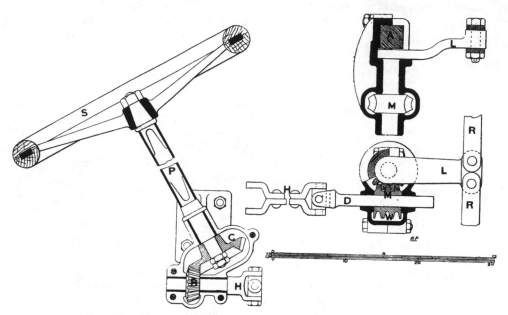

FIG. 80.—WOLSELEY 5-HP. CAR: DETAILS OF STEERING GEAR.

Steering gear

The chief details of the stearing gear mechanism are shown in detail in Fig. 80; S is the steering wheel attached to the steering pillar P, at the lower end of which is a bevel wheel G, gearing with the bevel pinion B on the spindle H. This spindle is coupled to the spindle D, from which motion is transmitted to the T piece L through the worm gear M. The links R R couple the T piece L to the arms on the steering wheel pivot pins. It will be noticed that the links or connecting rods R and the forked connector H are made of rolled steel bar, thus avoiding possible flaws in forgings if used for the purpose.

Chapter III

THE WOLSELEY, DE DION BOUTON, AND JAMES AND BROWNE PETROL CARS

ALTHOUGH not belonging to the light class of cars already dealt with, it is expedient to follow the description and illustrations of the 5-HP. Wolseley with those of the 10-HP. car, as some of the details, such as the steering gear, road wheels, rear brake gear, and spring hangers, are common to both except as to dimensions. The general arrangement is the same of both cars, except in so far as the difference in the motors introduces variation.

The 10-HP. car is illustrated by Figs. 80a to 87, in which Fig. 80a is from a photograph of a car with a tonneau body of usual pattern, but the body may be of any desired type from that shown, or bodies similar to the landau or brougham forms may be fitted.

Fig. 81 is a side elevation of the car, and Fig. 82 is a plan of the underframe and running gear partly in section.

The engine has at M M two cylinders 4·5 in. diam., the stroke being 5 in. **Engine** These are connected to cranks in the same line and placed so that they have a good bearing centrally between them, the end bearings being of great length, like those in the 5-HP. engine, but longer. The carburettor c is placed centrally between the two cylinders but above them, the air being drawn in through the filtering vessel A. The admission valves, mounted on the cylinder heads, seen on either side of the carburettor c, are connected by flanges to a breeches piece, flat slide throttle valves, and their grids being interposed between the central flanges which couple this breeches pipe to the passage from the carburettor. The throttle valve is in two parts, one of which is actuated by a small lever on the steering column and a rod connexion therefrom, the other being connected by rod and bell crank to the pedal which actuates the countershaft brake, this throttle valve being so set that the first part of the downward movement of the pedal gradually closes the throttle, but never completely shuts off the gas even when the pedal is hard down for obtaining full effect of the brake. When this brake is put hard on the driver must declutch, as the brake pedal does not do this, the connexion which is usual in many cars not being adopted. The connexion of pedal brake and throttle valve is extremely convenient for driving in traffic, and declutching by the brake

73

Interconnection of pedal brake and throttle valve

pedal would often be inconvenient. The engine is not governed, but as it will run down to a very low speed on the pedal throttle, the running is more economical than when running by brake and declutching with an engine running on the governor. The simultaneous operation of pedal brake and anticipating throttle secures a natural sequence of the necessary operations of control in traffic, namely slowing down by throttling, followed by putting on brake and conversely release more or less of brake accompanied by increase of gas admission and engine speed.

Construction of pedal brake

The pedal brake drum B′ is provided with a deep internal flange to carry water, though it is not often that it is found necessary to use the brake so much as to make the water cooling important. The brake is shown in detail on a large scale in Figs 86 and 87. In these the brake drum

Fig. 80a.—The Wolseley Long Wheel Case 10-HP. Car with Tonneau Body, 1903 Type.

D on the spindle s is surrounded by the wood blocks w attached to the steel band B, $\frac{1}{16}$ inch thick, the ends of the band being pivoted at c upon the vertical arm of a bell crank lever and near F, at the end of it, the lever itself being suspended and pivoted at P by a link L pivoted, hinge-like, at B to the aluminium extension G of the gear case. Connected to the top of the arm c of the bell crank lever is the rod R actuated by the pedal, the spring o returning the lever when the pedal releases it. From the edge elevation it will be seen that the levers are double and well arranged. Between each wood block a small distance piece F is fixed. The brake is adjustable

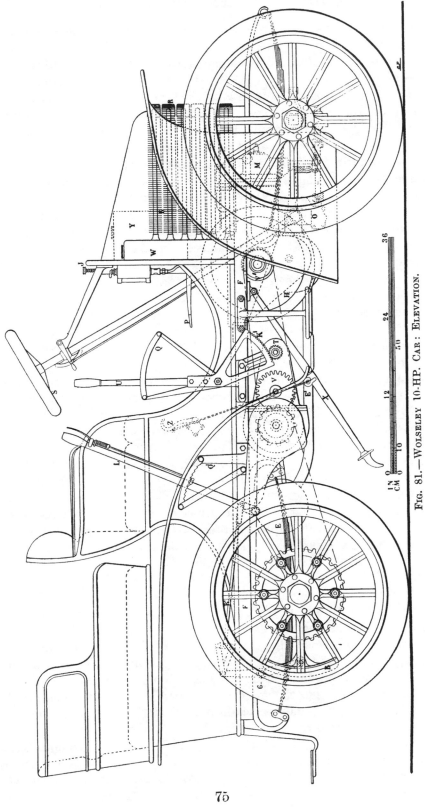

FIG. 81.—WOLSELEY 10-HP. CAR : ELEVATION.

by the three alternative pin holes in the levers c and by the screwed end of the rod in the piece R.

The arrangement of the gear changing apparatus in the 10-HP. is not the same as in the 5-HP. It will be gathered from Figs. 81, 82, 83, the gear-changing lever U is fitted with a toothed quadrant K′, which actuates a pinion T on the outer end of the spindle which carries the grooved **10-HP. change-speed gear** cylindrical cam K. The movement of the lever U to either of the four positions for the four speeds forward or the backwards positions for reversing rotates the cylindrical cam, so that the striking forks, see J′, Fig. 83, which have projections engaging in the cam grooves, are moved to any of the positions necessary to place the wheels 1 1, 2 2, 3 3, 4 4, alternatively into gear, as described with reference to Fig. 82. Part of the cylindrical cam K and one of the striking forks J′ are shown in Fig. 83, the main purpose of which is to illustrate the automatic secondary clutch of the speed-changing gear. **Object of secondary clutch** The object of this is to secure complete freedom of the spindle N and the gear wheels from the inertia effects due to the weight of the rotating parts after the main clutch is out.

As shown in the plan, Fig. 82, when the main clutch H is taken out by the pedal connexions, like those shown by Fig. 76, and the cam K moved by the lever U, so that none of the gears are in mesh, there are several parts running free, falling in velocity, and to be accelerated again when by the further movement of the lever the gears are again in mesh for some other speed. These parts are the internal part of the clutch H and the connected driving pinion v′, the chain on v′ and v and the spindle N, with all that is on it. By means of the arrangement of secondary clutch shown by Fig. 83 the mass to be thus accelerated is reduced to little more than that of the spindle and the gear wheels upon it inside the case.

On the spindle carrying the grooved cam cylinder K is a second cam with one groove K′, in which the roller on the piece L engages. The piece L surrounds the sleeve I, on which the driver chain wheel v is fixed. This sleeve is capable of movement on the boss fixed on the end of the spindle N and towards N. On this boss is a square-toothed part c c′, which engages normally with similar teeth on a steel ring fixed to the inside of the pinion v, the teeth c of the clutch engaging with the corresponding teeth c′ of the inner ring as shown.

When the cams K and K′ are rotated, the first action is the movement of v inwards, disengaging the clutch c c′. The main clutch, pinions v′ and v and the chain upon them, are then free. Further rotation of the cams engages one of the pairs of gear wheels, and any difference of speed **Operation of change-speed gear** between the wheels on the spindles N and G only involves changing by acceleration or retardation the speed of the spindle N and its wheels. The clutch c c′ then engages, and the chain pinions v v′, chain, and inner part of the main cone clutch are then accelerated. The driver then allows the main clutch to engage in the usual way. This is a very ingenious device for avoiding the heavy shock which the inertia either of motion or of rest of the parts between the gear on the sprocket pinion shaft and the cone

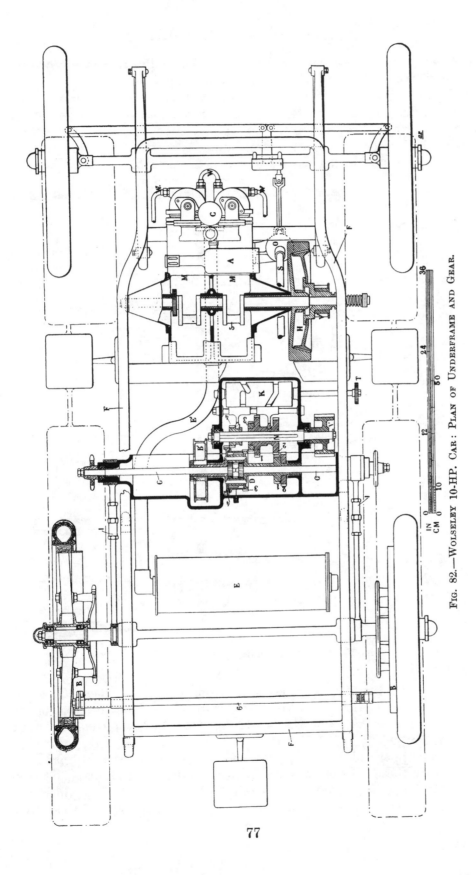

Fig. 82.—Wolseley 10-HP. Car: Plan of Underframe and Gear.

77

clutch would otherwise put on the gear and spindles, stresses too often under-estimated, and very easily and often excusably so.

Figs. 84 and 85 illustrate the shoe brakes, which operate on the internal brake ring fixed on the driving wheels. Fig. 84 is a side view of the brake gear, and shows also the spring hangers for the hind axle springs. In this F is a part of the channel steel frame to which the spring hangers D D′ in one double-ended piece are attached. The blocks B are pivoted

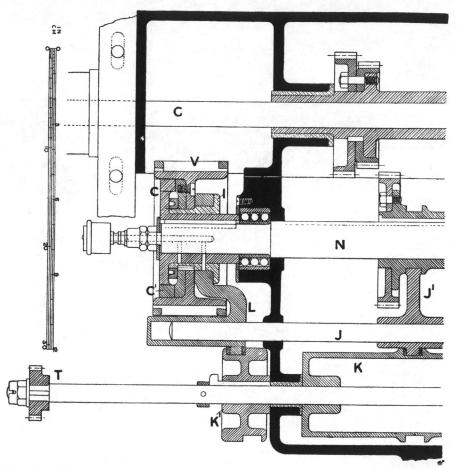

FIG. 83.—WOLSELEY 10-HP. CAR: SECONDARY SPEED-CHANGING CLUTCH GEAR.

on the ends of a bar P′, which goes across from one to the other. This bar is suspended by the double-armed lever H pivoted freely on the cross bar P, and by the arm J′ keyed on the bar P, Fig. 85. The arm J is also keyed upon the bar P.

To the upper ends of the levers H and J are attached pairs of links K K′, and between these pairs of links are pivoted crosshead pieces T T. Between these crossheads is pivoted a connecting rod socket end, into which the rod B′ is screwed with a left-hand thread. At the other end of this rod is a corresponding socket screwed right-handed, this socket being pivoted to the brake lever L.

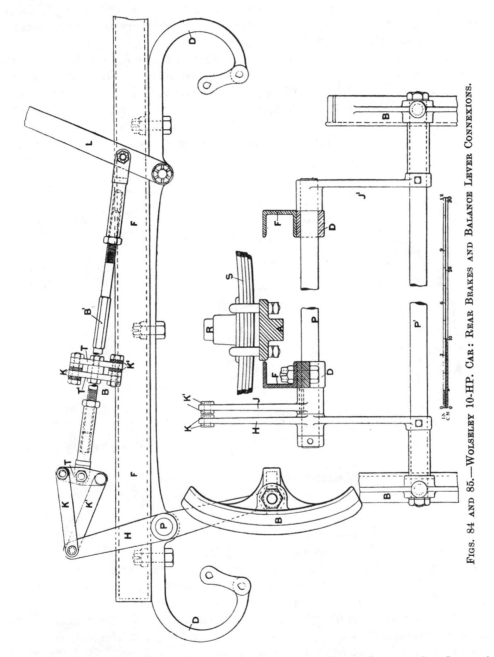

FIGS. 84 AND 85.—WOLSELEY 10-HP. CAR: REAR BRAKES AND BALANCE LEVER CONNEXIONS.

Now it will be seen that the brake lever pull on the crossheads T T is equally imparted to the links K and K′, the links K putting on the brake block immediately below the lever H, and the links K¹ putting on the brake through the lever J, bar P, and lever J′, the same pressure being exerted on each, even if the brake blocks are not at equal distances from the brake

Equalisation of braking effort

79

ring when off, or even if the pull on the brake lever causes a small torsional spring of the bar P. The two views, Figs. 84 and 85, have been made to overlap to save space, but it will be seen that part of the spring is in its proper place with relation to the frame F and hangers D D. The axle is seen at A, and the rubber chock block at R.

Since the above drawings were made the main axle A is no longer forged with ears for clip bolts to connect to the springs. The axle is instead turned out of a solid bar of rather high carbon Vickers-Maxim steel, with rounded shoulders where reduction in diameter occurs, the form of clip and clip bed for the springs being modified to suit. This no doubt forms an excellent and trustworthy axle, and at the same time conforms to modern

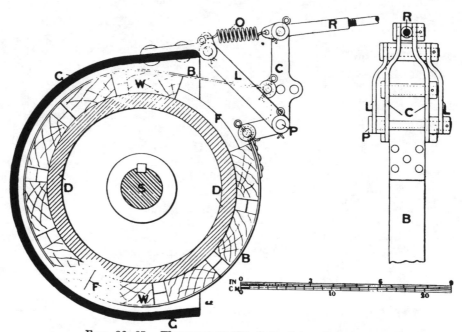

FIGS. 86, 87.—WOLSELEY 10-HP. CAR: PEDAL BRAKE.

Design of axles practice of machine tool work in preference to forging when possible. For the axles of some of the larger power racing cars Mr. Austin has adopted the locomotive side rod practice of milling out the sides of rectangular steel bars. He thus gets strength and lightness with facility for using rolled bar sections, and leaving metal on where wanted.

For making the forked ends of the fixed part of the steering axle, he puts a saw-cut about six inches down the ends of a bar of steel, and then opens the ends out to form a ⟶⟨ , and thus forms the axle from rolled steel bar with very little forging and no welding work.

The Wolseley cars present many features of much interest, and are the most noteworthy example of really successful cars with horizontal engines driving direct to parallel shafts without any bevel gear.

THE WOLSELEY MOTOR VEHICLES

The 20-HP. cars are of the same general arrangement as the 10-HP. except that the four-cylinder engine carries a fly-wheel and cone clutch at each end of the crankshaft and two silent chains from the crankshaft ends to the first motion change gear shaft.

The success of these cars is due to the thoroughly good design, material and workmanship of all the parts, and design does not mean merely appearance or simply fitness with regard to function or to durability, but to those manufacturing conditions which must be considered when things have to be well made without any unnecessary cost for labour. The best material should in any case be used, but these cannot be well afforded unless manufacturing conditions are considered and followed up with ability. Where this is not done by men able and interested, the product can only be sold at a profit while high prices are obtainable.

In the most recent (March, 1904) of the Wolseley cars, several detail modifications have been made. One of these is the substitution for the **Internally expanding brakes**

FIG. 87A.—A WOLSELEY 20-HP. CAR WITH TONNEAU BODY.

internal sector-shaped brake shoe, as shown in Figs. 84, 85, by a new form of expanding brake in which a split ring is, as in many friction clutches, expanded into the surrounding ring to secure the frictional or braking hold. Brakes of this form are shown by Figs. 130 to 132, and Fig. 234, pp. 117 and 223.

The Wolseley Company have also fitted their carburettors with a supplementary valve, by means of which an approximately uniformly carburetted charge is obtained through a considerable range of speed.

One of the 20 HP. cars is illustrated by Fig. 87a. It is a handsome, powerful car, fitted with a four cylinder engine, double clutch, and two driving chains from engine crankshaft to countershaft. The frame of this car is now of the pressed steel type.

Since these cars were brought to a standard pattern, the Wolseley Company has made a new light car of 6½ BHP., the main features of which are designed on the lines of the larger cars, but it is fitted with a live back axle driven by a single chain running on a differential gear, and a pinion on the end of a spindle in a three-speed gear box, which also gives a reverse motion. The three speeds and reverse are operated by the one lever.

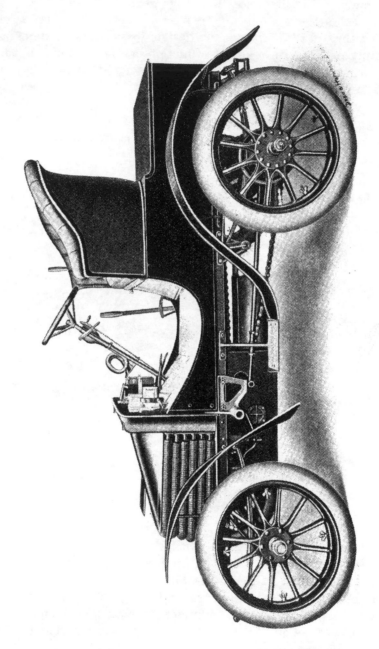

FIG 88.—6·5-HP. WOLSELEY LIGHT CAR, 1904.

6·5 HP.
Engine
The general arrangement of this car is shown by Figs. 88, 89, and 90, which are from photographs, and by the details, Figs. 91 to 93. As in the 5 HP. car, the motor is horizontal and has a single cylinder. It is 4·5

in. diameter and the stroke is 5 in. ; the normal speed is 800 revolutions per minute, at which speed it gives 6·5 BHP. As in the other cars no governors are fitted, and by the use of the throttle valve the speed may be

FIG. 89.—6·5-HP. WOLSELEY LIGHT CAR—ELEVATION OF CHASSIS.

varied from a minimum of about 300 to a maximum of about 1,000 per minute.

The valves are placed as in the 5 HP. motor, at the head of the cylinder they are therefore easily accessible. The exhaust valve is of cast iron, with a steel stem screwed into the valve when hot, the valve shrinking upon

83

it. The form is as shown in Fig. 91, and it gives no trouble and is very durable.

The commutator is driven by a worm wheel on the end of the crank-

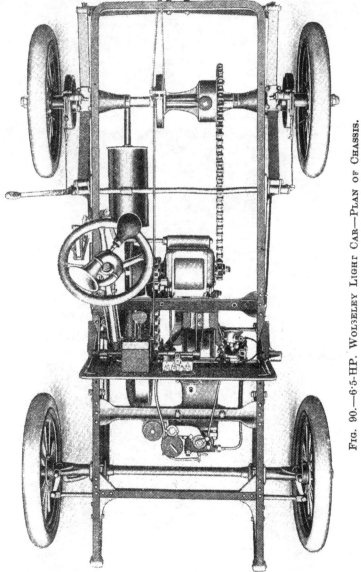

FIG. 90.—6·5-H.P. WOLSELEY LIGHT CAR—PLAN OF CHASSIS.

shaft. It is of the form shown in Fig. 92, in which the wiper A pressed lightly by the small spiral spring s makes an easy sliding or wipe contact with the conducting pieces in the ring of the commutator and accommodates itself to surface without detrimental friction.

Ignition is on the high tension jump spark system, with coil and accumulator as in the other Wolseley cars.

The carburettor is of the ordinary float feed type, admission of vapour

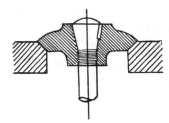

FIG. 91.—THE WOLSELEY EXHAUST VALVE.

being controlled by a throttle valve, moved by a small hand lever on the steering pillar as in the other cars. The lubrication is also practically **Lubrication** identical with that of the large cars. Pipes are taken from a multiple sight feed drip lubricator to each bearing which requires continuous lubrication. The lubricator is fixed on the dashboard and the oil gravitates to each bearing through separate pipes at a rate adjustable at each drip pipe. The oil bath splash lubrication system is not depended

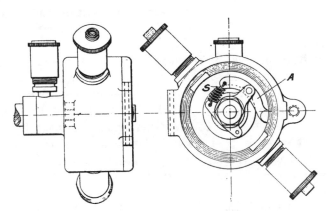

FIG. 92.—THE WOLSELEY COMMUTATOR.

upon for the engine, suitable drip pipes and holes being provided to lead the oil to the connecting rod bearings and pistons. If the supply of lubricant has been correctly adjusted, there will be no accumulation in the crank chamber; but if by over-lubrication such an accumulation occurs, then the excess oil should be drained away.

The gear box is filled with a heavier oil, and only requires occasional replenishing, and the slow moving and less active bearings and pins are lubricated with grease through Stauffer lubricators.

The gear box is separate from the crank chamber of the engine, though attached to it, and the gear gives 7, 13, and 20 miles per hour with normal engine speed, and a maximum of 25 miles. The clutch is similar to those in the other cars, but it is upon the projecting end of the first motion spindle of the change-speed gear. The power is transmitted through a Renold silent chain from a pinion upon the engine crankshaft to a similar pinion fixed to the female part of the clutch, both of which may be moved slightly along the shaft for disengagement or engagement with the male part of the clutch fixed upon the gear spindle.

Live rear axle

The back axle, Fig. 93, is illustrated by a part sectional view and that part not shown to the right of the differential brake drum c' takes the same form as the left-hand end of the axle. The driving chain wheel J is fixed to and drives the differential casing by means of the same bolts that hold the parts of the casing together, and the one centring spigot serves two purposes. The steel axle tubes B are fixed in the long sleeve extensions of the differential casing, and it will be seen that one of these sleeves, that on the right, has cast with it the brake drum c'. The wheel gauge is 4 ft.

Spur wheel differential gear

$0\frac{1}{2}$ in., and the centre line of the car is, therefore, between the chain wheel J and the brake drum c'. In common with the other Wolseley cars, the differential gear is of the spur-wheel type, and any objectionable end thrust such as occurs with bevel gears is avoided. The centre pinions of this gear are formed solid with the driving shafts A, and by adopting this method of construction the number of loose or separate parts is reduced. The outer ends of the shafts or driving axles A are coned and the road wheels are upon these parts, secured by sunk keys and castle-headed hub cap nuts with split pins. It should be pointed out that the road wheel hubs take their bearing upon the outer surface of the tubes B, and that these tubes carry the weight of the car and resist road shocks, the centre shafts A serving only as driving shafts. Inspection of Fig. 93 will show that the driving shafts are smaller in diameter than the bore of the tubes B, and are supported at their outer ends at the wheel hubs and at their inner ends in long bearing bushes close up to the differential pinions. Except when the differential gear operates there is no relative motion between A and B, the latter parts revolving in bearings G, situated immediately below the leaf

Spherical bearings

springs S, which carry the rear part of the car. The bearing shell is at its centre part of spherical form and a certain amount of movement or flexure of the shaft may occur without meeting with resistance from frame connexions or causing the frame and attached parts to undergo unnecessary deformation. The spherical bearing also secures advantages in the erection of the car in that the different parts and bolts, when tightened up, do not necessarily in any way set up twisting stresses, the bearings permitting the axle to accommodate itself to the desired positions of the springs and frame. The position of the longitudinal frame members of channel section is shown at F. A detail is given at Fig. 93 of an alternative form of axle bearing. Two rings of balls H H are used, spaced equally each side of the bearing G, and intensity of compressive stress on the ball race surfaces is thus reduced. Attention may be directed to the construction of the gear box and road wheel bearings as shown by Figs. 77, 79, 82 and 83. In some cases it will be Design of ball bearings observed as many as three adjacent rows of balls are used and for the road wheels two sets of two adjacent rows. When the spindles or shafts are stiff, and the deflection is negligible, this design of ball bearing should prove extremely satisfactory.

The brakes differ from those described on the 5-HP. car. The pedal brake acts on the drum c′ on the rear axle, and the side hand lever brakes are also band brakes acting on drums fixed on the driving wheel naves, one of these being shown at c.

The frame is of light channel steel bars, and is carried on semi-elliptical

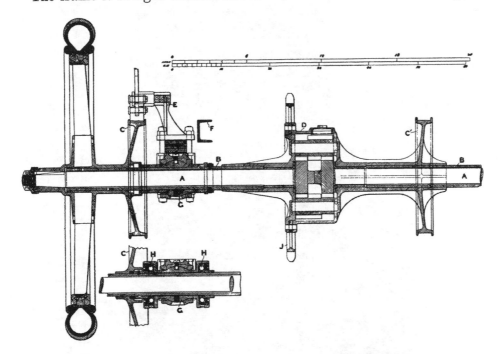

FIG. 93.—6·5-HP. WOLSELEY LIGHT CAR—BACK AXLE.

springs, the back ends of the rear springs being carried by the ends of a transverse spring.

The jacket water-cooling arrangements include a radiator of the Wolseley type of very ample cooling surface, and a circulating pump as in the other cars worked by the half-speed or camshaft of the motor.

The car [1] is one of the best of all the many light cars which have been brought out at a moderate price during the past year. Its driving gear is of the most efficient kind, the car is well made throughout and is simple and accessible. There have been small changes in details of construction but no essential change of design has been found necessary, and there is now a large number of these cars in use.

A modification [2] of the engine design that has not been shown here

[1] *Automotor Journal*, May 7th, 14th, 21st, 1904.
[2] *Motor Traction*, June 22nd, 1905.

relates to the arrangement of the camshaft and the exhaust valve operating lever. The camshaft is now arranged below the crankshaft and parallel

FIG. 94.—THE DE DION-BOUTON LIGHT TONNEAU, 10-HP.

to it, and is driven through spur instead of bevel gear wheels. The side shaft and bell-crank valve operating lever shown by Fig. 75 has been replaced by the parallel shaft within the crank chamber, and a simple rocking lever pivoted in the crank chamber side transfers the motion given to it at one end by the exhaust cam to the exhaust valve spindle at its other end. Laminated leaf springs are also used instead of the coil springs formerly used for the exhaust valves. These changes in arrangement and design were earlier adopted for the larger engines.

THE DE DION-BOUTON LIGHT CAR

MM. de Dion et Bouton have within the past two years produced some light cars differing materially from the voiturettes of which they made

FIG. 95.—THE DE DION-BOUTON DOUBLE CYLINDER ENGINE 10-HP.

so many, as illustrated in vol. i., pp. 255–6. The engine has been placed in front under a bonnet, as seen in Fig. 94, instead of at the back; the single cylinder engine has given place to a double cylinder engine of different arrangement; and the gear has been modified to give a third speed. One of these of 8 HP. has been described with reference to Figs. 16 to 24, pp. 20–25 in the present volume. Another departure by the same firm is illustrated herewith by engravings, showing the 10–12 HP. car with double cylinder engine.

In this type, the De Dion type of gear has given place to sliding speed-change gear, with free-wheel devices to prevent the shock usually resulting from change of speed and to facilitate it.

Fig. 94 gives an exterior view of the car, and Figs. 95 to 104 illustrate the double cylinder engine, the new arrangement of gearing, and the new form of cone clutch employed with it.

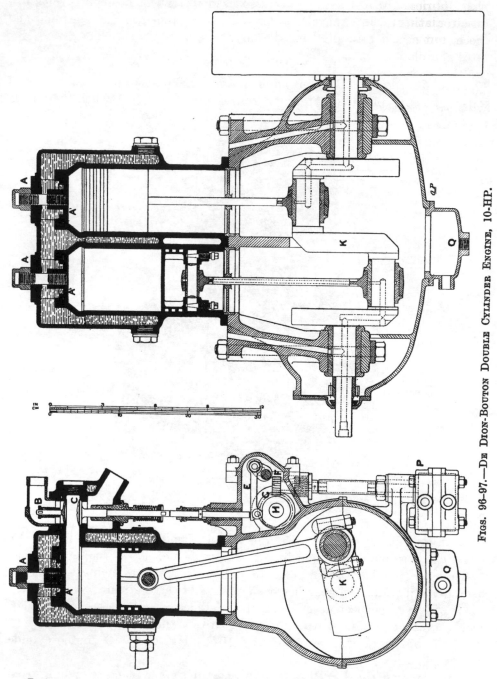

FIGS. 96-97.—DE DION-BOUTON DOUBLE CYLINDER ENGINE, 10-HP.

Method of lubrication   It is not necessary to describe the engine at length, the chief dimensions of which can be found by means of the scale giving metric and English measures ; but references may be made to some of the features. Among these are the simple and direct character of the compression and port spaces, the very complete system of lubrication to all the journals, including the ring lubrication of the camshaft, shown in Fig. 98, and the gear-driven oil-circulating pump P, Figs. 95 and 96. The two cylinders are cast in one piece, but equal thickness is observed in nearly all parts, and water circulates completely round them, so that no distortion of form is likely to occur by rise in temperature. The top of the cylinders is moreover freed from internal stresses by contraction in cooling when cast by coring out the top, which is afterwards filled in by a removable plate, held in place by a stud, and closed nut at A A. The water space is large, and both valves are easily accessible by the removal of one clamp bar and the connexion to the carburettor pipe.

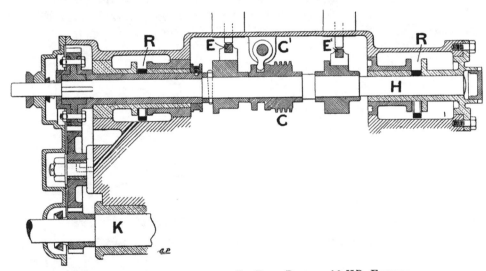

FIG. 98.—CAMSHAFT OF THE DE DION-BOUTON 10-HP. ENGINE.

The oil pump P is worked by a worm G on the camshaft gearing with a wheel F on a vertical spindle seen in the vertical transverse section of the engine, Fig. 96. The worm G is also seen in the angularly taken section on the line of the crank and cam shafts K and H respectively, Fig. 98. The passage to the pump is connected to the pocket Q at the bottom chamber, in the top of which, not shown, is a strainer. From the delivery side of the pump, connexion is made to passages leading to the crankshaft bearings, seen in the section, Fig. 97, one of the pipes for this being seen in Fig. 95. It will be noticed that even the stud spindle of the intermediate spur wheel for driving the camshaft is provided with an oil passage. *See* Fig. 98.

The admission valves are automatic, the exhaust valves being raised by the De Dion-Bouton pivoted lifter arm E, which can be adjusted by the pivot arm so as to be raised earlier or later and for a longer or shorter interval, thus governing on the exhaust.

Oil circulating pump

Governed exhaust

91

For starting purposes, the compression pressure may be relieved by moving the cam spindle H by the finger pivoted at G and working in a groove in the boss of the worm G. The short spindle on which the finger is fixed is seen in Fig. 95.

The form of sparking current make and break employed is seen in Fig. 95, and separately in Fig. 99. In the latter, H is a block upon which are mounted the platinum pointed contact screws G and G^1, with which contact is alternately made by the spring blade F, held by the rocking fork B, pivoted at C, and rocked by the double cam A A^1, against which the fork ends D and E respectively bear. The block is bushed where the cam spindle runs in it, and two spring-held pins S and S^1 press into a groove on the spindle, and keep the block in position. The block is held in position by the lug R, and may be covered by a box held by the studs K K^1. The current is led to the adjustable contact screws by the binding screws, T T^1.

The new form of sliding wheel change-speed gear is shown by Figs. 100 to 103. Fig. 100 is a vertical section of the gear box and gear, one-fourth

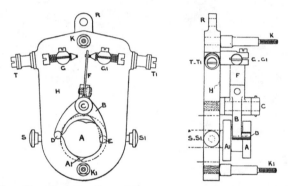

FIG. 99.—THE DE DION MAKE AND BREAK COMMUTATOR.

full size. C is the spindle driven by the engine, and on it are the sliding pinions gearing alternatively with the free-wheel spur gear on the driving spindle D, on the end of which is the bevel pinion E gearing with the bevel wheel, which drives the back axle, by means of the De Dion-Bouton universal joint-rods. The slowest speed is obtained with pinion P^1 gearing with wheel P, and the second, third, and fourth speeds, by the engagement of S^1S, O^1O, and Q^1Q, respectively, the reverse being obtained by the interpolation of a wheel between P^1 and R, the latter and the fourth speed wheel Q not being free wheels.

Rachet gear wheel clutches The driven wheels for the three speeds are made with side cheeks N, bolted together with the spur ring firmly held between them, and running freely on the spindle with the ratchet wheels formed on the shaft and pawls V between the cheeks.

By the use of these ratchet wheels, it is held that in changing from a higher to a lower speed, as in hill climbing, the difference between the speed at the moment of the shaft D and the speed proper to the gear to be put

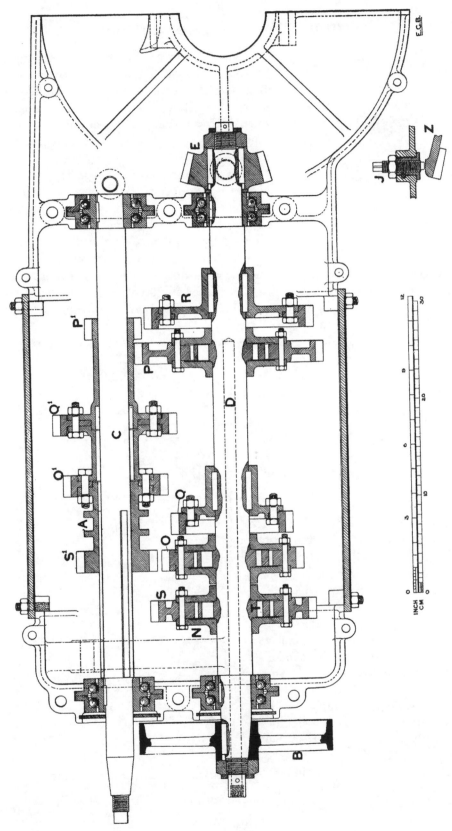

Fig. 100.—The De Dion-Bouton Four-Speed Change Gear.

93

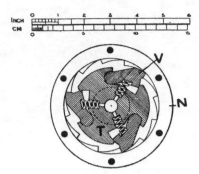

FIG. 101.—DE DION FREE-WHEEL
CHANGE-SPEED GEAR.

in mesh is made up by the overrunning of
the shaft D, until it has fallen to the speed
of C, and the utmost inertia shock felt by
the teeth is that due to putting the free
wheel into motion or checking it. Both
spindles run, it will be seen, in ball bearings
of large size.

Figs. 102 and 103 are respectively vertical and horizontal sections of a part of the
gear box. They show in part the apparatus for sliding the change wheels on the
spindle C by the fork A on the sliding tube U,
Fig. 103 engaging with the groove A, Fig. 100.

On the end of the tube U is a boss with a downwardly projecting finger,
which comes into contact with the flat end of the lever F, on the other end
of which is a pivoted grooved tongue which, when U is at its extreme position,
shifts the reversing pinion into position.

Separately, in Fig. 100, is shown a set screw J which is inserted in the bevel
wheel case, opposite the pinion E,
for preventing the deformation or
shifting out of pitch position of
the bevel wheel Z under exceptional stress.

Fig. 104 shows the novel arrangement of pressure-balanced
or self-contained clutch employed
with inwardly engaging cone.

The fly-wheel is fixed to a disc
K[1] on the end of the crankshaft K
by bolts, which also hold a disc
having a long tubular boss on
which the male portion of the
cone M slides. Within this boss
slides the headed tube Q, carrying
a ball race corresponding with
that on the boss of the cone M.
Formed on the end of the central
tube shown in black section and
fixed to K[1] is a tubular projection
screwed at its extremity inside
and outside which is taper and
slotted to form four horns, between which the four-way head Q
slides. A central nut holds a
collar for retaining the clutch
spring, and outside it is a

**Balanced
clutch**

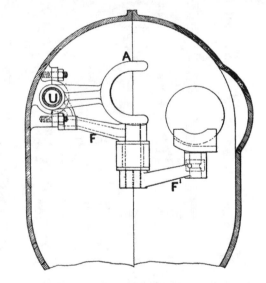

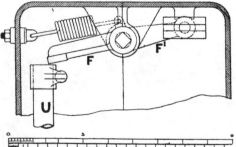

FIGS. 102, 103.—DE DION-BOUTON CHANGE-
SPEED GEAR.

94

large screwed ring, or pinching nut. Through a short range, therefore, the spring pressure is adjustable by this interior cylindrical nut and the exterior pinching nut. Through the centre of the spring and fixed at the inner end of the tube Q is a bolt S with a nut on the end. This bolt and nut does not in ordinary work perform any part, but it limits the separation of the spring and inner tube when the pinching ring nut is taken off. The spring therefore in pushing the cone into the fly-wheels bears against the collar fixed inside the inner end of the tube Q, and against the collar on the outer end of the bolt S, held in place in the adjustable externally screwed

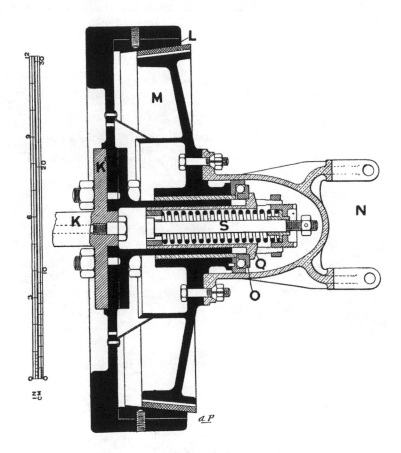

FIG. 104.—DE DION-BOUTON CONE CLUTCH.

collar by the pinching nut. When the clutch is pulled out of gear by a pull exerted through the universal joint N, the ball bearing O comes into play, and receives the thrust of the cone in motion instead of at rest.

Another form [1] of clutch is now being used on some of the De Dion cars in which a flat annular disc is gripped by toggle-pressed friction surfaces. It is not, however, as simple as the Weston clutch as used on some cars.

[1] See *Automotor Journal*, No. 208, p. 1537.

MOTOR VEHICLES AND MOTORS

The James and Browne Cars

A car possessing many features of interest is illustrated by Figs. 105 to 118. Fig. 105 shows it as fitted with a tonneau body, but it is made with various forms of body, including the brougham.

It is of English design throughout, mainly due to Mr. F. L. Martineau and Mr. T. B. Browne. It is propelled by a horizontal enclosed engine with outside cranks instead of the usual bent cranks, the transmission gear being all spur wheels, no bevels being used.

The car illustrated is of 9 HP. An 18 HP. four-cylinder car of the same general design is also made, the modifications being only in the water-

Fig. 105.—James and Browne 9-HP. Petrol Car.

cooler and in the engine so far as the use of four cylinders enforces change of arrangement.

The wheel base of the 9-HP. car is 6 ft. 6 in., the gauge being 4 ft. 1½ in., the wheel base of the 18 HP. car being 7 ft. 9 in., and the gauge the same as that of the 9 HP.

The frame is of channel steel with the channel placed outside. The whole of the machinery is connected directly to this frame, which it helps to stiffen as against racking stresses, there being no diagonal stays or ties.

The engine has cylinders 4 in. bore and 6 in. stroke in both the 9-HP. and 18-HP. type. The inlet valves are 2 in. in diameter and automatic, and the exhaust valves 1½ in. in diameter, with a lift of ⅜ in., and operated by a cam.

THE JAMES AND BROWNE MOTOR VEHICLES

The cylinders and water jackets are cast in one piece, and great care has been taken with the design so that the walls of the cylinder and valve ports shall be all of equal thickness, and all corners rounded so that there

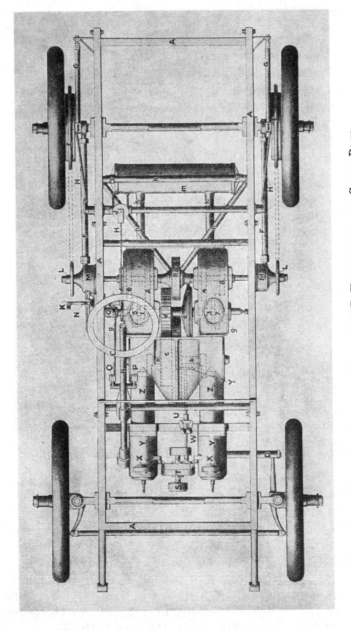

FIG. 106.—JAMES AND BROWNE 9 HP. UNDERFRAME AND GEAR: PLAN.

may be no warping due to heat. The ignition plug is placed in the centre of the cylinder head.

The engine speed is controlled by a governor actuating a throttle valve, and this valve can be opened or closed at will positively by a handle on the steering column, so that the speed of the engine can be varied from that permitted by the governor through a very wide range.

The inlet valves (Fig. 107) are held in their places by stirrups hinged so as to fold downwards, after slacking back a central set pin which presses on a dome cover fixed to the inlet pipe.

Position of valves

The inlet pipe is fixed at its other end to the carburettor by a central bolt, so that by loosening this the whole inlet pipe can be rotated round this bolt and allows of the withdrawal of the inlet valve.

The exhaust valve z is directly under the inlet valve, and by taking off the exhaust spring and withdrawing a cotter which holds the exhaust valve stem proper to a tubular extension on which the spring pulls, the whole valve can be removed through the inlet port.

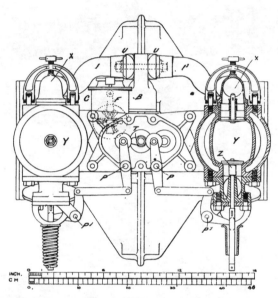

FIG. 107.—JAMES AND BROWNE VALVE CHAMBER AND MOTION.

The combustion head is so shaped that when the piston is at the position of full compression and the valves in place, the combustion space is as nearly as possible hemispherical, there being no port spaces or pockets.

This is shown by Fig 107, and is an advantageous arrangement which is possible with the horizontal engine. In the centre of this illustration will be seen the half-speed cam T, which alternately actuates the rocking levers and links which give motion to the bell crank levers, pivoted at P^1

Springs in tension unsatisfactory

through rocking arms pivoted at P. Formerly the valve operating levers were connected by one spring in tension, holding both valves on their seats. Tension springs have, however, been found very liable to fatigue and fracture, and now one spring in compression is used to hold each valve down. They are ingeniously connected to the valve stem between a cotter and a carrying bracket, which forms a pivot for the valve-lifting levers P^1, and the springs are easily put in or taken out. The valves are large, and there is ample room round them for ingress and egress.

A float-feed carburettor is employed, not shown on the drawing, but placed at C, as indicated in Fig. 108, immediately over the cam box T in Fig. 106. It is formed to receive by spherical union joints the two pipes $i\ i^1$, conveying the carburetted air to the inlet valves, the union joints being kept on their faces by the one bolt previously referred to. The engine runs normally at about 500 revolutions per minute. It runs under, i.e. the outward stroke is made when the crank is describing the lower half of its circular path. The vertical resultant of the angularity of the connecting rod during a working stroke is therefore upward, which tends to reduction of wear of cylinder and piston, and helps in the spread of the lubricant. The crank chambers being closed, the pistons act to some extent as pumps, and the cases being fitted with release valves acting at very small pressure, there

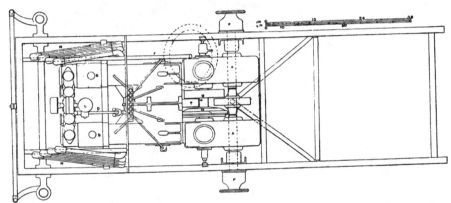

FIG. 108.—JAMES AND BROWNE 9-HP. CAR: UNDERFRAME COOLER AND LUBRICATOR PLAN.

is more of vacuum effect than pressure, and this, it is claimed by Mr. Martineau, draws in the oil and helps in its distribution.

A plan giving an external view of the whole of the frame and running gear is given in Fig. 106, from which it will be seen that the cylinders Y and crank chambers Z are carried by a strong spectacle plate connecting the side frame members, the chambers Z being of aluminium, so that although large they are not heavy. They are connected firmly to the gear-boxes $f, f,$ which at the back end are supported by a strong tubular frame stay. The easily removable inlet valve covers are seen at X, and between the cylinders at this point is the contact make and break at S, driven at half-speed by the small spindle carrying the governor W. The same spindle, it will be seen from Fig. 109, is fitted with a small chain-wheel for driving the pump at about twice the speed of the engine.

Support of engine and gearing

The transmission commences at the buffoline ring spur-wheel, fixed to a projecting ring on the fly-wheel, on each side of which, as shown by the dotted cross lines on the shaft in Fig. 109, is a very long bearing. This spur-wheel gears with a similar wheel on the clutch countershaft, which also carries the sliding gear pinions on squared parts. These pinions drive the corresponding wheels on the sprocket pinion shaft, in the centre of which is the differential gear within a casing, which forms the drum for the pedal brake. The relative position of all the gear is clearly seen in Fig. 109, which also shows that each half of the divided differential gear shaft is carried in bearings with the gear wheels between them, the motion of the shaft being conveyed by a form of Oldham coupling to the short sprocket pinion spindles in the long bearings at M (Fig. 106).

The clutch-shaft ends run in adjustable centre bearings at the ends, the clutch being moved by a rod through the centre of the shaft actuated

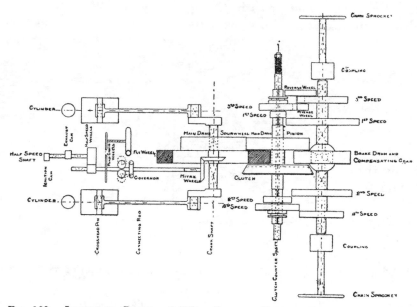

FIG. 109.—JAMES AND BROWNE 9-HP.: DIAGRAM PLAN OF TRANSMISSION GEAR.

by a quick thread cylindrical nut Q, partially rotated by the pedal O and connecting rod therefrom. The clutch is separately shown by Figs. 110, 111, and 112, to be hereafter described.

As shown in Figs. 106 and 109, the movable gear pinions are all in the neutral position, and in Fig. 109 are seen the grooves on the bosses of the third and fourth speed pinions, into which the shifting forks engage. These forks are actuated by a cylindrically-grooved cam, not shown, partially rotated by the speed-change lever. The reversing is effected by the interposition of a spur-wheel and pinion acting on the third-speed driving pinion and first-speed spur-wheel on the differential shaft.

The clutch is of the self-contained cone type, but of original design, as shown by Figs. 110, 111, and 112. In these, Fig. 110 is a perspective view showing the actuating levers seen complete in the end elevation, Fig. 111, and in section in Fig. 112. On the clutch shaft A is fixed the disc F, on the boss of which is a sliding flanged collar K. The disc F carries four fulcrum joints,

J, on which are pivoted four levers I, the ends of which bear against the flange of the collar K. In the centre of the shaft A is the sliding rod M,

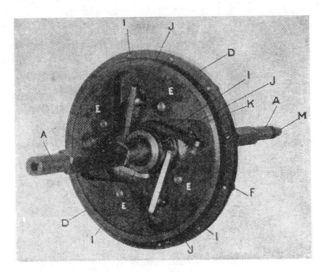

FIG. 110.—CLUTCH OF JAMES AND BROWNE CAR.

which by the action of the pedal pushes against the key L in a slot in the shaft and fixed in the collar K. When collar K is thus moved its flange **Special design of clutch**

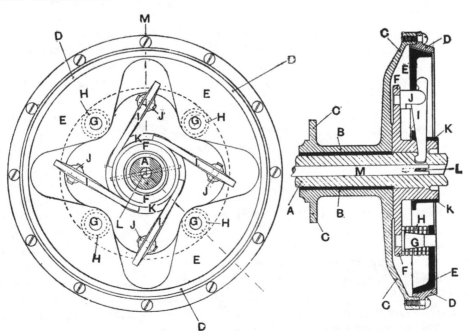

FIGS. 111 AND 112.—CLUTCH OF JAMES AND BROWNE CAR.

presses against the ends of the long arm of the levers I, and their short ends bear against the face of the internal cone disc E, which is carried by

101

the pins G fixed on the disc F, and push the cone inwards, and thus the exterior part D C of the clutch fitted with a bush B, which runs loose on the shaft A. On the boss of the clutch at C is a flange, to which the main driving spur-wheel d (Fig. 106) is fixed. The pressure by which the male part of the cone is pressed into the female part is determined by the four springs H on the pins G. As shown in Fig. 112, the lever I appears to enter the shaft A. This is a consequence of showing the lever complete instead of cutting it and showing it partly in section, as an inspection of Figs. 110 and 111 will show. Owing to the large leverage on the levers I, the clutch is easily operated, and a very small movement is sufficient to declutch. The coned friction surfaces are metal to metal.

The water tank contains four gallons of water, and is fixed inside the bonnet on the dash board, and the radiators are arranged on each side, so that the water from the pump passes through the cylinders, then through the radiators, and so to the tank at the top, the water being thus sufficiently cooled to avoid formation of steam. The water passes direct from the bottom of the tank to the circulating pump.

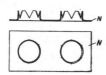

Radiator construction

FIG. 113.—JAMES AND BROWNE WATER COOLER.

The radiators are very originally made up. They are formed of tube 1 in. diameter, and there are 30 ft. of tube on a 9 HP. car, some of them being formed as shown in Fig. 113. The radiator ribs or gills of the most recent forms of the cooler each connect two tubes alternately horizontal and vertical, and so hold the whole radiator together as one mass. The arrangement is clearly shown by Fig. 113, in which the gills N are shown separately and on the tubes V. The gills are set sufficiently far apart to ensure the free passage of the cooling air, and they are thus much more efficient than when placed so close together as to entrap the air, as is often done.

In the 18 HP. car the radiator and tank are combined in one and placed in front of the car. The arrangement of the gills is the same as on the 9 HP. car, but the tubes are vertical and connect the upper and the lower tanks together. The total capacity of the tanks and radiator is 5½ gallons, **Quantity of cooling water** and there are 82 ft. of tube in the cooler. The lower tank is divided into three compartments, and the water from the engine jackets enters the two outside compartments and flows up the tubes (so condensing any steam there may be) into the upper tank, thence down the centre tubes to the central compartment of the lower tank, and so back to the pump. The radiator tubes with the gills upon them are electro-plated with tin by the Martineau method.

The countershaft brake consists of a pair of embracing shoes or straps on the drum R, pulled together by two tension rods and held as against rotation in either direction by the tubular stays h h (Figs. 106 and 108), attached to the longitudinals of the frame with a stretcher cross-tube, forming a triangular brace.

The main brakes are shown in detail by Figs. 114 and 115. They are internal brakes with shoes 5 acting within the rim formed with the sprocket wheel 1. The brake shoes 5 are carried by the flat suspending links 6 and 7, pivoted to the top of the arm 4, clamped to the rear axle, and prevented from rotating by the tension bar 2 connected to the back of the car frame. The brake is applied by the brake lever K (Fig. 106), which pulls the rod H, on the forked end of which is an equalizing pulley J, round which is an equalizing chain pulling on short lever arms acting on the side rods H H. These are connected to the lever 9 of Figs. 114 and 115. which acts on the toggle levers 8, and so forces the shoes forcibly into the rim.

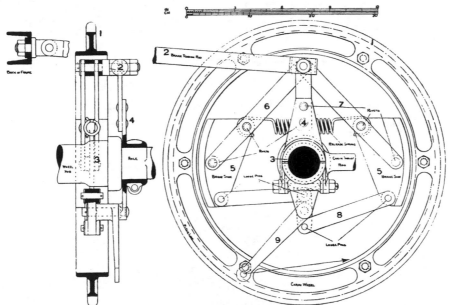

FIGS. 114 AND 115.—MAIN BRAKES: JAMES AND BROWNE CAR.

The steering gear is of simple worm and wheel type.

The pump employed for circulating the cooling water is shown by Figs. 116 and 117, and is of the centrifugal vortex vane type driven by a light chain. The water enters centrally at H at one side of the fan, and escapes tangentially into the passage of increasing diameter to the escape at E. The two sides of the case are alike in form, so that the pump may be put together for right or left hand. A pet cock is fitted at A to permit the escape of air and to test the working of the pump.

The general arrangement of the lubricator connexions to the main lubricator E at the back of the dashboard is seen in plan in Fig. 108, and a section of the lubricator at any one of the sight feed drops is shown by Fig. 118. A row of seven needle valve lubricators, as here shown, is formed on one holder. The oil enters any or all of them by the round hole seen **Lubrication** at the upper part of the main casting, and may be cut off from all by one tap. Each lubricator is separately adjustable, and is turned on or off by

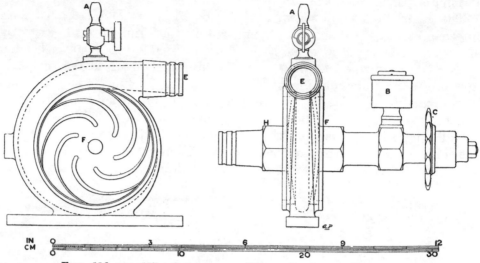

FIGS. 116 AND 117.—CENTRIFUGAL PUMP: JAMES AND BROWNE CAR.

the thumb cap at the top of the needle valve, which allows the oil to drop into the glass tube chamber. The short piece of glass tube is seated in recesses filled with a woodite ring.

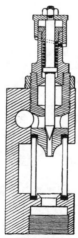

FIG. 118.—JAMES AND BROWNE LUBRICATOR.

The 18 HP. engine has four cylinders side by side, the gear arrangements being similar to those of the car illustrated. An 8-HP. light car,[1] weighing about 9 cwt., has been recently introduced. A two-cylinder horizontal engine with the cylinders oppositely placed is arranged with the crankshaft and gear box across the car or parallel to the rear axle. A single chain is used to transmit the power from the engine to the live axle. These cars have been very thoughtfully worked out and developed, and are of great merit.

[1] *Autocar*, July 1, 1905.

Chapter IV

THE BENZ, BENZ-PARSIFAL, AND DECAUVILLE
CARS

THE history of the Benz cars of the earlier types, and the pioneer work of Herr Carl Benz, were given fully in vol. i., and the forms of cars made by Benz down to 1900 were given. Since that date very large numbers of the Benz cars have been seen in use in this country, and almost every other country which has any roads. In deference to the experience in favour of the front position of the engine, the Benz Co. made some modified types of car with the engine in front, but still with the belt transmission from the engine to the first-motion spindle. With these cars some good records have been made here and abroad.

A 15 HP. Benz phaeton of the 1902–3 pattern is illustrated by Figs. 119, 120 and 121. These cars are of considerable interest to those who still favour a belt drive, and satisfactory experience causes many to believe in this method of transmission.

In the car illustrated a horizontal double cylinder engine is employed, the pistons being $4\frac{7}{8}$ in. in diameter and $4\frac{7}{8}$ in. stroke. The engine is rated as 15 HP. at normal speed, of about 600 revolutions per minute. The engine is throttle governed by hand. A spray-maker float-feed carburetter is employed, and ignition is effected by high tension methods. The circulating pump seen at S is gear driven. The frame is of wood strengthened by a flat bar along the inside.

In the engravings, Figs. 120 and 121, the two cylinders are seen at G G, supplied by one Benz carburetter at C through the pipe F.

On the crank-shaft is a wide pulley 17, which drives the fast and loose pulleys V and U, giving motion to the intermediate countershaft, on which are the sliding pinions within a gear box, and to the wheels on the sprocket countershaft 12.

The change-speed wheels are moved by a fork, seen on rod 20 in the plan, Fig. 121, on which is cut a rack, not shown. In this rack gears a pinion indicated at 18 by a dotted circle. This pinion is on a vertical spindle in the gear box, and is actuated by a similar rack seen in dotted lines in the plan, and partly shown by the rod 19 in the elevation, moved by the rod 22 and

105

Change-speed gear and method of control lever 7. The pedal 16 applies the countershaft brake 11, and the pedal 3 shifts the belt to the fast pulley, where it is held by a catch. The pedal 8 applies the side brakes, and releases the belt fork catch, and this belt is then returned to the loose pulley by a spring. The side brakes are also independently applied by a side lever, and held by it so as to release the foot from the pedal and hold the car when standing on a hill.

The reversing is effected by a movable interpolating pinion shifted by the lever 9 and connexion 21, which are seen in the elevation, Fig. 120. The throttle valve is controlled by the lever I, and connexions to the valve at H, and the advance or retard ignition by the lever P and connexions to the lever J on the make and break commutator.

FIG. 119.—BENZ PHAETON, 15-HP.

A petrol tank is placed at A with a cock on the supply pipe immediately beneath it, and another stop-cock at B on the footboard. Both cylinders exhaust into one pipe K and one silencer L.

A water tank is placed at Q, immediately over the cooler at R, the pipe connexions 25 from the cylinders delivering into the tank and from the radiator by pipe 26 to the pump, and thence to the bottom of the cylinder jackets. The main lubricator is placed at W, a battery at W, and an induction coil at N.

Since this car was made another type of car has been turned out of the Mannheim works known as the Benz-Parsifal car. In this car vertical en-

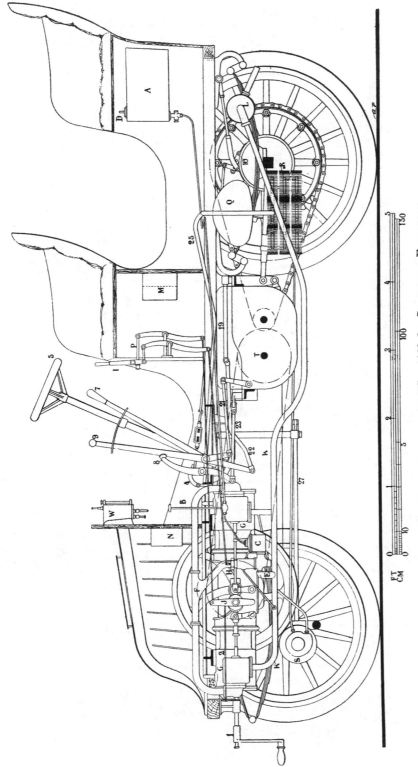

FIG. 120.—BENZ PHAETON, 15-HP., 1902-3 : SECTIONAL ELEVATION.

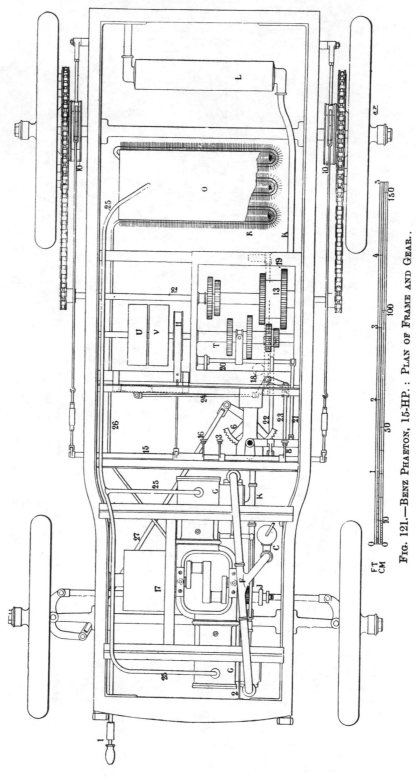

Fig. 121.—Benz Phaeton, 15-HP. : Plan of Frame and Gear.

108

gines are used, and the general arrangement of the gear is that of many manu-facturers, who work on what may be generically called the Panhard-Levassor system, but with modifications. In fact, the Benz firm has ceased to make the belt-driven cars, and in deference to fashion are giving up what

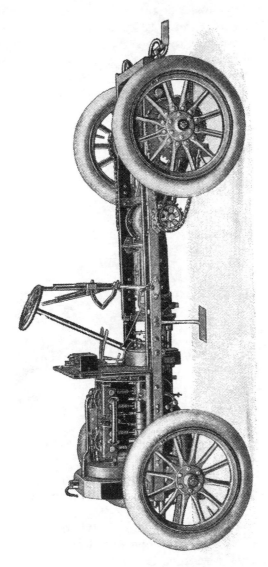

Fig. 122.—Underframe and Running Gear of 16–20 HP. Benz-Parsifal Car.

they themselves are satisfied by long experience produces a most satisfactory soft running, easily maintained car, namely, a belt-driven car.

The Benz-Parsifal Car

Some of the features of the car made towards the end of 1903 are shown by Figs. 122 to 125. Fig. 122 is a perspective view of the underframe and running gear of a 16–20 HP. car, and Fig. 123 a front view. The frame is made of pressed steel longitudinals and cross members, of channel section, and about 4 mm. thick, and a central depth for the 10 HP. car of 4·25 ins.

The engines of the 8 and 10 HP. cars have each two cylinders. The 8 HP. cylinders are 3·54 ins. (90 mm.) diameter, and 4·34 ins. (110 mm.) stroke ; the 10 HP. are 4·75 ins. (120 mm.) and 3·94 ins. (100 mm.) stroke. The engine of the 16 HP. car, to which some of the accompanying engravings relate, has four cylinders of the same size as those of the 8 HP. car, which are said to give 10·5 HP. on the brake.

Fig. 124 shows the engines of the 16–20 HP. car partly in elevation and partly in section. It will be seen that the inlet and exhaust valves are on opposite sides of the cylinder heads, and are all mechanically operated by exactly the same means. They are all easily accessible by the removal of one bridge nut over each pair of valves. The gearing by which

FIG. 123.—FRONT VIEW OF 16–20 HP. PARSIFAL CAR.

the two camshafts are driven from the pinion on the end of the crank-shaft is of large pitch and width, and made of vulcanized fibre shrouded with gun metal.

Magneto ignition   The ignition is effected by a magneto-electric machine of the Bosch type running at the same speed as the engine, and driven by gear and lathe driver. One end of the armature winding is earthed to the engine frame, and the other end passes to an omnibus wire across the top of the cylinders, to which switch lever connexions are made to the insulated anvil bit by the make and break palet in the compression space operated by the ignition gear cams, the make and break at palet on anvil being the same in principle as that described with reference to the Mors gear in vol. i. p. 178, and Simms' engine, p. 289.

Fig. 126 shows the Parsifal gear box of a car fitted with the four-cylinder engine, nominally of 16 HP., but capable of giving 20 HP. on the brake.

110

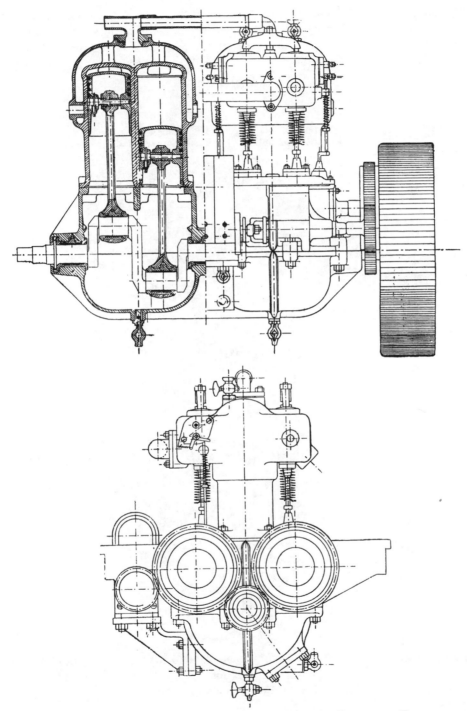

Fig. 124.—Benz-Parsifal 4-Cylinder 16–20 HP. Engine, Sectional Elevation
and End View.

111

Only one bevel wheel is used on the transverse sprocket pinion shaft, the reversal being effected by the interposition of the pinion seen near the largest spur-wheel in the left-hand end of the gear box. The gear, as will be seen, gives four forward speeds. The spindles in the gear box run in ball

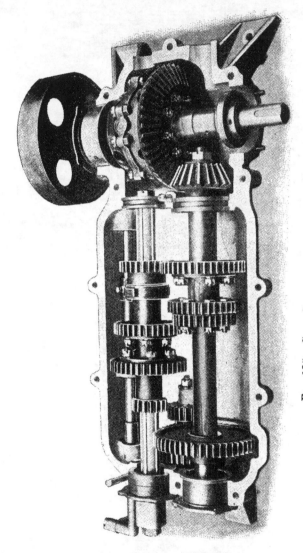

FIG. 125.—GEAR BOX OF 16–20 HP. PARSIFAL.

Ball bearings to gear spindles bearings, and the spindles and gear are of usual dimensions, although, as in many cases, the spur-wheels in the box appear to bear but small relation in dimensions to the bevel pinion and wheel they drive.

THE DECAUVILLE CAR

The Decauville cars are of the live axle type. They were until recently made with round tubular framing, as shown by the engravings Figs. 127

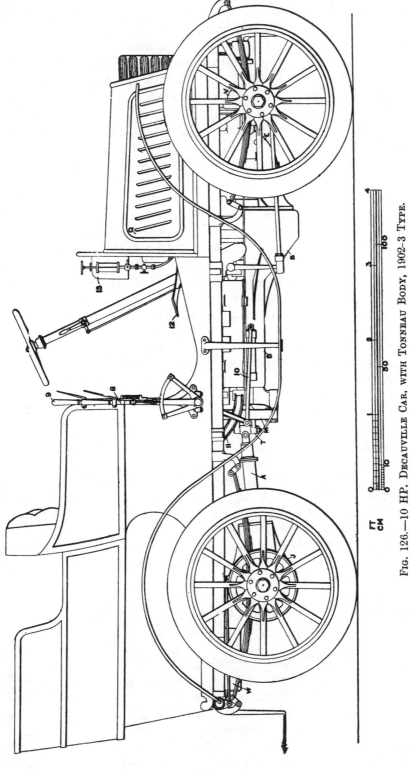

FIG. 126.—10 HP. DECAUVILLE CAR, WITH TONNEAU BODY, 1902-3 TYPE.

113

8

to 133, and although the frame is now of the pressed steel plate form as shown by Figs. 129 to 131, the principal features of the design of the car are the same.

A side elevation of the 10 HP. tonneau body car is given in Fig. 126, which shows the general design and disposition of the parts shown more fully in the vertical and horizontal sections, Figs. 127, 128. A feature of the design is the construction of the engine crank case and the gear case, including the enclosure of the fly-wheel and clutch, practically in one piece, so that the parts are not only well protected, but the shaft and gear bearings **Engine and** remain in true alignment unaffected by the frame. The part which includes **gear box** the gear case and fly-wheel case is marked by B B, and it is firmly connected by a large flange to the part forming the crank chamber E. All the main parts of the machinery may thus be said, so far as the frame is concerned, to be made in one piece like the bed frame or plate of a combined engine and dynamo. This bed is carried on the frame by short arms projecting from the side members of the frame and from the transverse member behind the gear case When firmly bolted to these short arms the machinery bed completes them as transverse members, and the arms cease to be cantilevers depending on torsional resistance of the side frame tubes, so long as the bolts are kept tight. The frame tubes are quite straight, the longitudinals and cross tubes being well connected, but no diagonal stays are inserted to prevent racking strains in horizontal directions. Such stays are seldom used, although they are necessary when the sides are only connected by cross members at right angles to them. This requirement is well met by the new frame, Fig. 131.

The engine of the 10 HP. has cylinders 110 mm. or 4·32 in in diameter and a stroke of the same dimension, and its normal speed is 1,000 revolutions per minute. The cranks O are at 180 deg. The upper part of the cylinders, which are cast together, is well jacketed with large water space, and a centrifugal pump U gear driven at about 2,500 revolutions per minute, maintains a good circulation through the jacket and the Loyal small tube cooler R. The inlet and exhaust valves are superposed, and are of large size, readily accessible, as seen in Fig. 127, and the piston bodies are of ample length. The inlet valves are of the automatic type. The governor is on the crankshaft at V, and the commutator is on an extension of the end of the camshaft. The governor acts on the exhaust valves.

The clutch C is of the self-contained type, the clutch spring x being placed between the fly-wheel and the inner cone so that no end thrust on the crankshaft or gear spindle occurs.

The change-speed gear is of the sliding type, or Levassor system, and gives three speeds with a through drive on the full speed. Four speeds are fitted to the most recent of the cars. The lowest speed is obtained with the pinion 1 on the square spindle N gearing with the wheel 7 on the upper spindle, which by the pinion 5 gears with the wheel 2 on the squared part G of the lower spindle. This, by means of the universal joint T, spindle A, with a 20-tooth bevel pinion at its outer end, gives motion to the 52-tooth

114

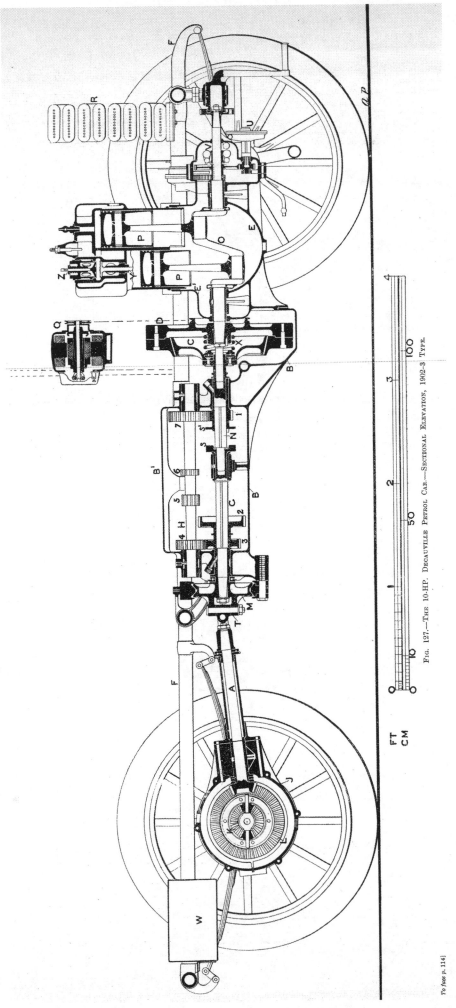

Fig. 127.—The 10-H.P. Decauville Petrol Car.—Sectional Elevation, 1902-3 Type.

[To face p. 114]

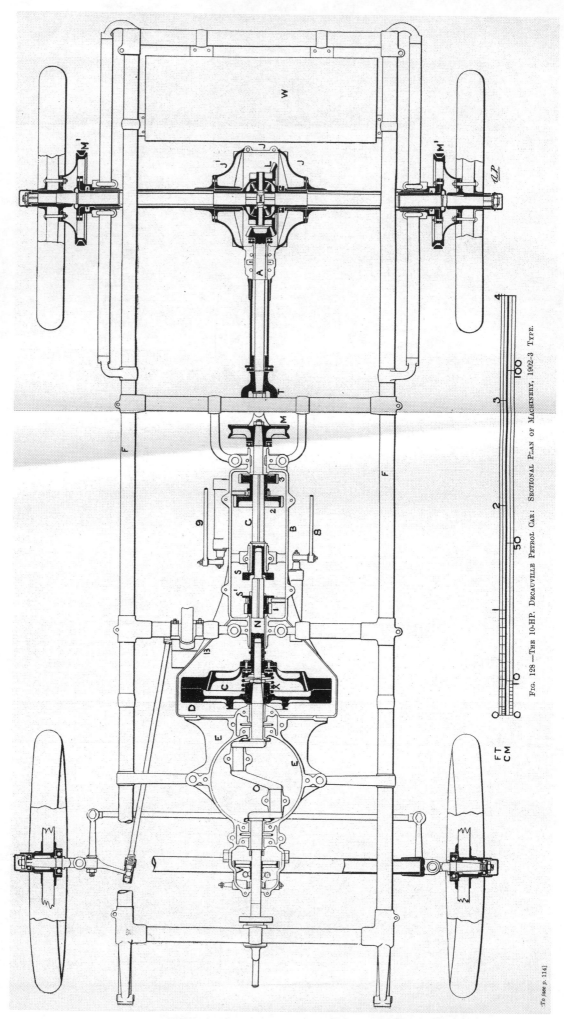

FIG. 128.—THE 10-HP. DECAUVILLE PETROL CAR: SECTIONAL PLAN OF MACHINERY, 1902-3 TYPE.

[To face p. 114]

THE DECAUVILLE MOTOR VEHICLES

bevel wheel L on the differential gear K on the live driving axle. With these gears the driving wheels, which are 32 in diameter on the tread, run at 85 revolutions per minute.

The second speed is obtained with 1 and 7 in gear as shown, and with the pinion 4 gearing with the wheel 3, the revolutions of the driving-axle are 168 per minute.

Operation of change of gear

To the third speed the pinion 1 is taken out of gear with 7, and the claw clutch s' upon it slid into contact with the claw clutch s on the inner end of the spindle G, thus connecting the spindle N which runs in the hollow end of the spindle G with the latter, and forming one spindle, so that the jointed spindle A runs with the crankshaft, all the wheels in the gear box being put out of gear. The sliding forks for moving the pinions 1, 2 and 3 on the lower shaft are shown in Fig. 128, the fork sliding rods being carried in long tubular projections from either side of the gear box, and moved by rods 8 and 9 connected with the hand lever 8. The reversing is effected by putting cog wheel 2 as far forward as it will go, and thus brings it into gear with an intermediate pinion, not shown, but which is driven by 6, and moved into place by a separate lever outside the lever marked B in Fig. 126.

The pedal brake acts upon a drum M with a v-shaped periphery, and with half circle cast-iron brake bands, the brake being very well designed, but it is difficult to get any of these brakes to wear uniformly so that their application does not throw a side pressure one way or the other upon the gear spindle carrying the drum. The side brake drums M', Fig. 129, have also the v-shaped periphery with a flat bottom to the v, and they are compensated so that both have equal pull upon them.

The driving axle is shown in plan in Fig. 128 as carried in plain bearings in the ends of the tubes forming the outer part of the axle, and projecting therefrom and carrying the wheels. The axles as made in this way, although common on live axle cars, have been found to be liable to breakage between wheel and bearing, and the Decauville cars have now the tubular part of the axle extended beyond the spring, so as to form a fixed axle upon which the driving wheels revolve. To do this the wheel naves have to be a little larger, but the central spindles which rotate and give motion to the wheels are relieved of the load, and have upon them only torsional stress. To enable them to drive the wheels they are extended through to the outer end of the fixed tube, and are there provided with a large claw head, which engages with corresponding claws or teeth in the nose of the wheel nave. The arrangement is shown by Fig. 132, in which A is the tubular axle carried beyond the springs and forming the axle on which the wheels run, as seen on the left hand. At the right-hand end the wheel is shown off the axle, and the parts are separated. B is the case containing the differential gear. C is the squared end of the driving axle, which being free in the tubular fixed axle receives only the torsional stress of driving. At the outer end D of each driving axle is the claw-shaped disc head, which fits into the notches or claws in the wheel nave nose E. When

Live axle design

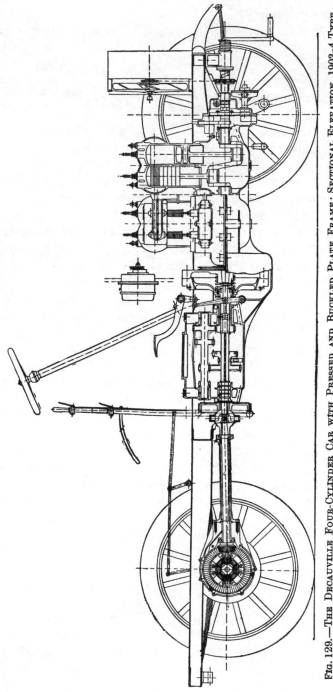

FIG. 129.—THE DECAUVILLE FOUR-CYLINDER CAR WITH PRESSED AND BUCKLED PLATE FRAME: SECTIONAL ELEVATION, 1903–4 TYPE.

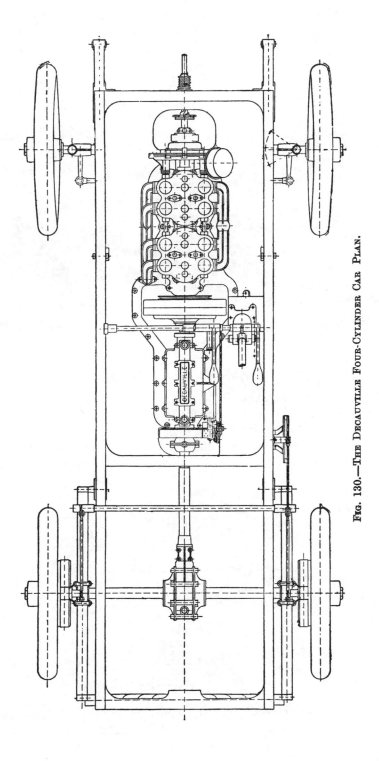

Fig. 130.—The Decauville Four-Cylinder Car Plan.

117

the wheel is in its place the nuts G are screwed on the end of the axle, and set by a set screw. The propeller rod C D is then put in place and the nose cap F screwed on. Instead of the plain bearing on the end of the axle a ball bearing is now used.

FIG. 131.—THE DECAUVILLE BUCKLED PLATE FRAME.

Dished plate frame The new form of frame originated by the Decauville Co. is shown by Figs. 129 to 131, the latter from a photograph of the underside of the chassis. It shows clearly the large buckled plate for the engine base and gear box, and the way in which the side members of the frame are entirely

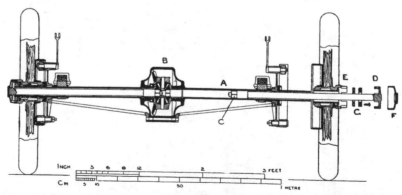

FIG. 132.—THE DECAUVILLE FIXED LIVE AXLE DRIVE.

prevented by the plate riveted to them from moving parallel with each other by racking stresses set up by running over bad roads, gutters, and rails. The frame is shown as carried upon the two usual side springs and a rear cross spring. This will also be seen in the plan of the 1903 car, Fig. 130.

118

Fig. 133 gives a separate view of the method of driving the circulating pump. The view is a vertical section through the front part of the crank chamber and shows the gear for driving the camshaft as well as the pump. On the crankshaft is a pinion which drives a wheel centrally above it.

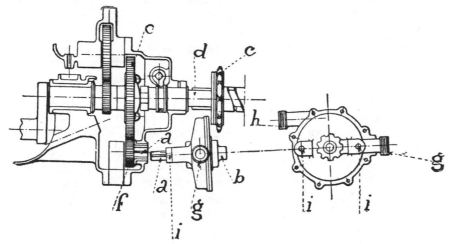

FIG. 133.—ARRANGEMENT OF THE DECAUVILLE PUMP AND GEAR.

This gears with the wheels on either side of it for driving the camshafts for the inlet and the exhaust valves. On the end of the crankshaft is the gear wheel *c* driving the pinion *f*, on the spindle of which is a socket *a*, with which the square end of the pump spindle, also marked *a*, fits, and is easily

Water circulating pump

FIG. 134.—A DECAUVILLE FOUR-CYLINDER 16 HP. TONNEAU CAR, 1903-4 TYPE.

dismounted if necessary. On the extremity of the crankshaft is a chain wheel for driving the fan at the back of the cooler, as seen in Fig. 129.

Fig. 135 is from a photograph of four-cylinder car of the 1903–4 type. The arrangement of the electric ignition apparatus is very complete, and will be made clear by means of the diagram of connexions shown in Fig. 136.

The combination of accumulators, contact-breaker, induction coil and sparking plugs, forming what was until quite recently the only high tension system in use, is employed. The internal arrangement of the trembler induction coil and the construction of various forms of sparking plugs were fully described in chapter xx. of vol. i, and it will be assumed that the readers of this volume are already acquainted with the design and operation of these parts.

Electric ignition system The diagram, Fig. 136, has been made to show the connexions used with the four-cylinder 16 HP. car, but the system in a simpler form applies equally to the 10 HP. car. It will be seen that in addition to these parts already mentioned, and which are in general use, there is a small dynamo connected in parallel with the accumulators through a speed-controlled switch G S. S I is the usual switch employed to connect or disconnect the accumulators, and in this case the dynamo also, from the rest of the system when the car is in or out of use. The duty of the dynamo is thus the same as that with the " Mors " car. See pp. 176 and 374, vol. i.

The construction of the small two-pole shunt-wound dynamo is shown by Fig. 135, and its position on the car is indicated at Q in the longitudinal view in section, Fig. 127. It is fixed in a recess cut for it in the dashboard, and the pulley or belt-wheel end projects sufficiently in front to be driven by a round or small V-belt from a wheel cast with the engine fly-wheel. The light circular metal cover at the commutator end of the dynamo projects through the dashboard, and may be seen facing the driver and just below the lubricator box 13 in Fig. 126. The armature spindle T, Fig. 135, is supported in one long bearing, cast with and fixed by its flange F to the combined frame and pole-piece ring H, and it may be seen that the oil reservoir containing the ring lubricator J is also secured by the same set-screws that are used for F. The armature spindle is reduced in diameter at the pulley end and a gun-metal or bronze bush is fixed on the reduced portion by means of the spring washer and nut shown. Lugs formed with this bush carry the pins upon which the bell-crank arms of the little centrifugal governor turn. A spring connects the outer ends of these arms and causes the inner ends of the bell-cranks to press the thin sliding friction collar against the outer face of the pulley-wheel boss, and the pulley drives

Construction of dynamo the armature spindle by frictional contact with the inner collar fixed to the bearing bush and the outer sliding collar. When a certain predetermined speed of rotation has been reached the governor balls move outwards under the influence of centrifugal force, and the pressure on the friction collars being relieved the pulley may slip. The maximum speed at which the armature revolves is thus fixed independently of the speed at which the engine occasionally runs, and the voltage of the current delivered by the dynamos is therefore similarly limited.

The armature and commutator are both carried on the piece K, and the tubular portion on which the armature is secured completely houses the bearing inside it, and any oil that finds its way along T and out at K is delivered ultimately to the pocket P, and thence by a drain hole to the oil reservoir. There is, therefore, no possibility of the armature or field windings being affected by oil leakage.

At low speeds the switch bridge arm G is, as shown, out of contact with the two screws S, but when the motor has speeded up to about 600 revolutions per minute the governor arms D move outwards and spring the bridge G into contact with the screws S.

The second of these screws is behind that shown in Fig. 135, and the points are bridged across by the strip piece attached to but insulated from the spring arm E. Referring to the wiring diagram, Fig. 136, the switch S I

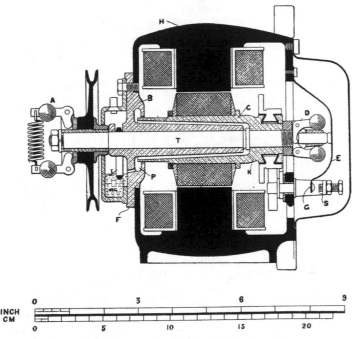

FIG. 135.—THE DECAUVILLE DYNAMO ELECTRIC MACHINE.

being closed, completing the accumulator circuit, the engine may be started, current for ignition being supplied only from the accumulators. When the motor has speeded up, the governor-controlled switch G S automatically closes, and the dynamo and accumulator are both connected to the $+$-terminal of the primary circuits of the four coils A^1 to A^4 and, to use the electrical term, are in parallel. The live ends of the high tension coils P^1 to P^4 are connected to the ignition plugs similarly figured, the return being made through earth to the starting or low tension ends of the secondary coils at T. The dynamo is so designed that when the normal motor speed has been reached and G S is closed, the voltage is a little higher than that of the accumulator, so that the dynamo not only supplies the ignition system but also delivers a small continuous charge to the accumulator, or is ready to do so as soon as the accumulator voltage drops a little from the usual causes. The accumulator is in this way kept fully charged, only supplies current for ignition when starting the motor or when running quite slowly, and if the dynamo at any time unexpectedly fails, the accumulator is in a condition to supply the necessary current until the dynamo has been put in order. Failure of the dynamo is very unlikely to occur, but it may be detected at once by means of the ampère meter shown in the dynamo circuit and placed in a conspicuous position, and which would not show its usual indication or any indication at all. Recently the design of the dynamo has been still further perfected; the driving pulley is now fixed direct to the armature spindle, as shown in Fig. 127, and the friction drive dispensed with. The armature and field windings are so proportioned that when the motor is

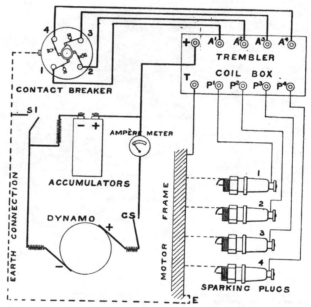

FIG. 136.—ELECTRICAL CONNEXIONS OF 16 HP. DECAUVILLE CAR.

running at more than its normal speed no increase of voltage results, and when the motor is intentionally made to race the ampère meter does not indicate more than 4 ampères.

I have devoted so much space to the Decauville car because of the originality of the design of the small car, though it was defective as to strength and material in several details, and because of the many meritorious features of the design of the larger cars, which are the Phoenix of the light original.

A few of the larger cars of 40 HP. have been constructed with fixed rear axle and transmission of power to the road wheels by side chains, but these are in the minority, and the designs here illustrated with live rear axle and bevel gear are typical.

Chapter V

THE MILNES-DAIMLER AND THE MERCEDES CARS

ANOTHER example of the rapid modern development of design, though on well known general lines, but with numerous details of much importance, is the car known here as the Milnes-Daimler, illustrated by Figs. 137 and 138. This car was made at the Marienfelde Works, and it, with further developments when these works were amalgamated with the Cannstatt works, became well known as the Mercedes car, one of which, in the racing form, and having an engine capable of giving about 60-HP., won the race for the Gordon-Bennett cup in Ireland on July 2, 1903, with M. Jenatzy as driver.

The car, as illustrated by Figs. 137, 138, of the 1902 type, ceased to be made after the middle of last year. It had a steel plate reinforced wood frame with straight longitudinals set over the front springs, and therefore within the width of the back springs, which are carried by projecting hangers. These outrigger hangers are subject to a torsional stress which is visited upon the side frame members. The twisting stress on the frame is, however, minimised at the back by forging strong wings on the hangers, which extend a considerable distance under the transverse member of the frame.

Offset spring hangers

The machinery was carried by an angle steel underframe F F, connected to the transverse front member of its main frame, and at the back to the transverse member a little in front of the back axle as seen in the plan, Fig. 138.

The motor was of the four-cylinder Cannstatt design, the cylinders M M being cast in pairs.

The admission valves were of the automatic type, and ignition was effected by a rotary magneto machine I, and internal make and break apparatus.

A single float-feed carburettor was employed at A, the air aspiration tube H is passed between the pairs of cylinders, and had a mouth partly embracing one of the exhaust pipes.

A separate exhaust pipe E E was used for each pair of cylinders, the pipes being carried separately to the silencer S.

123

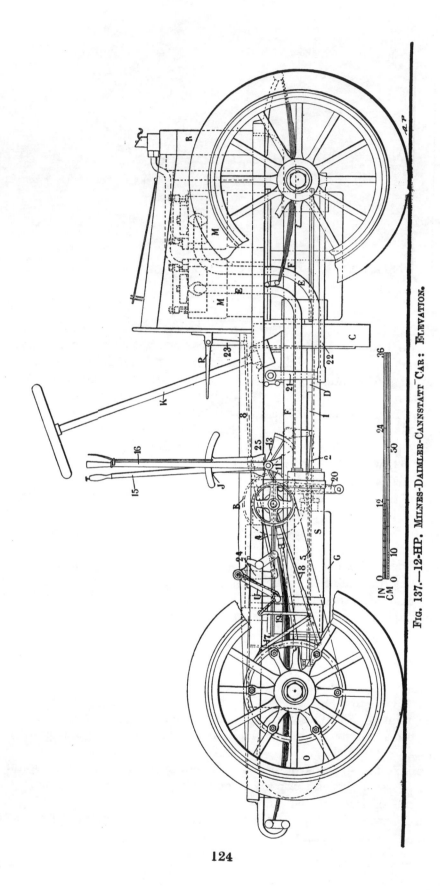

Fig. 137.—12-HP. Milnes-Daimler-Cannstatt Car: Elevation.

124

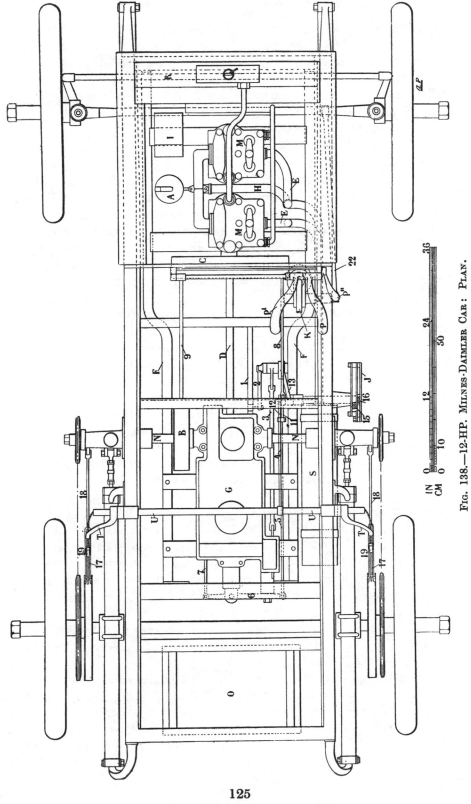

FIG. 138.—12-HP. MILNES-DAIMLER CAR: PLAN.

125

MOTOR VEHICLES AND MOTORS

The cooling water was carried in the cooler or radiator R of the honey-comb type, the air being sent through the cooling tubes by the air pressure produced on the face by the movement of the car, assisted by the action of an exhausting air propeller at the back of the cooler.

The power of the motor was imparted to the transmission gear by a cone clutch at C and shaft D connected to the spindle in line with it through the gear box G to the thrust bearing at the back, and where it can be operated for liberating the clutch by the pedal P′ and connexions to the lever 6.

The transmission gear was of the sliding gear type with bevel wheels on the upper shaft for rotating the transverse sprocket pinion shaft N. The sliding gear for changing speed was operated by the hand lever 16 on the quadrant J, and on the spindle which carries the toothed sector 13. This sector rotated a pinion carrying a short bell crank, one arm of which was connected to the rod 2. The end of this rod was connected to a stud pin by which the striking rod 1 was moved and thereby the gearing.

Reversing was effected by interposing a wide pinion moved by the same sector, and by the other arm of the bell crank moving the rod 3, the wide pinion gearing with the low speed wheel and pinion not then themselves in gear. Referring to details in the engravings it will be seen that the clutch is moved by the pedal P′ and by the brake lever connexions, both being connected to the rod 5, Fig. 138, jointed to one end of a lever 6, pivoted on the end of a strut 7. Near the end of the lever 6, where connected **Operation of clutch** to the rod 5, is a telescopic box containing the spring by which the clutch is forced into engagement by pull on the lever 6, and thereby on the thrust bearing on the main or central spindle in the gear-box which is a continu-ation of the shaft D.

There are two brakes, one on the drum B actuated by the pedal P on a spindle, at the back of the dashboard, on which is a lever arm connected to the rod 9, Fig. 138.

The side brakes 17 are actuated by the hand lever 16 in the outer part of the quadrant J, and on a spindle which carries a short lever arm, to **Design of brakes** which is connected the rod 4. This rod pulls a lever 24 on a tubular spindle U, Fig. 138, through which is threaded a steel wire cord which, through the agency of the bent levers T, actuates the levers 17, 18, 19 of the Panhard and Levassor type of brake gear connected to the brake bands, an un-desirable form of gear not now used on any cars.

It will be seen from Fig. 138 that when the brake pedal P is depressed the clutch pedal P′ is also actuated, though the former can be released without engaging the clutch.

At P″ is a small pedal for controlling the governor, so that for short periods when starting, or when in traffic and for passing, the speed of the engine may be accelerated.

The steering pillar K actuates a worm on a toothed segment, on the spindle of which the lever arm 21 is fixed for operating the steering wheels by the connecting rod 22, connected to the steering arm on the off jointed steering axle.

126

Fig. 139.—Mercedes Racing Car 60-HP., 1903 Pattern.

127

This car was brought out in 1901, modified in 1902, and was a very fast car. Two cars of this type ran in the Glasgow trials of 1901, one a 12 HP., and one a 16 HP., the latter gaining a gold medal.

FIG. 140.—A 40-HP. MERCEDES TOURING CAR, 1903-4 TYPE.

The Mercedes car, which is the outcome of these cars, is illustrated by engravings representing the 28, 40 and 60-HP. sizes, and some of their parts by Figs. 139 to 150.

FIG. 141.—40-HP. MERCEDES FRAME AND RUNNING GEAR.

Fig. 139 is of the same design as that which, driven by M. Jenatzy won the Gordon-Bennett cup in Ireland.

Fig. 140 shows one of the 40-HP. touring cars, 1903 pattern, one of the first lot made with the pressed steel side frames.

Fig. 141 shows the off-side of the frame and running gear of the 40-HP. car of a few months later date.

Fig. 142 shows the same underframe and running gear from the near side with the bonnet removed.

Fig 143 is from a photograph to a larger scale showing in plan the engine, gear box and underframe of the 40-HP. car.

Fig. 144 is a view of the underframe, engine and running gear of the same car as seen from below.

Fig. 145 is a longitudinal view to scale of the 40-HP. car.

Fig. 146 shows the Mercedes carburettor.

FIG. 142.—40-HP. MERCEDES FRAME AND RUNNING GEAR.

Fig. 147 shows the differential gear and sprocket shaft and ball bearings of the 40-HP. car.

Fig. 148, gives in section and side elevation the construction of the rear axle and side brakes of the 40-HP. car.

Fig. 149 shows the construction of one of the forms of pivoted steering axle used in these cars, the end A of the axle shown as fixed upon a tubular axle being now made of I section.

Fig. 150 is from a photograph of the near side of a 28-HP. touring car of the same make and design.

These cars have obtained a deservedly high reputation for smoothness of running and for good workmanship.

They are made at Cannstatt and at Vienna, and the F.I.A.T. Cars built at Turin may be looked upon as in some respects the same car. They represent in most particulars all that is generally considered best practice in the vertical engine and chain driven type of car, successful running in

Fig. 143.—40-HP. Mercedes Frame and Gear from Above.

130

Fig. 144.—40-HP. Mercedes Frame and Gear: View from Underneath.

131

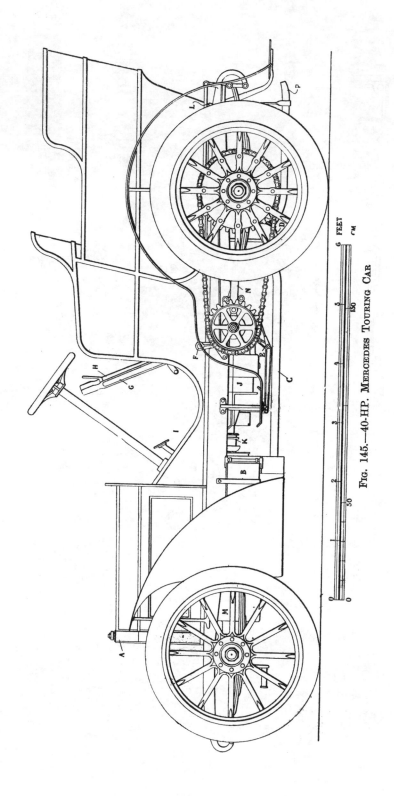

FIG. 145.—40-HP. MERCEDES TOURING CAR

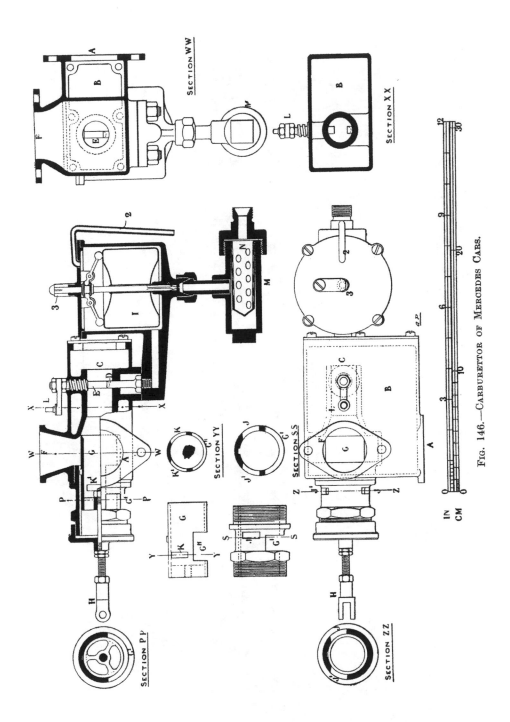

FIG. 146.—CARBURETTOR OF MERCEDES CARS.

133

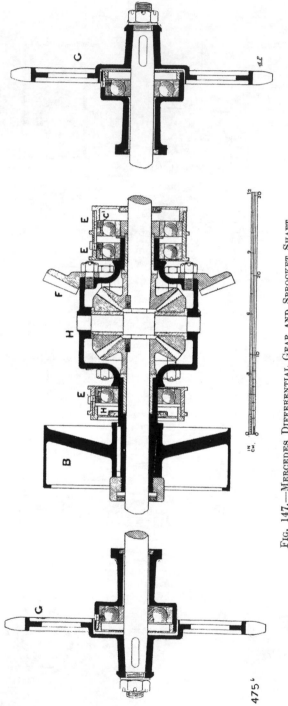

FIG. 147.—MERCEDES DIFFERENTIAL GEAR AND SPROCKET SHAFT.

134

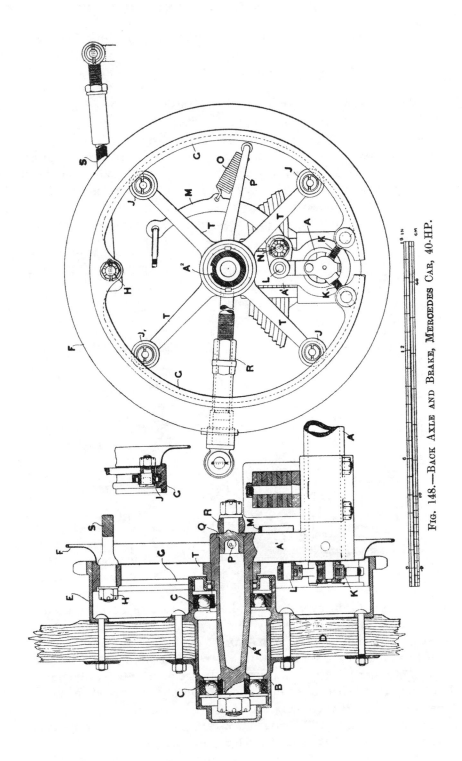

Fig. 148.—Back Axle and Brake, Mercedes Car, 40-HP.

trials and races having contributed most to this opinion. It has yet, however, to be seen whether lengthened usage will confirm this opinion and whether greater accessibility and simplicity of parts will not be looked for.

The pressed steel frame which these cars introduced is now in general use by all the largest makers. In the 60-HP. cars these frames are about $4\frac{1}{4}$ in. deep in the middle, $1\frac{3}{4}$ in. wide, and nearly $\frac{3}{16}$, or about 4 mm., thick. **Frame construction** When the heavily stressed connexions to the frame have a good foot or base, and are well attached by judiciously placed and well closed rivets, these frames may be more satisfactory than the wood frames to which Messrs. Panhard and Levassor adhered so long, but many of these frames have been sent out by some firms with badly designed connexions for

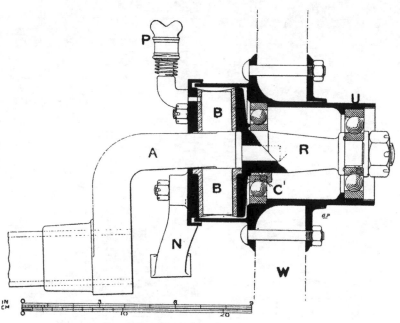

FIG. 149.—MERCEDES STEERING AXLE.

sprocket spindle bearings spring hanger brackets and transoms, and the imperfection will show itself by the loosening of the fastenings and wearing away at the holes in the thin frame metal.

The wheel base of the 60-HP. car is 8 ft. 9 in., and the wheel gauge the old standard of 4 ft. 8 in., the wheel base of the 40-HP. being 7 ft. 10 in but for cars with side entrance bodies, the wheel base is longer. The springs in all these cars are of great length, and hence their effectiveness; those of the 1904 cars of 18-HP., said to be capable of 28 BHP., are, for instance, 4 ft. 8 in. long for the back axle.

The general arrangement of the machinery of the car will be readily gathered from Figs. 143, 144 and 145.

The engines have four cylinders, and mechanically operated inlet valves of large diameter, but not of the three-seated type used two years ago. The inlet and exhaust valves are now similar and interchangeable. The mechanically moved valve is preferred, as it does not chatter on its seat as many automatic valves do, and therefore makes less noise; its economic value has, however, to be proved taking into consideration not simply a slight saving in fuel, if any, or a slight increase in the power of a given engine, but the extra cost and extra number of **Variable lift to inlet valves** moving and wearing parts, and in most cases some loss of accessibility. The valves have a variable lift and variable period effected by the increase or decrease in the length of the valve-pusher rods on which are nuts moved by a rack, the outside of the nuts forming toothed pinions. Thus

FIG. 150.—A 28-HP. MERCEDES TOURING CAR, 1903 TYPE.

the longer the rod the earlier it opens the valve, and the later it closes it. The exhaust valves have flat faces and seats.

The 60-HP. engine has cylinders 5·51 in. or 140 mm. diameter and 5·9 in. or 150 mm. stroke.

The 40-HP. engine has cylinders 4·72 in. or 120 mm. diam., and 5·51 in. or 150 mm. stroke, and the 18-HP. has cylinders 90 mm., 3·54 in. diameter, and 120 mm., or 4·72 in. stroke.

All these engines have the low tension magneto ignition apparatus,[1] the period of ignition being variable. The inlet valve camshaft is provided with cams for working the tappet, which is normally in contact with the insulated current-carrying stem inside the combustion space. At the proper moment it is pulled out of contact by a rod actuated by a cam and a lever resting on the cam, and carried so that the cam may reach it

[1] See photographs in *Automotor Journal*, vol. viii. pp. 805, 825.

sooner or later according to its position.   The system of ignition is practically the same as that shown by Fig. 204, p. 189.

The carburettor used is shown by Fig. 146, in which the large upper figure is a longitudinal section, and the bottom central figure a plan.   Air enters at A into the box B, and thence into the tube C past the jet D, which may be closed by the screw E, and out to the engine at F.   The passage to F is controlled by the cylindrical throttle valve G, moved by the governor connected to the rod H.

As shown, the throttle valve is closed as far as it does close, and the jet is closed, but a small quantity of air is admitted under these conditions at the slots J, and thereby into similar slots K in the throttle valve, and thence

Carburettor at the end of the throttle valve, see plan, into the passage F.   The passage of air to the cylinders is thus never completely cut off, and a weak mixture may pass until the petrol is shut off by the screw-stop and lever L.

Petrol enters a filter N in the case M, and thence to the float chamber, which has an overflow pipe 2.

Some slight modifications have recently been made in this carburettor, so that the variation of the quantity and strength of mixture is more nearly automatic through a greater range of speed.

The transmission gear includes change-speed gear of the sliding wheel

Water cooled brake drums and pinion type, giving motion to a bevel wheel, gearing with a bevel wheel F Fig. 147, on the differential gear box on the sprocket spindle, which also carries a pedal brake drum as seen in Fig. 144, and at B, Fig. 147.   A similar brake drum is also carried on the forward end of the second motion shaft, these drums being respectively 9·45 in. and 7·87 in. diameter in the 60-HP. and 40-HP. cars.   Water is admitted to the interior of these drums when the brakes are applied on the 60-HP. cars.

The clutch is of the cylindrical expanding form, metal to metal, of cast-iron.   The internal expanding part or split ring carries at one side of the split a pivoted lever, which with a short jointed toggle spreads or opens the two ends ; the other end of the lever projects toward the shaft and bears

Types of clutches upon a cone of rounded or approximately hemispherical form.   When this is pushed toward the fly-wheel by the clutch spring, the lever is pushed out of a radial towards a tangential position, and the clutch ring expanded. A similar form has been used on some of the Charron, Girardot and Voigt cars.

A coil clutch is also used on the Cannstatt cars for some of the sizes, and it is put into operation by means similar to those described for the expanding cylindrical clutch.[1]

It was on these cars that the honeycomb type of cooler was first used as a development of the condenser-tube and tube plate form previously used by the same makers.   This cooler is now usually made up of a large number of thin $\frac{5}{16}$ in. tubes, expanded and squared at the ends, and with

Honeycomb cooler only about $\frac{1}{32}$ in. between them for the flow of the water.   These ends are all soldered together, and the batch of tubes surrounded by a light

[1] *Autocar*, February 27, 1904, p. 287.

strip-brass frame.   A 60-HP. car is fitted with a cooler having about 1,600 tubes about 6 in. long, $\frac{5}{16}$ in. diameter at centre part, thus giving a total cooling surface of about 67 square feet.   The fly-wheel spokes act as fan blades of an exhausting fan to draw air through the cooler tubes.

The speed-change gear wheels were in the 1903 cars moved by the rotation of a large disc side grooved or face cam.   This system has been discarded, and lever and rod and fork arrangement is again used, the hand lever being arranged to move in a treble slot quadrant, one of the slots being of half length only for the reversing motion, a switch lock on the lever itself being arranged to prevent the reverse motion position being taken accidentally.[1]   Formerly a latch was pivoted on the quadrant.

All the bearings throughout the transmission gear are of the large ball non-adjustable type, as seen at E E in Fig. 144, the balls running in shallow grooves in the sleeve ring and shell, a removable piece c′ being inserted in the shell to admit the balls.   This type of ball bearing is also used on the bearings of the sprocket wheel C and on the heavy Daimler omnibuses and (Marienfelde) wagons. **Ball bearings**

The form of ball bearing on the driving wheels on their axles is shown in Fig. 148, the axle A^2 being formed in one steel forging with the arm A^1 formed with the socket into which is fitted the tubular axletree A.   The latter arrangement has now been discarded in favour of a steel forging of I section, as already mentioned for the front axletree.   The axle itself is hollow and has fitted in it the screwed and pinned pin P which carries one end of the radius rod R.

The side or main brakes are also shown in detail in Fig. 148.   They are operated after the Panhard and Levassor fashion by a steel wire rope threaded through the cross tube carrying the brake lever.   The expanding brake ring G, 15·4 in. diameter, is internally grooved to receive the supporting and guiding rollers J at the ends of the arms T.   Pivoted **Side brakes** on the side of the axle stem at N is a lever M, the short end of which at L carries a double-jointed piece, to the lower joint of which are pivoted the adjustable spreader screws K.   The expanding ring G is prevented from rotating by the rod S pivoted at H, and a spring O on the fixed arm P returns the lever M, and allows the contraction of the ring G when the pull on the arm M ceases.

The ring G expands into the drum which forms the sprocket wheel E attached by bolts to the wood spokes of the wheel.   It will be noticed that effective means are adopted as shown at H, Fig. 147, and at other similar places in it, and on the axle in Fig 148, for keeping out dust by a soft spongy filling in a deep groove.

The steering gear consists of a long nut on the steering column, the nut sliding in a slotted arm much as formerly used in locomotive brake gear. The steering axle is shown by Fig. 149.   In this A is the fixed part of the axle, now forged in one piece with a channel steel axletree instead of the tubular central part shown. The end of the piece A fits nicely in the jaw

[1] *Automotor Journal,* vol. viii. p. 354.

forged in one with the axle R, the pin B B being a driving fit in the part A in which it is further secured by a pin. As shown in the drawing, the bushes in which B fits were inserted by making the jaw in two parts divisible in the vertical centre line of B B, the outer part next the axletree being bolted on two of the bolts also holding the steering arm N; this jaw is now made solid, and the bushes inserted from top and bottom and pinned, the pin B, as stated, being inserted after the end of the axletree is in place. The ball bearings are again seen here, the balls being relied upon to keep the wheels on by the slight groove in which they run.

Exhaust is used to maintain the necessary pressure in the petrol and lubricator tanks.

The wheels, or rather the tyres of the 60 HP., are 920 mm. by 120 mm., and 910 by 90 mm., and those of the 28 HP. 870 by 90 mm., all the same diameter.

Chapter VI

THE PANHARD AND LEVASSOR CARS AND ENGINES

THE Panhard and Levassor carriages described and illustrated in detail in vol. i. pp. 137–157, were what were then racing cars of 6 HP. of the 1898 type, and a four cylinder 12-HP. car which ran in the Paris-Bordeaux race and made an average speed of 30 miles per hour. The engines were of the Phœnix pattern of Daimler general design, and that of the 6 HP. had two cylinders of 91 mm. or 3·59 in. diam., and 127 mm., or 5 in. stroke. They were fitted with the hit and miss governor acting on the exhaust valve, and were excellent engines. They were, however, superseded by a form of engine called by the makers the Centaur ; why, it is difficult to say. In this engine the hit and miss exhaust lifting gear was dispensed with and throttle governing adopted. It gradually became a very successful engine, and has been the model upon which many others have since been made. It has more recently been modified to a small extent, the chief alteration being the use of mechanically operated inlet valves, but it must be noticed here as an important type, and as representing a useful stage in the evolution of such engines.

The engravings Fig. 151 show one of the four cylinder engines of this type, made when the ignition tubes were still being fitted, to the end of 1902, and beginning of 1903, and when electric ignition was being also provided, the ignition plugs being inserted at c, Fig. 151.

These were the first engines which the firm fitted with the governed throttle valve carburettor shown by Fig. 153, and in which the inlet valves were placed in the lanterns A, Fig. 151, with each pair held down by one bridge and bolt B E. Many of these engines are running and are much liked, although the cars themselves are out of date and ignition tubes discarded. The exhaust valve covers are similarly held down in pairs by one bridge and bolt, Y Z, as in the earlier Coventry Daimler engines. The exhaust pusher rods are forked to carry the rollers running on the cams and forked at H at right angles, to straddle the camshaft F and form a guide to the rod. These features and the governed throttle carburettor

141

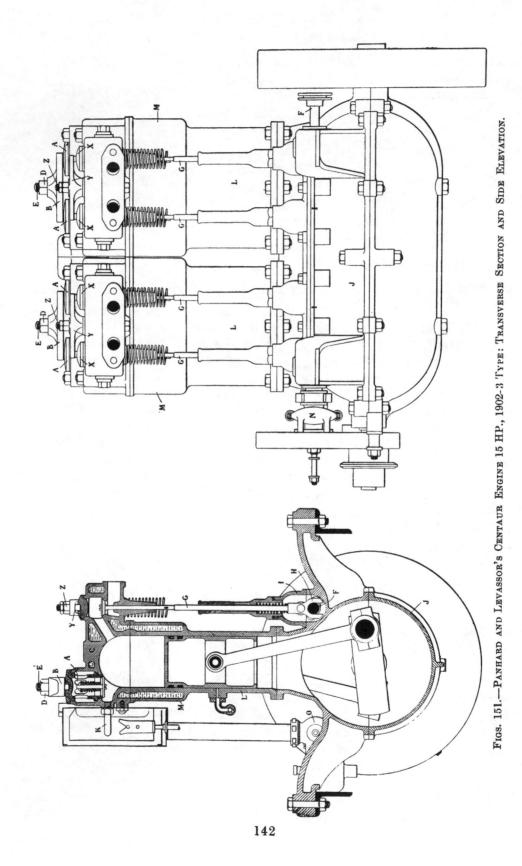

FIGS. 151.—PANHARD AND LEVASSOR'S CENTAUR ENGINE 15 HP., 1902-3 TYPE: TRANSVERSE SECTION AND SIDE ELEVATION.

142

were made the subject of a patent, but no serious attempt has been made to show that the subject matter was sufficient.

The governor N, Fig. 151, as shown in the plan, Fig. 152, is connected

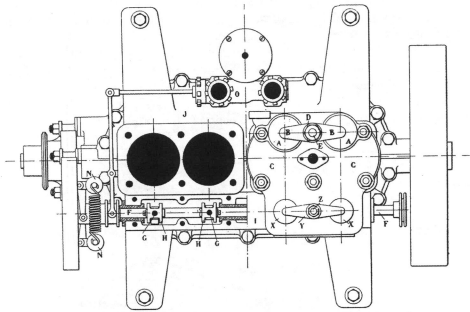

FIG. 152.—PANHARD AND LEVASSOR'S CENTAUR ENGINE: PLAN.

to the throttle valve rod c of Fig. 153, showing the carburettor. At the end of this rod is a piston throttle valve a. In this carburettor air enters at j, passes through the short tube l by the jet e to the annular space

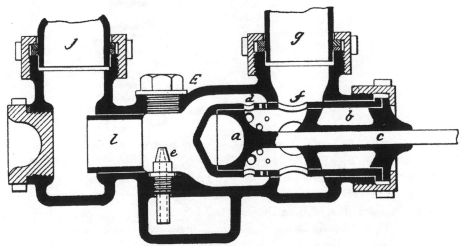

FIG. 153.—GOVERNED THROTTLE CARBURETTOR OF CENTAUR ENGINE.

surrounding the throttle valve cylinder, thence through the holes d into the valve cylinder, and away through the holes f to the pipe g connected with the inlet valves.

In the position shown, the maximum quantity of air is being admitted, the engine standing or running with full load below the speed at which the governor enters into action. When the governor comes into play the air is gradually cut off and the larger holes at d may be closed, and later the holes next in size. The smallest holes leave just sufficient air passage to allow the engine to continue working on no load.

It will be seen that this arrangement is nothing more than an elaborate throttle valve, and accomplishes nothing more than would be effected by a butterfly valve in the pipe g or by a perforated plate or other form of throttle. As a throttle valve it worked very well through a short range of speeds, but, as will be seen, the velocity of the air past the jet diminished

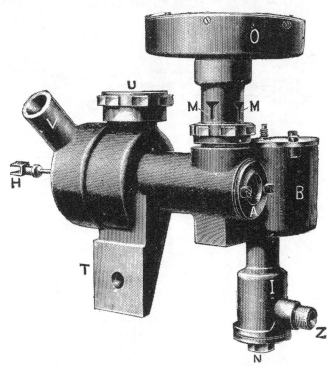

FIG. 154.—PANHARD AND LEVASSOR'S KREBS CARBURETTOR.

rapidly with the closing of the throttle valve, and it and the suction on the petrol soon ceased to be sufficient to give any petrol, and hence the quality of the mixture as well as the quantity fell off. Since this date the modification known as Krebs' carburettor has been made, which in brief is equivalent to taking out the plug E, lengthening the tube l to a little past the jet, and **Kreb's** putting a spring air admission valve in the place of the plug. The result **carburettor** would be that with the engine running slowly the whole of the air required to fill the cylinders would pass to l without opening the valve at E, the velocity being sufficient to induce the petrol to leave the jet. As the velocity of the ingoing air increased with the increase of the speed of the engine the valve at E would open and more air enter, but as some would enter without passing through l the suction on the jet remains proportional to the total quantity of air, i.e. the quality of the mixture remains unaltered.

The newer form of carburettor designed by Commandant Krebs, of the Panhard and Levassor firm, is shown by Figs. 154, 155. Fig. 154 is an exterior view in which B is the float vessel and v water connexion to hot-water jacket. In Fig. 155 K is the cylindrical supplementary valve, which moves downward against the resistance of the spring R.

The valve is attached to the stem, which is fixed in the small disc surrounded by the diaphragm leather Q. The slight vacuum formed by the suction of the pistons acts upon this diaphragm, some holes in the piece

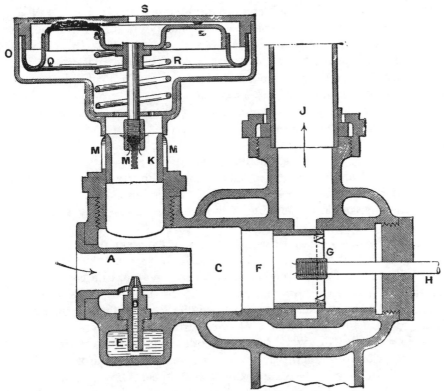

FIG. 155.—PANHARD AND LEVASSOR'S KREBS CARBURETTOR.

through which the valve stem passes permitting communication between the carburettor and the diaphragm chamber. In the small cylinder in which the valve K works there are formed a series of slots or ports, M M, wider at the top than at the bottom ends. When the engine is standing or running slowly the piston valve K stands at its highest position, and the air inlet ports M are covered. The whole of the air admitted then passes through the tube A past the jet. Up to a speed of say 200 revolutions per minute the velocity of the air past the jet is not sufficient to give too high a flow of petrol from the jet. At a higher speed, however, the suction increases, and more petrol is sucked out of the jet at each pulsation, and as

145   10

the speed increases the suction, pulsations follow each other so closely that the momentum of the liquid becomes sufficient to make the flow a continuous one. At this point and for a considerable range of further increase the petrol is in excess of that required to maintain a sufficient strength of mixture, or rather it would be were it not for the suction on the diaphragm, which lowers it and the valve K, and admits more or less air, according to the speed and suction, through the ports at M M. The air which enters in this way prevents the increase in velocity of the air through the tube A past the jet D, and so prevents unnecessary increase in the flow of petrol and richness of the mixture. It will be seen from this that the movement of the piston valve must be sympathetic, easily affected by increase or decrease in the suction and corresponding air velocity. It will also be seen that when the engine speed passes the point at which the mixture is right for a little above the slowest speed, and at which the point of continuous flow of petrol is reached, that a considerable increase in the inlet of air is necessary for a given increase of suction. To meet this requirement the ports at M are almost of a T-shape, so that at the first part of the downward motion of the valve, a larger area of opening per degree increase of speed is presented than by the succeeding descent caused by further increase in engine speed. With a further increase of speed more air is admitted by further descent of the valve, but not at so rapid a rate of increase as at first. The velocity of air inrush at the tube A may then increase, but the flow of the petrol having through a considerable range now been continuous, its rate of increase of flow decreases, a limit being imposed by the size of the hole in the jet piece.

So far the action of the carburettor has only been dealt with independently of the effect of the use of the throttle valve, but this valve modifies the requirements of the supplementary air inlet valve, and makes them less onerous. When the engine is running at its slowest, this valve nearly closes the passage to the pipe J to the engine, only sufficient opening being left to give a sufficient velocity of air through the tube A to induce that petrol flow which will give the weakest burnable mixture under the conditions of ignition and combustion in an engine cylinder. For a considerable increase of what may be called the lowest speed, brought about by the opening of the throttle valve, the velocity of air past the jet will not be great enough to give too high a mixture, but with further opening of the throttle valve and increase of speed the additional increase of air inlet must be considerable for the reasons above given.

From the foregoing it will be seen that the whole requirement resides in the provision of a sufficiently sensitive supplementary air valve, which will give a quick and gradually decreasing rate of increase of air inflow with increase of speed of engine, the supplementary air being admitted so that the air past the jet varies but little from the minimum to the maximum rate.

This has been done in various ways, and may be done by other and very simple forms of carburettor and valve.

Behaviour of carburettor

THE PANHARD AND LEVASSOR MOTOR VEHICLES

Fig. 156 is from a photograph of one of the Panhard and Levassor 8-11 HP. cars with a landauette body. It has a three cylinder engine with bore and stroke dimensions of 8o and 120 mm., or 3·15 and 4·73 inches respectively, and the normal speed is about 800 revs. per minute. The inlet valves are automatically operated, but those of the larger

FIG. 156.—PANHARD AND LEVASSOR LANDAUETTE WITH 8-11-HP. THREE CYLINDER ENGINE, 1905 TYPE.

engines are now mechanically operated. The early difficulties with the three cylinder engine have been overcome and the performance of the car is as consistently good as that of the other Panhard carriages.

The most noticeable changes which have been made in the past year in the construction of the Panhard and Levassor cars have been less of their initiative than of agreement with that of others. Pressed steel frames, longer wheel base, equal size of all wheels and various changes in the form of bodies to suit altered fashion and better knowledge of the most convenient. Detail changes in the gear-change levers, mechanically operated inlet valves, high tension magneto ignition, and in the mode of control of the governor, have been made ; and the gear box has been altered so that the change-speed wheels lie in the same horizontal instead of vertical plane as the main spindle. These, however, are no advances in principle, and could not be shown without drawings, which in this case I do not possess. The exterior appearance of the recent cars is so much like that of many other cars that it is not necessary to produce here views of the other types now made, of 15–20, 24, and 50 HP.

Chapter VII

THE BROOKE CARS

A N interesting and original car of English design and construction is the Brooke car, made by Messrs. J. W. Brooke & Co., of Lowestoft.

It was one of the first to be fitted with a three-cylinder engine, and it has always been fitted with the Estcourt form of natural circulation cooler, thus avoiding the use of a circulating pump. More in deference I believe, to fashion than to any other influence, Messrs. Brooke & Co. are now

Fig. 157.—A 12-HP. Light Brooke Car.

making a four-cylinder car with honeycomb cooler and circulating pump, and with a modified form of their carburettor.

These cars are made in three standard powers, one a 12-HP. three-cylinder engine car, which is illustrated by Figs. 157 to 162, and a 15-HP. with four cylinders. The third type is of 35-HP., and has made its appearance this year. A four-cylinder engine is used, the bore and stroke dimensions being 5·5 and 6 inches respectively, and the stated power is developed at 620 revs. per minute. A carburettor of the form described with reference

149

to Fig. 165 is used with this engine, and a clutch of the metal to metal expanding type.

Fig. 157 is from a photograph of one of the 12-HP. cars fitted with a spider seat, and Fig. 158 shows a similar car with a closed tonneau body or limousine with canopy top and wind guard, the frame and arrangement of the running gear being the same as that shown by Figs. 159 to 161, upon which any form of body may be fitted.

The frame, as shown, is of channel steel filled with ash, but may be of the pressed steel channel section, which is used in the 12-HP. light car.

FIG. 158.—12-HP. BROOKE LIMOUSINE, 1902.

The frame is strong and well stayed as against racking stresses by gusset plates shown in the plan, Fig. 159.

The engine is shown by Fig. 162, the sequence of cycles or firing strokes being as indicated by diagram 5, Fig. 311. From this it will be seen that a much more uniform turning effort is obtained with the **Three cylinder engine** three-cylinder than with the two-cylinder engine. The idle period between working strokes is reduced from 180 to 60 degrees, and the third cylinder thus adds considerably to smoothness of running.

The cylinders of the 12-HP. engine are 3·62 in. diameter, and the

150

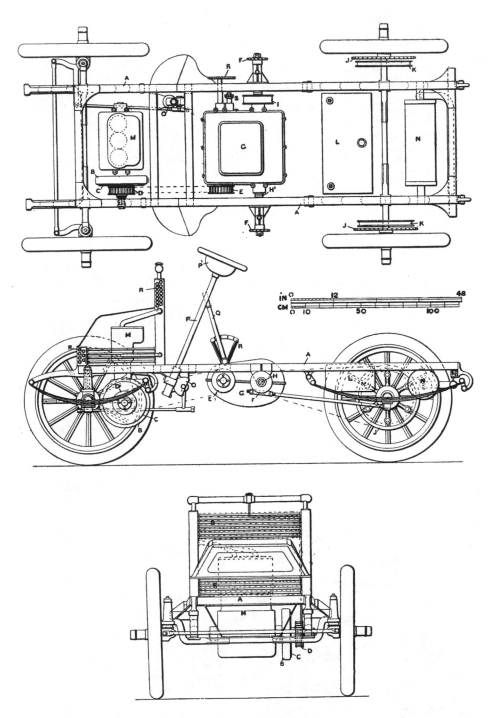

FIGS. 159, 160 AND 161.—THE BROOKE 12-HP. UNDERFRAME AND RUNNING GEAR:
ELEVATION, PLAN AND END VIEWS.

151

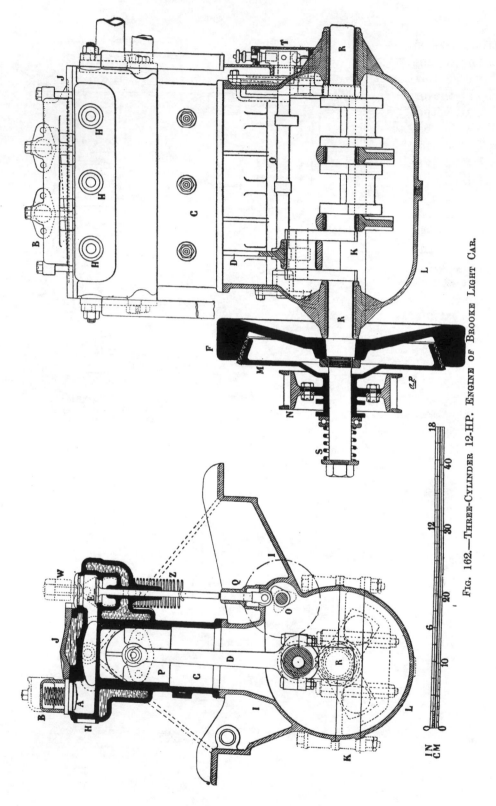

FIG. 162.—THREE-CYLINDER 12-HP. ENGINE OF BROOKE LIGHT CAR.

152

stroke 4·75 in., the normal revolutions being 900, at which speed it gives on the testing stand 14 BHP.

Fig. 162 shows the valves as being on opposite sides of the cylinder, as was the practice last year. They are, however, now all on the one side and all actuated as is shown for the exhaust valve E by one camshaft O. The valve covers are all held down on their seats by one clamping bar W with one set screw to each cover.

The crank chamber is horizontally jointed together in three parts with one of the joints at the crankshaft centre line. By removing the lower part of the crank chamber, the connecting rod ends may be got at, and, when necessary, the crankshaft, connecting rods and pistons may be taken out when the middle section of the crank chamber is removed.

The provision of a bearing at each side of every crank dip removes the objection which obtains with many multiple cranks.

The Brooke carburettor used with the 12-H.P. engines has, since the beginning of 1902, differed from those of any other makers. It is a combined carburettor and governor, and was at first made as a governor only, to govern by the variation in the exhaust delivery pressure with variation of engine speed. In this form it is shown by Fig. 163, in which the flange

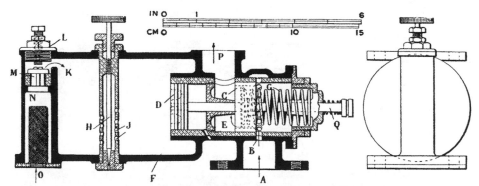

FIG. 163.—THE BROOKE GOVERNOR OF 1902.

A was attached to the carburettor, and the supply pipe to the engine at P. A shunt pipe from the exhaust pipe was connected at O, the exhaust gases passing through a filter there, and then through a valve M into a chamber at K. The rise in frequency of exhaust discharge, and thereby virtual pressure in the chamber with increase in engine speed, caused the outward movement of the piston D, to which was attached the piston throttle valve C. Normally the spring G kept the throttle valve open as shown, the carburetted air then passing through the holes B, thence through the valve to P. With increase of engine speed, the valve was pushed more or less over the holes B. For setting the apparatus, the needle escape H in the tube J was used. The pin Q, normally kept where shown, acted as accelerator i.e. when used it prevented the action of the governor.

This form was discarded in favour of one which is still used, in which the chamber F and parts M, K, H, J were replaced by a circular diaphragm

153

instead of the piston D. This was connected by a rod to the throttle valve C. Part of the guide piece E was cut away, so that there would be open communication to the diaphragm, which then, through the partial vacuum caused by the suction in the cylinders, acted by atmospheric pressure on the throttle valve.

It then acted as a throttling governor; the greater the degree of vacuum caused by the suction of the pistons, the more were the holes B covered by the throttle valve. This arrangement [1] was, moreover, combined with the float valve chamber and jet, and is the form now used in the 12-HP. cars.

Another form, illustrated by Fig. 164, which Messrs. Brooke have used, includes the diaphragm J, which, being larger than the piston throttle valve G, is more powerfully acted upon by the atmospheric pressure, and

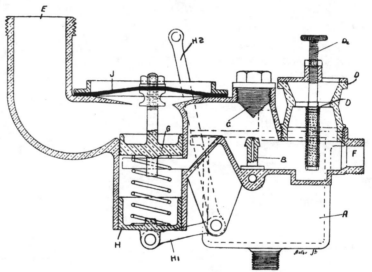

FIG. 164.—THE BROOKE GOVERNING CARBURETTOR, 1903–4.

a supplementary inlet valve D^1, admitting supplementary air to the air on its way to the carburetting jet B, is employed. The strength of the mixture is thus governed by the supplementary air valve, and the diaphragm controlled valve is used as a governor. The depression of the diaphragm and movement of the throttle valve are checked by a spring, the resistance of which may be increased for acceleration purposes, by the levers H^1, H^2 acting on the piston H. This governor carburettor thus differs from the Krebs, the diaphragm not being used to actuate the supplementary air admission, and the latter being made prior to the passing of the main air supply to the jet instead of after as in Krebs'. It has recently been decided to discontinue the use of this form of governing carburettor, and to use that shown by Fig. 165. Consideration of Fig. 164 shows that the

[1] *Automotor Journal*, February 21, 1903, p. 200.

delicate movement obtainable by a diaphragm was misplaced in using it
for actuating a throttle valve, while the supplementary air admission was
not sufficiently delicate or variable through sufficient range. The relative
positions and areas of the inlet F, jet B, and passages were, moreover, not
well selected with reference to the processes of carburation and variable
demand.

In Fig. 165 shows the governing carburettor recently made and now

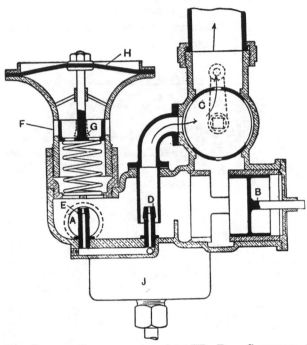

FIG. 165.—THE BROOKE CARBURETTOR FOR 15-HP. FOUR-CYLINDER CAR, 1905.

used with the four-cylinder cars. Air is admitted to the carburettor
through the part A, drawing petrol from the spraying jet E out of the
float chamber J, and passes through the throttle-valve passage, the
opening of which is regulated by the piston valve B, operated by the
engine governor, and through the hand-regulated rotary throttle valve
C to the induction pipe. Additional air is admitted through ports F,
when the engine is running at its higher speeds, the piston valve G,
regulating their opening, being operated by the suction of the engine
on the flexible diaphragm H connected to it, a spring pressing the valve
G upwards normally closing the ports. When the car is at rest and
the engine running light the throttle valve C may be turned to the position
shown in the engraving, the mixture then being drawn through the alternative passage past the jet piece D into valve C and through a small opening
into the suction pipe, as indicated by the arrows, to the engine. With
this carburettor the speed of the engine when running free may be reduced

New type of carburettor

to 100 revolutions per minute, the inductive air rush in the petticoat pipe round the jet D being sufficient to ensure a rich mixture.

Transmission of power by chains

The power is transmitted from the engine to the first motion spindle A in the gear box, Fig. 166, by a Renold silent chain running on the toothed pinion N fixed on the centre of the clutch cone M, Fig. 162.

This chain runs on the similar pinion E, as seen in Figs. 159 and 160 and Fig. 166, on the spindle A, upon which are the three chain pinions carrying chains driving the wheels 1, 2, 3 on the sprocket shaft H. On the ends of this shaft are the sprocket pinions, as seen in Figs. 159 and 160.

The wheels, 1, 2, 3 run loose on a sleeve H^2 on their shaft, this sleeve being connected with the differential gear box L. The sleeve is continued

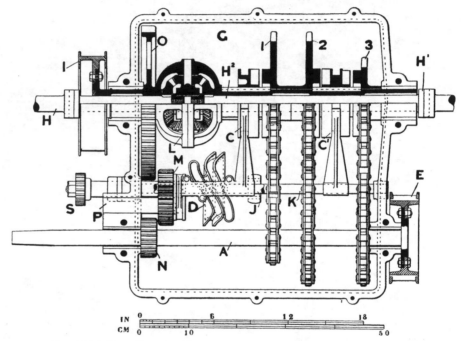

FIG. 166.—THE BROOKE TRANSMISSION AND SPEED-CHANGE GEAR.

on the other side of the gear box, and there carries the brake drum I. This brake drum is now carried inside the gear box, as in Fig. 170, where it is fitted accurately with brake band or shoe pieces and runs in oil, so that it is always in good condition.

The wheels 1, 2, 3, Fig. 166, are selectively used and put into gear by the movement of the grooved cam D, rotated by the speed lever quadrant acting on the pinion S. The cam D is on a spindle between those carrying the clutch fork arms C and C^1, and it acts on pins and rollers, shown by dotted lines, one actuating the clutch lever C in the top of the gear box, by which the slowest speed and the reversing pinions are thrown into gear; and the other moving the clutch lever C^1 on a spindle at the bottom of the box, by which the second and highest speeds are put into gear.

156

The cam D and boss of the arm C also perform the reversing by moving the two pinions M into mesh with the pinion N and the wheel O, two pinions being necessary in this case because the driving spindles A and H rotate in the same direction, and the pinion N could not be got into gear with O.

The chains are roller chains of 1⅛ in. pitch, run smoothly and quietly, and the mechanical efficiency of the transmission is high. It is, I believe, the only car in which sprocket chains are used in the gear box instead of the usual small teeth spur-wheels and pinions.

Use of chains in gear box

The water cooler used with this car is practically the same as that illustrated in vol. i. p. 342. It is an excellent cooler, light and effective, and requires no pump or pump gear.

The cars fitted with the four-cylinder engines, shown in section and end elevation by the diagrams Figs. 167, 168, have the engine placed with

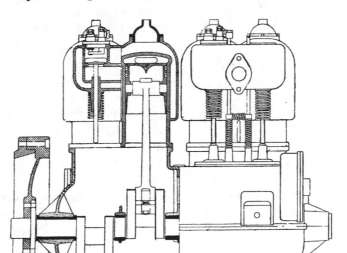

FIG. 167.—FOUR-CYLINDER ENGINE OF 15-HP. BROOKE CAR.

the crankshaft fore and aft. The inlet and exhaust valves are on opposite sides of the cylinders, all mechanically moved, two camshafts being employed, but the camshaft which works the inlet valves also operates the low tension igniter pallet, a Simms-Bosch magneto machine driven by gear from the inlet valve camshaft being used.

The igniters, Fig. 169, are simple and strong, and are placed directly over the inlet valves, the anvil pin being insulated and the rocking spindle carrying the pallet being constructed much as in the Cannstatt engines. The arm on the outside end of the rocking spindle is wide, so that the cam or finger on the top end of the vertical spindle, which can be varied as to time of lift and partial rotation, does not miss it at any vertical position.

Two systems of ignition

The negative wire from the magneto is as usual earthed, and the

157

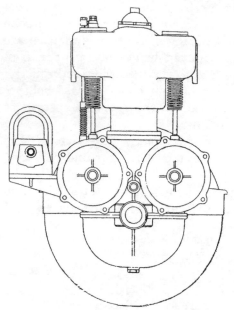

FIG. 168.—BROOKE 15-HP. ENGINE.

positive carried by insulated wire to an omnibus plug bar, into which are fitted the four plugs for the four cylinders. The magneto runs at the speed of the camshaft.

A rotative commutator and trembling coils with ignition plugs are also used, the commutator being driven by a chain from the inlet valve camshaft. Both systems are controllable by the driver by switches on the steering column.

The cylinders are 3·7 in. or 94 mm. diameter, and the stroke 4·52 in. or 115 mm. The normal speed is 900 revs., at which it is stated to give about 20 BHP., and about 25 BHP. at 1,500 revs. If these figures be correct, it is clear that a critical point of maximum power per 100 revolutions is reached soon after 900 revolutions are exceeded, the mechanical efficiency falling off rapidly after this speed is passed.

Engine lubrication Pressure-fed lubrication is adopted for the engine and for the petrol feed. A good arrangement of drain plugs is shown by Fig. 167 in the bottom of the crank chamber. If the centre plug is unscrewed, the whole of the oil may be drained away, or any water that may have accumulated can be first drawn off. When the driver wishes to ascertain whether the quantity of oil in the crank chamber is sufficient or not too little he can do so by unscrewing the plug to the left, when any excess of oil may be run off until the oil remaining in the crank chamber is level with the upstanding boss into which the plug is screwed. If the quantity is already too little, oil may be fed into the crank chamber until it commences to overflow. The provision of this or other means of definitely knowing the state of lubrication of the engine cannot be too highly commended.

The water in the honeycomb cooler is circulated by a pump driven by gear from the exhaust camshaft. The use of the pump in this car does not in any way indicate imperfect working of the natural automatic circulation in the 12-HP. car.

The transmission gear, Fig. 170, is the same in general design as that shown by Fig. 166, but the second motion shaft L is driven by a bevel pinion K^1 on the continuation of the clutch shaft and bevel wheel L^5 fixed on the shaft between the two larger chain wheels L^2 and L^3, instead of the silent chain drive shown by Fig. 162. Except that the bevel wheel and pinion run inside the oil-tight gear case,

FIG. 169.—BROOKE LOW TENSION IGNITER.

158

while the chain is outside, the advantage would appear to be with the chain arrangement.

The wheel base of the 12-HP. car is 7 ft. 6 in., and the wheel gauge 4 ft. 6 in. The wheels are all 34 in. diameter, the tyres being 870 mm. by 90 mm., or 34·26 in. by 3·54 in.

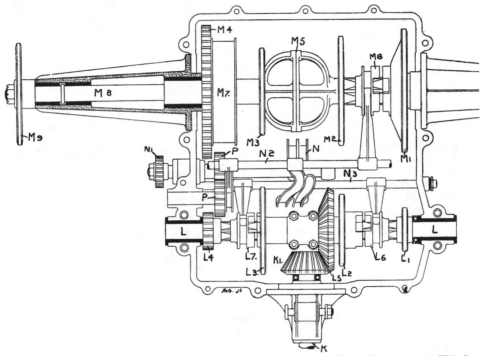

Fig. 170.—Transmission and Speed Change Gear of Brooke Four-Cylinder 15-HP. Car.

These dimensions are the same for the 15-HP four-cylinder car, except that the wheel gauge is 4 ft. 8½ in.

The Brooke three-cylinder car worked exceedingly well throughout the 1903 Reliability Trials of 1,000 miles, and was almost the lowest in fuel consumption.

Chapter VIII

THE MORS CARS

Air and water cooling

THE reference to the two systems of cooling adopted in the two sizes and types of Brooke cars recalls the further system of combined air and water cooling used by Mors for two- and four-cylinder cars at the time when the drawings for this book were commenced. The heads of the cylinders were water cooled, and the barrels cooled by circular radiator ribs or flanges like those still used on the cylinders of motor cycles. For the 1903 cars the cylinders became liners of an aluminium case forming a water jacket between it and the liner.

Fig. 171 is an elevation and Fig. 172 a plan of a 15-HP. four-cylinder Mors car so cooled, and as made until a little more than two years ago, when the cooling ribs were discarded in favour of water jacket.

A large number of these cars were made, and some of the prizes in some of the most remarkable racing and trial events were carried off by them, including the world's record for the flying mile in 1902, when a speed of 78 miles per hour was reached.

The frames F of the cars were of ash reinforced by inside steel flitch plates deepened in the upper portions as shown by Fig. 171, the engine and gearing being carried on a secondary frame T. The wheel base of the car shown was 6 ft. 2in., the wheels being 34 inches and 30 inches respectively.

Low tension magneto ignition apparatus was used, the magneto being of the Mors design and make, and the system similar to that common in France long before this date, a system which, with modifications to suit application to motor vehicles, has become more common since.

Separate throttle valves to each cylinder

Four separate pipes from the carburettor N led to the inlet valve chambers, the admission being controlled by the governor Z acting through the lever 11, vertical spindle 21 and rod Y coupling up the four throttle valve arms like that marked 20, Fig. 172. At the outer end of the governor lever 11, was attached a connexion to a pedal by which the action of the governor could be delayed or prevented for acceleration purposes. Four separate exhaust pipes joined at 26, and thence branched to the two ends of the silencer E.

Hardly any change has been made in the general arrangement of the

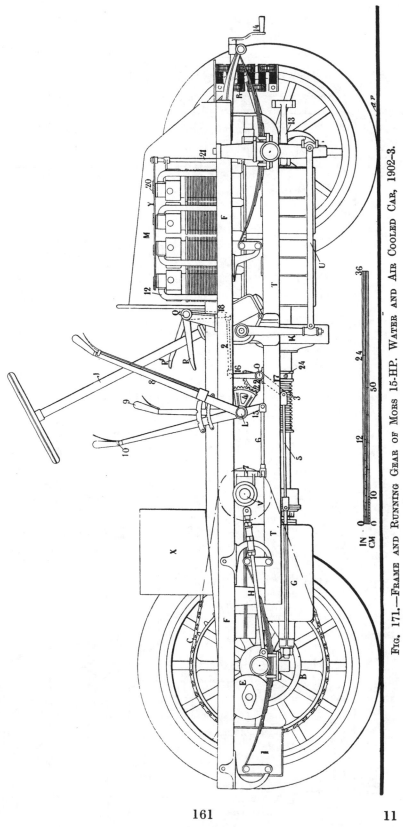

FIG. 171.—FRAME AND RUNNING GEAR OF MORS 15-HP. WATER AND AIR COOLED CAR, 1902-3.

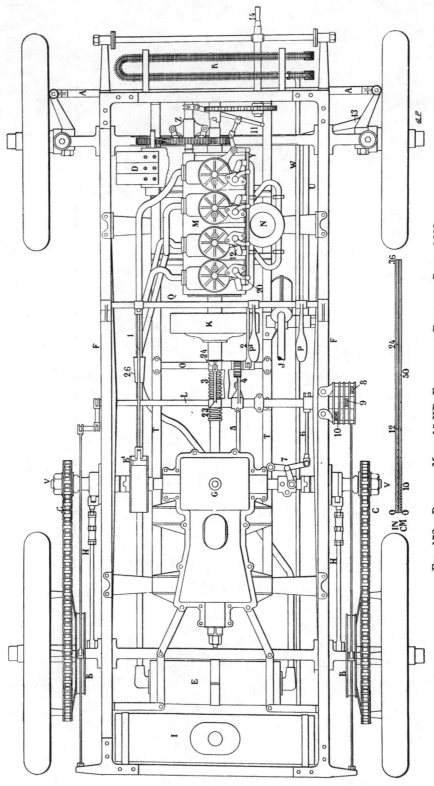

Fig. 172.—Plan of Mors 15-HP. Frame and Running Gear, 1902.

162

Mors cars to date. The transmission gear of the 1903 type is illustrated by Fig. 173, in which F is the engine fly-wheel, and D the aluminium male part of the clutch which slides on the shaft E and is kept in position by the spring C, a five-collar thrust bearing being provided at the rear of the gear box for this shaft. It has, however, little or nothing to do in the ordinary sense as a thrust bearing, as the pressure of the clutch spring is not put upon it as thrust. The spring C bears against a fixed collar upon the shaft, and the end of the clutch centre sleeve. In the end of the shaft in the fly-wheel centre is inserted and fixed by a pin a short piece with a large collar head. This has upon its rearward face a ball race, and a similar race is carried by a boss fixed on the fly-wheel. This takes the thrust due to the spring, and it is only when the clutch is held out of action while the engine is running that the ball thrust bearing comes into action. The change-speed wheels are on the main shaft E, the upper or secondary shaft carrying the fixed gear and the bevel pinion by which the bevel wheel on the differential motion is driven. The pairs of wheels and pinions are marked

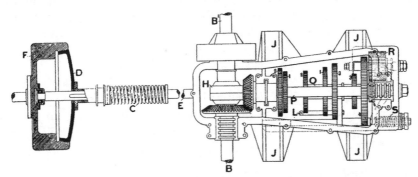

Fig. 173.—Transmission Gear of Mors 15-HP. Car, 1902-3.

1 to 4, and their positions will be readily traced. The differential shaft B B¹ is fitted at its ends with universal driving pieces, as seen in Fig. 172, which drive the short sprocket spindles running in long bearings attached to the frame sides.

The change wheels are moved by the hand levers 9 and 10, the one being on the sleeve on the spindle L, and the other on a tube outside that, the side brake lever being fixed to the spindle itself. The sector 4 gearing with the sector 22 on the spindle O is moved by the lever 10, and thus the rod 5 connected to the rod L, Fig. 173, in the gear box. The lever 9 is in Fig. 171 shown behind lever 10, but should be in front of it. It operated the reversing gear. As now made, however, it is dispensed with, one lever performing the whole of the gear changes. By pushing the lever to beyond the out-of-gear point on lowest speed, the end of the boss of the fork pushed by the rod L comes into contact with the lever pivoted below the thrust bearing and seen in dotted lines in Fig. 165. Further movement of the hand lever pushes this pivoted lever against the resistance of the spring S, and the other end of the lever pushes the reversing pinion R into gear with the pinion 3 and wheel 4. The pedal P operates the brake B[1], and P[1] moves the clutch out of gear by pulling on connecting rod 2, arm 16, and corresponding forked lever 24.

The steering gear, clearly shown by Figs. 171 and 164, consists of the worm and segment at the foot of the steering column, the lever K, universal joints, connecting rod U, and cross connecting rod A. A rod connecting the ends of the two front spring hangers was used to provide a fixture for the bearing of the starting handle 14, but this rod may be commended as a desirable strengthening stay, though it is seldom used.

The 80-HP. Mors cars have also been but little changed in design since 1902, the 1904 racers being in some respects exceptions. In the smaller cars the ignition apparatus has been modified in various details, including the method of timing the ignition, but the Mors cars were always in advance of most others in the neatness and lightness of this gear, some types of which will be illustrated elsewhere. It need hardly be said, perhaps, that the latest engines have mechanically moved inlet valves.

One of the 80-HP. 1902 cars, with a boat-shaped body, made a record speed at Welbeck Abbey, driven by the Hon. C. S. Rolls, who subsequently altered the body to a bonnet and a seat, and won a speed race with it in the Phœnix Park speed trials last year. After that it was fitted with a touring body and sold to the Duke of Manchester.

The change-speed gear in the 1903–4 cars has been modified so that the second motion shaft lies alongside instead of above the main shaft, and a second bevel wheel has been put on the sprocket or countershaft, so that on the highest speed this shaft is driven direct by the bevel gear instead of through bevel gear and the change-speed spur gear.[1]

[1] *Automotor Journal*, December 26, 1903, p. 1394 ; and *Autocar*, December 20, 1903, p. 626.

Chapter IX

A DESIGN FOR A BELT-DRIVEN CAR

BELT-DRIVEN cars have fallen into disrepute ; or, at all events, have gone out of fashion. In many cases the belt failed, not because of an inherent defect in the belt system but because it was defectively carried out. There are still many belt-driven cars running on the Benz and Maybach system, but this is only because they are in the hands of careful and sympathetic users. In almost all cases the belt pulleys were too close together and some of the pulleys too small. The belts were generally narrow. All these causes conspired to throw a very much higher stress upon the belts than would ever be attempted in machine shops and mills where changes in weather had less effect than they had on the belts of most cars. It is not impossible, however, that the comfortable silent running of belt-gear may yet again be used on small cars, and if this be so a description of a car designed when the belt car was going out of fashion may be of interest.

Figs. 174 to 195 illustrate the car designed by the author, made by Messrs. Hayward, Tyler & Co., and burned to ashes and scrap in the fire which occurred in their works in the autumn of 1903, soon after the first tests of the car had been made.

The object of the design was to use a belt of much greater length than had been possible with other designs, larger pulleys, only one shift of belt for four speeds, solid through one piece axle and a light car. *Features of design*

Fig. 174 is a side elevation, and Fig. 175 a plan of the car. The frame is shown as made of hickory $1\frac{1}{4}$ in. $\times$ $1\frac{3}{4}$ in., capped by a bar of $1\frac{1}{4}$ in. $\times$ $\frac{3}{8}$ in., and soled with a bar of $1\frac{1}{4} \times \frac{5}{16}$ mild steel, connected by $\frac{1}{4}$ in. steel bolts at 6 in. centres, except where bolts for fittings took their place.

The rear part of the frame rests upon the springs through intermediary pivot brackets ; the springs are carried by hangers which at the rear are formed at the ends of the separate underframe F F pivoted at E on a shaft which slides in the bracket D. The forward parts of the springs B are carried in hangers fixed to the underframe F. Towards the back part of the latter are fixed bearings pivoted transversely for carrying the shaft G on which are carried the pulleys H, H^1, J, J^1, and which is the main or drawing axle. The pulley H is driven by a belt U running on the pulley N, driven by the

165

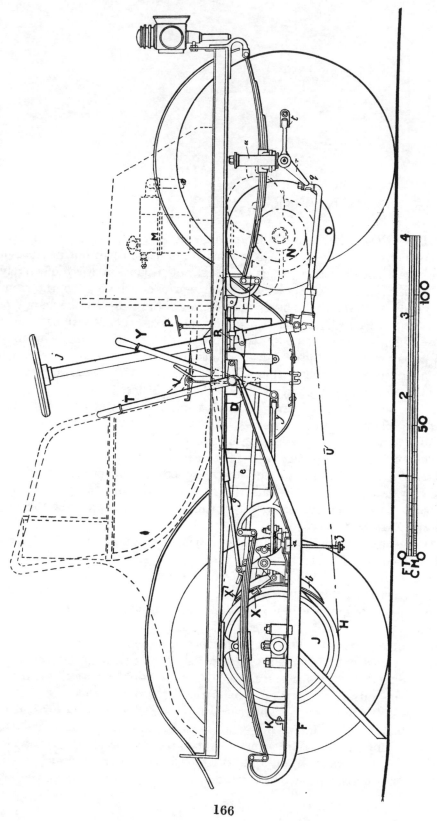

FIG. 174.—DESIGN FOR A BELT-DRIVEN CAB, 6 HP.: ELEVATION.

166

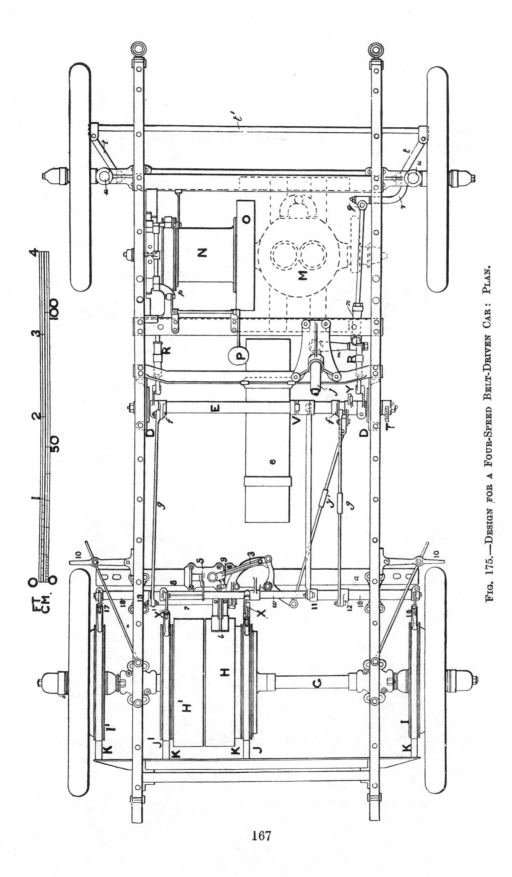

Fig. 175.—Design for a Four-Speed Belt-Driven Car: Plan.

167

Operation of clutch bands motor M, and carried upon an extension of the motor shaft. The clutch bands J and J¹ are alternatively actuated by the pedal v, which is pushed forward or backward, and thus, by a rod connecting the lower end of the pedal lever with an arm 11 on the solid spindle w, moves the bell-crank arms at X and X¹, by which one clutch band is tightened when the other is loosened. These parts are shown to a larger scale on Figs. 176, 177, which also shows

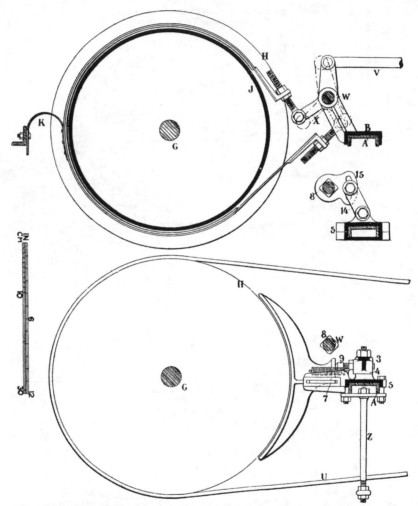

FIG. 176.—BAND CLUTCH FOR SPEED-CHANGE GEAR ACTUATED BY PEDAL AT V.
FIG. 177.—CENTRAL PULLEY BRAKE AND BELT SHIFTER FOR THIRD AND FOURTH SPEEDS.

the bracket providing a bearing for the shaft w, and the transverse bearer a on which are placed the belt-shifting gear, and at its extremities the tyre brakes hereafter to be described.

Belt shifting gear For shifting the belt U the lever Y, which, like the pedal v, is pivoted on the outer tubular part of the shaft E, carries at its lower end the jointed end y of the rod y^1, which moves the bell-crank lever pivoted at 2, Figs. 179, and 180, and through the rod or link 3 moves the slider 5, by the lower part of which the belt-shifting fork z is carried.

The transmission and change-speed gearing is shown partly in section and partly in elevation by Fig. 178. It comprises a central loose pulley H, which carries the small gear wheels or pinions B, C, D, E, on short or stud spindles A. These gear wheels mesh with corresponding wheels or pinions fixed on the shaft to be driven R, and with other pinions carried on the bosses of loose wheels or pulleys H¹, J¹, or, J on either side of the central pulley. The central loose pulley being driven the side pulley H¹ may be held by the band-clutch J¹ to give one speed to the shaft G, or the wheel J may be held to give another speed.

For instance, if the pulley H¹ be held by the band-clutch on J¹, the gear wheel E will roll round on I and carry with it the pinion D in the same

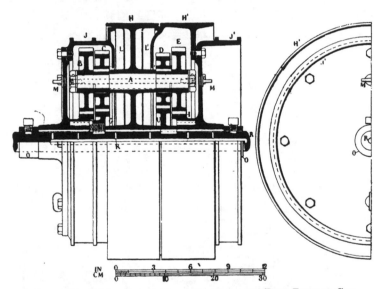

FIG. 178.—FOUR-SPEED DRIVING GEAR OF BELT-DRIVEN CAR.

direction as that of the pulley H ; correspondingly, if J be held by band-clutch the gearing on its side will give motion to the shaft R through the pinion G at a different speed, due to the relative dimensions of the gearing. If the belt be shifted so as to run on H and H¹ at the same time, the gear will be locked and a third speed obtained, namely the highest. By running the belt on H¹ and holding the pulley H, the pinion I will drive the wheels E, D, giving motion to wheel U and shaft R at another or fourth speed.

When the belt is, by pulling rod y^1, thrown fully over on to the pulley H¹, the set screw head 9 in the link 3 pivoted at 4 comes into contact with the piece 7 pivoted above F, Fig. 180, which carries the brake shoe piece b, Figs. 174, 175, and B, Fig. 180, which holds the pulley H for obtaining the third speed ; these parts are also shown by Fig. 177. To prevent the slider 5 from being moved and then thrown over to H¹ while the band-clutch is holding J¹, the shaft w is made square at 15, Figs. 179, 180, and at 8,

Change-speed gear

169

as shown by Fig. 176, and on the slider piece 5 is carried a piece 15, in which
is a square hole capable of sliding over the squared part of the spindle ;
before the slider 5 can be pushed over, the pedal v must be centrally placed,
so that neither clutch-band is holding, and in that position the square

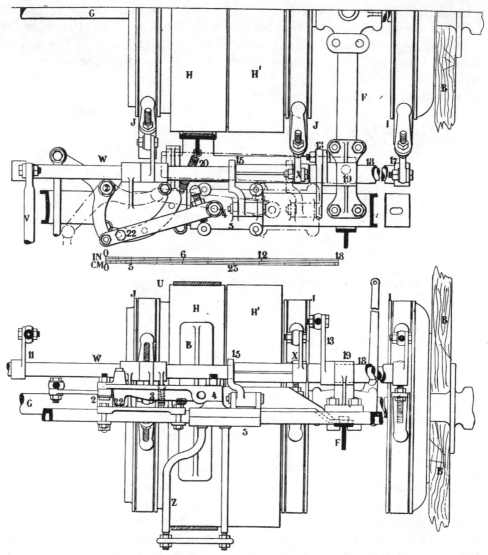

FIG. 179.—BELT-SHIFTING AND SPEED-CHANGING GEAR: PLAN.
FIG. 180.—BELT-SHIFTING AND SPEED-CHANGING GEAR FROM THE FRONT.

hole of 15 and the square at 8 will coincide, so that the slider may be pushed
over opposite H[1]. At 22 in the link 3, Figs. 179, 180, is a socket in which
is a spiral spring, which presses upon a short round-nosed bolt, which engages
with slight depressions in the quadrant-shaped part of the fixed bracket
over which the socket or boss 22 passes when the bell-crank pivoted at 2,

Fig. 179, passes towards the position shown in dotted lines. When it engages with the depression shown by dotted line near 22 the belt will be half on H and half on H¹, giving the highest speed ; when it is in depression, shown by full line, the belt is on H¹, and H is loose ; further movement to the position shown by dotted lines forces the piece b upon pulley H, and thus gives the third speed, the spring 20 pulling the piece b back again when the belt is removed from H¹.

On the two ends of the spindle w, Figs. 175, 179, 180, is an exterior tubular spindle 18. These tubular ends are supported in the brackets 19

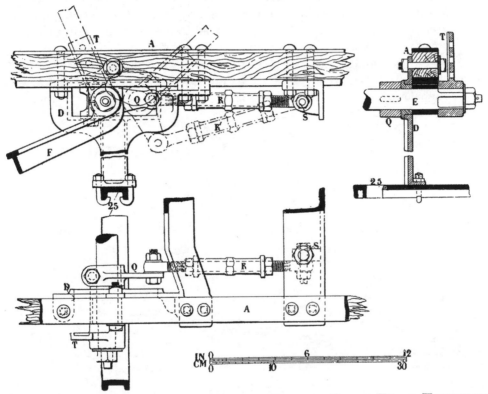

FIGS. 181, 182, 183.—SIDE ELEVATION : PLAN AND SECTION OF PIVOTED END OF UNDERFRAME OF BELT-DRIVEN CAR.

and carry the brake-operating bell cranks 17 and 16, Figs. 174, 175, and 180. These are operated by the rods $g\ g$ and arms $f\ f$ on the tubular part of the spindle E, actuated by the pedal f^1, Fig. 174.

Fig. 181 is a side elevation of the pivoted end of the underframe and of the means and apparatus for adjusting its fore and aft position with relation to the upper frame.

Fig. 182 is a plan, and Fig. 183 is a section through the pivot eye at the end of the pivoted underframe pieces F through the sliding bearing carrying the pivot spindle, and through the bracket supporting it. In this A is the upper frame, to which is attached the bracket D, which is ex-

Adjustment and relief of belt tension

171

tended downwards to carry the crossbar 25, upon which are fixed the steps as seen in Fig. 1. On the pivoting spindle E is fixed an arm Q and a lever T, seen also in Fig. 174, to the end of the arm Q is attached the adjustable link R, pivoted at its front end at S. With the lever T, arm Q, and link R in the position shown, the pivoted underframe is in the backward position with the driving belt tight upon the pulley H and the arm Q and link R being both inclined slightly upwards they remain in that position. When it is desired to relieve the driving belt of the tension upon it, the lever T is thrown over to the position shown by dotted lines, and the arm Q pulls upon link R, which takes the position shown at R¹ ; the whole of the under-gear is thus pulled forward, the brass E¹ sliding in the bracket D. For the adjustment of the tension on the belt, the link R is made adjustable as to length by means of the nuts and screws shown.

When the underframe is thrown forward in the manner described, the tyre brake shown in part in Fig. 175, but for clearness omitted in Fig. 174, is brought into action ; this brake, is shown by Figs. 184, 185, 186. In

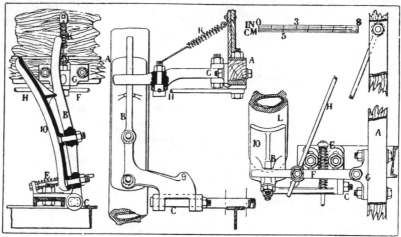

FIGS. 184, 185, 186.—TYRE BRAKE GEAR.

Tyre brakes these is part of the T-steel pivoted underframe, and the crosspiece seen at *a* in Figs. 174 and 175, which carries the gear-shifting apparatus ; the ends of this crossbar are extended as seen in Fig. 175, and upon these ends are fixed the pivot brackets, to which are pivoted at C the arms B upon which are bolted the brake shoes 10 ; attached to the upper frame A is a bracket carrying the arm F pivoted at G, but held in the desired position at or near a right angle with the frame A by means of the tension rod H pivoted to the frame and adjustably fixed in the arm F by means of the clip bolt *j*, shown in section at Fig. 185.

When the underframe is pulled forward, as already described, the tyres L of the driving wheels are brought into contact with the brake shoes 10, which are prevented from moving forward by contact of the arm B with the arm F.

DESIGN FOR A BELT-DRIVEN CAR

To prevent shaking or rattling movement of this brake tackle, and inertia effects, a compression spring is placed at E, and a tension spring at K.

Reversing is effected by the apparatus shown in Figs. 175, 187, 188. In these O is the motor fly-wheel to which is attached an extension B of the motor shaft, upon which runs loosely the driving pulley N; the end of the shaft B is carried in a bearing fixed to the bracket C; to this bracket also are attached small brackets supporting the spindle on which is fixed the finger F and the cam G; suspended from the upper frame A is a forked lever H, which engages with a cap J in which is fixed one end of a rod K passing through a hole in the spindle B, and thereby a key engaging with a central pinion on a feather on spindle B, the inner end of the pinion forming a clutch which engages with the claw clutch boss on the pulley N; partly

Reversing gear

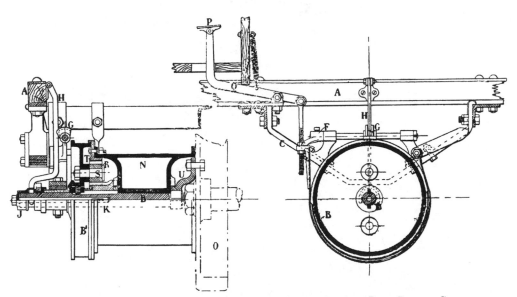

FIGS. 187, 188.—DRIVING PULLEY AND REVERSING GEAR OF BELT-DRIVEN CAR.

encircling the pulley B¹ is a brake-band P, supported in front upon a bolt carried in part by a bracket fixed to the frame and in part at the foot of the bracket which carries the spindle of cam G; at the rear it is carried by an adjustable suspending rod on the end of the lever Q actuated by the pedal P. For all ordinary forward running the parts above described are in the position shown; when it is necessary to reverse, the pedal P is depressed, and the finger F partially rotates the spindle E and pushes the cam G against the forked lever H, and so pulls the rod K and the pinion out of contact with the clutch boss of the pulley N; continued depression of the pedal P pulls the band tightly upon pulley B¹ and prevents its rotation; the central pinion then drives the pinions R which gear with the internal gear wheel T, and thus drives the pulley N in the direction opposite to that of the rotation of the operating motor. The pinions R, of which there

173

were three, are carried upon a stud spindle s fixed in a boss in the band-clutch pulley B^1. The spindle B is carried in a boss U, with square hole, fixed to the fly-wheel O.

The steering spindle extends downwards within a tubular pillar j,

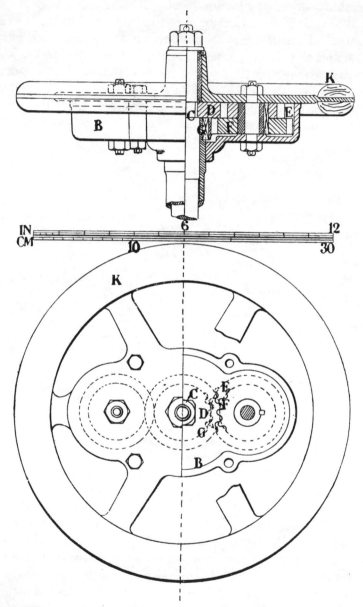

FIGS. 189, 190.—STEERING LOCK OF BELT-DRIVEN CAR.

which is adjustably held within a bracket fixed to transverse members of the frame, to the arm m, pivoted to the rod n^1 at n and at q; at q this rod embraces the axle-steering arm r, which is extended to form the arm t

DESIGN FOR A BELT-DRIVEN CAR

connected by a connecting-rod t^1 to a similar arm t on the near front wheel axle, the steering axles being pivoted at u u ; at n at one end of the rod n^1 is a socket containing spiral springs for absorbing the more severe shocks which would otherwise be delivered to the steering-wheel spindle. In order that the steering may be irreversible, I employ the epicyclic self-freeing gear illustrated by Figs. 189, 190 : the wheel K turns freely upon the upper part of the spindle ; to the under part of the steering-wheel centre is attached a box B, which turns freely upon the tubular steering pillar. Upon a square c is a pinion D gearing with wheel E, integral with a pinion F, the pinion F gearing with a wheel G fixed on the top part of the tubular steering pillar j.

Epicyclic steering gear

With this epicyclic gear, turning the steering wheel in either direction causes the pinion F to roll upon the fixed pinion G, and thereby the wheel E rotates the pinion D on the steering spindle ; when, however, any tendency

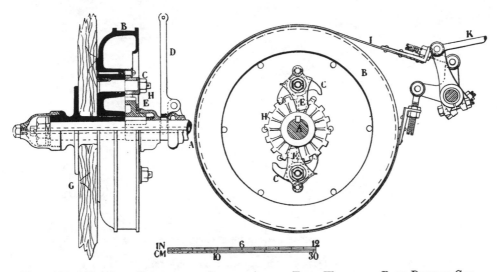

FIGS. 191, 192.—SIDE BRAKES AND DOUBLE-ACTION FREE WHEEL OF BELT-DRIVEN CAR.

originating in the road-steering wheels to rotate the steering spindle occurs, motion then originating at D so far as the hand-steering wheel is concerned causes the gear wheels D, E, F, G, to lock. This locking device would, in some cases, simplify the steering-arm arrangements, but it is not as good as the worm and nut, or worm, nut and quadrant arrangements used in various cars. No differential gear was employed, but in its stead the double-action free-wheel arrangement shown by Figs. 191 and 192. In this A is the axle, part of the end of the bearing being shown broken off as A, Fig. 191. On the axle is the nave of one of the driving wheels, the inner part of which is formed by the brake drum B on which are two bosses for pins carrying the pawls c, which engage with the square tooth ratchet wheel H fixed on the axle. On the face of each pawl is a steel spring plate E normally free of the slightly raised ribs on the face of the ratchet wheel. A light spiral spring on a pin outside the pawl pivot keeps the pawl normally in the

Double-action free wheels

175

position shown for forward running. When it is necessary to run backward, the driving gear being reversed by the method described, a light lever D pivoted on the outer end of the axle bearing is pulled toward the body of the car. This lever is forked at the lower end and bears against a cup seen in Fig. 191, inside which is a light spring between the cup bottom and the ratchet wheel boss. Pulling the lever towards D pushes the lip of the cup against the spring plate, which then bears on the ribbed face of the ratchet wheel. This, by slight frictional hold, tilts the pawl so that the teeth C, previously free, come into gear, and the wheels are driven backwards.

Fig. 192 shows the band brake gear on the end of the tubular spindles on the ends of the shaft w, seen in Fig. 175, and in Figs. 179 and 180.

The reversing gear employed on the driving pulley N worked satisfactorily for changing from forward to backward motion of the car, and it

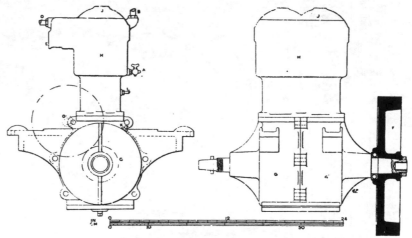

FIGS. 193, 194.—8-HP. FORMAN MOTOR USED ON BELT-DRIVEN CAR.

could be used as a declutching and pedal brake for running in traffic, but one oversight was made in its design. In putting the forward motion out of gear by pulling the central claw-ended pinion out of gear with the claw-faced boss of the pulley N, no inertia forces came into play, and the reverse motion was gradually brought into operation by the band clutch on the pulley B¹, Figs. 187 and 188.

Modified reversing gear

When, however, the forward motion was again put into gear by releasing the band clutch and letting the central pinion claw clutch into position as shown, the pulley N had to be put into motion at the speed of the engine. Although this speed could be lowered at the moment to about 250 revs. per minute, the inertia of the pulley N was sufficient to bring so large a moment of resistance into play, that the spindle B was twisted between the pinion and the square at the inner end. If the pulley had been made of aluminium, the inertia force would have been much lessened, but this would have introduced a difficulty as to the form of the claw clutch. The

176

difficulty was entirely removed by boring out the fly-wheel rim to an angle of 15 degrees, and fixing a clutch cone on the inner flange of the pulley N. The teeth of the claw clutch pinion in the centre were taken off and a loose collar put in to fill the space. The end cap of the spindle B was altered to carry a spring to press the male part of the cone clutch home. The cone clutch did the forward driving, and the central pinion and reversing gear and band clutch worked as before, and the pulley N was put gradually into motion by the cone clutch in the usual way. The reversing gear could thus always be used as a pedal brake.

The engine used on this car was a throttle-governed 8-HP. of the Forman Motor Co., but driven at nominally 650 revs. instead of 900. At this speed it was found to give 6 BHP. for short periods. It had cylinders 90 mm. or 3·54 in. diam., and 100 mm. or 3·93 in. stroke.

Figs. 193, 194 show the outline and dimensions of the engine and

FIG. 195.—8-HP. FORMAN MOTOR OF BELT-DRIVEN CAR.

supporting arms, and Fig. 195 is from a photograph of the engine used. The commutator was taken from the position shown, as it would have been very difficult of access, and was modified and driven by a chain on the dashboard.

The cooler used was one of the Albany Co.'s apron form, placed high on the dashboard, the bonnet being a very low one. The unfortunate end to the car was very disappointing, but recent developments in light cars by English makers have removed the disappointment except in so far as lengthened experimental running of the car would have been of interest.

From the external view of the 16 HP. engine shown by Fig. 195A, it will be seen that in many respects the arrangement is similar to the 8 HP. engine. Four cylinders cast in pairs are used and the crank chamber is divided horizontally at the crankshaft centre line. The upstanding end

FIG. 195A.—16 HP. FOUR-CYLINDER FORMAN ENGINE.

of the governor operated lever is connected to a throttle or throttle valves not visible in the illustrations.

Recent types of Forman engines differ in external appearance from those referred to, but some of the changes are not essential and more nearly concern improvements in general arrangement and in mechanical detail.

Chapter X

THE RICHARD-BRASIER, CLEMENT-TALBOT, AND DE DIETRICH CARS

THE Georges Richard cars, now the Richard-Brasier cars, comprise within less than three years an epitome of the salient changes which have been made by many constructors.

Well known as a cycle manufacturer Georges Richard began with a voiturette, one of which is illustrated by Fig. 197, and has ended with an 80-HP. racing machine which, driven by M. Théry, wrested the Gordon-Bennett Cup from Germany for France on June 17, 1904, and again won this trophy for France with a racer of nearly 100-HP. in the recent Gordon-Bennett Race in Auvergne. The engine used for this racer has four cylinders, 160 mm. or 6·3 inches diameter with a stroke of 140 mm. or 5·51 inches, and its stated horse-power at 1,200 revs. per minute is 96.

The power is transmitted to the driving wheels by side chains.

Richard entered four different cars in the Paris-Berlin tourist car race in 1901: one, the voiturette, Fig. 197, with a 4-HP. single cylinder engine and belt-driven gear, and called a Poney car; one a similar car with an 8-HP. engine; one a double-cylinder chain-driven car of 10 HP., as shown by Fig. 198, and one other. Two of the voiturettes reached their journey's end "sans accident ni incident."

These voiturettes did not become well known in England, although they were useful little well-designed cars, the general arrangement being similar to the three-speed belt-driven Vivinus car, and the two-speed Hutton car. It was, however, of better design as to the three-speed change gear, and as to the belt pulleys, which were of 8-in diameter, and of the same size on engine and countershaft. The pulley centres were nearly 4 ft., so that the belt was much better treated than it was in the Benz cars of the same period. There were numerous points in which these voiturettes were ahead of the time. Many were made and used in France in 1902, and in 1903 the air-cooled engine was replaced by one with water jacket and gear-driven circulating pump. One of these cars ran in the 1,000 miles reliability trial of that year. The engine was 5½ HP., at 1,200 revs. per minute; wheel base, 5 ft.; wheel gauge, 3 ft. 8 in., and tyres 28 in. diameter on wood wheels. Some trouble was experienced with cylinder overheating, but none with the belt.

Belt-driven voiturettes

179

The live axle of the car, Fig. 197, was carried by central bearings close to the differential gear box as well as at the ends, a braced channel iron axletree securing the distribution of the load. The main bearings [1] under the springs were carried by pivot pins, so that the spring of the axle would not cause them to grip crosswise, they were proper shaft bearings, four diameters long and provided with ring lubrication. The central bearings, which had smaller load, were adjustable of the Bown pattern.

The change-speed gear was simple and of very ingenious arrangement. The three gear pinions, through which the power at the different speeds was

FIG. 197.—5-HP. GEORGES RICHARD BELT AND GEAR "PONEY," PARIS-BERLIN TYPE.

transmitted, were all in one piece with a hollow spindle which ran on a fixed spindle with ball bearings at the two ends. For two speeds these wheels came into operation by sliding a double pinion on the primary shaft. For the third, or highest speed, this double pinion was slid further in the same **Gear design** direction, when it became a clutch for the loose wheel previously driven by the secondary gear. Thus on the highest speed the drive was direct or without the intervention of the secondary gear, though the latter remained in gear when no longer doing work. The principle of the arrangement, namely, having the final transmittor on or forming part of the primary spindle, was subsequently used by Richard in his belt and chain-driven cars,

[1] *Le Chauffeur*, July 25, 1901.

THE RICHARD-BRASIER MOTOR VEHICLES

Fig. 198, of 1901, and again in the live axle bevel gear driven cars, like the 1902 cars, Fig. 199, and the similar cars in 1903, with change-speed lever at the side of the car instead of on the steering column, as in 1902. The gear design was in fact the forerunner of many developments. The arrangement made it possible to get the gear into a very small box, as may be gathered from Fig. 199, which is from a car that did good service in Tunis and Algeria for the photographic corps of the Touring Club de France.

The 1903 cars were similar in almost every detail to the 1902 car, except the arrangement of the lever connexions for operating the change-speed gear, and in the use of magneto ignition apparatus. Automatic inlet valves centrally over the top of the cylinders were used, and in some a circulating pump was used, but generally, as in the 1904 cars, natural or

FIG. 198.—GEORGES RICHARD BELT GEAR AND CHAIN CAR, 10 HP., 1901 TYPE.

thermosyphon circulation was employed. The frames in all cases were tubular with a secondary underframe to carry the engine, gear case, and other fittings. The engine of the 10-HP. car had two cylinders 4·12 in. diam., and 3.94 in. stroke—the speed being high.

The low tension ignition apparatus was of the same form as that used in the Gordon-Bennett Cup car of 1904. It is illustrated by the diagram, Fig. 200. Pivoted to the top of the pusher rod A is a finger B which presses, when A lifts, against the short end of the rocking arm C, which rests normally against stop S, and moves on a spindle K, the inner end of which, inside the combustion chamber, carries the pallet or hammer D. This hammer is normally apart from the inner end of the insulated anvil or terminal piece E, as shown by dotted lines. It is moved by the arm C, through the interven-

Igniter actuating gear

181

FIG. 199.—10-HP. GEORGES RICHARD GEAR-DRIVEN TOURING CAR, 1902 TYPE.

182

tion of a coiled spring on the spindle K. The lower part of the trip finger B has an inclined surface near the pivoted end, and adjacent to this is a cam F pivoted on a fixed pin. This cam can be adjusted as to position with reference to the trip finger, from the driver's seat, by the lever G. When the pusher A is raised by the cam H the finger B raises the hammer terminal D into contact with the insulated terminal E, continued lift of B only acting on the coil spring and the spring J, but in this continued lifting the tapered surface of the trip finger is pushed past the cam F, and the finger is thus, according to the position of the cam, sooner or later caused to leave the end N of the lever C by taking the position shown by the dotted lines ; at this instant the distended spring J asserts itself, pulls the lever C back to its original position, and makes a break of circuit at the velocity of recoil of the spring, aided by the spring at the back of the lever C.

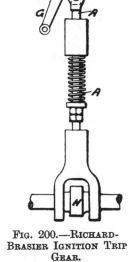

FIG. 200.—RICHARD-BRASIER IGNITION TRIP GEAR.

It will thus be seen that the period at which the finger B is tripped from the end N of the lever C, and, therefore, the period of recoil and of ignition, depends on the position of the cam F, which is adjustable. It is noteworthy, however, that in the Gordon-Bennett car only two positions of the lever on the dashboard for this adjustment were made, namely, a late position for starting the engine, and an early position for the engine after starting ; the driver could thus make no mistake as to the best position of his sparking lever, and he was relieved of all consideration with regard to it.

Positive clutch

The Richard-Brasier clutch is of the pressure balanced cone class, but for the racing cars it is also provided with means for making it positive after the car is fully in motion, so that no slipping can occur, Fig. 201. Over the central boss of the male cone is a ring A carrying six steel pins B, the inner ends of which may, when the cone is engaging, find their way into any six of thirty-six holes in a flat ring C fixed to the fly-wheel web. On depressing the clutch pedal these pins are first withdrawn, and in Fig. 201 are so shown, continued motion withdrawing the clutch cone D. The device is said to work well and cause no difficulties, and it may be easily understood that with large powered cars running under race conditions all prevention of slip may be very valuable. Whether this positive part of the clutch be used or not, the design by which the thrust is self-contained is one of the best made.

The Richard-Brasier 1904 cars are all fitted with live driving-axle and bevel driving-gear, the frames being of the now popular pressed plate steel, but with a supplementary under-frame to which the gear box is rigidly attached. The clutch shaft is connected to the main shaft in the gear box by a short double universal joint spindle, so that no slight want of alignment can affect the proper working of the clutch or free running of the shafts, and the block of the double joint can be easily taken out for access to the clutch. The engines of all sizes from 8 HP. to 24 HP. are now fitted with mechanically opened valves, magneto ignition apparatus, and with the Brasier carburettor similar to that used in the Gordon-Bennett car. In this carburettor two petrol jets are inserted in the induction tube radially, nearly touching each other, and pointing upwards at an angle of about 40°. The points of these jets [1] are placed within the apex of a parabolic constriction of the induction pipe, at a point where the air velocity will at the same time be greatest, and where the triturating action will be supplemented by the surface rubbing action produced by the form given to the tube; air is admitted some distance below the constriction referred to, and also by hand adjustment between the jets and the connexion to the engine. No supplementary air valve, as in what are known as the automatic carburettors, such as the Krebs already referred to, is employed. It is assumed that the two jets pointing towards each other, like the jets of certain gas or acetylene burners, produce what has been likened to a liquid butterfly, easily vaporized by the rushing air; inasmuch, however, as the petrol is brought in by the inductive action of the air current the jets will not impinge upon each other, but on issuing from the jet orifices will follow the direction of the air which induces the motion.

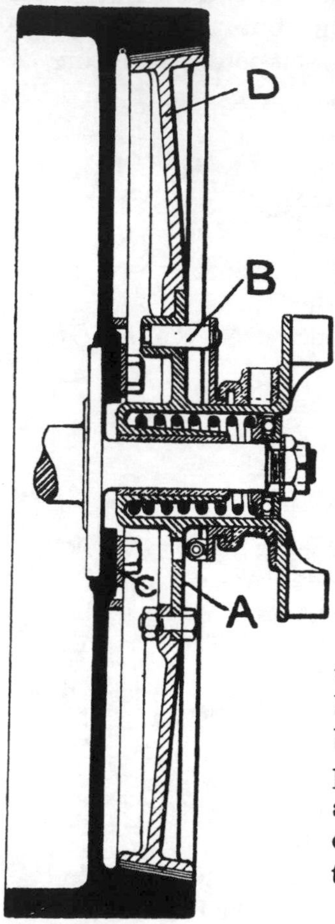

The 1904 cars are made in four sizes, namely, 8, 12, 16 and 24 HP., requiring the use of only two sizes of cylinders, pistons, connecting rods and some other parts. They are as follows —

Double jet carburettor

Fig. 201.—Richard-Brasier Friction and Positive Clutch.

Declared HP.	8	12	16	24
Number of cylinders . .	2	2	4	4
Diameter of pistons . . .	{ 85 mm. / 3·35 in.	104 mm. / 4·09 in.	85 mm. / 3·35 in.	104 mm. / 4·09 in.
Stroke of pistons . . .	{ 100 mm. / 3·94 in.	100 mm. / 3·94 in.	100 mm. / 3·94 in.	100 mm. / 3·94 in.
Weight of chassis, approx. .	{ 500 kilog. / 1,103 lb.	625 kilog. / 1,378 lb.	700 kilog. / 1,543 lb.	750 kilog. / 1,653 lb.
Diam. of wheels	{ 760 mm. / 29·9 in.	800 mm. / 31·5 in.	870 mm. / 34·26 in.	870 mm. / 34·26 in.

[1] *La Vie Automobile*, June, 1904 ; *Horseless Age*, July 27, 1904, p. 87.

THE RICHARD-BRASIER MOTOR VEHICLES

The 24-HP. car is also made with a longer frame for large body, and the weight of the chassis becomes 800 kilog., about, or 1,764 lb., and the wheels are increased to 34·5 in.

The advantages which attach to making these four cars with only two sizes of cylinder and only one length of stroke cannot be over-estimated from the mere works point of view, but they are almost equally important to the purchaser. It is an instance of the many which shows how carefully every point in the design of these cars has been considered.

The dimensions of the 8-HP. car remain as given above, but the particulars of the larger cars as made in 1905 are as follows :—

Declared HP.	12	16	24	40
Number of cylinders . .	4	4	4	4
Diameter of pistons . .	78 mm. / 3·07 in.	85 mm. / 3·35 in.	104 mm. / 4·1 in.	125 mm. / 4·92 in.
Stroke of pistons . . .	110 mm. / 4·33 in.	110 mm. / 4·33 in.	120 mm. / 4·72 in.	130 mm. / 5·12 in.
Weight of chassis, approx. .	762 kilog. / 1,680 lb.	788 kilog. / 1,736 lb.	825 kilog. / 1,820 lb.	1016 kilog. / 2,240 lb.
Diam. of wheels . . .	810 mm. / 31·9 in.	875 mm. / 34·5 in.	875 mm. / 34·5 in.	920 mm. / 36·2 in.

With the four-cylinder engines and the range of powers given it has become necessary to adopt more varied cylinder diameters and strokes and the advantages of few standard sizes are lost.

The Richard-Brasier engines develop their stated powers at 1,300 revs. per minute, and 1,400 for the 8 HP. engine.

Some of these cars have not done best in the Automobile Club 1,000 mile trials, but the best cars may easily lose marks in such trials when the system of marking is not sufficiently tempered with judicial discretion.

THE CLEMENT AND CLEMENT-TALBOT CARS.

The Clement car, so called after the name of the designer, is one of those which has rapidly changed in details of design during the past two years, and during that time has attained much popularity as a well-designed light car presenting numerous details which represent some of the best points of foreign practice. The frames of the 1902 cars and 1903 cars were of wood with steel reinforcing plates, but those of the 1904 cars are of pressed steel plate similar to those illustrated on pp. 116 and 129.

The engine of the 1902 car was looked upon as an example of neat design with well-arranged valve seats and ports, large water jacket and easy water circulation. Like all engines of this type, however, it had one serious defect. One of these engines is illustrated by Fig. 202, with cylinders 85 mm., or 3·35 in. diameter, and 108 mm., or 4·25 in. stroke, and of 8 HP. It was, of course, fitted with automatic inlet valves, and the exhaust was governed as to period of opening and closing by resting the pusher rod on the top of a rocking lever A, the pivot of which was moved in the direction of length

185

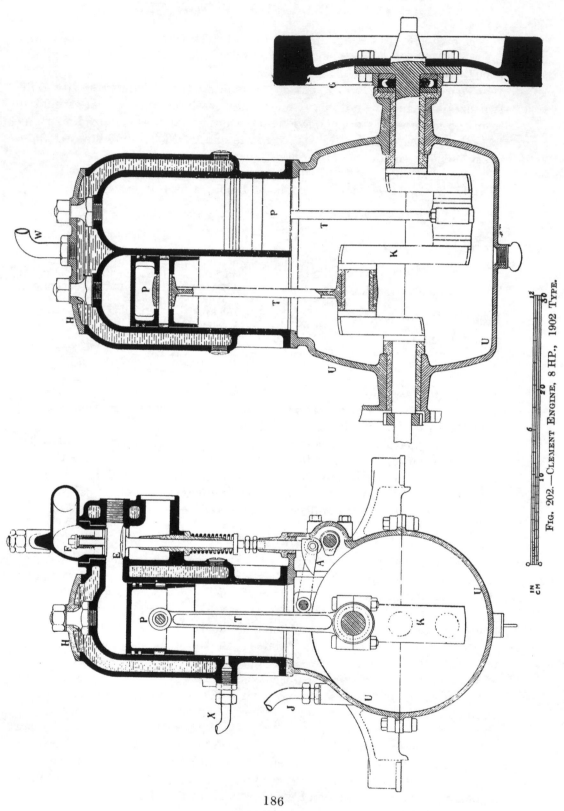

Fig. 202.—Clement Engine, 8 HP., 1902 Type.

186

of the lever, so that the cam met it earlier or later as was desired. The cylinders were cast in pairs with a minimum thickness between them, and there was no room for a bearing between the two cranks, the long central web form being employed. This practice has been given up for a very good reason. The torsional stress on the crank pins on the end of the crank web K are exceedingly heavy for a comparatively small torque effort at the fly-wheel. With a bearing in the centre the torsion stress on the crank-

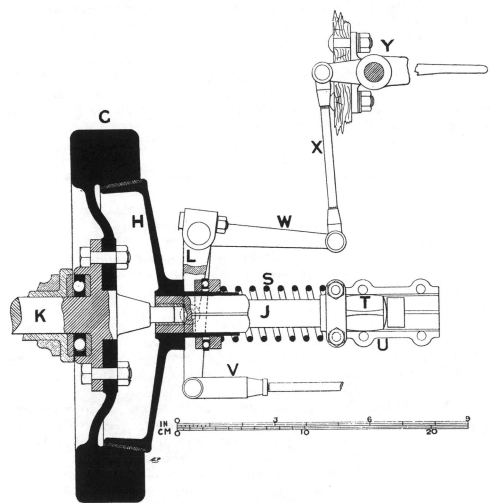

FIG. 203.—THE CLEMENT CLUTCH, 1902 TYPE.

pin is almost nil, without it the torque effort transmitted to the fly-wheel by the remote cylinder involves torsional stresses on crank-pins and central arm which are of a very ambiguous character, and, in fact, the crank of this form though it has been largely used, is a mechanical wrong and should never be used where the piston impulse is considerable and sudden.

Design of crank shafts

In the new form of Clement engines this form of crank has been discarded in favour of one with a central bearing.

The clutch, Fig. 203, used in this car was novel in its arrangement of operating gear and connexion to the driving shaft, but the spring pressure introduced, as was then common, a thrust which had to be met by a ball bearing outside the crankshaft bearing, and at the gear box. The clutch shaft J was centred and carried on the end of the crankshaft, and transmitted power to the gear box spindle by a squared end T in a divided sleeve U, one end of which embraced the gear box spindle. Between the end of that spindle and the end T a short distance piece was inserted. By taking off the sleeve U, this distance piece was easily removable, and then the clutch cone and its spindle and spring could be taken out.

Improved clutch   Since this clutch was brought out, two changes have been made. In the 1903 cars, a segmental cylindrical substitute for the cone was used. Each of four segments was mounted on a guide spoke and was forced outward into the rim of the fly-wheel by a spring in the usual place, but acting on the inner ends of four toggle levers. This clutch has since been discarded, although it introduced no end thrust, and the cone clutch like that illustrated again adopted, but the clutch spindle, instead of being bored out and fitting on, and being carried by a very small reduced end of the crankshaft, is now housed in a socket on the disc end of the crankshaft and of ample size.

The two larger four-cylinder cars as constructed this year are fitted with multiple disc clutches enclosed and running in oil.

The transmission and change-speed gear of the 1904 Clement-Talbot car is of the type which gives a direct drive to the propeller shaft on the fourth or highest speed, the arrangement being of the lathe back-gear type, with a claw clutch on the main spindle for the fourth speed. The secondary spindle is vertically above the main spindle, while in the Clement car, the main spindle is at the top. Plain bearings of good length were used throughout, a ball bearing only being used to take the main clutch thrust, but ball bearings have since been adopted for the gear spindle bearings. The arrangement of the gear is of the type introduced in the Argyll cars, the main spindle being divided for the purpose.

The ignition mechanism in the Clement car is the ordinary high tension, but in the Clement-Talbot cars magneto low tension apparatus is used. It is similar to that adopted by several continental makers, and is shown by Fig. 204, in which E is the insulated anvil terminal inside the cylinder, D the pallet or make and break terminal, not insulated, on the inside end of the spindle K. Upon the outer end of spindle K is fixed the lever C.

Magneto ignition gear   The short end N of this lever is hit by the head B of the pusher rod A by the force of the recoil of the spring S, when the cam H allows it to drop. Normally the striker terminal D is separate a short distance from E, but when the cam H raises the pusher rod A, the spring J asserts itself and pulls the terminal D against the insulated terminal E, thus making contact and completing electric circuit. At the same instant the rotation of the magnet armature is approaching the point of maximum inductive effect, and immediately following this the pusher rod A is dropped by the cam H, and in its fall it strikes the end N of the lever C, breaking the circuit at the velocity of recoil of the spring s. To vary the period at which the head B strikes N and thereby gives rise to the ignition spark, the forked rocking lever L pivoted at M is moved by the lever O towards or from the cam H, and the moment of fall of A is thus respectively delayed or hastened. The lever L is forked and the roller fork on the end of the pusher rod slides between two ears connected by the rubbing piece on the cam.

A very simple form of carburettor was formerly used in the Clement cars as shown by Fig. 205. This, like most carburettors then in use, acted very well through a somewhat limited range of quantity of air carburetted. When designed and adjusted for the quantity of carburetted air required for an engine at normal working speed, the inspiration effect at the jet J was insufficient at low speeds. A cap with holes in it to correspond with two holes L was used to adjust the quantity of air entering, and the suction at the jet; but it required a supplementary air supply, so that the carburettor being set for low speed, the extra air required at high speed could be in part supplied by the supplementary inlet, in such proportion that the aspiration effect on the jet should only increase in accordance with but not necessarily quite in proportion to the increased quantity of air.

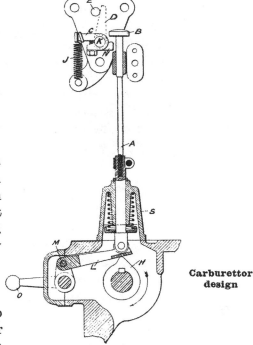

Carburettor design

FIG. 204.—CLEMENT-TALBOT MAGNETO IGNITION SPARK GEAR.

In the 1904 Clement carburettor [1] this has been obtained by means of a petticoat cylinder over the jet. The latter stands up about 2 in. above its base, and has a screw thread cut on its exterior to give a long path for petrol not carried off by the air current to trickle down and be exposed for evaporation. The petticoat is surmounted by a light piston with V-shaped ports in it, and this lifts with the increase of the partial vacuum formed by the engine pistons as speed

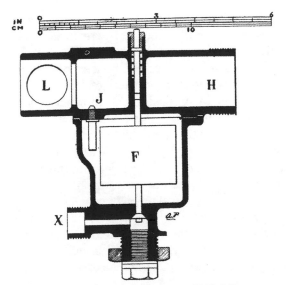

FIG. 205.—CLEMENT CARBURETTOR, 1902-3 PATTERN.

[1] *Automotor Journal*, June 4, 1904, p. 672.

MOTOR VEHICLES AND MOTORS

increases. The petticoat tube pipe being conical, the large end of the cone downwards, the rate of increase of velocity of the air round the jet with increase of engine speed decreases as the cone rises, and at higher speeds the piston rises and admits air to the engine which has not passed the jet. The piston, in fact, forms the supplementary valve which is to be found now in nearly all carburettors except the Richard-Brasier. It is not a simple carburettor, but the parts are nearly all machine work. It is illustrated in Chapter XXXIII as the Rover carburettor and the design is due to Mr. E. W. Lewis.

The Clement-Talbot 1904 carburettor[1] is a modification of the Krebs, but the supplementary air valve has no diaphragm, and the throttle is an ordinary butterfly valve. It is jacketed with hot water from the cylinder jackets, but calls for no separate description.

The final power transmission is by bevel gear through a spur wheel and pinion differential to the live axle which runs in ball bearings. These wheels are fixed on the rotating divided axle, and not on the fixed tube as in the Decauville, Orleans and some others.

The front wheels also run on ball bearings on the steering axles.

The cars are made in four sizes with engines of the following dimensions:—

```
10–12 Nom. HP. 2 cylinders ·88 mm. 3·46 in. diam.140 mm. 5·51 in. stroke.
12–14   ,,    ,, 4   ,,  ·80 ,, 3·15 ,,   ,, 120 ,, 4·72 ,,    ,,
16–20   ,,    ,, 4   ,,  ·95 ,, 3·74 ,,   ,, 130 ,, 5·12 ,,    ,,
24–30   ,,    ,, 4   ,,  ·105 ,, 4·14 ,,  ,, 140 ,, 5·51 ,,    ,,
```

These cars have been, and are being, made in large numbers, and a very large new factory has recently been built in West London for their manufacture in England.

THE DE DIETRICH CARS.[2]

Among those cars which have gained an excellent reputation by reason of their general good design and sound construction, few are now better known than the De Dietrich, made at Lunéville in France. As with a few others, which, if judged by successful performance, might be here described, only a brief reference can now be made without repetition of either illustrations or references to parts or particulars of design common to a large number of cars. Four types are now made with four-cylinder engines of 16, 24, 40 and 60-HP. of similar design. The frames are of pressed steel, and the use of the underframe supporting the engine and gear box visible in Fig. 205A has been discontinued, the engine being carried on the frame longitudinals by brackets cast with the crankchamber, and the gear box is hung at the front and back by bearer arms and lugs from the suitably spaced transverse frame tubes. By avoiding the use of the underframe, and providing for the greater departure from the alignment of engine and gear-box and shafts, that then occurs, by the use of flexible couplings or universal joints wherever relative movements by frame flexure may take

Underframe abandoned

[1] *Autocar*, July 2, 1904, p. 6.
[2] *Automotor Journal*, July 1, 8, 15, 22 and 29, 1905.

place, some weight may be saved and construction simplified without any added disadvantages.

The general arrangement of the transmission gear remains essentially as indicated by Fig. 205A, the bevel differential gear being, as usual, in the gear

FIG. 205A.—24 HP. DE DIETRICH CHASSIS, 1903 TYPE.

box, and the transmission of power from the divided differential shaft is through side chains to the rear road wheels. A single gear-changing lever is used, the different pairs of wheels being engaged or meshed by progressive sliding of the gear-wheels on the first motion shaft. In accordance with good present practice the equivalent of the feathers or keys which transfer the drive to the sliding group of wheels is obtained by milling or tooling them out of the solid shaft. An example of this form of shaft is given by **Gear shafts** Figs. 376 and 377, p. 436, its section being shown at B in the latter figure. Such a shaft costs no more to make, and its form and that of the wheels upon it is preferable to the squared shaft and wheels with square-holed bosses.

A feature of interest in the De Dietrich cars is the use of a separate or stationary clutch spring so arranged that its effort for engagement of the clutch is delivered through a fork and collars to the female clutch cone, **External clutch spring** the end thrust from the spring being taken by a ball thrust bearing near the front end of the engine shaft. As an arrangement permitting the use of a stationary spring, it is preferable to those designs in which the spring effort is delivered to the tail end of the first motion shaft, and in which the movement of the shaft in the gear box becomes necessary when engaging or disengaging the clutch. The clutch spring may be seen in a vertical position in Fig. 205A, in front of and a little to the right of the centre line of the gear box.

In recent cars ball bearings are used at all points in the transmission gear and at the road wheels. The engines, though of considerable interest as examples of skilful design and of careful detail arrangement, necessarily are similar in many respects to others of the four-cylinder type. The four-throw crankshaft is carried in three bearings, with the cranks between bearings at 180 degrees, and with no cranks at 90 degrees. The fly-wheel is fixed to the shaft by a flanged coupling and it is made as an exhaust fan to assist the belt-driven fan behind the honeycomb type of radiator now used with these cars. The fan blades are cast as short fly-wheel arms between the outer rim and the centre or clutch drum part of the wheel.

The camshaft has fixed upon it the inlet and exhaust valve cams, and by an arrangement of sliding features or keys moving in grooves in the exhaust cams the compression may be relieved. The inlet valves are operated through vertical push rods, pivoted levers, and short tappet stems in contact with the inlet valve spindles and in the manner indicated by Fig. 259, but with differences in detail.

Low tension magneto-ignition gear is fitted to these engines and a camshaft at the opposite side of the crank chamber to that occupied by the valve operating camshaft gives motion through spur-wheels to the magneto, **Timing magneto ignition** and through cams to the push rods which actuate the igniters. By a sliding sleeve and spiral keys at that end of the camshaft which is driven, the time of ignition and the position of the magneto-armature are simultaneously varied. The means adopted for timing the ignition admit of the parts being enclosed and copiously lubricated, and they can, moreover, be conveniently made of such sizes that the trouble incidental to the wear frequently occurring with the small and too delicate parts of some high tension magnetos and their attached gear are avoided. The igniters are simple and substantial and no spring operated quick break or trip gear is used with them.

THE DE DIETRICH MOTOR VEHICLES

The speed of the engine is controlled by a throttle valve hand-operated, **Throttle valve connexions** and there are connexions between the throttle valve and the clutch pedal and an accelerator pedal in order that when the clutch is disengaged racing of the engine may be reduced, and so that if desired the throttle valve may be foot- instead of hand-operated.

The carburettor is of the float-feed spraying type, so designed that **Carburettor** movement of the sliding throttle valve sleeve simultaneously varies the area of the opening to the engine induction pipe and the area of the ports through which air, not passing through the mixing chamber or past the jet, may enter. The streams of fresh and carburetted air oppose each other and their early admixture is encouraged. The relative areas of the petticoat pipe around the jet and the fresh air ports having been experimentally determined a near approach to constancy of quality of the mixture received by the engine may be obtained.

The engines are lubricated on the splash-bath principle, the added **Lubrication** lubricant whilst the engine is running being supplied by a sight feed Dubrulle lubricator and through branch tubes to the lower part of each cylinder barrel. Oil may be fed to these tubes by a force pump connected with the lubricator if deemed necessary, and to enable the driver of the car to exercise his judgment as to the quantity of oil in the crank-chamber, and to add to, or reduce it, a filling hole or lubricator, and drain cocks and gauge glasses are fitted to the crank-chamber.

The cylinder diameters of the four engines of the stated powers are **Engine dimensions** 104, 120, 130, and 150 mm., or 4·1, 4·72, 5·12 and 5·9 inches respectively, and the strokes are 120, 120, 140 and 170 mm., or 5·12, 5·12, 5·51 and 6·7 inches respectively. The normal speed is 800 revs. per minute.

The arrangement of the brakes may be gathered from inspection of Fig. 205A, the 1903 type of car, but the rear brakes are now of the internally expanding and enclosed type.

There are in the De Dietrich cars evidences of careful attention to the many small points which contribute much to successful behaviour, and an absence of multiplication of parts the duty of which can be more simply performed.

Chapter XI

THE MAUDSLAY CARS

A CAR presenting several distinctive features is that made in several sizes by the Maudslay Motor Company.

Fig. 206 is a side elevation of the car as made with a rectangular steel tube frame filled tightly with ash, and a secondary underframe to carry the engine and transmission gear.

It is a chain-driven car with cone clutch, and with the second motion spindle above the prime mover shaft in the gearbox.

Disposition of valves and camshaft

The engine is illustrated by Fig. 207, the transverse vertical section being through one cylinder. The inlet valves D are automatic, and the exhaust valves E are worked by rocking levers N actuated by cams on the spindle Q seen in the sectional part of Fig. 206. The spindle Q is driven by a worm-wheel on a vertical shaft K, gearing with a worm-wheel, the upper part of the vertical shaft being jointed so that the whole camshaft may be tilted so as to leave easy access to the exhaust valves. This will be seen more clearly by reference to Figs. 208, 209.

The governor seen at R, Fig. 207, is on a tubular spindle, also on the shaft K, driven by screw or spiral gear from and at half the speed of the crankshaft. It operates a horizontal spindle I in the induction pipe, on the end of which is a disc throttle valve at 6, Fig. 206. The fork and lever and the end of the spindle moved by the governor is seen on the left hand end of Fig. 209.

Forced lubrication

The jacket is of large proportions, the water entering at w^1 and leaving at w, separate inlet and outlet to each cylinder, though the three cylinders are cast in one. In some of the engines the jackets are in part made of aluminium panels. A system of forced lubrication is adopted, a small pump being used to force the oil through the crankshaft (see Fig. 206), and to all other main bearings.

The cylinders are 5 in. diam., and the stroke 5 in., the normal speed being 800 revs. per minute. The valves are shown as on the line of one diameter in Fig. 207, transverse to the engine, but they are now set on a diameter in the direction of the length of the engine, so that all the valves may be worked mechanically and directly by the cams ; but all the valves are still in the cylinder head.

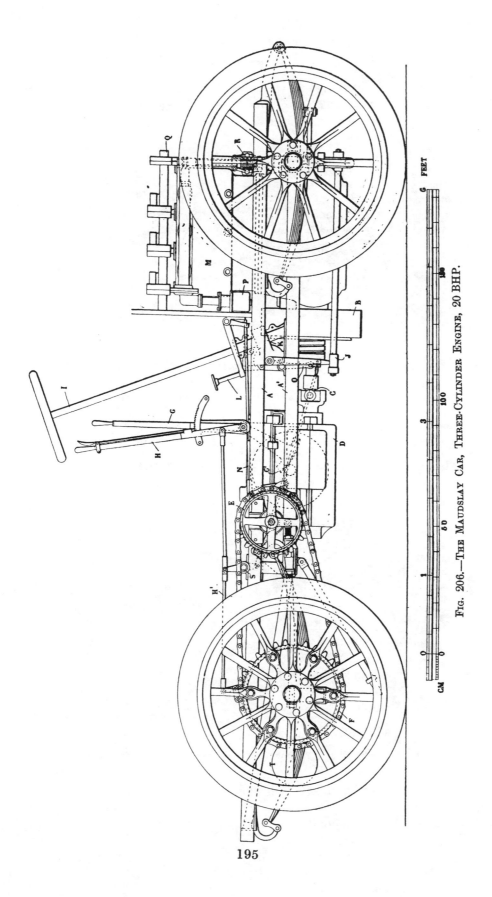

Fig. 206.—The Maudslay Car, Three-Cylinder Engine, 20 BHP.

195

The cranks are, of course, set at an angle of 120°, but the order of firing is as though they were at 240°, giving a much better distribution of effort than when the cycle sequence is in the order of the cranks at 120°.

FIG. 208.—MAUDSLAY ENGINE VALVE GEAR: CAMSHAFT IN WORKING POSITION.

This will be seen by reference to the chapter on "Sequence of Cycles," diagrams Nos. 4 and 5 of Figs. 311 and 312, p. 358.

Perspective views are given in Figs. 208 and 209 of the valve gear of

FIG. 209.—MAUDSLAY VALVE GEAR: CAMSHAFT RAISED FOR ACCESS TO VALVES.

Accessibility of valves the engines with positive motion inlet valves, the first showing the gear in working position, and the second showing the camshaft tilted up on its pivoted legs, the corresponding lugs being released by loosening the three-hinged bolts shown.

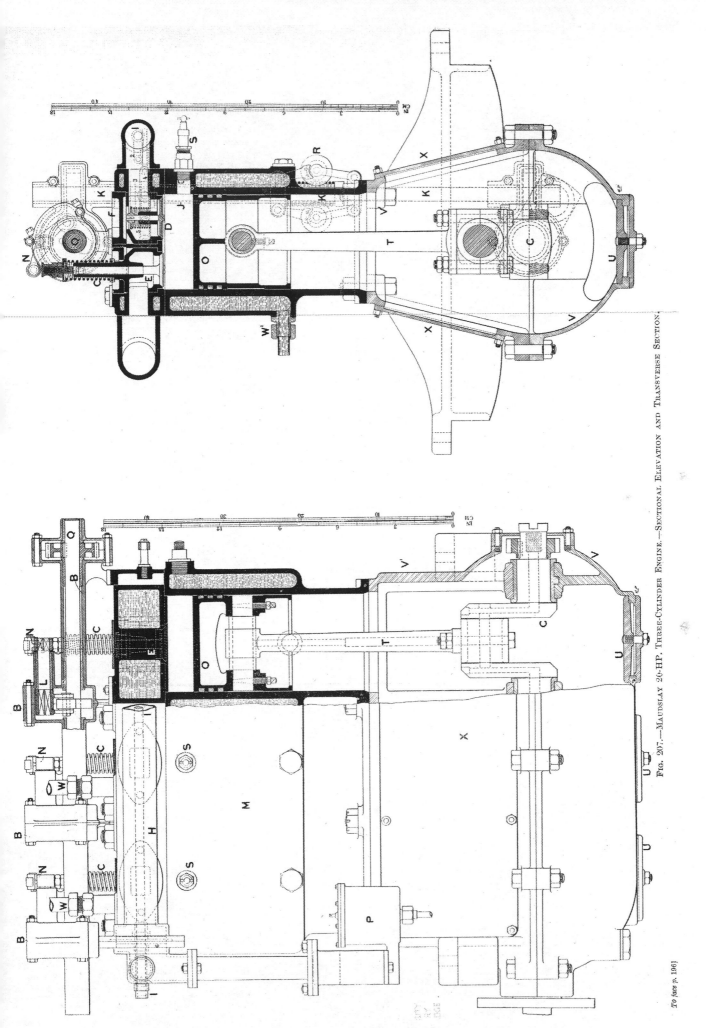

FIG. 207.—MAUDSLAY 20-HP. THREE-CYLINDER ENGINE.—SECTIONAL ELEVATION AND TRANSVERSE SECTION

To face p. 196]

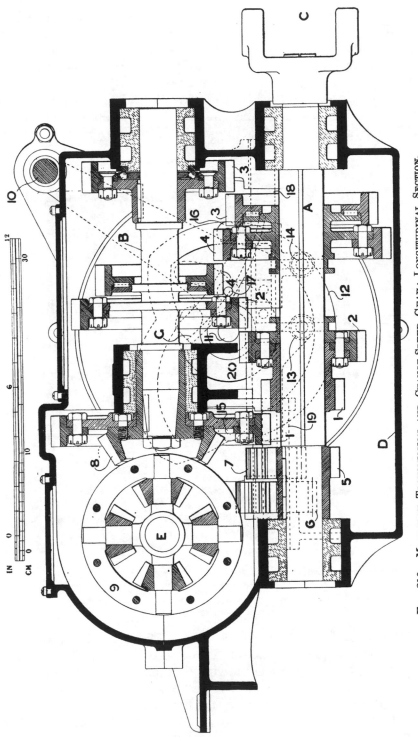

FIG. 210.—MAUDSLAY TRANSMISSION AND CHANGE-SPEED GEAR: LONGITUDINAL SECTION.

197

The transmission and change-speed gear is very substantial and well designed. It is shown by Figs. 210 and 211, which are respectively a longitudinal section along the spindle and differential gear centres : and a transverse section at the centre of the disc cam by which the gear-shifting forks are moved.

The engine drives the lower shaft A by the universal joint C. Four speeds are given by the wheels and pinions which are marked in pairs, as 3, 3 and 1, 1. The forms of the cam grooves,[1] 12 and 20, are seen in dotted

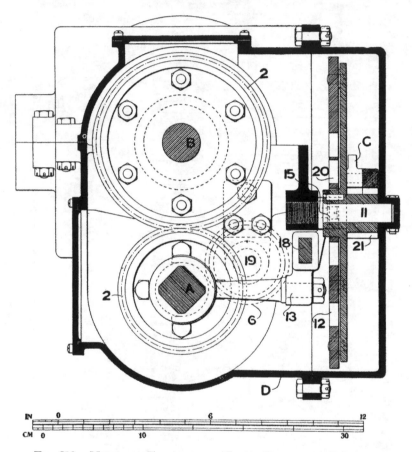

FIG. 211.—MAUDSLAY TRANSMISSION GEAR : TRANSVERSE SECTION.

Method of gear changing

lines in Fig. 210, and they appear in section in Fig. 211. The cam disc is rotated by a toothed quadrant seen in dotted lines in Fig. 206, and moved by the lever G. The quadrant gears with the pinion 21, Fig. 211, on the stud spindle 11. The fork arms 13 slide on the guide bar 18. Projecting from the slide, and on upwardly projecting lugs, are three pins with rollers on their ends which engage with the two grooves, one of the pins being fixed to the guide bar itself, which also moves for changing the position of pinion 6

[1] *Automotor Journal,* vol. viii. p. 269.

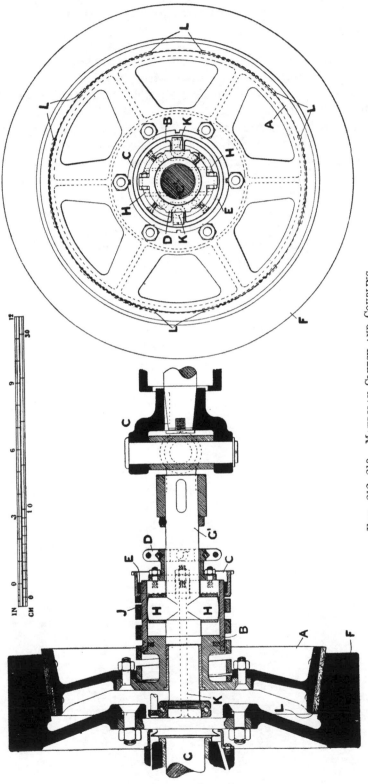

FIGS. 212, 213.—MAUDSLAY CLUTCH AND COUPLING.

for reversing. One of these pins and rollers is seen in dotted lines at 15 in Fig. 211, and one of the forks is seen fixed into the slide at 13.

A ball bearing is used on the shaft G to take the horizontal resultant of the drive of the bevel pinion 8, which works the bevel wheel on the differential 9 on the sprocket shaft.

The clutch, Figs. 212, 213, is of the self-contained type giving no horizontal thrust to the crankshaft or to the gear-driving shaft.

On the end of the crankshaft G is a ball thrust bearing at K. This is connected by two bolts K K to lugs on a ring E, which bears upon the volute spring J. Fastened to the boss of the cone is a cylindrical centre enclosing the bush blocks of the tee-piece H H of the short flexible coupling rod G^1. Connected to the block in this cylinder is the pedal-moved sliding ball thrust bearing D. Pulling this away from the fly-wheel compresses the spring and releases the cone.

Chapter XII

THE DAIMLER CARS

THE Daimler Company was one of the first established in England, and it has the longest English record of good cars, which have performed well. In recent years it has departed considerably from the practice in design of the Cannstatt Daimler Co., from which it originally obtained its models. Under the works management of Mr. Martin the designs have in many points received important modifications, and the cars now made have regained a position among the leaders similar to the corresponding position held by Daimler cars in 1899.

In the 1903 1,000 mile durability trials these cars did exceptionally well, two gold medals being awarded them, one for a 22-HP. touring car, and the other a 22-HP. shooting brake or omnibus, the first obtaining 2,871 marks, and the second 3,000 for reliability, the maximum being 3,000.

During the summer of 1904 these cars gained several of the first prizes in keenly contested hill-climbing and other trials, as at the Nottingham hill trials, the Yorkshire, and the Bexhill, in August of that year.

The company is confining its manufactures at present almost entirely to the larger sizes of cars.

Fig. 214 is from a photograph of one of the cars made for King Edward VII; and Fig. 215 has a similar underframe, but has a body like that of the car referred to as having won the gold medal.

They are both of 22 nominal HP., with cylinders 105 mm., 4·13 in. diam., and 130 mm., or 5·11 in. stroke, the normal speed being 720 revs. per minute.

The frame of the car, Figs. 216, 217, as made in 1902–3, was of wood reinforced with inside steel flitches, as shown in Fig. 217. This has been slightly modified for the 1904 cars of 28 HP. by stopping off the wood part of the sills of the frame at the front spring hangers and continuing the steel flitch into the front dumb irons or spring hangers, these hangers being split to receive the flitch which is then firmly rivetted in the split dumb iron. From about 2 ft. behind the back hanger of the front springs to the front end of the slit in the dumb irons is a piece of channel steel $\frac{3}{16}$ in. thick rivetted to the flitch plate, and with it rivetted in the slit fork of the dumb

Continued use of armoured wood frame

201

irons. This method of construction gives a strong and sufficiently flexible frame, and the channel steel being on the inside of the flitch and therefore within the wood sills makes the frame narrower at the front than at the back by the thickness of the wood, or say twice 2 in.

FIG. 214.—THE KING'S 22-HP. DAIMLER CAR.

Figs. 216, 217 show the 12 HP. as made in 1902 with three speeds, the same car with 14-HP. engines having in 1903 been provided with four speeds, all other main features remaining unaltered.

The engines, as will be seen, were carried on a separate short under-

202

Fig. 215.—22-HP. Daimler Touring Car.

frame, the gear box rigidly connected to the transverse casing of the sprocket shaft being fastened to the main frame at the rear of the car. An inspection of Figs. 216 and 217, in which the parts are all correspondingly marked, will make lengthened description unnecessary. The propeller shaft F from the engine drove the squared spindle in the lower part of the gear box on which were three sliding pinions, the smallest pinion 12 gearing with the largest wheel on the upper shaft for the lower speed and with the wheel next forward through the intermediary of an idle pinion always in gear with it for reversing. Plain bearings were used throughout with ring lubrication. The leading dimensions will be gathered from the scale provided in the engraving.

Fig. 218.—12-HP. Daimler Engine : Carburettor Side.

The engines of this car are shown by Figs. 218, 219, in detail by Figs. 220, 221, from which it will be seen that high tension electric ignition apparatus was employed with tube ignition also. The cylinders were 86 mm. or 3·39 in. in diam., and the stroke 100 mm., or 3·94 in. Automatic admission valves were used, the camshaft for the exhaust valves being enclosed, the hit and miss gear formerly used on all the Daimler engines being discarded. Throttle valves were.in some cases fitted to the breeches pipe over the inlet valves of each pair, as seen in Figs. 218, 219 and in other engines of later pattern ; one throttle valve was used as seen at J K, Figs. 220, 221. The cranks were made in two pairs at 180°, each double crank being of the Z pattern with a long web without bearing between the cylinders,

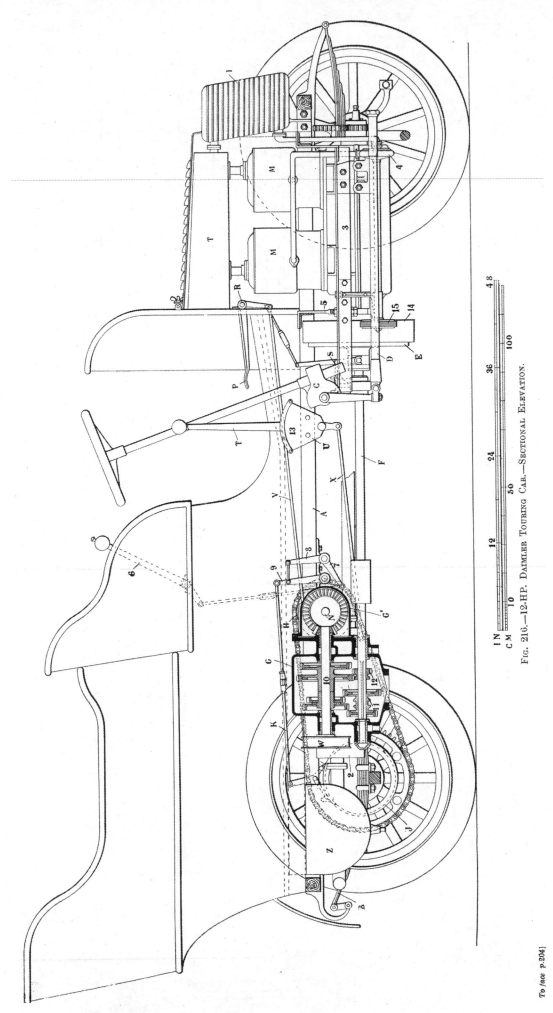

Fig. 216.—12-H.P. Daimler Touring Car.—Sectional Elevation.

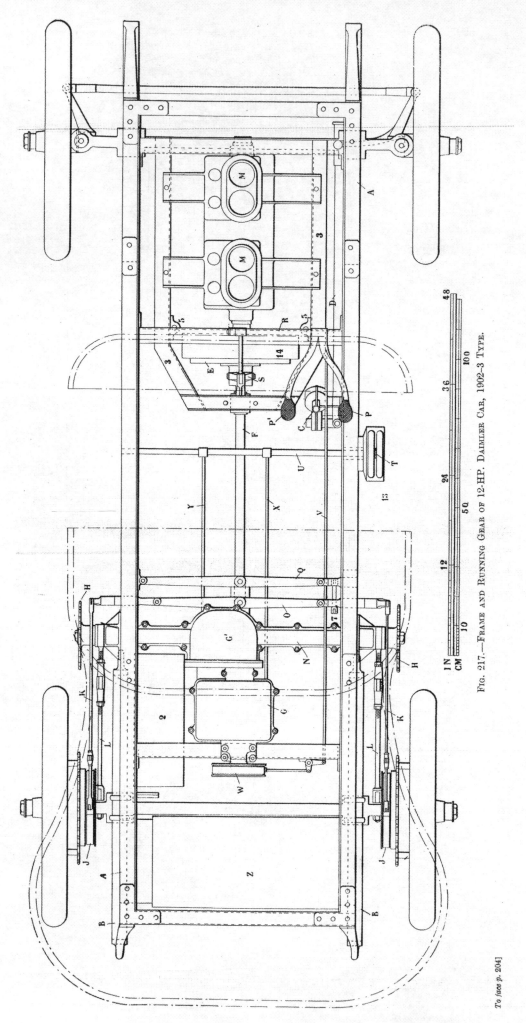

FIG. 217.—FRAME AND RUNNING GEAR OF 12-HP. DAIMLER CAR, 1902–3 TYPE.

To face p. 204]

a form of crank not to be recommended. The lift of the exhaust valves was very readily adjustable by the adjusting pieces and locknuts E of the pusher pieces. The carburettor used was of the well known Daimler pattern, illustrated by Fig. 221A. The size of the opening at the top of the jet C was adjustable at F by means of the needle valve shown, and the position of the air flow restricting neck at E could be varied with regard to the height of the jet and in relation to the position of the atomising cone D. Additional air not passing the jet entered through adjustable openings S, and the carburetted mixture entered the engine induction pipe at A. The warmed air entered the chamber B by an opening not shown, but behind the jet C.

Fig. 219.—12-HP. Daimler Engine: Valve Gear Side.

The upper portion of the mixing chamber is, as shown, jacketed. A drain cock occupied the hole shown in the boss at the bottom of the float chamber through which accumulated water, stale petrol, or petrol could be taken.

The 1904 engines of the 18-HP. Daimlers have cylinders 95 mm. diam. 3·74 in., and stroke of 130 mm., 5·11 in.; the 28-HP. have cylinders 110 mm. or 4·32 in. diam., the stroke being 150 mm. or 5·9 in. The normal speed of the 18-HP. is 800 revs., and that of the 28-HP. 750 revs. All the valves are on one side of the cylinders, and are mechanically moved by a single camshaft. The cranks are in one plane or at 180°, the sequence of the cycles being 1, 2, 4, 3, which, as will be seen by reference to Nos. 6 and 7 of diagram, Fig. 311 and diagrams, Figs. 312 313, page 358, gives a much more equable turning moment than when the four cranks are in

205

two planes or at 90°. The cylinders are cast in pairs, the head of the combustion space being cast in one with the cylinders, but the jacket is cast open, and the top covered with an aluminium cover plate common to two cylinders.

The camshaft is placed outside of the crank chamber, and the cams move rocking levers which have rollers resting on the cams and adjustable pusher pieces acting on the ends of the valve stems. These parts are easily accessible in contrast to those of some of the foreign designs, and although not completely covered in, the cams run in an oil trough.

Short ports

With the object of keeping the valve port spaces in the combustion space short, the valve stems are set at an angle divergent from the cylinders to meet the outer ends of the rocking levers over the camshaft, which is farther than usual from the cylinder centres.

All the valves are very easily taken out by removing plugs over them, there being no pipes to remove for the purpose, the exhaust and the induction pipes both being on the valve side of the engine.

Features of design

The engine is fixed direct to the channel steel parts of the main frame by arms cast on the upper part of the aluminium crank chamber, no supplementary under-frame being used, and the gear box is separated from the engine and is carried by the back part of the frame with the sprocket shaft bearings, a long propeller shaft F with a form of sleeve joint giving freedom for self-adjustment for frame flexure, being employed. This arrangement makes it easy to suit the machinery to any length of frame by simply lengthening or shortening the propeller shaft and the rods V, X, Y. The sleeve referred to nearly fits the squared end of the propeller shaft and the first motion shaft in the gear box, and permits the end on movement of the clutch spindle when the clutch is put out of gear.

Ignition is of the high tension type with one coil and trembler, and one low tension rotary timing contact in a distributor box on vertical spindle driven by bevel gear from the camshaft end, the distributor being above the engine cylinders where it is easily accessible. The distributor contains the four distributing contact pieces connected to the ignition plugs by the high tension wires ; the wiring is thus simple and the wires short.

Compound throttle valve

The carburettor[1] is of the float-feed jet type, the warm air passing over the jet in a small passage to a cylindrical throttle valve, which controls the outflow from carburettor jet to cylinders, and at the same time the ingress of cold air admitted to the warmed carburetted air coming from the jet passage. The cold air inlet is provided with a supplementary valve, so that the carburettor is of the automatic kind through a considerable range. Pressure feed is used, obtained by a shunt from the exhaust pipe to a distributing box, to which pipe connexions are also made to the lubricating oil reservoir, hand air pump and pressure gauge. It is difficult to see the use of a pressure feed to a jet, the flow to which is controlled by a float valve unless it be merely convenience as to position of petrol tank.

---

[1] *The Autocar*, vol. xiii. pp. 132, 164, 200.

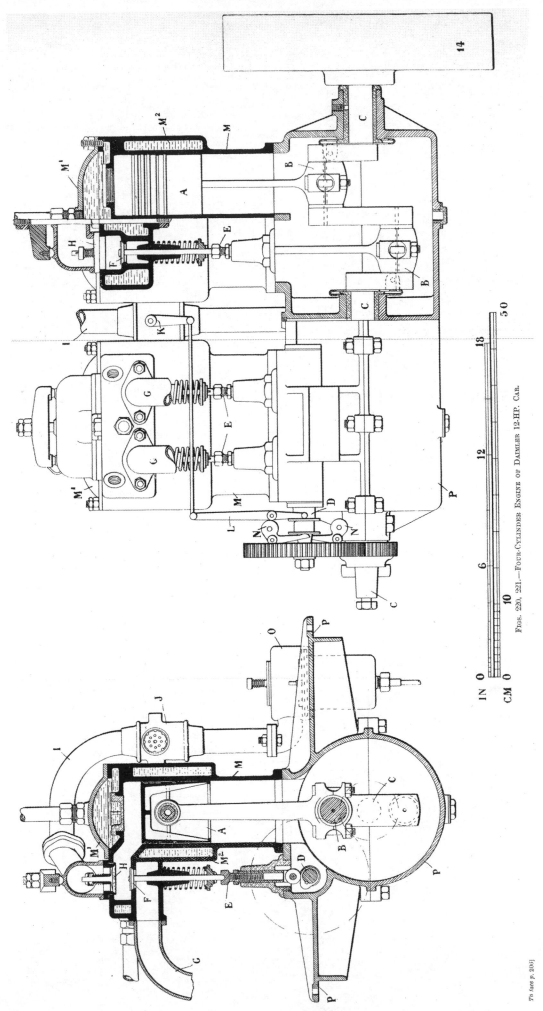

FIGS. 220, 221.—FOUR-CYLINDER ENGINE OF DAIMLER 12-HP. CAR.

[To face p. 236]

The gear box is of the design shown in Fig. 216, but is fitted with gear for four speeds instead of three.

One hand lever T, as in Figs. 216, 217, in a double slot quadrant 13, serves to move the two forks of the two pairs of shifting pinions. The lever T is on a tubular spindle U, which slides on a central solid spindle. This permits the lever T to be passed sideways from the outer slot to the inner one through a break in the separation between the two slots. When **Operation of change-speed gear** the lever is in the central position shown all the wheels are out of gear. The two forks are worked by separate rods X and Y when the lever is in the inner or the outer slot of the quadrant. This method of gear changing gives a definite position for each speed and the gear changes may be made with more certainty than is frequently the case with closely notched quadrants of the usual form. Detail changes have been made in construction but the gear is now operated in the same way.

For reversing a second lever is used, working on the inward side of the quadrant. This lever and its connexions move an intermediate idle pinion into gear with the lowest speed pinion and wheel when the change-speed lever is in the central position. This arrangement of gear gives short strong spindles, not likely to bend or twist, or bind in the bearings, which are all ring lubricated. All the toothed rings of the gear wheels

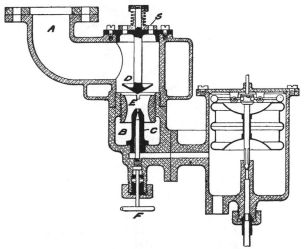

FIG. 221A.—THE DAIMLER CARBURETTOR, 1902–3 TYPE.

and pinions, except the smallest for lowest speed, are renewable on the sleeves.

The construction of the clutch differs from that of most other makers. There is no spring on the clutch shaft. On the back of the clutch cone boss is a ball thrust bearing and collar. This collar is forced toward the fly-wheel by a forked lever, the upper part of which is pulled by a spiral spring on the top of the frame. This spring is adjustable even when the engine is running. The arrangement is very simple and adjustment very easy, but it involves the use of this ball thrust always working. As, however, the total thrust pressure is small compared with that which such a bearing can carry indefinitely, the objection which often obtains with ball bearings does not apply. It is necessary, however, that the parts carrying the ball bearing be very accurately erected centrally with the fly-wheel.

There are four brake drums, two on the countershaft ends inside the sprocket pinions, instead of one drum on the end of the second motion spindle in the gear box as shown in Figs. 216, 217; and two on the sprocket wheel drums as shown. All have outside bands lined with steel, and all are provided with pivoted link compensating connexions of the same kind as those described with reference to κ κ, Figs. 84 and 85. The side brake lever is seen at 6, Fig. 216.

The radiator shown in Fig. 216 had ribbed aluminium tube plate boxes forming vertical ends which were connected with horizontal tubes. The cooler now used has vertical tubes fastened at the bottom in a shallow, narrow, parallel copper tank, and at the top in the under side of a ribbed aluminium vessel supported by side castings of the same metal which are also formed to carry the front side lamps. The water is circulated by a centrifugal chain-driven pump, the spindle of which is properly carried by a bearing on each side of the chain sprocket. The water passes from the pump to the jacket at a point near the exhaust valves, and thence it finds its way round the cylinder head and out at the top on its way to the cooler. In the 22-HP. cars the coolers had 110 ft., and the 12-HP. had 60 ft. of Clarkson tube. The 28-HP. car of to-day has 72 feet of closely gilled tube, $\frac{7}{16}$ inch external diameter, and the 30–40 HP. car has a cooler of the same general dimensions, but with four instead of three rows of staggered vertical tubes, having a length of 95 feet. A fan at the back of the coolers adds to their capacity by accelerating the air flow.

The engines of the Daimler cars [1] as now made are among the most efficient constructed anywhere. Their dimensions, power, and approximate speeds are—

TYPE.				DIMENSIONS.			
HP.	28	36		{110 mm. or 4·33 inches	bore.
Revs.	750	1,000		{150 mm. „ 5·90 „	stroke.
HP.	30	40	51}	{124 mm. „ 4·88 „	bore.
Revs.	650	900	1,000}	{150 mm. „ 5·90 „	stroke.
HP.	35	45	61}	{134 mm. „ 5·28 „	bore.
Revs.	600	900	1,000}	{150 mm. „ 5·90 „	stroke.

Interconnected throttle valve and ignition timing

The control of these engines by means of a single hand lever above the steering wheel possesses the advantage of simplicity. The period of ignition and the throttle valve opening are simultaneously varied, an arrangement that appears to be satisfactory, though better results must of course be obtained by careful use of independent rather than interconnected throttle and ignition levers.

The body of these cars is hinged at the back ends of the frame and held in front by catches. The body can thus be easily raised for access to the parts beneath.

[1] *Automotor Journal*, vol. x. pp. 160, 221, 255, 282.

Chapter XIII

THE HUMBER CARS

A TYPE of car differing from those at present illustrated is the Humber car with the trussed tubular frame shown by Figs. 222, 223 and 224. The first is from a photograph of the 8-HP. car, and the second shows the frame for the 12-HP. car.

FIG. 222.—12-HP. HUMBER TONNEAU CAR.

From Figs. 223, 224, it will be seen that the main frame tubes c are very strongly trussed by the tension rods B B, which also by their angular position firmly maintain the position of the lower supplementary frame D D, by which the gear box is carried. Figs. 225 and 226 show the engine of the 12-HP. cars of the 1902–3 patterns. Fig. 225 is an elevation of the valve gear and exhaust pipe side of the engine, of which Fig. 226 is a trans-

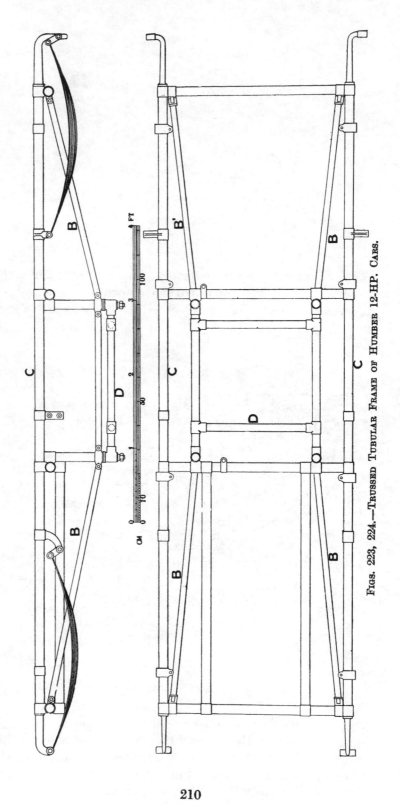

Figs. 223, 224.—Trussed Tubular Frame of Humber 12-HP. Cars.

210

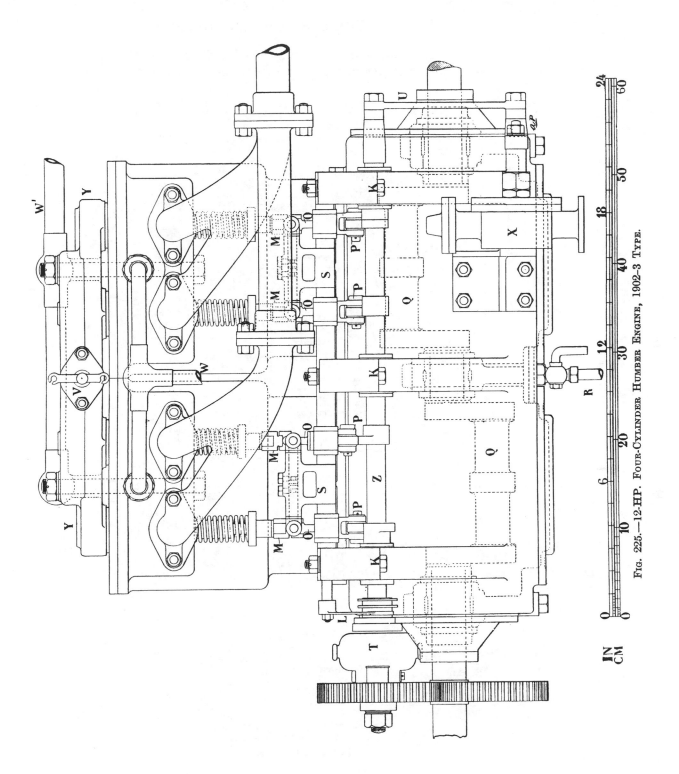

Fig. 225.—12-HP. Four-Cylinder Humber Engine, 1902-3 Type.

211

verse section.   The cylinders are 90 mm., 3·54 in. diam., and the stroke
102 mm., or 4 in., the pistons being now flat on the top instead of dome-

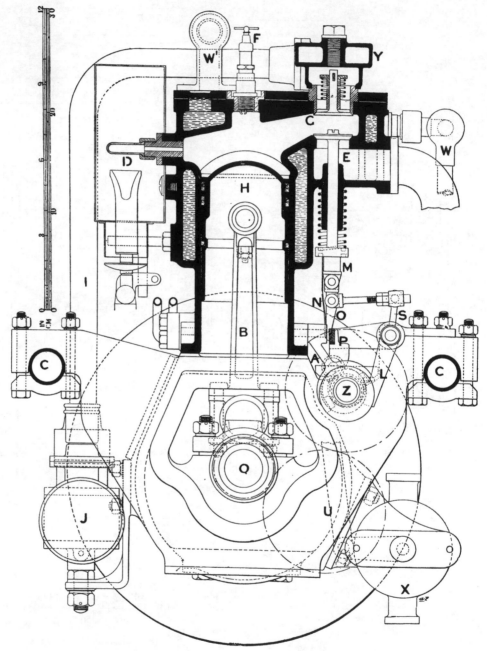

Fig. 226.—12-HP. Humber Engine.

shaped as shown, the compression space being correspondingly altered.
The diameter was increased to 3·62 in. early in 1903. The engine, as shown

was fitted with the hit and miss gear for operating the exhaust valves and governing by them, as in the earlier Daimler engines. Rocking levers P actuated by cams on the shaft Z lifted the knife-edge finger O pivoted on the end of the pusher rod M, this finger being lifted by the steel edge at P

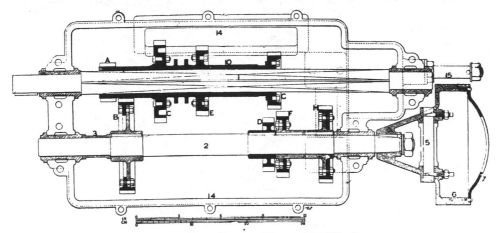

FIG. 227.—GEAR BOX OF HUMBER 12-HP. CAR.

when the engine was running within its speed, but missed it when the governor T caused the arm L to rise on the larger part of the collar on which its end rested, and so tilted L and S, and by the forked links pivoted at N pushed the finger O out of contact with P. Within a groove on the collar

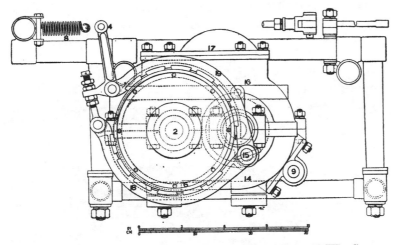

FIG. 228.—BACK END VIEW OF HUMBER GEAR BOX, 12-HP. CAR.

actuated by the governor was a forked lever A, Fig. 226, pivoted on a small spindle above that letter, which, by a hand-worked connexion, prevented the trip action from taking place when it was desired to run the engine above normal speed. It will be seen that the engine was provided with ignition tubes D. This was discontinued in 1903, and only electric ignition fitted,

213

the exhaust valves being worked direct by a camshaft within the gear case and throttle governing adopted. It will be noticed that the cranks Q were all in the same plane, each pair of cylinders being on one wide crank. The circulating water was admitted near the valves at w and taken off to go to the cooler by the pipe w^1 from the flat aluminium top of the jacket.

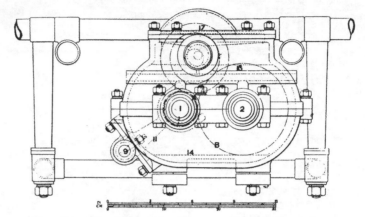

FIG. 229.—FRONT END VIEW OF HUMBER GEAR BOX.

The cooler shown is of the Loyal type, but hinged on the circulating pipe ends in the form of trunnions, so that the cooler may be turned down out of the way for easy access to the engine.

The circulating pump was a semi-rotary worked by a rod u from a crank-pin on the end of the camshaft. The exhaust pipes were of large size coupled by two breeches pipes to one outlet.

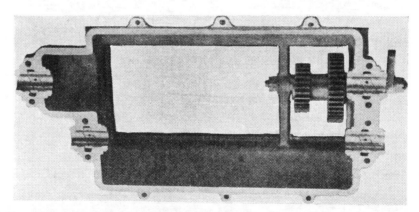

FIG. 229A.—GEAR-BOX TOP AND REVERSING PINIONS—12-HP. HUMBER CAR.

The carburettor J was of the Longuemare type of the time placed very low, and connected by the pipe I to the pipe Y over the inlet valves.

A cone clutch of the pressure balanced type, as used in the 1904 cars, see Fig. 230, drove the first motion shaft in the gear box through a double-jointed propeller rod, Figs. 227 to 229. This shaft was square, and on it

214

were the four change-speed pinions and wheels which alternatively geared with the four on the second motion shaft 2. A spindle in the gear box carried a pair of idle pinions, see 17, Fig. 229, of better size than usual for insertion between the pinion A and wheel B for reversing, the pinions being on an excentric part of the spindle which carried them, as seen in dotted lines in Fig. 229 and in Fig. 229A. Partial rotation of this spindle put the reversing motion into or out of gear.[1] The fork seen at 11, Fig. 229, which

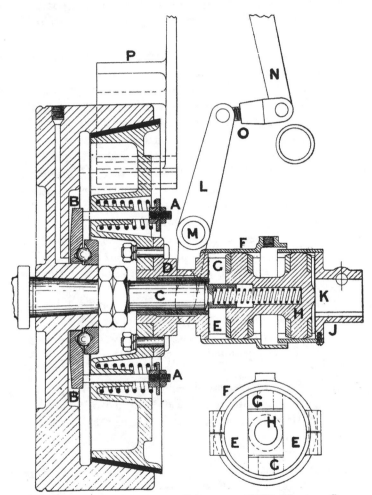

FIG. 230.—FLY-WHEEL AND CLUTCH OF 12-HP. HUMBER CAR.

engaged with the groove between the central pair of change-speed wheels C and E, Fig. 227, was moved by the sliding spindle 9 in the same figure.

On the back end of the second motion spindle was a universal joint 5 for the propeller rod which drove the bevel pinion on the differential gear on the live back axle. This pinion was cut out of the solid forged with the spindle which ran in ball bearings. The differential was of the spur pinion type.

[1] *Cantor Lectures,* 1903, by the Author, Fig. 29.

Gear.box suspension

The gear box was fixed to the transverse tubes marked D in the plan, Fig. 224, and the whole quadrilateral under frame could be dropped without other disturbance by taking off the nuts on the bottom ends of the vertical truss struts.

A brake drum 6, Fig. 228, was fixed on the back end of the second motion shaft, the brake band being of malleable cast iron pivoted at 15, and worked by a lever 4, the brake holding in both directions.

The side brakes on the driving-wheels were also of the external band type held by a horn extending from the spring pad bracket on the end of the tube surrounding the live axle.

The 12-HP. clutch, as will be seen from Fig. 230, which is one-fourth full size, is easily adjustable as to pressure by four springs on the studs, which are fixed in the back disc B of a ball thrust bearing on the boss of the fly-wheel. The aluminium cone is carried on an extension C of the crank-shaft by a bushed steel boss D, which has a jaw E extending into the cover F and forming the propelling guides of the brasses G of the universal joint H between the engine and the gear box. Similar jaws are formed on the boss J, which is fixed on the first motion shaft in the gear box. The spring inside the joint piece H keeps the universal joint lightly in contact always with the spindle at K. The jaws as made fit the sides of the ends of the piece H, and so prevent any angular movement in the pivot blocks. Only motion in the plane of the jaws can, therefore, take place, and half the usefulness of the universal joint is lost, it is not universal, and if the driving and driven shafts be not in true alignment this coupling does not procure ease of running. The forked lever for working the clutch is seen at L pivoted at M. It is moved by the lever N connected to the spindle moved by pedal, adjustment being made by the screw joint O, which should have a lock nut at O.

Freedom of universal joints

The gear box was one of the very long type, which compares unfavourably with some of those previously herein illustrated, which are very short and much lighter.

The cars were made in several sizes, including a 20-HP., which did some remarkably high-speed running in 1903 at Dublin after the Gordon-Bennett race, an 8½-HP. light car, a 14-HP. touring car, formerly the 12-HP., and the 25-HP., formerly the 20-HP.

The general design and main features of the 10-HP. engine of a 1904 type car are shown by Fig. 231. The cylinders are 4 in. diameter, and the stroke 4 in., the cylinders being cast in one pair with the jacket open at the top and covered with one aluminium plate. The cranks are at 180° with a single central web, the shaft and pins being 1½ in. diameter. Although only designed last March, it will be seen that automatic inlet valves are used, the exhaust valve being lifted by a pusher rod with an adjustable cap. The pusher is guided by a detachable guide, but the guidance is not carried down to near the roller as it should be. The crank case separates at the centre of the main bearings, as is the practice with most makers, but it is very desirable that the bearings should be carried by the upper part of the case, so that the lower part may be taken

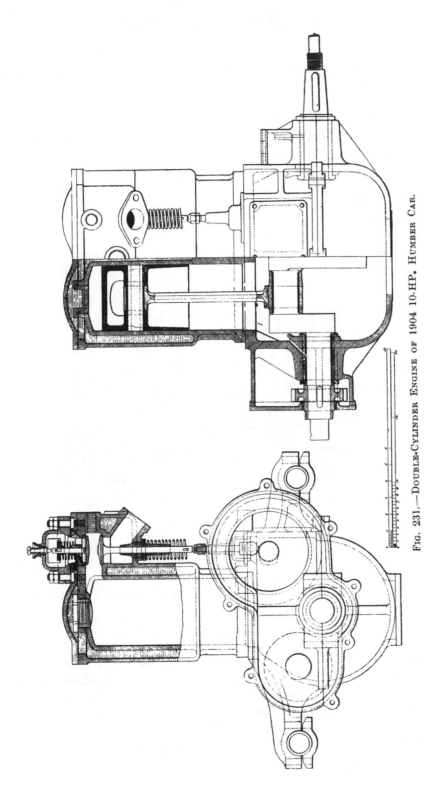

FIG. 231.—DOUBLE-CYLINDER ENGINE OF 1904 10-HP. HUMBER CAR.

217

off for access to connecting rod ends without dropping the crank and connected parts. The engine is of good substantial simple design, but for
reasons given at page 357 concerning cycle sequence, and elsewhere concerning crank stresses, I would prefer to have a crankshaft, type No. 2
of Fig. 312. All the bearings are of bronze lined with white metal.

The Humber Co. are also making two sizes of very small cars known
as the Humberette of 5 HP., and Royal Humberette of 6½ HP.

The latter car is illustrated by Fig. 232. It is a live axle car driven
by a single-cylinder engine of the Humber make of de Dion-Bouton type,
with a cylinder 3⅞ in. diam and 4 in. stroke, giving its power at 1,100
revs. per minute. It is provided with three speeds, 7, 15 and 25 miles
per hour at normal speed of engine, and low-speed reverse. The engine is
water-cooled, and circulation is maintained by a gear-driven pump.

The frames are tubular.

FIG. 232.—THE 6½-HP. ROYAL HUMBERETTE.

The wheel base is 5 ft. 3 in., and wheel gauge 3 ft. 6 in.

The Humberette is fitted with two speeds only, and its engine runs at
1,500 revs. per minute.

Small car prices The wheels of the Humberette are tangent-spoke cycle type, those of
the Royal Humberette are Hancock or artillery wheels, with 28 in.
2½ in. pneumatic tyres.

The two cars of Beeston make are respectively 160 guineas and 140
guineas, or 150 and 125 guineas if of Coventry make. This gives some
idea of the prices of small popular cars in midsummer, 1904.

The Humber Co. have now ceased to make any of these cars and the
three types [1] now constructed by them are all equipped with four-cylinder
engines. Figs. 232 and 232B are reproduced from photographs of the
8–10 and 16–20 HP. Beeston Humber cars, and the third type the 8–10 HP.

[1] *The Autocar*, Feb. 11, 1905.

THE HUMBER PETROL MOTOR VEHICLES

Coventry Humber, not here shown, retains some of the features of design of the earlier cars, notably the tubular frame already described.

The engines have four throw, three bearing crankshafts and the cylinders are separately cast with the mechanically operated inlet and exhaust valves at opposite sides and operated by separate camshafts. The leading particulars of these engines are as follows—

Type.	Bore.	Stroke.	Revs.
8–10 H.P. Coventry Humber	80 mm. or 3·12 in.	95 mm. or 3·75 in.	1200
8–10 H.P. Beeston ,,	77 mm. or 3·03 in.	108 mm. or 4·25 in.	900
16–20 H.P. ,, ,,	95 mm. or 3·74 in.	125 mm. or 4·92 in.	900

FIG. 232A.—8-10-HP. BEESTON-HUMBER CAR—1905.

The last of these engines is controlled as to speed by a governor arranged to slide the inlet valve camshaft lengthways. The cams have a varying profile and by end sliding of the shaft upon which they are fixed a variable lift is given to the inlet valves. The action of the governor may be opposed by an accelerator pedal or independently by a hand lever below the steering-wheel. Both the Beeston cars are fitted with the high tension magneto and accumulator and coil systems of ignition, either of which may be put in service through a two-way switch.

Variable inlet valve opening

Duplicate ignition systems

The car shown by Fig. 232A, when fitted with the two-seated body, weighs 12 cwt. and the larger car, Fig. 232B, with a body to carry five passengers, weighs 17 cwt.

In the external appearance of these cars there is noticeable similarity, particularly as regards the fore part and the distinguishing form of the dash. The radiators used are of good design, having the neat appearance of the now frequently used honeycomb type without its inherent weakness of construction. The cooling surface is made up of a large number of thin

219

flat tubes connected to and framed by the outer rim or tank space of the radiator. Pressed steel frames are, it will be observed, used in preference to the well-designed tubular type now only used for the Coventry car, and the ends of these frames are produced and curved downwards for the reception at front and rear of attachments to the spring ends.

Fig. 232b.—16–20-HP. Beeston-Humber Car, 1905.

The nature of the transmission gear has already been sufficiently described by reference to the other cars and the same system of change-speed gear, propeller shaft and bevel gear with live rear axle is retained for these recently introduced cars.

Chapter XIV

THE BRUSH, AND WILSON AND PILCHER CARS

THE Brush Electrical Engineering Co., foreseeing the probable use-fulness of motor cars for connecting up tramway lines with outlying sources of traffic, commenced the manufacture of cars in 1902 on Panhard lines. They were at first fitted with Abeille engines and with continental axles and springs.

The company has since made other engines on similar lines, and recently some engines of 16 HP. and 30 HP., with mechanically operated inlet-valves

A 16-HP. car of 1903 type is illustrated by the accompanying engravings, Fig. 233 being a perspective view of a tonneau touring car, and Figs. 234, 235, being respectively an elevation and plan of the frame and running gear of the same car. Fig. 236 is a front view of the car. As seen from this view the engine is fitted with valves mechanically moved by a duplication of the exhaust valve camshaft.

The frame is of the wood and steel flitch build like the Panhard, and the engine is carried on a suspended subsidiary underframe. The wheel base is 8 ft., and the gauge 4 ft. 8 in.

The engine has four cylinders 3·94 in. diam., and 5·5 in. stroke, and it runs normally at 900 revs. per minute. It is of the same design as the engine shown by Figs. 237, except that the admission valves are placed on the side of the cylinders in chambers like those of the exhaust valves E, and mechanically worked by precisely the same means, instead of the central inlet valve shown. In Figs. 235, 236, the exhaust valves are at O and the inlets at P, their cams being driven by the duplicate wheels N, N¹. The method of actuating the inlet valves is a simple one very generally used, but the central position for the inlet valves I, as used in the other engines, is a good one. The descriptions which have been given of other engines will have made the reader familiar with the points conveyed by an examination of Figs. 237. The external design of the engine is precisely the same as that of the two-cylinder 8-HP. engine shown by Fig. 238, duplicated and with the crank chamber lengthened to suit, and the support-ing arms H placed between each pair of cylinders. The carburettor C D also is placed between the two pairs of cylinders instead of between two

221

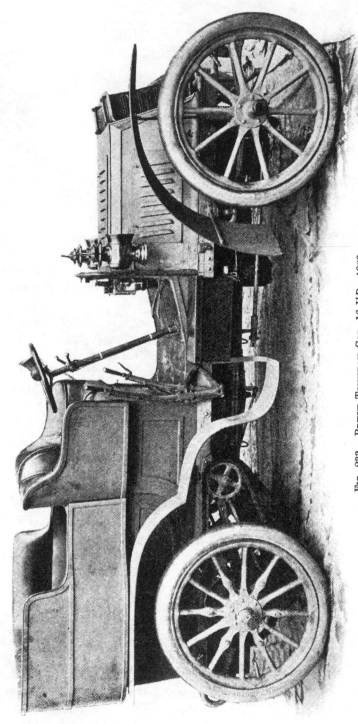

FIG. 233.—BRUSH TONNEAU CAR, 16-H.P., 1903.

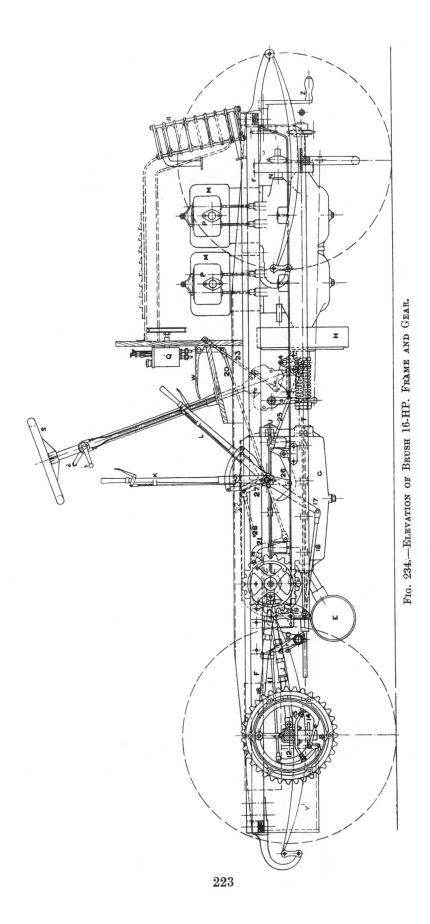

FIG. 234.—ELEVATION OF BRUSH 16-HP. FRAME AND GEAR.

223

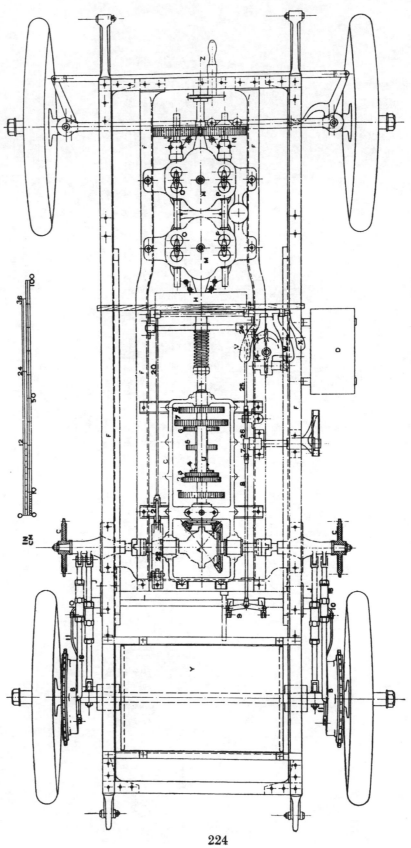

FIG. 235.—PLAN OF BRUSH 16-HP. FRAME AND GEAR.

224

cylinders, and the two induction pipes and throttle valves serve two cylinders instead of one.

The clutch, as now made, is of the cone type. The cone is of aluminium fastened to a steel sleeve which fits over an extension of the crankshaft which is screwed at the end to receive a nut. About three-fourths of the length of the sleeve is bored out large enough to take a central spiral clutch

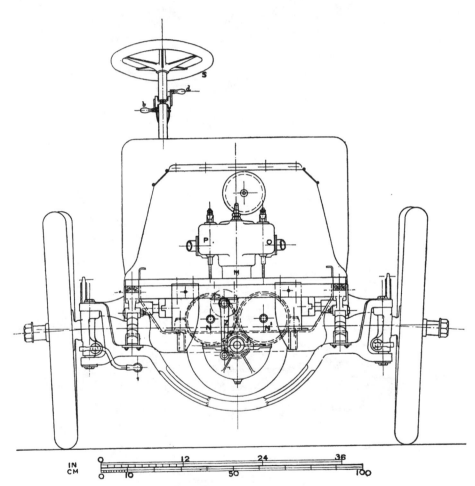

FIG. 236.—FRONT ELEVATION OF 16-HP. BRUSH CAR.

spring, which sits on the shoulder at the bottom of the large part of the bore, and the crankshaft extension passes through all. The pressure on the cone is regulated by the nut which bears with a washer on the outer end of the spring. Claws are formed on the end of the sleeve for a claw clutch drive to the gear box. The clutch is thus of the self-contained type,[1] and is very

[1] *Autocar*, vol. xii. p. 235.

like the Richard-Brasier clutch, Fig. 201, and it drives the lower and first motion spindle T of the gear box direct. The clutch is worked by

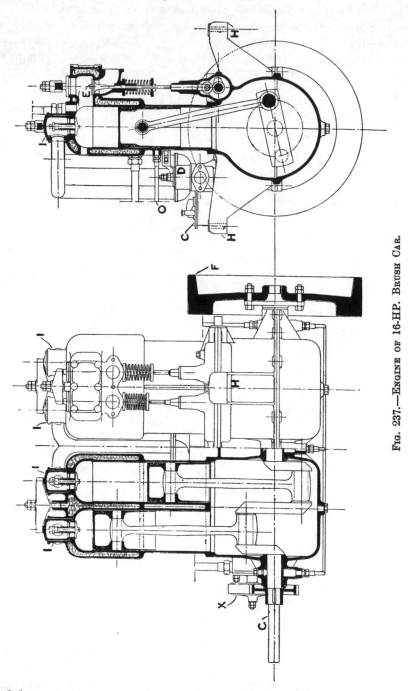

FIG. 237.—ENGINE OF 16-HP. BRUSH CAR.

the pedal v, connecting rod 23, and fork lever 24. It is also taken out of gear by the side brake lever by a downward extension with a roller acting on the bent lever 26 and the rod 25.

226

THE BRUSH PETROL MOTOR VEHICLES

As will be seen from the plan, the second motion spindle of the gear drives the countershaft by bevel gear, this shaft driving by Oldham couplings the two short sprocket spindles which run in bracket bearings, strutted by distance rods J from the main axle. In some cars ties are inserted from the front of these bracket bearings to a point of the frame about 15 in. in front of them

FIG. 238.—VIEW OF 8-H.P. ENGINE OF BRUSH CAR.

The gear gives four speeds, the shifting pinions on the lower spindle being moved by the lever L, downward extension 17, rod 18 connected to a sliding rod at 19 with forks in the bottom of the case.

The countershaft brake 22 is actuated by the pedal w, rod 20 and lever 21.

The side brakes of the internal expanding type are actuated by lever K, lever 27 (Fig. 234) and rod 28, marked 7 and 8 in Fig. 235. The rod is connected to a whippletree bar, or crosshead, 9 (which is in one piece, not two as appears in Fig. 235), the ends of which are connected to two levers, one on a solid spindle across the car, and the other on a short tubular spindle upon this as seen in the plan Fig. 235. The ends of these spindles carry bent levers 10, 10, connected by rods 11, to the bent lever 12, Fig. 234, by which the two semicircular segments of the brake are spread into the brake drum B. The two brake segments are pivoted at the top on a distance rod 16 pivoted on the sprocket-bearing brackets. The crosshead 9 seen in the plan gives the compensating action by which the pull on the brakes on both sides is equalized.

Internally expanding side brakes

The cooler is of the Loyal type, the circulation being kept up by a gear-driven pump. The British Company have now given up light car manufacture, but are building motor omnibuses.

The Wilson & Pilcher Car

A car which has been from the first and remains unlike any other is the Wilson & Pilcher, of which Fig. 239 is a plan, 1902–3 type. It is a live axle car with two bevel wheels on the differential gear, one for going ahead and the other astern. It is a car which comprises a great many ingenious details, few of which, however, have received from other makers that form of flattery which is said to be the sincerest. It is fitted with a four-cylinder horizontal engine, with cylinders placed transversely in the frame, the cylinders being 3·5 in. diam. and 3·75 in. stroke. The crankshaft is thus placed longitudinally in the car, and drives direct through the gear box shafts, all in line, for all speeds. The engine, gear box and rear axle bearing frame are all on a built-up subsidiary frame partly formed by crank case and gear box and bevel drive case, but running from end to end. The cylinders 5, 5 are not put in pairs but are staggered, those on one pair of cranks being on opposite sides. The cranks are in one plane and at 180 deg., the cycle sequence being as in No. 6 diagram, Fig. 311, p. 358. The fly-wheel 2 is in front of the car, has dished spokes and is of large diameter. Numerous modifications in the details of the engine have been made for the 1904 cars, which it is not necessary to describe here.[1]

Features of design

The commutator 10, the oil pump 9, and water pump 8 are driven by the bevel and spur gear shown. The carburettor is seen at 7, and the admission valves at 6,6. The steering segment is in the box 22. At 23 is a pedal lever by which the rod 36 is pulled for working the friction clutch 11, and at 39 is a pedal lever for working by rod 32 the brake at 15. At 24 is a lever, also seen in section, for working the reversing lever 14 by the rod 31, the reversing being done by throwing one bevel wheel out of gear with the bevel pinion and the other in. All the gearing is now made with beautifully cut curved spiral or helical teeth. At 25, 26 are cross levers for working the change-speed gears by rods 33, 34, 35, 37. At 11 is an ordinary cone

[1] *Automotor Journal*, vol. ix. pp. 463, 492, 518.

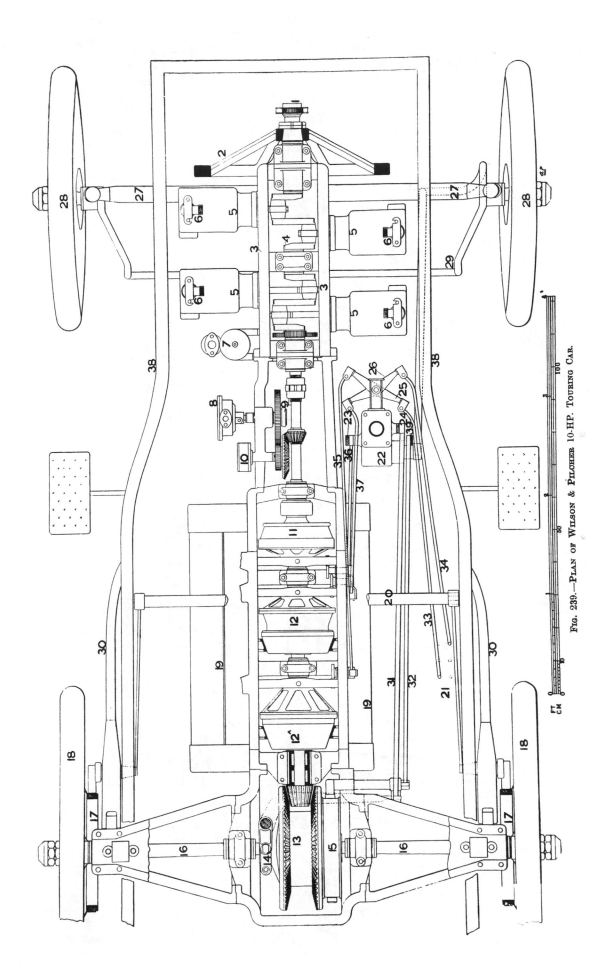

Fig. 239.—Plan of Wilson & Pilcher 10-HP. Touring Car.

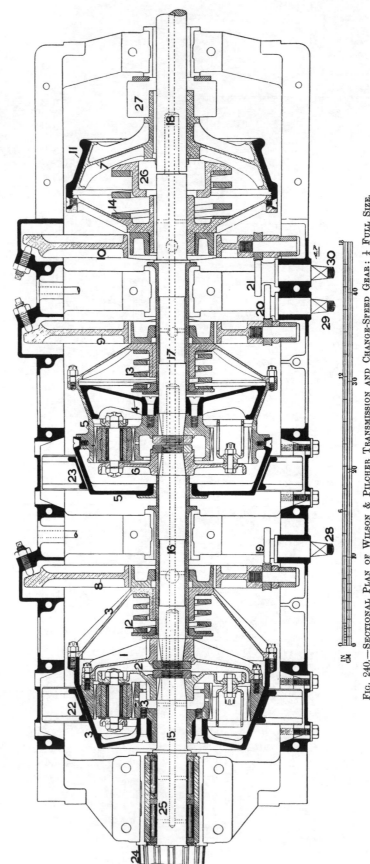

FIG. 240.—SECTIONAL PLAN OF WILSON & PILCHER TRANSMISSION AND CHANGE-SPEED GEAR: ¼ FULL SIZE.

clutch, 12 and 12A being speed-gear clutches. These are shown in detail in Fig. 240. At 20 is a cross shaft for working by rods 21 the band brakes 17 on the driving wheels. The silencer is seen at 19. Although a circulating pump is used natural circulation is also available from a tank above the engine to a Clarkson cooler beneath the front of the car.

Some modifications in the lever and lever connexions have been made in the 1903-4 car, by means of which the cross levers 25 and 26 have been replaced by modified connexions from a side lever in a four-way slotted quadrant.

As shown in Fig. 240 the propeller shaft is in three pieces, 15, 16, 17, the part 18 being a continuation of the crankshaft or driven by it. The male part 7 of the simple clutch is fixed on the shaft 18. The female part

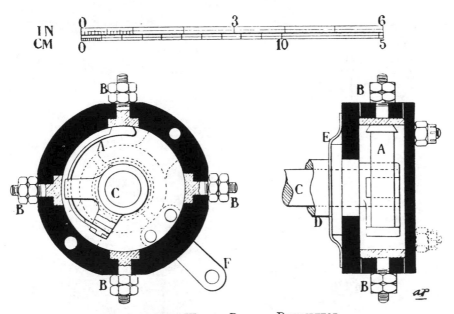

FIGS. 241, 242.—WILSON-PILCHER DISTRIBUTOR.

11 slides on the shaft 17, and is pressed on 7 by the spring 14. Behind the female part of the clutch is a lever 10, one end of which bears against a set screw, and the other against a cam 21 bearing against a roller on a pin in the end of the lever. Partial rotation of the cam spindle 30 forces the lever 10 towards the clutch through central pivot pins shown in dotted lines. This movement compresses the spring a little, and pushes the female cone 11 off the male cone 7. Everything is now out of gear, the engine driving only the shaft 18, but when this clutch is in and the other clutches in the position shown the carshaft 15 and bevel pinion 24 are driven at the highest speed, a direct drive as though the transmission shaft were all in one piece.

To obtain the lowest speed the black section cone 5 is pushed by cam 20 and lever 9 into a fixed conical ring 23 seen in section. This action liberates cone 4 by pushing 5 off it. Cone 4 carries on its boss a spur pinion of an

Construction of gear

231

epicyclic train, the stud spindles of the pinions of which are fixed in the disc 6. Rotation then of shaft 17 gives to shafts 16 and 15 a speed proportional to this gearing. Now by cam 19, lever 8 and cone 3, the black section cone 3 is pushed off the cone carried by the spider 1 and into the fixed cone 22. The slow speed given by the central set of gear is now further reduced by a similar reduction.

Four speeds are thus obtained, one with every rotating clutch at work and the whole of the contents of the transmission shaft revolving as one piece, i.e. the highest speed ;

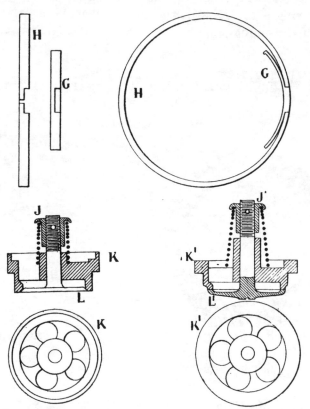

FIGS. 243, 244.—PISTON RINGS AND ADMISSION VALVE SPRINGS OF WILSON-PILCHER ENGINE.

Second, by declutching the central cone 4, fixing the cone 5, and thus putting the epicyclic gear into operation.

Third, by liberating clutch 5, allowing 4 to operate, and then fixing clutch 3, and causing its reducing gear to drive.

Fourth, by fixing both cones 5 and 3, and so backing the slow speed gear of 5 on low speed gear of 3.

The commutator or distributor used in the Wilson & Pilcher car is shown by Figs. 241, 242. Fixed on an ebonite box is a boss D in which runs the spindle C carrying a two-armed boss fitted with a spring contact piece A. The end of the spring has a little stud rivetted in it, and it is slipped into a slot or housing in the end of one of the arms. The centre of spring rests in a saddle at the end of the other arm, and by spring pressure bears against the contact pieces B in turn. The distributor can be partially rotated on the spindle c for adjusting the period of ignition by the arm F.

The admission valves used in the engines are seen in Figs. 244. The point in them is that they have springs which have at their upper part two or three coils close wound so that the pin which keeps the nut J in place cannot get out.

The piston rings H fitted with tongued springs G, which also serve to prevent the rotation of the rings, are shown in Fig. 243.

In the 1902 reliability trials of the Automobile Club an 8-HP. Wilson & Pilcher car worked very well indeed, gaining for reliability full marks on four days out of six, and nearly full on the other two, a total of 1,787 out of a possible 1,800. It was, however, slow on the hill-climbing tests. A 12-HP. car with an engine having cylinders 3·75 diam. and 3·75 stroke, running at 900 revs. per minute, entered the 1903 trials but retired.

The endeavour to obtain a straight through drive is praiseworthy, but it is obtainable in much simpler ways, although these other ways do not combine with them the shockless changing of speeds which is obtainable with the gear described.[1] It is, however, very nearly so with the Argyll gear, which, for instance, is very simple in construction and use.

[1] *Autocar,* vol. xiii. p. 102.

Chapter XV

THE VELOX AND MARSHALL-BELSIZE CARS

The Velox Car

A CAR was made in two sizes last year by the Velox Co., which though no longer made possessed some points of interest, albeit the greater part of the car was a copy of the Humber. The 12-HP car had a four-cylinder engine with cylinders 3·5 in diam., 4 in. stroke and ran at a normal speed of 900 revs. per minute.

It had a trussed tubular frame, and the clutch was exactly the same as the Humber. The gear box [1] was, however, arranged with the second motion shaft carrying the fixed gear wheels above the main shaft carrying the sliding pinions, the latter shaft being made in two parts so that on the fourth speed a direct drive through from engine to driving axle was obtained. A sliding rack worked by a toothed segment on the side lever spindle, and a segment in the gear box, was used to work the forks instead of a sliding rod in the lower part of the box, the advantage being that the spindle carrying the segment for working the rack was above the level of the oil in the box and not so liable to drop oil on the ground. The propeller shaft was both between the engine and gear box, and between gear box and axle drive, see Fig. 245, made of the form shown by Fig. 230, but the central part of the jaws was cut away so that the joint pieces H could move in all directions.

A feature of the car was the spring drive between gear box and rear axle, as shown by Figs. 245 and 246.

In Fig. 245 N is the gear box with the main shaft at w, on which is a drum Q which served as pedal brake drum, and as half the case for containing the springs Q4, Fig. 246, inside the inner part Q1 of the case. Bolts Q2 alternately fixed in one part of the case and free to move in slots in the other part, permit relative motion of the two drums through a small range and act as abutments for the springs Q4. Thus the torque of the shaft w **Spring cushioned drive** was transmitted through the springs to the propeller shaft s, but its behaviour in practice has not induced any other maker to adopt it, and it may easily be seen that the spring might cause a very objectionable hammer action

[1] *Automotor Journal*, vol. viii., 1903, pp. 29, 55.

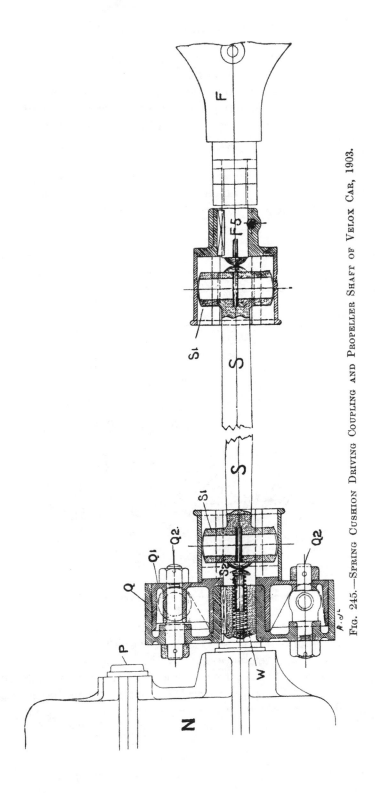

Fig. 245.—Spring Cushion Driving Coupling and Propeller Shaft of Velox Car, 1903.

through recoil between clutching and declutching, and when playing with brake and throttle in traffic and on changing gradients.

The pedal brake is clearly shown in Fig. 246. Two cast-iron segments R R are pivoted at R3 to a suspending arm R4 and carried by it and an adjusting screw R5 so that the return movement of the rod R7 always causes the two band parts R, the one to lift off the drum, and the other to drop clear of it.

The steering pillar and gear are shown by Fig. 247. The steering spindle v3 is carried within a tubular pillar v6, and at its lower part rests on an adjustable step screw, so that wear at the shoulders of the worm can

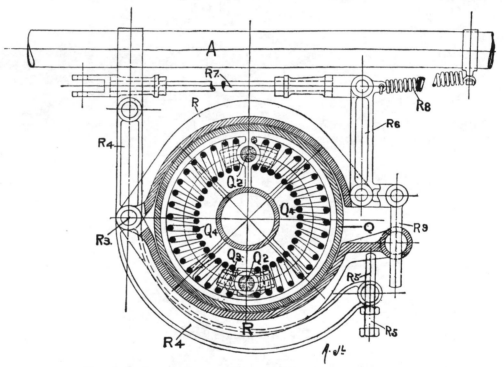

FIG. 246.—SPRING CUSHION DRIVING COUPLING VELOX CAR.

be taken up. The worm gears with a renewable segment rivetted to a lug formed on the segment and steering arm spindle. The steering gear box v1 is clamped at any desired angle by the clamp v, the upper part of which is fastened to the frame A by a clamp. The upper part of the tubular pillar is bushed, and on the upper part of the steering spindle is a tube v4 which fits in this, and may be set at required height by clamp lugs v7, feathers in v3 fitting in featherway in v4.

THE BELSIZE CAR

Another car of English make brought out in 1902, and containing numerous points in detail which have not been lost upon other makers, is the 12-HP. two-cylinder Belsize car, illustrated by Figs. 248 to 250.

236

THE BELSIZE PETROL MOTOR VEHICLES

Like several other cars brought out a short time after the drawings for this book were put in hand, it has been modified in several details, and amongst others it has since been fitted with a three-cylinder engine and of

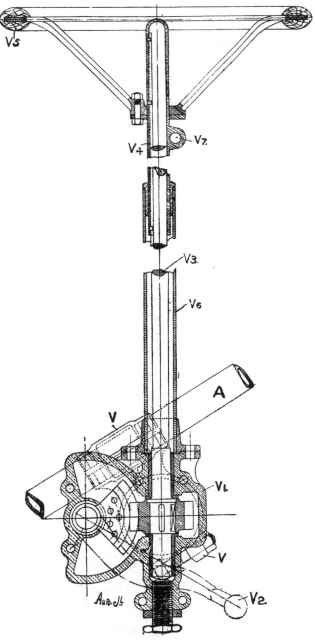

FIG. 247.—STEERING PILLAR AND GEAR OF VELOX CAR.

16 to 20 HP., with mechanically moved inlet valves on one side of the cylinder, the exhaust valves being on the other. The inlet valve cams are stepped so as to give a variable degree of inlet controlled by hand. The

237

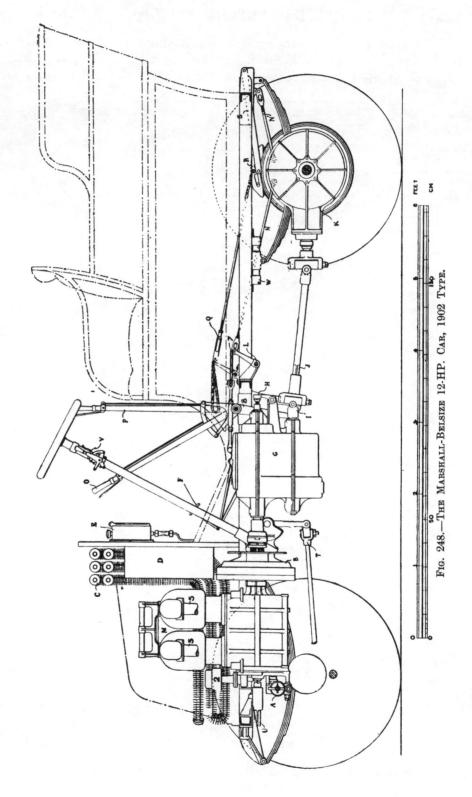

Fig. 248.—The Marshall-Belsize 12-HP. Car, 1902 Type.

238

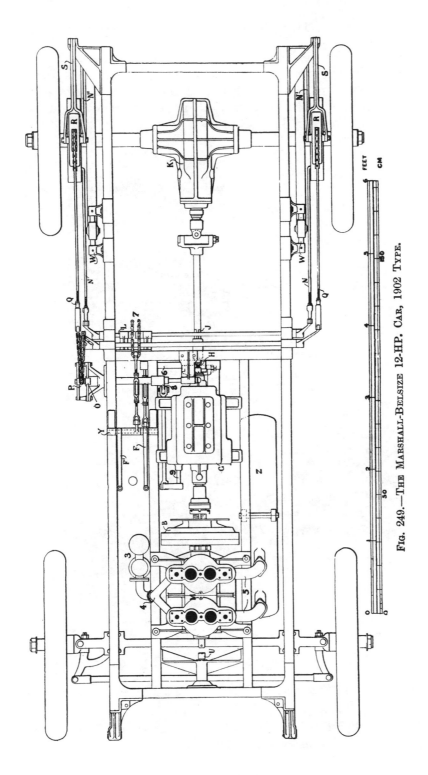

Fig. 249.—The Marshall-Belsize 12-HP. Car, 1902 Type.

239

throttle valve consists of two thin cylinders, one sliding over the other, one worked by the governor and the other by hand. The bottom of the crank chamber can be removed for access to the interior parts without dropping the crankshaft. The ignition apparatus includes a single trembler coil and a high tension distributor.

FIG. 250.—ENGINE OF BELSIZE 1902-3 TYPE 12-HP. CAR.

The car illustrated is made with a channel steel frame, to secondary longitudinals and transverse members of which the engine and gear box are fixed.

The longitudinal and the end transverse members of the frame are connected by well designed gusset brackets of malleable cast iron, which also provide for spring hanger, brake and other connexions.

240

The engine had cylinders of 100 mm., or 3·94 in. diam., the stroke being 110 mm. or 4·34 in. These dimensions were subsequently increased to 4 in. diam. and 4½ in. stroke, and the engine made with one, two, three or four cylinders, giving 6, 12, 18 and 24 HP. at 1,000 revs. per minute.

The engine of the 1902-3 type car is shown by Fig. 250.

The inlet valves were automatically operated and made easily accessible by single stud clamping bridges. They were at the top of the cylinders, the exhaust valves being worked by rocking levers giving easily adjustable period of opening. The engine of the 15 to 20-HP. 1904 car has three cylinders, and has mechanically operated inlet valves, and the cooler is of the multitubular square front vertical type assisted by a fan. The carburettor 3 is of the Longuemare type.

The commutator 2, in the design shown by Fig. 248, is on a vertical spindle driven by enclosed bevel gear in a crank chamber extension, and is well proportioned and accessible, and it may be clearly seen in Fig. 250.

The circulating pump A is also attached to the engine and driven by enclosed gear in the crank chamber direct from the crankshaft.

The cooler or radiator of the 1902–3 car surrounded, and in part covered the bonnet, the water tank being in front of the dashboard under the cooler tubes, where they passed over from side to side of the bonnet. The cooler had a large superficial area as may be gathered from Fig. 248.

The clutch B was and is of the self-contained or balanced type, the inner cone being pressed inwards towards the fly-wheel for release. It was operated in the car shown by the pedal F, which had a downwardly projecting arm connected by a shackle-adjusted rod, Fig. 249, to the lower arm of the lever H, the upper part of which pushed the spindle carrying the inner cone of the clutch.

The gear box G contained gear for three speeds and reverse of the lateral sliding type operated by the lever O, arm 8, and connecting rod to the slider rod 9, on the end of which in the gear box was the shifting fork Reversing is effected by putting the slow speed gears into relation with an idle pinion.

The power was transmitted from the lower gear spindle by the universal joint I and propeller shaft J to the bevel pinion, driving a bevel wheel of good size on the differential gear in the case K forming the central part of the casing of the live axle. The gear box has been modified for the 1903-4 car, so that on the high speed there is a direct drive through from engine to bevel wheel on the live axle.

There were two brakes, both on brake drums on the wheel naves. One was an internal brake expanding into the drum actuated by the pedal F¹ on a spindle held at Y, an upward arm on which carried a shackle-adjusted rod and small pulley on which was a chain 7. The ends of this chain were attached to two arms pivoted at L, Fig. 248, on transverse spindles, the ends of which carried arms, Fig. 249, to which are attached the rods N N connected to one end of each of the expanding rings, the other ends of these

Compensation of brakes

rings being held by adjustable rods N^1. The chain 7 being free to move on the small pulley gives an equal pull on each brake.

The side brakes were operated by the lever P, which transmitted the pull by a chain, one end of which was connected to an arm and the rod Q on the off side of the car, the other end being connected to an arm on the cross shaft L, and transmitting motion to an arm and the rod Q on the near side, both pulling on the front ends of the brake bands R on the outside of the brake drums. The other ends of the brake bands in the design illustrated are held by the forked stays S S. The steering gear is of the type in which a nut, through which the threaded end of the steering column passes, gives motion to a rack and the toothed quadrant with which it gears, and so through the usual lever arm and connecting rods to the steering wheels.

The front springs were carried in front by cast malleable dumb irons, and at the rear in shackles in the ordinary way, and the back ends of the rear springs are held in jaws formed with the corner castings as already described.

The front ends of the rear springs are carried by slippers sliding on a rod carried by the bracket pieces w.

The Belsize Motor Co. is now making a popular car of small size known as Belsize Junior.

This little car has pressed steel frame longitudinals and a 6 ft. 3 in. wheel base and 4 ft. gauge. The transmission consists of a gear box close to the engine. From the end of the first motion spindle, a chain drives a central countershaft. From this countershaft is a chain to the outside of the differential gear box on the live axle. The gear gives three speeds and a reversing motion, and the second motion spindle does not run when on top speed. The engine has a single cylinder 4·5 in. diam., and a stroke of 5 in., and it gives, it is said, 7 HP. at 1,000 revs. per minute.

Both valves are mechanically moved, and are easily accessible. The throttle and ignition control levers are brought to the top of the steering wheel, the ignition being of the trembler and wipe contact type.

One end of the crankshaft projects through the gear box, and on its end carries a cone clutch which drives a spur pinion sleeve otherwise loose on the shaft. On the end of this sleeve is a claw clutch face, and next a double-faced claw clutch which can engage with this sleeve or similar one on the other side of it. This latter carries the sprocket pinion by which the countershaft is driven. On either side of the shaft referred to is a spindle carrying sliding spur-wheels which can be brought into gear to give the first speed and reverse and the second speed. The sliding clutch on the crankshaft and the sliding wheels on the two parallel spindles are moved by forks which receive motion from one cam plate.

Separately adjustable chains

The countershaft is carried in ball bearings, and adjustable radius rods connect it with the driving axle. Each chain can be adjusted separately.

The cooler is of the honeycomb type, water being circulated by a toothed wheel pump.

THE BELSIZE PETROL MOTOR VEHICLES

The total weight of the car is given as 8 cwt.

The 12-HP. Belsize Junior has a two-cylinder engine with bore and stroke of 4 and $4\frac{1}{4}$ inches respectively, developing its rated power at 1,000 revs. per minute. The transmission gear is like that of the 12-HP. car already described, but with detail modifications, and the change-speed gear of this and the other cars now constructed is arranged with the direct drive on the highest speed and with the second shaft and its gear wheels then disengaged and stationary. The 18–24 and 30–40 HP. cars of 1905 have three- and four-cylinder engines with similar leading dimensions of $4\frac{5}{8}$ in. bore and 5 in. stroke. A car equipped with a six-cylinder engine is also now constructed, the cylinders having the dimensions of the 12-HP. Belsize Junior.

The inlet valves are, as mentioned, operated mechanically through push rods and pivoted levers in the manner adopted for the exhaust valves and illustrated by Fig. 250, the inlet valves and camshaft being on the opposite side of the engine.

Variable lift to inlet valves

By means of a camshaft arranged to slide longitudinally and cams with a varying profile, a variable lift may be given to the inlet valves in the manner already described for some other cars. The control of the engine by this means is by hand lever and connexions.

With all these cars a single trembler induction coil is used supplying the sparking plugs in turn through a rotary high tension distributor, a similar rotary switch or contact maker being required in the low tension circuit.

Among other changes common to many cars as now made is the adoption of pressed steel frames, the use of the multitubular or honeycomb design of radiator, and the considerable extension of the wheel-bases to accommodate more roomy and luxurious bodies.

Many of the interesting detail features of the Marshall Belsize cars are retained in these more recent types,[1] the greatest changes being those affecting external appearance.

[1] *Automotor Journal,* Jan. 28, Feb. 4, 1905.

Chapter XVI

THE NAPIER CARS

FEW cars are more widely known than the Napier cars, not only as touring cars of several sizes with from two- to six-cylinder engines, but as racing cars. Originally made on exactly the Panhard and Levassor lines, but in many respects of stronger build, they have gradually developed into a distinctive character, and in various trials and races have often taken the highest places. The Napier racing cars have been well known since one of them, driven by Mr. S. F. Edge, won the Gordon-Bennett Cup in 1902 in the Paris-Vienna run, when all the continental cars one after another retired through troubles of one kind and another, which were greater than those which attacked the Napier, or were not so vigorously or persistently combated.

Figs. 251 to 253 illustrate the frame and running gear of the 9-HP. car like that supplied to the Right Hon. A. J. Balfour.

The frame F is of the wood and steel flitch pattern with a supplementary underframe H carrying the engine M and gear box G, as in the Panhard car of the time.

The engine drove the first motion spindle, which was a continuation of the shaft clutch cone shaft, the secondary spindle above the primary having upon it the fixed gear wheels.

The sliding pinions V, Fig. 252, are moved by the lever L, connecting rod 10 and slider rod 12, having upon it the gear-moving forks, one as indicated between the figures 12 and 10 on Fig. 252 for the reversing pinion.

The countershaft was driven by the bevel pinion and wheel, see Fig 253, and this drove by claw-coupling ends at O the short sprocket spindles in bracket bearings fixed and stayed to the frame as seen in Fig. 252. The claw couplings were not, it will be seen, universal joints as in the Brush car, but were only means of driving the sprocket spindles in such a way that the countershaft was easily removable.

Clutch and brakes interconnect

The clutch pedal P acted on the rod 9 jointed to the lever 11, Fig. 251, and pulled the clutch out of contact against the resistance of a spring in the cylinder in front of the lower forked end of the lever 11. The side brakes also put the clutch out of gear, by causing the end of a short arm 11, Fig. 252,

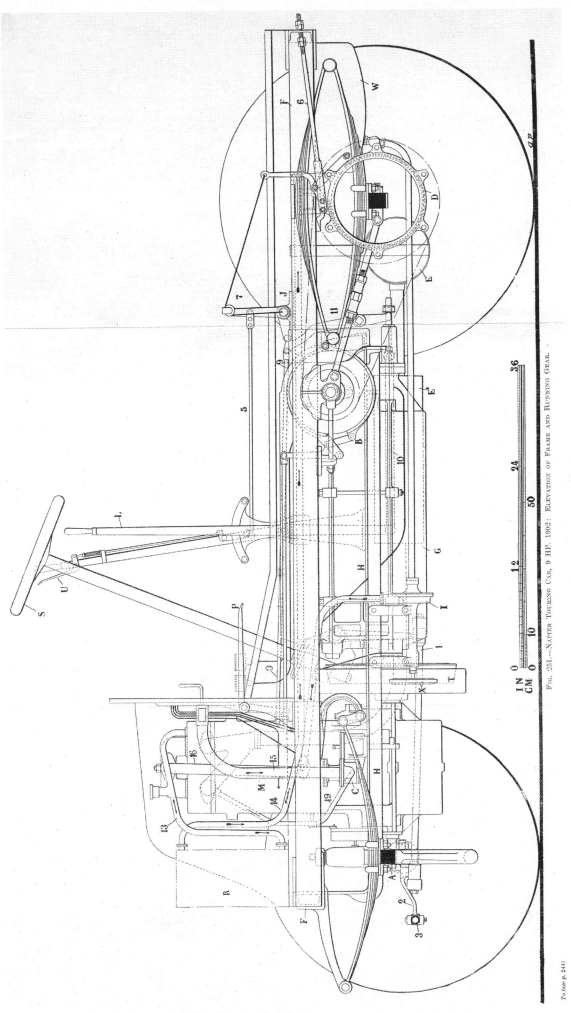

Fig. 251.—Napier Touring Car, 9 H.P. 1902: Elevation of Frame and Running Gear.

To face p. 244]

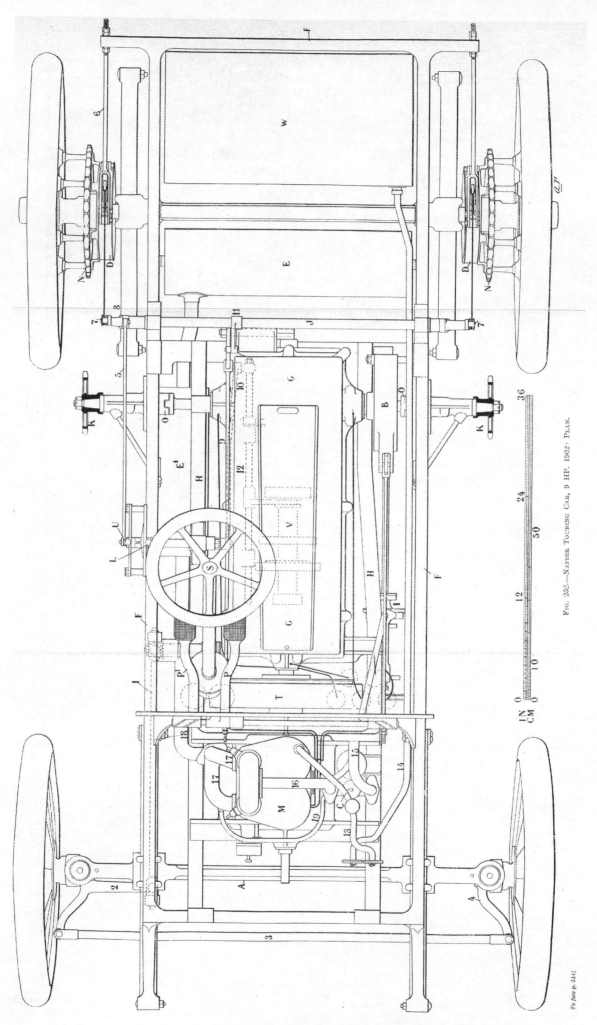

Fig. 262.—Napier Touring Cab, 9 HP, 1902 · Plan.

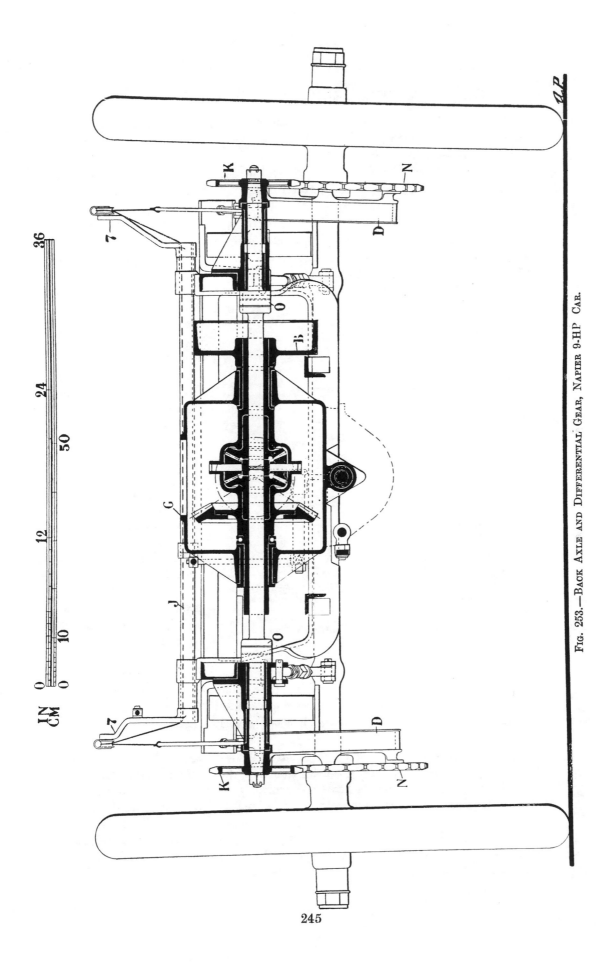

IN
CM

FIG. 253.—BACK AXLE AND DIFFERENTIAL GEAR, NAPIER 9-HP CAR.

on the brakeshaft J to press against the hollow curved back of the upper part of lever 11 of Fig. 251, a small roller being carried at the end of the lever 11, as seen in the plan.

The pedal brake B was actuated by the pedal P[1], and the connecting rod and lever shown in Fig. 251 pivoted to the forked lugs at the end of the brake band at B.

The carburettor, Maybach type, is seen at C, the air admission pipe 15 being carried up to the dashboard, where there was an adjustable cap, a shunt pipe 19 from the exhaust pipe being used to supply hot air to the jacket of the carburettor nozzle chamber.

The connexions to the Loyal water-cooler are shown by the figures 13, 14, and the arrows on the pipes between cylinders, cooler R, pump I and water tank W.

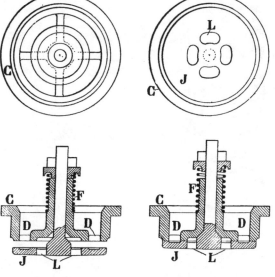

FIGS. 254.—TRIPLE-SEATED INLET VALVES, NAPIER ENGINES.

The side brake bands D were operated by the compensating wire rope threaded through the hollow brakeshaft J and over small pulleys at 7, as in the Panhard cars.

The lubricating pipe connexions are shown by a number of black lines from the Stauffer lubricators shown in dotted lines, to the several bearings.

The automatic inlet valves used in the Napier cars are shown by Figs. 254. They have, as will be seen, three seats and three inlet edges or orifices, two formed at the edges of the passages D D, and one by the passages not shown in the sections but seen at L in one of the plans. This valve gives about double the area of inlet for a given lift that is given by the ordinary single-seated valve, and is a useful adaptation of the similar valve which was formerly used for safety-valve purposes.

Triple seated valves

The Napier 15-HP. four-cylinder car of 1904 is shown by Figs. 255 to 262. From these drawings it will be seen that the frame is of the armoured wood type, that the engines have mechanically operated inlet valves, that the water cooler is of the honeycomb type with fan behind it, and that in general it represents the practice of midsummer, 1904. The frame F is supplemented by an underframe F1 formed of light unequal channel steel

Frame construction

attached to the main frame by an outwardly bent end as seen in the plan, and an upturned splayed end fixed to the front transverse member of the

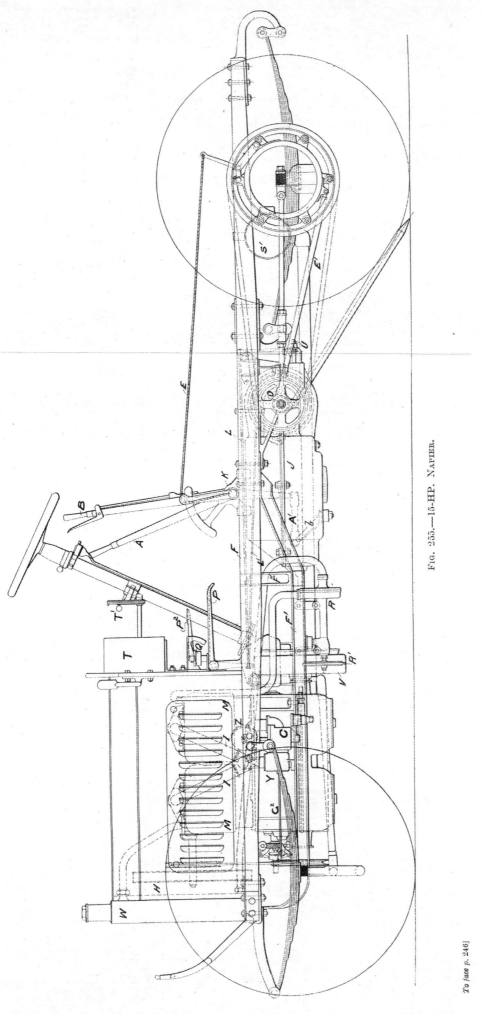

Fig. 255.—15-HP. Napier.

The material originally positioned here is too large for reproduction in this reissue. A PDF can be downloaded from the web address given on page iv of this book, by clicking on 'Resources Available'.

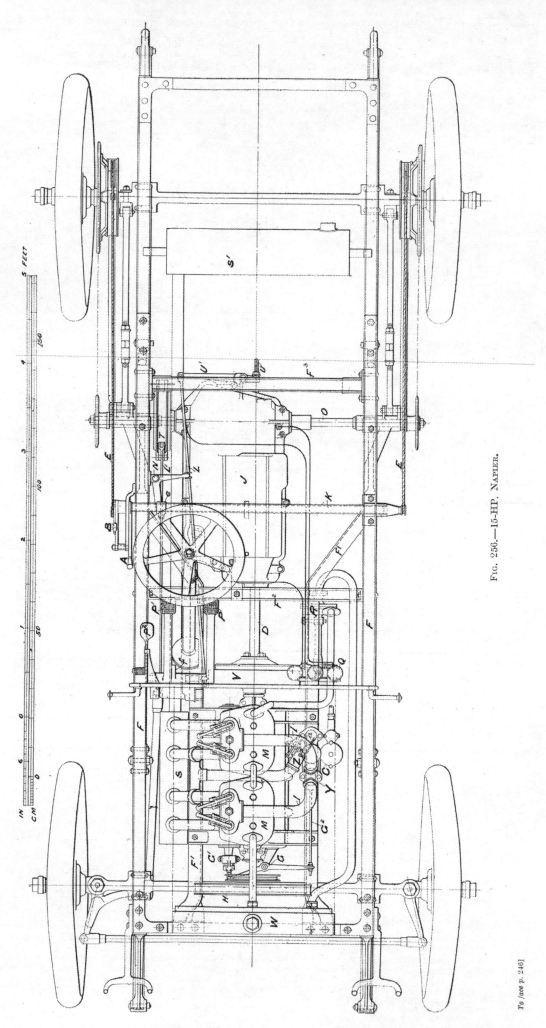

FIG. 256.—15-HP. NAPIER.

main frame. This underframe carries the engine, and in conjunction with the transverse member F2 carries the front end of the gear box J, which has a three-point suspension, two of the points being its bearings on the countershaft O. The side members of the frame are well connected by the transverse members F2 and F3, and by the end pieces which have corner strengthenings.

The engine has cylinders 3·5 in. diam., and 4 in. stroke, the normal speed being 900. The 18-HP. engine has four cylinders having the same

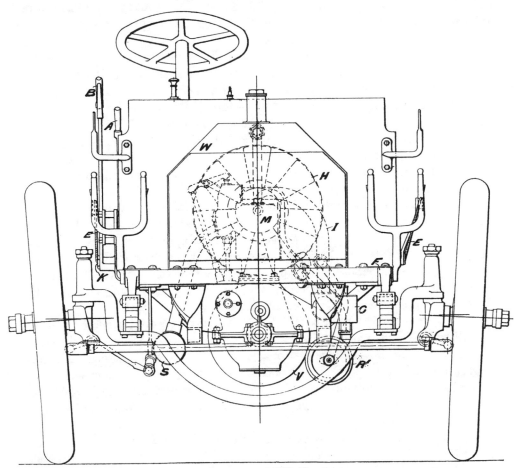

FIG. 257.—15-HP. NAPIER.

dimensions, but the stated speed is 1,200 revs. per minute. Both a high mean pressure and high mechanical efficiency are required to enable the declared powers to be obtained.

As will be seen from Fig. 259 both valves are on one side, the inlet valve being worked by a rocking lever L pivoted at P, pushed by a rod Q, moved by a cam on the half-speed shaft B, similar to the cam for the exhaust valve. It will be seen that the fork x on the bottom end of the pusher piece carrying the roller moved by the cam B is well guided so that the side

247

pressure resulting from the push of the cam has no effect on the pusher piece. Adjustment when necessary of the rod Q actuating the rocking lever L of the inlet valve is made at Q^1, and a spring at L^1 maintains a slight pressure always on the rod, so that it does not leave either the cam pusher stem or the outer end of the rocking lever. The connexion of the inlet pipe from the carburettor is clearly shown in Figs. 255 and 256. The piston, as will be seen, is of the flat-topped form, and the port to the combustion space is direct and short.

The cylinders being cast in pairs have their centres at a minimum distance, but owing to the endeavour to keep the engine case as short as possible the distance between the centres of the crank-pins is still less, and hence the necessity for the one-sided form of the big ends of the connecting rods. If a little more length could be given to the crank-case this might be avoided. The water jacket is cast complete with the cylinder heads, and the pipe connexions between jackets, pump and cooler are clearly seen in the elevation and plan. The water tank shown in the 9-HP. 1902 car is, it will be seen, dispensed with. Splash lubrication is depended upon for the engine, but Stauffer lubricators are used for the main bearings, three of the lubricators being shown at Q in the plan, and the pipe connexions shown by black lines.

Ignition is by the high tension accumulator and trembler coil method, one coil only being used, placed at T, Fig. 255, the arrangement of commutator being that recently introduced on the Napier cars. The advance or retardation of the ignition period is effected by the handle at T1. The commutator or distributor is driven by a chain from the fly-wheel boss, not shown, running on sprocket pinions on the spindle projecting through the dashboard, on the box T, and inside the bonnet.

System of ignition

The apparatus[1] is similar to that for the six-cylinder engine, which consists of a distributor with six cam points and one contact blade and screw points for making and breaking the primary circuit between battery and coil. Driven at the same speed is a second distributor which is in the secondary or high tension circuit of the coil. The high tension current is delivered to a rotating centre piece, on which is a single distributor plate which alternately approaches and leaves in its rotary path six corresponding contact plates which are disposed equally around a normally fixed ebonite ring. Contact on the primary circuit interrupter is broken before contact of the secondary circuit pieces ceases. Although contact pieces for the high tension circuit are mentioned it should be noted that they do not come into actual contact, but the moving piece only approaching rather than actually rubbing the fixed contact pieces.

Although the apparatus would work without the interrupter in the circuit between battery and coil, that arrangement would be objectionable for three reasons. Firstly, the continuous action and passage of current to the primary and continuous action of the trembler; secondly, because of the production of the high tension spark on approach and departure of

[1] *The Autocar*, vol. xii. p. 480.

the rotating high tension contact piece to the fixed contact pieces, and thereby the burning of these contacts; and, thirdly, as a result of the second great irregularity in the actual period at which the spark would occur at the plugs. I have spoken of fixed contact pieces, but the ring carrying them is, like the contact blade on the primary circuit interrupter, movable within a small range for the purposes of varying the period of ignition. These parts are interconnected and the time interval between contacts is regular. The rotating low tension contact-making cam and the high tension

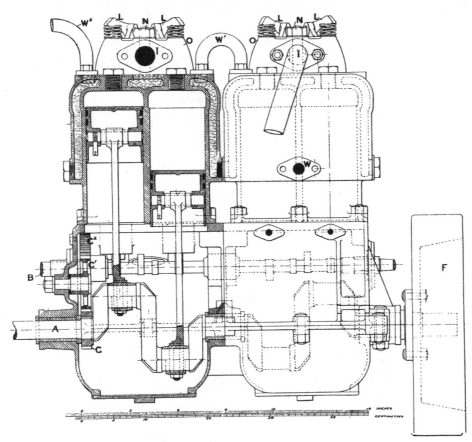

FIG. 258.—NAPIER 16-HP. MOTOR.

distributor arm are also driven together through gearing, and by these means the possibility of imperfect timing of ignition is reduced.

The sequence of the cycles in the cylinders, and, therefore of ignitions, is shown for the cranks at 180°, and at 60°. in the diagrams Nos. 11 and 12, Fig. 311, page 358; but a modification in the sequence of firing with cranks as shown may be made which will modify the connexions between the high tension distributor and the ignition plugs.

The sequence in the four-cylinder car here illustrated is also shown on the diagram above mentioned, but it will be noticed that the sequence

249

ERRATUM.

Fig. 258, p. 249: *for* 16-HP. Motor *read* 15-HP.

of the cranks and the cams, and the relative positions of the inlet valve levers as shown in the drawing, require correction, the relative position shown being only for convenience of the drawing.

The carburettor seen at c, Figs. 255 and 256, is of the simple float-feed kind, but is provided with Napier's air admission control valve, which is actuated by a diaphragm, one side of which is in communication with the jacket water by a pipe not shown. Inasmuch as increase in the speed of the engine increases the speed of the pump R, driven by the friction wheel R1 running against the fly-wheel V, it gives increase in the pressure of the water

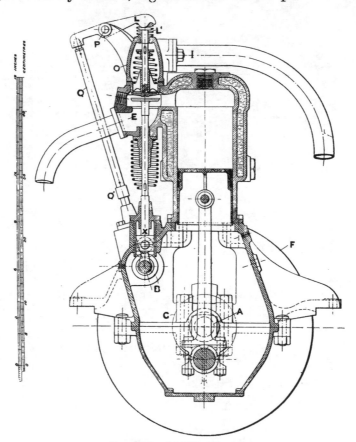

Fig. 259.—15-HP. Napier.

delivered by the pump, the pressure on the diaphragm, and, therefore, the opening of the throttle valve increases with increase in speed of the engine. **Hydraulic air regulator** The diaphragm is seen at z, and in the plan the short lever on the butterfly valve used, pushed by the central spindle of the diaphragm, can be seen. The short lever on the throttle valve is returned to its position by a spiral spring *a*. This valve then acts as a throttle of the air inlet to the carburettor. The air inlet pipe passes between the two cylinders, and is provided with a wide mouth contiguous to the surface of the silencers by which the air is heated. As well as the diaphragm controlled inlet valve, a throttle

250

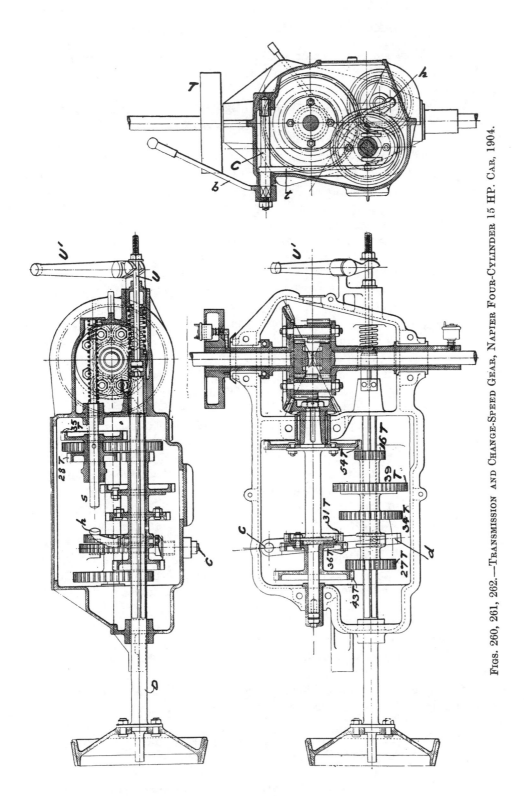

FIGS. 260, 261, 262.—TRANSMISSION AND CHANGE-SPEED GEAR, NAPIER FOUR-CYLINDER 15 HP. CAR, 1904.

251

valve at Y between carburettor and the pipe supplying the cylinders is used. This is actuated by the arm G from the governor G1, and the connecting rod G2.

Another form of the hydraulic air regulator [1] has been used by Mr. Napier, the diaphragm moving a piston supplementary inlet valve, but that illustrated is the more recent.

Transmission to the gear box is by a simple cone clutch, the spindle D of which is continued through the gear box to a tube containing the spring by which the clutch is forced into the fly-wheel. The method by which the spring is brought to bear upon the clutch and the means for relieving it, by pull on the cheese-headed rod U, when actuated by the pedal rod L, and fork-ended lever U', are similar to those used in the earlier forms of Coventry-Daimler and Napier cars (Figs. 65, p. 99, vol. i.).

Operation of change-speed gear The sliding wheels and pinions are as heretofore on this shaft, but a different arrangement is adopted for the reversing gear, and a spur-wheel differential is used as seen in Figs. 260 and 261. These shifting speed-change wheels are moved by the lever A, Fig. 255, on the shaft K, the lower end of the lever having a spherically jointed connexion A', attached to the bent lever b which moves the spindle C, Figs. 260 to 262, on which is the fork-carrying arm t, which is curved round as seen in Fig. 262, the gear-shifting fork being pivoted on t at its first bend. The arm t terminates at its second bend in a circular or cheese head h. In the positions shown none of the gear wheels are in mesh, but progressive movements forward of the speed change lever A bring the pairs of wheels 27T and 43T, 34T and 36T, and 39T and 31T into gear, for the second, third and fourth speeds forward respectively. Similarly, from the position shown, a movement of the lever A backwards puts the wheels 16T and 54T into gear for the first speed forward and a second movement of A in the same direction moves the four gear wheels on the squared shaft D still further to the right, the end h of the curved arm t having meanwhile come into contact with the end of the reversing wheel-carrying spindle s, and moved it and the wheels 28T and 35T, which are fixed to it, as far to the right as the bearing boss, in which s runs, will permit. In this position wheels 16T, 35T, 28T and 54T are in gear, in the order given, and so rotate the shaft carrying 54T in the required direction for reversing or travelling backwards. When any of the four forward speeds are engaged h is out of contact with the spindle s, and under pressure from the spring shown on the right hand end of s, keeps it and the two wheels upon it in the position shown, and wheel 28T therefore out of engagement with wheel 54T. By means of this arrangement the spindle s and its wheels are normally stationary, and are only active when the reversing gear is in operation.

The pedal brake band T is in two parts pivoted at the back and worked by the pedal P[1] and rod L[1]. The same pedal also releases the clutch through the action of the lug f which moves the clutch pedal P.

The side brakes are worked by the lever B, tubular shaft K and wire

[1] *Autocar*, vol. xi. p. 529.

rope E on the Panhard plan, but the brake is simple, strong and double acting, the rotation of the band being prevented by the radius rod E[1].

Mr. Napier has made some cars[1] with six-cylinder engines which are of much interest, and a large racing car of the same type of about 80 HP. The latter car is provided with only two forward speeds, the engine being relied upon to meet all intermediate requirements. The clutch of these cars, unlike that of the 15-HP. car illustrated, is provided with three adjustable springs used in such a way that the pressure on the cone is self-contained within the clutch after the manner of the clutch shown in Fig. 230. The inlet valves are actuated in the same way as in the 15-HP., Fig. 259. The cylinders are made without jackets, and an extremely light jacket afterwards formed over the cylinder by electrolytically depositing over a plumbago coated wax core a sufficient thickness of copper, the wax being afterwards melted out after the manner of the bronze casting makers of the Delhi district. The circulating pump is of larger size than is commonly used and runs at a slower speed. Each jacket is separately supplied with its cooling water, and warm water is supplied, as in the other car, to the carburettor jacket, one carburettor being used for the six cylinders which are supplied in pairs by three pipes, the pipe admitting a supplementary supply of air to the carburettor being carried up to the dashboard, where there is a hand control.

Electrolytically formed jackets

The ignition system is as already described. The weight of the racer car is under 1,000 kilogrammes to meet Gordon-Bennett conditions. The car was sent to Homburg for the German race, though not actually used on that occasion.

The engine of the 30-HP. Napier car has six cylinders with a bore and stroke of 4 inches and its normal speed is 1,000 revs. per minute. The 50-HP. six-cylinder engine runs at the same speed and has the same stroke. but the cylinder diameter is 5 inches.

Six cylinder engines

The chassis equipped with these engines have pressed steel frames, but it is interesting to note that the 18-HP. car is built with a wood frame reinforced with steel flitch plates like that shown by Figs. 255 and 256. The continued use of such frames by the Napier Company and the Daimler Company of Coventry testifies to their satisfactory character, especially now, when the steel frames are so much in use and by many who pay as much attention to the dictates of fashion as to the requirements of design.

[1] *Automotor Journal*, vol. ix. p. 729.

Chapter XVII

SOME RECENT LIGHT PETROL CARS

The Swift Light Car

NOTEWORTHY success was achieved in the Hereford Trials last year by the two small cars of the Swift Company, as will be gathered from some results of those trials given on page 504. One of these cars also did well in the 1,000-mile trials of 1903, and received a silver medal.

Previous to the construction of these cars the Swift Company had entered the list of those who endeavoured to dispense with a differential gear so as to use a solid through axle, and to make a two-speed car with a direct drive on both speeds. A through axle is very desirable, but the difficulties are considerable, and the devices substituted have hitherto been attended with defects in operation and details inferior to the gear displaced by them. The free-wheel ratchet gear employed, moreover, adds to the operations for reversing.

Original transmission gear. The transmission gear used in this experimental car is of some interest as a type which has been tried and found wanting. It is illustrated by Fig. 263, which shows that it is very like certain forms of mowing machine gear, except for the parts necessary for reversing.

In a casing A were bearings for the main or driving axle, and for two spindles P and H, the former being driven by the engine. On it were two bevel pinions R and T, the first gearing with a large bevel wheel C giving the low speed and the latter gearing with the smaller bevel wheel D giving the high speed. Between the two bevel pinions was a double-faced claw clutch S, by which either pinion was put into gear. By means of a spur pinion O, and corresponding pinion N, the spindle H was driven and ran idle, except when reversing motion was required. The spindle H, by pinion 67, drove a corresponding pinion 67A on a sleeve carrying the bevel pinion 68. In ordinary running this sleeve, as well as the spindle H, ran idle, the sleeve being loose on its spindle. For reversing the claw clutch 61 was put into engagement with the end of the sleeve by the spectacle lever 60 pivoted at the back on the bracket 60A. This gave the reverse motion on the slow speed. With this arrangement there was no gear box, and a direct drive was obtained for both speeds. Both the driving pinions R and T were, how-

254

ever, always in gear, and the five pinions required for the reversing gear were also always in gear. There were seven pinions always running with only one doing positive work ; the driving of all the others was all lost work, and the simplicity of the car was only an apparent simplicity, and the cost of production much higher than with a simple gear box which will give three speeds and requires only one extra pinion for reversing.

An external view of the 7-HP. single-cylinder car which did so well at Hereford, and of the 7-HP. double-cylinder car is given in Fig. 264. This view is given to show the general design of exterior. It is a very comfortable two-seat car, silent and economical, the single-cylinder car being more economical than the double-cylinder and running as smoothly.

The frame is tubular, carried on leaf springs of sufficient length, supplemented as they are by small coil springs of the form formerly used in

FIG. 263.—TRANSMISSION AND SPEED-CHANGE GEAR OF SWIFT LIGHT CAR.

bicycle saddles, and placed between the tube frame and the bottom of the body frame. The wheel base is 5 ft. 10 in. of both cars, which were Nos. 14 and 26 in the trials at Hereford.

The engine of the No. 14 car was of Fafnir make and De Dion type, and had a cylinder 95 mm. 3·74 in. diam., and 110 mm. 4·32 in. stroke, and ran at about 1,100 revs. per minute.

The engine of car No. 26 was by White and Poppe, and had two separate cylinders 80 mm. 3·15 in. diam., and 85 mm. 3·35 in. stroke, with half-speed spindle for working the valves placed transversely to the crankshaft, below the cylinders and between the cranks. This spindle carries two cams, one on each side, actuating respectively the two inlet and two exhaust valves through the medium of a pair of bell crank levers.

A De Dion Bouton carburettor was used in both cars.

The change-speed gear of these cars gives three speeds, and is of the now almost standard type of small gear box with sliding change-speed pinions and reversing pinion, and through or direct drive on the highest speed to a jointed propeller shaft driving a bevel pinion gearing with a bevel wheel on the differential gear on a live axle which runs in large ball bearings.

As shown above, the change-speed lever quadrant was so proportioned and placed that the lever when full over forward was inconveniently low and far from the seat. This has been now modified, and the side brake lever now works on the same quadrant. I understand that the single-cylinder car at £175 will not be made in future, but the No. 26 pattern will be continued. The new 1905 car has a pressed frame and a two-cylinder White and Poppe engine of 7-HP., the cylinders being 80 mm. diam. 3·15 in., and 90 mm. 3·55 in. stroke. The crank is built up with crank-pins on the same line or at 360 deg. and with a bearing between them.

FIG. 264.—THE SWIFT 7-HP. LIGHT CAR, 1904–5.

In addition to the 7 HP. two-cylinder car the Swift Company are now making a 12-HP. car with a four-cylinder engine and with the same general arrangement of transmission gear.

The cars at Hereford weighed respectively 1,204 lb. and 1,302 lb ready for the road.

These cars, the Wolseley 6-HP. and the Siddeley 6-HP. cars, were awarded gold medals.

The Siddeley car, made by the Wolseley Company, was similar to that described on page 82, but had a seat of the same type as that shown by Fig. 264, and had a cooler of the Loyal type with a surrounding water container, the appearance being similar to Fig. 264.

THE CHAMBERS LIGHT CAR

A car presenting several novel features is the car designed by Mr. J. V.

Chambers, and made by Messrs. Chambers and Co., Belfast. It was one of the light cars entered for the Hereford Trials in August, 1904, but meeting with a slight accident, not likely to recur, it had to retire.

The engine is horizontal, having two cylinders fore and aft driving by a light chain the sprocket wheel on the combined differential and change speed gear on the rear axle. The arrangement permits the use of a light **Reduced load on chain** chain because the stress on it is not increased when the slow speed is put into gear, as the change gear is driven by the chain and not the chain by the change speed gear. This is a feature which does not apply to the final drive chain of any other car, though it obtains with the first motion chain of cars such as the Wolseley, represented by No. 4 diagram, Fig. 485.

FIG. 265.—THE CHAMBERS 7-HP. LIGHT CAR, 1904–5.

The car is illustrated by Figs. 265–269c.

The frame is made of channel steel, to which is riveted an inside flitch plate, and outside is fixed wood lining to the channel steel.

The cylinders M M of the engine are side by side, with the cylinders to- **Carburettor close to cylinder** wards the front of the car, as seen in Figs. 267, 268 and 269, the carburettor main casting being attached directly to a passage cast on the cylinder covers, into the centre of which are screwed the two ignition plugs. The two automatic admission valves on the top of and immediately over the centre of the cylinders have covers held in place by a one-bolt bridge, and the inspiration passage between carburettor and valves being formed with the cylinder covers is kept at a sufficiently high temperature to prevent the formation of snow, resulting from the refrigeration due to evaporation.

The exhaust valves are beneath the cylinders, and are operated by the two cam-worked side rods seen in Fig. 269, and levers on the rocking shaft carried in bearings connected to or cast with the cylinder cover casting. The position of the carburettor is easily adjusted, and the throttle valve carried in the admission valve chamber casting is operated from the centre of the steering wheel by connexions *a b* to the lever shown in Figs. 268 and 269.

The ignition is by battery, coil, and wiper distributor, which is on the end of the half-speed shaft, and is moved for varying the period of ignition by another handle in the centre of the steering wheel which moves a collar *d* and thereby a forked lever *c* (Figs. 267, 268), which gives partial rotation to a small spindle across the engine to the wipe contact-maker case. The cylinders of the engine are 3·125 in. diam., and 4·25 stroke, and the power is given as 7 brake-horse-power, or 6½ horse-power at 1,000 revolutions.

A recent change in cylinder diameter to 3·375 inches has increased the engine power to 8-HP.

The transmission from the engine crankshaft B is by a single chain

FIG. 266.—THE CHAMBERS LIGHT CAR WITH THIRD CENTRAL SEAT.

within the aluminium gear case G (Figs. 267, 268 and 269A), which forms a distance piece between the crankshaft and the live axle, which is further stayed in the horizontal direction by stay-bars T T (Fig. 268). The stay-bars are definitely fixed as to length, and are attached to the spring carrying pads on the axle, and to bosses X in the sides of the aluminium chain case. For adjustment as to distance between crankshaft and sprocket wheel S (Fig. 269B), on the main axle, the aluminium chain case is carried by a nut B^2 (Fig. 269A),which screws upon the casting B^1 (Fig. 269A), and is seen in Figs. 267 and 268, the gear chain casting being held on the nut B^2 by pinching bolt through the lugs E (Fig. 269A), seen also in Fig. 268. The driving gear and change speed gear is of an ingenious cyclic and epicyclic combination. It is shown in section in Fig. 269B, and its operating gear in Fig. 269A.

Adjustment of driving chain

As will be seen in Fig. 269B, this gear also comprises the differential gear E, the case for which is coupled together by the pins C C upon which the epicyclic speed pinions of the speed gear run.

There are four of these pins or bolts c and planetary wheels. The central pinion F has a sleeve extension, to which is keyed a large spur

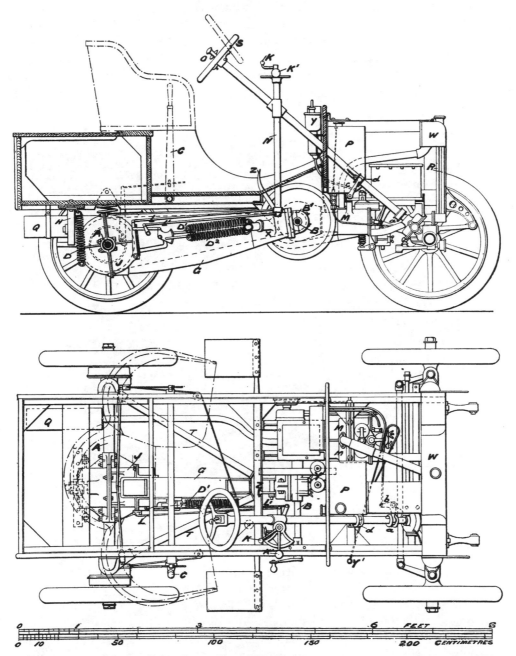

FIGS. 267, 268.—THE CHAMBERS LIGHT CAR 7 HP.—SECTIONAL ELEVATION AND PLAN, 1904-1905.

wheel B, which also carries clutch ring B¹. The central driving chain sprocket s runs free on this sleeve.

The internal gear ring formed with the wheel D has a large sleeve encircling the differential box and carrying a clutch actuating collar D². For convenience in describing the action of the gear the large spur wheels are lettered to correspond with the clutches they carry. Dogs O, O¹, shown in Fig. 269A on the spindle H, are provided to engage in the teeth of these gear wheels, and forked levers operated from the same spindle, engage with the clutch moving collars B² and D².

Operation of change-speed gear

To obtain the first speed, wheel D is held by bringing its dog O¹ into action, and at the same time moving B² and expanding B¹ clutch, thus

FIG. 269.—ENGINE OF CHAMBERS LIGHT CAR.

locking the chain sprocket with the central pinion F, and holding the internally toothed ring of the epicyclic wheel D.

For the second speed, wheel B is locked by its dog, and clutch D¹ engaged by collar D², B¹ B² being then out of action.

Third speed. Both dogs are disengaged and clutches B¹ and D¹ both expanded, thereby locking the driving chain sprocket to the differential box, and giving a direct drive from crankshaft to live axle.

When it is required to reverse, a small train of pinions on spindles M and N, shown in Fig. 269A, is tumbled into engagement with large gear wheels B and D, by the rod L¹ and the cam on spindle P. This train differentiates in opposite directions between these two gear wheels, clutch B^1 being used, both dogs and clutch D^1 being out of engagement.

The friction clutches are expanded by the semicircular pins G, which receive a partial rotative movement by levers L (Fig. 269B), when the sliding collars D^2 B^2 are moved inwards. A taper wedge and screw stop for effecting adjustment for wear of the expanding clutch rings is provided at s. The clutch rings are split at opposite diameters and connected by rings 1, 1. By the use of the dogs O, O^1 on the large spur wheels, the gear changing is quick and positive, the dog being introduced noiselessly at the moment of reversal of motion of its wheel, which is the free member of the crypto system.

The chain pinion on the engine has 9 teeth, which is rather small, and the pull on the chain with the engine doing 6·5 HP. at 1000 revs. per minute is 341 lb., the chain speed being 653·25 ft. per minute. The sprocket wheel has 36 teeth, and a $\frac{7}{8}$ in. chain is used. The central pinion F has

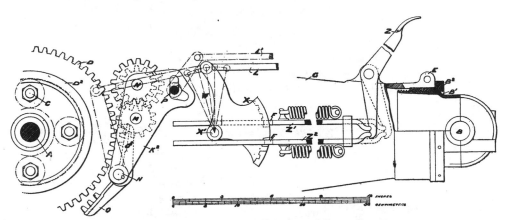

FIG. 269A.—SPEED CHANGE AND REVERSING GEAR CONNEXIONS.

24 teeth, and the internal ring 52 teeth, the idle pinions 14 teeth, the maxi- **Stresses in** mum pull on the pins C being about 560 lb., and on the teeth 280 lb. **gear**

There is a slight difference in the arrangement of the dog and clutch actuating rods as shown in Fig. 267 and in Fig. 269A. The latter is the most recent.

The dogs O, O^1 are placed opposite each large gearwheel, and are keyed to their spindle and actuated by a lever on the outside end of it. This lever is linked to a segment x as shown, and carried on to the hand gear K and K^1 (Fig. 268), by the rod L. The dogs only come into action at the extreme positions of the small hand lever K^1, shown in the neutral position in Fig. 268; the horizontal rods z^1 and z^2 (Fig. 269A), with tension springs, locking levers which allow the clutches to go into action, or are held out as the case may be by the segmental piece x, the latter and the notches F F on the rods z, z^1 forming a lock for the various positions indicated by the radiating dotted lines w corresponding with positions of the hand lever K^1 (Figs. 267 and 268). They are released by the pedal z, which must be pressed

261

quite down (with clutches out of contact) before the hand lever K^1 can be moved, thus acting as a safety device.

The rods z^1 and z^2 pass on either side of the segment x, and the notches F F permit the movement of the segment when the pedal is quite down by allowing the projections on the periphery of the segment to pass through them.

The reverse gear train is carried between two stiff steel plates K^2 (Fig. 269A), forming a frame pivoted on the spindle carrying the dogs previously referred to.

This gear is moved into mesh by a small crank P moved by the lever L^1, and the connexions to the handle K on the top of the steering column.

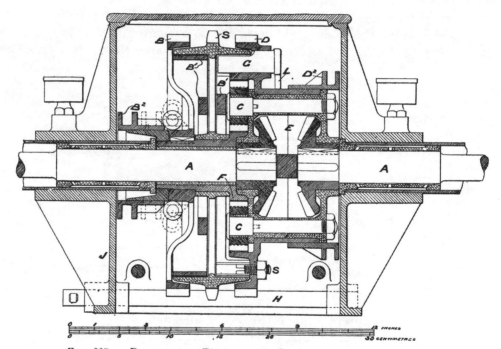

FIG. 269B.—DRIVING AND DIFFERENTIAL GEAR OF CHAMBERS LIGHT CAR.

This crank " throws " slightly past the dead centre, thus forming a rigid self-holding lock for the frame K^2 (Fig. 269A) when the gear is in action. The clutch B^1 is, as before explained, used in actuating this gear.

The chain is accessible by small sliding doors in the top of the chain case, and the whole of a long flat plate on the top can be removed when required.

The chain case with its triangular stays and back axle tubes form a complete structure, rigid in itself, but free to partake of the movement of the springs, differential spring movement being accommodated at B^1 (Fig. 269A). It can be readily uncoupled from the crank shaft and wheeled back from underneath the car for access to the parts more thoroughly when required.

RECENT LIGHT PETROL MOTOR VEHICLES

The shafts forming the rotating parts of the axle are provided with long bearings at A A of considerable diameter, but they are reduced in diameter between the inner and end bearings to a size sufficient for torque requirements, and so as to provide a little torsional elasticity to reduce the severity of sudden application.

The bearings are similarly designed at both ends, with good provision for lubrication, and the outer ends protected from mud or dust. All external stresses are taken by the tubes and casing, which are strongly stayed, and to which the chain case and triangular system of staying give great rigidity. The gear is silent, and easily manipulated.

The car is carried on spiral springs front and back, and a pair of strong spiral springs D (Fig. 267) relieve the driving axle of much of the insistent central weight.

The front axle is kept in place by two rods both jointed to the ends of the horns shown in Figs. 265 and 267, but the end of one is rigidly fixed to the axle, as seen in Fig. 267.

The steering is by worm and wheel, the steering wheel being pivoted on the top of the tubular column, so that it may be moved to the dotted line position shown in Fig. 267 for easy entrance and egress of the driver to and from his seat.

The jacket water is cooled by a natural circulation cooler having two rows of brass vertical tubes R connecting a top tank W and a small tank at their lower ends. The tubes are surrounded by top tank W and water legs at the sides, the small lower tank being free from the sides, and free to move up and down with the expansion and contraction of the tubes. The flow pipe to the cylinder jacket is connected to this (see Figs. 267, 268). The Hereford car cooler had 38 tubes $\frac{1}{2}$ in. diam., and 12 in. long. They are now increased to 14 in. long, and the number slightly increased by widening the tank, so that there is now approximately 1 sq. ft. per declared HP. of engine.

Amount of cooler surface

An ingenious arrangement for introducing a removable third or central seat is shown by Fig. 266, and by the dotted lines in Fig. 268, the two side seats being pivoted so as to take the position shown when the central seat is in use. When not in use a convenient receptacle for travelling odds and ends is placed in between the two seats as shown in Fig. 265.

Expanding internal brakes in drums on both driving wheels are employed, actuated by a pedal pulling equally on both by wire cord, and also by a handle C.

A 1-ton van with the same construction of gear has now been working some months quite satisfactorily. It has a double cylinder engine with cylinders 4 inches diameter and stroke 5 inches. For the change-speed gear, Fig. 269B, cast-iron clutch rings B and cast-iron clutch pulley S are used, and, as would be expected from experience in other branches of engineering, is more satisfactory than other metals for the purpose. In the van this transmission and change-speed gear is on a countershaft from which the road wheels are driven by side chains.[1]

Chambers 1 ton van

[1] A view of the van and gear is given in *Motor Traction*, July 27, 1905, p. 284.

Since the date of the Hereford Trials the coil springs then used have been abandoned in favour of laminated or leaf springs.

THE ROVER LIGHT CAR.

The 8 HP. light car,[1] of which two views are here given, is illustrated by Figs. 269c and 269D in elevation and plan with the body removed. In method of construction and in some essential features much originality is shown, and though the ingenuity displayed in the many small details cannot be described by reference to the general views a good general idea of the car may be gained by an inspection of them.

The engine crank chamber A, the casing around the clutch at B, and the gear box C, are connected together by flanged facings, and by reference to both views it will be seen that the Y arms L[1] are cast with C and upon its rear end. To the front of the engine crank chamber a forwardly pro-

FIG. 269c.—THE ROVER 8 HP. LIGHT CAR, 1905 TYPE. BODY REMOVED.

jecting bracket is fixed which through a connecting bolt takes a swivel bearing in a bracket attached to the centre of the inverted semi-elliptic spring above the front axle. There is visible in Fig. 269D a transverse bar E[1] to the outer ends of which the wood longitudinal frame bars of the body, 1¾ in. square, are connected. The rear springs are semi-elliptic and take their centre bearing on the brackets J[1] to which they are fixed. The length of the springs, 40 in., is exceptional. The body, it will be understood, is fixed to the transverse bar E at the front and is supported at the rear upon the side springs, suitable offset spring arms being fixed to the body frame.

The engine, clutch and gear cases, A, B and C, are virtually in one piece, swivel supported in front through the bracket mentioned upon the transverse spring and their weight is taken at the rear by connexion of

[1] The *Autocar*, Aug. 27, Sept. 3, 1904 ; May 13, 1905.

the arms L^1 to the body frame. The body, therefore, supports part of the mechanism, but it does not derive any support from what in other cars is the underframe. As far as possible the various parts have been designed to be independent of the support of an external frame, and it is evident, particularly from a study of the earlier design of this car, that it was intended that when assembled the parts should constitute an underframe independent of any useful connexion to the body. Exposition by practical trial, of inertia effects, however indicated the desirability of providing as fully as in the design shown by Figs. 269c and 269d, for relative movements of the parts, set up, when the car is travelling over irregular road surfaces, and for the relief of the tyres as far as possible from weight not spring borne.

From the gear box the power is transmitted by a propeller shaft to the bevel wheels and differential gear contained in the case E, at the centre of the rear axle. A torque bar to restrain the tendency of the case E and the connected parts to rotate about the axle centre in an opposite direction to that of the driving wheels is not used, but the propeller shaft is satisfactorily relieved of any bending stress by the use of the casing D with a bearing at the forward end.

The engine is of the vertical single cylinder type with both valves mechanically operated, and is provided with means of varying the amount of the opening of the inlet and exhaust valves, and the frequency of opening of the exhaust valve. The cylinder diameter is 4·5 in. and the stroke 5 in. *Variable frequency and amount of valve opening*
The cams are formed with a changing profile and may be moved lengthwise on the half time shaft through a shifting fork connected to a pedal. From an extreme position, in which both valves are given their maximum lift, the cams may be moved along the shaft, giving less and less movement to the valves by bringing their parts having lessening eccentricity or cam throw beneath the valve tappets. A position may be taken up in which, with small lift of the inlet valve, a cam to partially relieve the compression comes into operation, and in the extreme position the inlet valve receives no movement, but the exhaust valve is lifted towards the completion of each up stroke and remains open during every down stroke. The engine is thus *Method of control* caused to act as an air compressor driven by the car, and considerable braking effect results from the work of compression. By means of the pedal through which the longitudinal movement is given to the cam blocks, not only may the power developed by the engine be regulated, but the negative work or power absorbed by the engine when it is caused to operate as air compressor or air brake may be controlled. Additional means of control of the car, through Bowden wires operated by two milled discs at opposite points in the rim of the steering hand-wheel, comprise connexions to the throttle valve on the carburettor and to the contact breaker for timing the ignition. The normal engine speed is 900 revs. per minute.

The Rover automatic carburettor is described in Chapter xxxiii, dealing with carburettors, and it need not be here referred to.

Regular turning of the engine and reduction of vibration is assisted

Flywheel design

by the use of very heavy flywheels. The weight of the pair of wheels is 120 lb. and the metal is as far as practicable all in the rim, only a very thin centreing disc being required. The exchange of energy between the flywheels and crank shaft is effected through chocks cast on the rim, machined to fit the cheeks of the single-throw crank shaft. The latter and the half time shaft and first and second motion shafts in the gear box run in single row ball bearings.

Single disc clutch

The metal to metal single disc clutch is one of many interesting features introduced by the designer of the Rover cars, Mr. E. W. Lewis. The clutch runs in oil inside the casing B, and may be permitted to slip during engagement and disengagement. The necessary face pressure is obtained through levers, pivoted in the clutch casing or that part of the clutch which is driven by the engine, which multiply the spring effort five times.

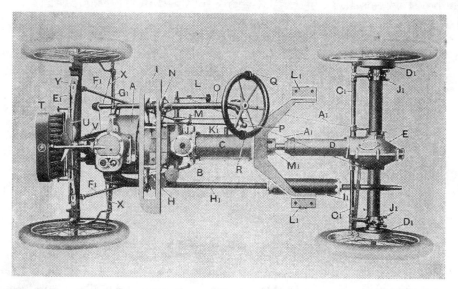

FIG. 269D.—THE ROVER 8-HP. LIGHT CAR 1905 TYPE. PLAN WITH BODY REMOVED.

The gear changes are effected in the usual way by a series of toothed wheels of ample dimensions which may be engaged by progressive sliding movement of the one set of wheels upon their shaft. Three forward speeds are provided and a reverse speed, and a direct drive is obtained on the top speed by engagement of a claw clutch, the second motion shaft being driven idly and thus permitting of a return to the low ratio gears without encountering stationary gear wheels. The gear shifting lever is at P below the steering wheel.

There are many other points in design to which only passing reference can be made, and among these is the use of flexible steel wire connexions, duplicated, between the lower end of the steering column and the steering heads. The duplicate connexions are left a little slack and do not transmit the steering efforts, but should one of those in service become disconnected or break owing to accident or improper use, then the duplicate connexion takes up the duty of that which has failed and the possibility of complete failure of the steering gear is thus rendered remote.

Aluminium has been freely used in the construction of the car, and there are evidences in various parts of thoughtful reduction of weight. With a body to seat two passengers the car weighs 9 cwt. and its overall length and width are respectively 9 ft. 6 in. and 4 ft. 8 in.

The diameter of the road wheels is 30 in. and the wheel base and track are respectively 6 ft. 6 in. and 4 ft.

A 6 HP. car [1] of similar construction weighs only 6 cwt. approximately and in it are embodied many of the characteristic features here referred to.

Two 16 HP. cars with four-cylinder engines competed in the Tourist Trophy race, September, 1905, and some reference is made to them in Chapter xli. There are points of similarity of design in all these cars.

[1] *The Autocar*, Jan. 28, 1905.

Chapter XVIII

SOME AMERICAN PETROL CARS

PROGRESS in the design and manufacture of motor vehicles in America has not been distinguished by any noteworthy advance upon the practice obtaining in either this country or on the Continent. It has however, led to two interesting results, both of which may be considered due to the natural process of elimination which attends the development of anything, and which has resulted, firstly in the production of a type of light motor-car universally recognized as American, and secondly in the adoption, for the larger and more powerful cars of much that may be readily recognized as of European origin.

Although commercial enterprise and energy has very largely contributed to the successful introduction of the large number of light cars to be seen in this country, the continued success of some of them sufficiently shows that low cost, satisfactory performance, and general convenience have been well considered by the manufacturers and much appreciated by the number of people who as owners have actual experience of them.

In the construction of these cars primarily designed to carry two persons, and provided with engines developing about 6 or 7 BHP., the American engineers have shown themselves possessed of the same ingenuity and capacity for invention which characterized the making and selling of the light steam vehicles described in·vol. i. of this book, chapters xxvi. and xxvii. It must, however, be pointed out that in many instances a more liberal use of more suitable materials has subsequently been found necessary.

It is noteworthy that recognition of the reliability of the petrol engine has had the effect of diminishing the number of those who had previously favoured steam cars, and has more largely added to the number of those who understand, or who are influenced by those holding the opinion that the oil engine already possesses advantages absent in the steam engine, and that improvement leading to further simplification and greater reliability will without doubt be more rapid with light cars equipped with the internal combustion motor, than with those employing steam as the power medium.

The attention paid to the construction of light cars suitable for short distance travelling, or of cars which are light, even when some of the requirements for longer distance travelling have been considered, has largely been brought about by the special conditions of limited distances over which good roads are to be found in America, and the very indifferent condition of the majority of the roads beyond town precincts, often mere tracks, having no made or maintained road surface. Experience with such roads has shown the necessity for the very sparing use of even the strongest materials in order that the completed road vehicle may be extremely light. That this result has been satisfactorily attained as regards horse-drawn vehicles is sufficiently indicated by the light four-wheeled carriages of American origin, many of which may be seen in this country. They have been found serviceable under very trying conditions partly because of their peculiar design and construction, but mainly because of the less destructive stresses set up in the vehicle itself, resulting from the reduction in weight and the consequent diminution of useless power expenditure on mutual destruction of vehicle and road.

The American light motor car or "runabout" has embodied in it some of the results of this long experience in light carriage work, and of those cars to be here referred to, the Oldsmobile is perhaps as well known,

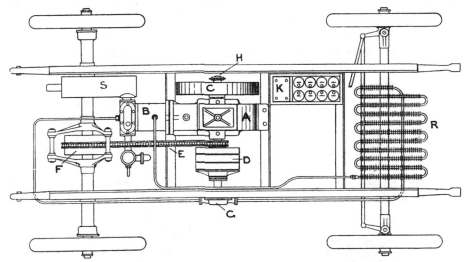

FIG. 270.—TYPE PLAN ILLUSTRATIVE OF AMERICAN LIGHT CARS.

and may claim precedence over others by reason of its early appearance in this country, and its distinctive features.

Fig. 270 is a block plan which may be considered representative of the general arrangement of a large number of American cars, some of which cannot be included in the "runabout" or voiturette class, because of the use of bodies having greater seating capacity, and of higher powered engines. Among those makers whose cars are better known, the horse power of which varies from 6 BHP. to a declared 10, are Messrs. The Oldsmobile Co., Cadillac, Elmore,[1] Ford,[2] Rambler,[3] and Pope Hartford.[4]

Character-istic design

[1] See *The Horseless Age*, May 18, 1904, vol. xiii.
[2] See *The Automotor Journal*, June 11 and 18, 1904, vol. ix.; *Horseless Age*, vol. xiv. p. 253.
[3] See *The Horseless Age*, April 20, 1904, vol. xiii.
[4] See *The Horseless Age*, April 27, 1904, vol. xiii., Jan. 25, 1905, vol. xv.

In all these cars the general disposition of parts is as indicated in Fig. 270 in which the single cylinder motor A B C is carried, either upon a rectangular angle steel frame, or transverse angle bars connecting the spring supported main members of the frame, and occupies a nearly central position. The engine normally runs at low speed, and the fly-wheel C, sometimes, and with advantage, carried at the opposite side of the crankchamber to that shown, is larger, and its accumulator and regulator power greater than is common with single cylinder motors. It is clear, however, that variations of the effort between fly-wheel C and driving chain wheel must be transmitted through the crankshaft when they are in the positions shown in the block plan, and the evil resulting from this setting up of racking stresses in the crankshaft is greater than that resulting from loss of balance or the additional bearing required when the fly-wheel is on the same side of the engine as the chain drive. Additional fly-wheel energy is, as might be expected, conducive to quiet running, the driving effort is more constant and therefore more merciful to transmission parts and tyres, and in a lower-powered car, the deliberate addition of fly-wheel weight may well prove economical and considerably more convenient than the use of a multi-cylindered engine to obtain in a different way the same advantages. Carried upon the extended end of the crankshaft is one of many varieties of epicyclic planetary gear D, by means of which a slow speed forward or backward may be obtained, or through a face clutch the whole gear may be caused to revolve at the same speed as the crankshaft, thus providing a high speed drive without any intermediate active gear between engine and road wheels other than the chain E on the occasionally active differential gear F.

It will be seen that there are tubular extensions of the open connecting frame each side of the differential gear, and within these fixed parts the driving shafts are housed, and run in either ball or roller bearings. The axle is, therefore, of the live type, but the driving shafts have to transmit power to the road wheels and in addition experience the road shocks. With the tubular sleeve construction, however, the centre driving shafts may receive some support close up to the road wheel hubs.

One of the drum surfaces provided by the speed gear D is as a rule utilized for a band brake drum, and a second brake is arranged with the brake straps, surrounding the surfaces of the differential gear casing F not occupied by the driving chain, or acting upon drums on each side of the differential but connected to the driving shafts. With some cars the only brake provided is that on the differential gear, the reversing gear brake strap, in connexion with the planetary speed gear, being considered sufficient for an emergency brake.

The motor may be started with the usual cranked starting handle, sometimes so arranged that it may be reached from the seat (as in the De Dion Bouton voiturettes) connected with a transverse shaft, chain wheel, and chain leading to the freewheel clutch H, on one end of the crankshaft. The other end of the crankshaft is sometimes supported in a bearing connected to the frame, but when, as shown in Fig. 270, an approximate balance is obtained by arranging fly-wheel and speed gear at opposite sides of the crank chamber such a bearing is not employed, and there is one place the less in the car likely to require attention and lubrication. A circulating pump G is also driven from one or other end of the crankshaft through some form of flexible coupling, and there are the simple connexions between cylinder jacket, pump G, and radiator R.

Other parts, such as petrol, and water tanks, the coil and cell box K, the silencer S, and the radiator R are arranged in whatever way may prove most convenient, and where they may be accessible without being unsightly.

The silencers in these cars are as a rule of very liberal dimensions though not necessarily heavy. The cylinder, heads, barrels and liners are generally separately renewable, particularly the first of these parts, and failure or wear does not then entail heavy replacement expenditure.

FIG. 271.—5 BHP. OLDSMOBILE LIGHT CAR.

These light cars are equipped with high tension ignition apparatus, dry cells being, as a rule, supplied in preference to the form of accumulator more generally used in this country.

The wide wheel gauge and comparatively long wheel base with rather high-pitched narrow bodies distinguishes them as American, and at once recalls the light horse-drawn spider vehicles, although the substitution of the motor for the horse has necessitated a reduction in wheel diameter.

THE OLDSMOBILE CAR.[1]

The general external appearance of this car is shown by Fig. 271, and the elevation and underneath plan of the same car with the body

[1] A description of this car appeared in the *Automotor Journal* of August 8, 15 and 22, 1903, to which readers may be referred for anything not dealt with in this brief description. For earlier references see the *Autocar*, September 1, 1902, and for a description of the 1904 type, February 6, 1904.

removed by Figs. 272 and 273, while details of some parts are given in the three following illustrations.

A distinguishing feature of the Oldsmobile is the use of very long side springs extending from axle to axle and performing the functions of carriage springs, radius bars and to some extent of frame. It possesses the merit of great freedom of movement combined with economy in parts and manufacture. To the under side of the small square angle steel frame K^1 upon which the engine and the body are carried, are bolted the long laminated springs K which may be well seen in Figs. 271 and 272. These springs are made up of six leaves, the lower one only being formed in one continuous length from front to rear axle, the others are for reasons of cost and weight only of just sufficient length for suitable connexion to the frame K^1 If

Freedom of spring carriage

Fig. 272.—5 BHP. Oldsmobile Car with Body Removed.

Fig. 272 be examined carefully the junction of the dummy wood packing piece at the centre, and the spring ends may be seen.

Undue liveliness at the front of the car is prevented by the addition of a double elliptic spring J, Fig. 273, interposed transversely between the front part of the body and the front axle. The lower half of this spring is connected at its centre with the short lever arm which gives motion to the connecting rod J^3, Fig. 273, and by the usual connexions to the steering pivots on the front wheels. The upper half of the spring has a short column fixed to it at its centre, this column J^1 having the tiller steering lever J^2 hinged to its upper end. Steering is effected, therefore, by movement of J^2 causing partial rotation or swivelling of J^1, spring J and the short lever to which it is connected at the lower side, as in the Ideal Benz cars. Movement by deflection of the spring J may take place without any tendency to movement of the steering gear. The underneath view of the car, Fig. 273, shows the relative widths of wheel gauge and frame, and the extent to which the front and rear axles project beyond the points of attachment of the springs. While this arrangement contributes to easy riding of the car over rough surfaces it must also be remembered that it involves unfavourable loading of the axles, and these parts must be strengthened, or reinforced as they have been in this car by the addition of truss rods or tension stays, below the tubular axle sleeves, visible in Fig. 273. The wheel base of the car is 5 ft. 7 in., and the gauge 4 ft. 7 in.

The cylinder dimensions of the 5 BHP. engine are $4\frac{1}{2}$ in. bore and 6 in. stroke, and when the car is running at a speed of 20 miles per hour with the high speed clutch engaged the engine runs at about 760 revolutions

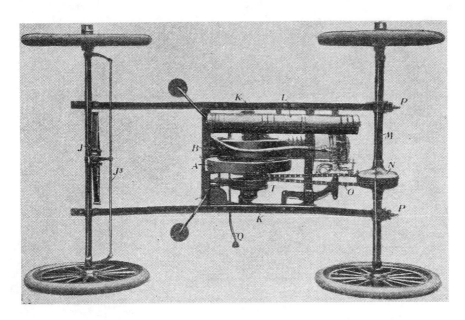

FIG. 273.—PLAN VIEW FROM UNDERSIDE OF 5 BHP. OLDSMOBILE WITH BODY REMOVED.

per minute. Its normal speed is stated to be 650 revolutions per minute, and at this speed the cylinder capacity is sufficient for the 5 horse-power developed.

The silencer L is unusually large considered with reference to the volume of the cylinder exhausting into it, and it is also effective. The use of a little more sheet steel does not add much to the weight of the car, and may be considered as justifiable weight expenditure when it results, as it has with the Oldsmobile, in very silent running of a car propelled by an oil engine. The connexion from exhaust bend to silencer by pipe D^2 may be seen in Fig. 272.

The back part or combustion head end of the cylinder is shown by Figs. 274 to 276, and from these illustrations it will be evident that the valve chambers are cast with and form part of the cylinder cover. From

Ample exhaust box capacity

the sectional views the extent to which this valve box is water-jacketed may be understood, but it should perhaps be mentioned that at the joint between cylinder and cylinder cover similar water ways face or match each other and permit a free flow of water between cylinder, barrel and valve box jackets. These passages and the barrel jacket which is cast with the cylinder have been omitted from Figs. 275 and 276, but their form does not need description, the external diameter of the barrel jacket being nearly equal to that of the cylinder flange.

That portion of the cylinder next to the water-jacketed portion B[1], Fig. 272, it will be noticed, is provided with radiating gills or flanges. Some

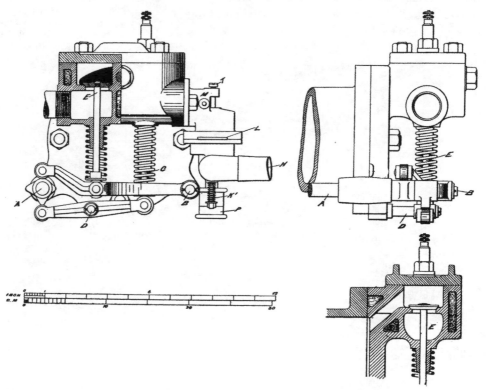

FIGS. 274, 275 AND 276.—5 BHP. OLDSMOBILE. ARRANGEMENT OF CYLINDER COVER, VALVES AND OPERATING GEAR AND CARBURETTOR.

of these are of questionable utility though possibly the first two, or perhaps three, may be sufficiently active to slightly supplement the cooling by water jacket. It may be seen by close observation that this effect has been considered, for the depth of the radiating gills decreases from a maximum at B[1] to a mere ridge at the junction of the barrel with crank chamber. The cylinder and part of the crank chamber are cast together.

Valves and valve operating gear

The arrangement of side cam shaft A, Figs. 274 and 275, skew gear driven from the crankshaft, and the valve lifting levers pivoted at B and D follows the practice of stationary gas and oil engine builders closely, and

274

should prove durable and satisfactory within certain limits of speed, for this more recent application.

Both inlet and exhaust valves are mechanically or positively operated and are of the same form and work in similar chambers, the only noticeable difference being in the length of the valve spindle guides, that on the right or inlet side, Fig. 274, being longer and more likely to remain gas tight, even if wear should take place in a guide so long in relation to spindle diameter. The position of the port connecting cylinder and valve box, shown black in the half sectional end view is more clearly shown in the lower sectional view through the exhaust valve chamber, and may be seen leading diagonally from the space above the exhaust valve down to the cylinder barrel. Removal of the top ribbed cover, held in place by four studs and nuts, permits of the valves being examined or got at for removal

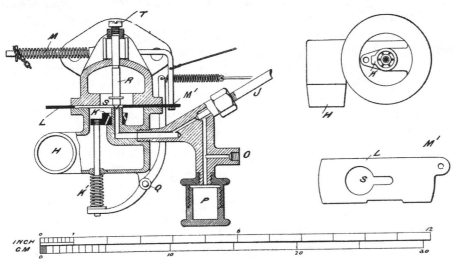

FIG. 277.—5 BHP. OLDSMOBILE. SECTIONAL AND PLAN VIEWS OF CARBURETTOR AND THROTTLE VALVE.

when required. The inlet valve cam is shown in engagement with the roller on the end of its lever pivoted at D, and its other end, also provided with a roller, is in contact with the collar piece on the end of the valve spindle C.

This end of the valve spindle works between the arms of the forked end of the exhaust valve lever which takes a long bearing the width of its fork, upon a long stud pin B, fixed in a lug cast for the purpose with the cylinder cover. The exhaust valve lever may be moved along its pin B, against a spring which normally keeps it away from the cylinder cover face, by means of suitable connexions under the driver's control, and it is normally in the position shown in Fig. 275, where the roller end of the lever arm can be seen above its cam, and vertically above the letter D. In the end view, Fig. 274, a small compression relief cam may be seen opposite the exhaust cam ; it is not opposite, however, on the cam shaft A longitudinally, and it is only when the exhaust lever is moved inwardly towards the cylinder

Relief of Compression

that it makes contact with both the exhaust valve and compression relief cams, the roller on the end of the lever being wide enough to allow of this.

The carburettor and throttle valve are combined and are illustrated by Fig. 277.

Petrol is supplied through a pipe connecting the supply tank and o, and its rate of supply to the inlet passage leading to the mixing chamber is controlled from the driver's seat by an extension of the needle valve spindle J. Connexion of the carburettor to the valve chamber is very direct, and is by the oval flange seen at the back of the sectional view, Fig. 277, and the complete carburettor may be seen in position to the right of Fig. 274. Suction from the engine causes a vertical up and down movement or fluttering of the valve stem R, and the amount of this movement may be adjusted until experimentally found correct by means of the set screw T. In this way the area of the petrol passage or jet may be varied but the rate of supply of petrol is, as already mentioned, controlled by the needle valve J. The form of the throttle plate L is shown by a separate detail, and the only opening remaining when the plate is in its closed position

Carburettor construction and action

is the narrow-necked part to the right of s. Enough air may then pass by the inlet opening H, and through the throttle opening to the mixing chamber to keep the engine running slowly when the car is standing. The means provided for moving the throttle plate hardly needs description. The plate is moved by means of the flexible connexion shown attached near the right angle bend of a rod that is attached to L at M¹. A spring M passing over the long leg of this rod after it has passed through and been guided through bosses in the inlet bend of the carburettor, tends to keep the plate L in its closed position. Movement in the direction of opening of the valve is through flexible connexion referred to, and a pedal to which it is connected. That is the opening of the throttle valve, and, therefore, the speed of the car is controlled by a foot operated throttle valve lever or accelerator pedal.

In cold weather if difficulty is experienced when starting the motor, carburation of the smaller volume of air required may be more certainly effected by means of the cup-shaped piece K surrounding and capable of sliding on the upper part of the petrol inlet pipe within the carburettor. When this piece K is moved upwards from the position shown to about the level of the plate L the opening s is nearly closed or is very much reduced in area, and most of the air entering the mixing chamber of the carburettor passes by way of the converging or concentrating ring of holes in K which may be seen in the plan view of the carburettor with the upper part and throttle plate L removed. As a result of restricted inlet area, a greater difference of pressure or greater suction is created by piston movement, at low engine speed of revolution, than would be the case if the area of air inlet was not reduced, and the supply of air issuing at high velocity from the holes in K impinges upon and intimately mixes with the petrol spraying from the jet, under suction influence. Movement is given to K when required through the vertical stem connected to it at one end, and in contact with the lower end of the curved lever pivoted at Q. Normally K is held in its lower position by the spring K¹, and the necessary pull at the upper end of the lever Q is obtained through a flexible connexion, and the spring there shown from a convenient position for control. The use of the spring pull makes it almost impossible for a careless driver to damage either K or L by exerting undue force, and while its application for this particular purpose is not perhaps of great importance, the provision of parts protected by such a method of operation is sometimes of the highest importance. As an instance of this, the method of operation of the 1903 Argyll light car change speed gear, described at pages 54 to 56, may be referred to.

The pocket P is for the collection and subsequent removal from time to time of any water, dirt, or impurities, of greater specific gravity than, and which may occasionally be found in, petrol.

The contact breaker used in connexion with the accumulator, induction coil, and sparking plug high tension system of ignition employed with these cars is illustrated by Fig. 278. Part of the crank chamber side wall T

Simple con- tact make and break parts

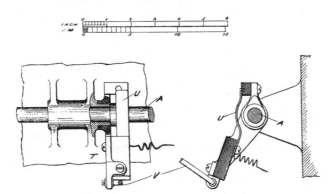

FIG. 278.—5 BHP. OLDSMOBILE. CONTACT MAKE AND BREAK.

is shown with a bearing cast with it in which the half speed, or lay shaft A works. Reference has been made to this shaft, and its continuation toward the cylinder end on the right is shown by Fig. 275, where its second bearing cast with the cylinder cover and the cams on its extreme end may be seen. One of the pair of skew wheels by which A is driven from the engine crankshaft is carried on that portion of A which projects through the bearing shown in Fig. 278 to the left. A light cast brass frame fits over a sleeve projection of this bearing on the right, and the frame may be rocked about this sleeve for purposes of advancing or retarding the period of electrical contact, and, therefore, time of ignition. This rocking movement is effected through the rod V which takes hold of an eye at the lower end as shown. Two ebonite or vulcanite blocks are fixed to the brass frame and to the lower of these is screwed the flat steel spring U, the same screw that holds U also serving as the point of attachment of the insulated wire connecting the primary winding of the induction coil to the contact breaker spring U. Normally the free end of U rests with spring contact upon the upper vulcanite block, but at each revolution of the side or half speed shaft A the contact cam shown upon it makes a wiping or sliding contact and completes the low tension electrical circuit. Except when contact is thus made by the cam, the spring U and the low tension wire connected to it are completely insulated, having contact only with the vulcanite blocks. This forms a most simple form of contact breaker, and one that is most unlikely to be a source of trouble by failure. The only improvement that might be easily carried out would be the addition of a contact so arranged that the movement of U by the contact cam would complete **Oil film resistance to low tension current** a direct return circuit, or direct connexion to engine frame, so that the low tension current need not encounter the high resistance of the oil films at bearing surfaces. Such a resistance as this may have a very marked effect on the character of the spark formed at the sparking plug, and hence on the satisfactory and economical running of the engine.

In Fig. 278 the contact cam has been shown of insufficient diameter at its circular part. It should be at least a quarter of an inch larger in diameter to provide a reasonable collar face to prevent the contact breaker frame from moving longitudinally off the bearing sleeve upon which it rocks.

No reference will be made to the speed change gear other than to state that it is of the planetary form, and involves the use of 11 wheels in two epicyclic trains. Road speeds may be obtained of 7·5 miles per hour for forward running, or about 6·6 miles per hour when the reversing train is in operation. When the high speed clutch is engaged the gear is locked and runs at the same speed as the crankshaft upon which it is carried, and, as already stated, the gear reduction at the driving chain wheels is such as gives a car speed of 20 miles an hour when the engine is accelerated to 760 revolutions per minute.

A change speed gear of somewhat similar form, that used with the Cadillac car, will be described, the number of speeds obtainable are the same, and the method of control almost identical. The use of a single lever with the Oldsmobile gear as compared with a lever and pedal with the Cadillac does not necessarily mean that the control of the former is easier.

An Oldsmobile car of 4 HP., as early as September, 1902, was entered for the reliability trials of the Automobile Club of Great Britain and Ireland, and was the only American petrol-engined car that took part in the trials. The performance of the 5 HP. and 6 HP. cars which took part in last September (1903), trials was such as led to the award of the Gold and Silver Medals in their class, the only awards made. The 6 HP. car which won the silver medal differed but slightly from the 5 HP. The engine ran at the same speed, and had a stroke of 6·25 in., and a cylinder diameter of 4·625 in.

Since 1902 these cars have gained largely in popularity, and have undergone some improvement in design, and, more particularly, improvement in the strength of some parts, and of the materials employed for them. The 5 HP. Oldsmobile weighs only about 7 cwt., and was one of the first light cars produced which met the requirements of a very large **Weight** number who wanted a car that could be easily handled, would go anywhere where a car may be expected to go, and would not be expensive in upkeep, or in first cost.

A larger car of 9 HP.[1] has recently been introduced which may be fitted with a removable tonneau. The external appearance of the car has been altered, and the characteristic curved dash by which the smaller cars were readily recognizable has been replaced by a bonnet with radiator water-cooler, petrol and water tanks in front, and the usual dashboard. The same form of long leaf springs is retained, but the arrangement of the front transverse spring is better, and consists of two semi-elliptic springs arranged back to back and shackled at their outer ends to the car frame and front axle respectively.

The general arrangement of the engine and transmission gear has undergone slight modification, and while the wheel gauge is still 4 ft. 7 in. the wheel base has been altered to 6 ft. 10 in.

The normal engine speed is 700 revs. per minute, the cylinder bore 5·69 in., and the stroke, 6 in.

As a result of the larger overall dimensions and heavier parts necessitated by the increased seating capacity and power the 9 HP. car weighs 17 cwt., which shows that from the error of too little they have now gone to the error of too much.

Radius rods are now fitted to the rear axle below the leaf springs, and by adjustment of their length provide also for adjustment of the driving chain.

The springs are free to slide in their brackets on the back axle, and are relieved from the driving effort.

This modification of the very simple arrangement employed in the light car is illustrative of the change of method rather than of dimensions which sometimes becomes desirable when a change in size or power of machine parts is contemplated. One of these 9 HP. cars with a cylinder described as of 5·5 in. diameter, and 6 in. stroke was awarded a bronze medal in the recent small car trials from Hereford, and one of the small Runabout cars, provided with a larger cylinder (5 in. × 6 in.) than usual, and of 7 BHP., began the trials, but through a mistake in coupling up the pump and jacket and cooler connexions had to retire from one of the trial runs, which put it out of competition.

THE CADILLAC CAR.

As already stated the Cadillac car is one of those whose arrangement may generally be said to closely follow that indicated in the typical block plan, Fig. 270. A full description is not necessary, but some of the details peculiar to the car are shown in Figs. 280 to 284, and an external view of one of the more recent types is shown by Fig. 279. The departures from the early design are not many, and perhaps the most noticeable is the

[1] *Automotor Journal*, August 13 and 20, 1904; *Horseless Age*, Sept. 7, 1904.

change in the formation of the front part of the car. The curved front or dashboard has given place to the flat dash and the more customary bonnet. The latter does not serve the usual purpose of enclosing and protecting the motor, but is used to contain the cooling water and petrol tanks, and provides some room for the stowing of any necessary tools and parts or accessories. A pressed steel frame is now employed instead of the L section previously in use, and some advantage obtained from a better distribution of metal, in side frame members of channel section, by variation in the depth from centre to ends, to suit the stresses that occur in such a structure as a car frame. The manufacture of such frames, or the use of them, has only become commercially possible by reason of the large demand that has been encouraged, and now exists, for them.

Convertible bodies
Fig. 279 shows a car with a " Surrey " body, but a change may be readily

FIG. 279.—6½ BHP. CADILLAC CAR WITH " SURREY " BODY AND PRESSED STEEL FRAME.

made to other forms if preferred. The back part is detachable, and either a tonneau may be substituted or a flat sloping back piece used, leaving the front seat available for two persons. Prior to the introduction of this recent design, semi-elliptic springs of peculiar form were used for front and back axles. These have been retained for the back axle, but a transverse spring is now used to carry the load on the front axle, and radius rods are used to determine its position, and to receive shocks conveyed through the road wheels, tending to longitudinal displacement of the front axle. The engine is carried upon two transverse frame members connecting the side pressed steel frames, and its position in the car may be judged from the fly-wheel which is visible in Fig. 279, the cylinder end being towards the rear of the car.

The side elevation of the engine, Fig. 280, enables the general construction to be understood. It rests on and is bolted to the frame cross members by bolts passing through them, and the bracket feet L L¹ cast with the crank chamber. The crankshaft is sufficiently long to take the fly-wheel on one side, and the planetary speed gear drum on the other side of the crank chamber, and the extreme end of the crankshaft at the speed gear side drives the jacket water circulating pump, which is independently fixed to the car frame, by driving pins which are the equivalent of a flexible coupling. These driving pins are fixed in a nut, which is screwed on the end of the crankshaft for the purpose of adjusting the high-speed gear clutch, and may be seen in Fig. 284, lettered L.

Engine construction

The separate semi-sectional view of the engine cylinder, Fig. 281, shows the method of connecting the cylinder barrel and valve head castings and the thin sheet copper water-jacket C to each other, and to the crank chamber. The water-jacket C has its end flanged, and a separate ring M,

Thin Copper Jacket

FIG. 280.—6½ BHP. CADILLAC CAR. SIDE ELEVATION OF ENGINE AND CHANGE SPEED GEAR.

like the loose flange on a pipe, serves the double purpose of holding the water-jacket C against the flange on the cylinder barrel, and both these parts up to the machined face of the crank chamber, by means of the long studs and nuts shown. In some engines the Cadillac Co. are now casting the jacket with the cylinder in the usual way. It will be seen that the cylinder barrel has a spigot fit at its point of connexion with the crank chamber which with good machining ensures these parts taking up correct relative positions in one direction. The attachment of the valve head casting to the cylinder is by means of the right and left hand threaded nipple which is formed internally of hexagonal shape in order that a suitably-shaped key may be inserted through the hole occupied by the sparking plug fitting. An end view of this nipple is given in the sectional view across the valve chamber, Fig. 283. The nipple is first screwed into the cylinder, the valve head then screwed into place with the copper jacket between its joint face and that of the cylinder, and the final tightening up effected by rotation of the nipple which holds the three pieces firmly together. A dowell or steady pin in the cylinder, and corresponding holes in the copper water-jacket flange and valve head casting determines the correct position of the latter when the threaded nipple is being tightened up.

The drain cock o allows any excess oil which may have collected in the cylinder as a result of over lubrication, to be drained off, and by means of the cock N the water may be run out of the cylinder jacket.

Both inlet and exhaust valves are positively worked, the former being in the upper, and the latter in the lower part of the valve chamber. Quite at the back end of the valve chamber is a machined recess into which

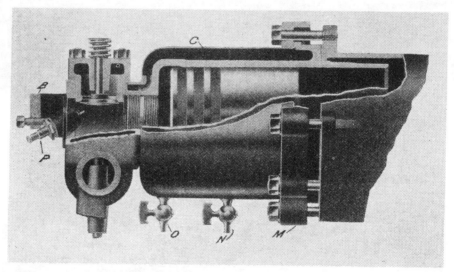

FIG. 281.—6¼ BHP. CADILLAC. PART SECTION OF ENGINE CYLINDER AND VALVE CHAMBER.

fits the disc part of the special sparking plug fitting used with these cars, and which has a metal to metal joint, the fitting being held in place by a strap and set screw, in the same way that the covers are held in place over the valve chambers with some engines.

Special form of sparking plugs   Special sparking plugs of simple and substantial construction are used and two are required, the insulated portion of each consisting of a plain central wire conductor passing through the mica insulation. The two plugs, when in place, stand at an angle to each other, and with the disc plate into which they are screwed, and their points approach, and by bending them the necessary or correct spark gap may be obtained. Apart from any longer life these plugs may have due to their strong construction, each pair is also the equivalent of two sparking plugs, but the lead to and return wires from the plugs, both being insulated, require the use of two conductors between coil and sparking plug. In the event of insulation failure at whichever may happen to be the high tension side of the sparking plug, all that need be done is to change over the connexions to the plugs, that plug which was faulty for the high, being, of course, serviceable for the low tension terminal for the return conductor. An adapter disc capable of receiving any ordinary standard sparking plug may be used by those who either have no spare special plugs, or prefer the more usual form, and recognize the fact that sparking plug troubles are not always due to insulation failure.

Fig. 282 illustrates the means provided for removing either of the half bearings without disturbing the crankshaft when renewal or adjustment is required, or, when an overhaul is desirable, how the crankshaft may readily be lifted out. In the illustration the cast iron cover plate and bearing cap has been removed with the upper half bearing, and the lower half has

FIG. 282.—6½ BHP. CADILLAC. DETAIL OF CRANKSHAFT BEARING.

been partly turned out, and shows how it can be got out when necessary. The holes in the cover plate, it will be noticed, are slotted, and this allows the plate to take up whatever position is required after adjustment of the bearings.

The lock-nutted setscrew shown in the upper crank chamber cover takes any upward thrust from the crankshaft bearing cap, the screws which hold the cover plate in position being only called upon to make the joint with the crank chamber. The crankshaft is 1 75 in. diam. at the main journals.

The cylinder diameter and stroke of the engine are both 5 in., and the engine is usually described as of 6 5 BHP. when running at 750 revolutions per minute.

Cylinder dimensions

A car similar to that shown by Fig. 279 was entered for the Automobile Club's small car trials from Hereford, and described as of $9\frac{1}{2}$ HP. The engine dimensions have not been increased, and it must be concluded, therefore, that the cylinder capacity in the engine of the earlier car was very liberal for the normal $6\frac{1}{2}$ HP., at which it was rated, or that in the Hereford trials it was intended to run at a much higher speed. It had however, the misfortune to break its driving chain, and was thus forced to retire.

With a gear ratio between the chain wheels of 3·75 to 1 at this engine speed the car will have a road speed of 16·75 miles per hour. Accelerated to 900 revolutions per minute at the engine the car would then be travelling at 20 miles per hour. Lubrication of piston, connecting rod, and other parts inside the crank chamber is by oil supply from the glass drip lubricator seen entering the top of the crank chamber, and by splash distribution assisted by oil channels and collector ring at one side of the crank cheeks.

The main crankshaft bearings are further lubricated by Stauffer grease cups and the usual semi-solid lubricant.

The circular pocket cast in the crank chamber, and inside which the larger of the timing gear wheels runs, sufficiently indicates the position of the cam shaft.

Design of valve gear

There is only one cam upon it, and that gives motion to the pusher rod P, which at its other end is in simple contact with one arm of the bell crank lever K through which the exhaust valve receives its movement. The compression of the exhaust valve spring keeps the pusher rod in thrust between the bell crank lever arm K and the roller jaw piece that is always in contact with the cam. P is a plain length of rod without any connexions or fastenings upon it, and may be pulled out or pushed into place very readily when the engine is being examined or assembled.

Outside the crank chamber, at one end of the cam shaft, is the contact breaker, which is like that illustrated in connexion with the Oldsmobile. At the other end, and visible in Fig. 280, is an eccentric and cover strap D^2, which gives motion through the rod D, and in a way about to be described to the inlet valve.

Referring to Fig. 283 an arm G^2 is carried upon the outer pin-formed end of M, which is rectangular in section, and is pivoted at its inner end at the position of the flat-headed screw N. A small movement of M about its pivot may occur, which movement is dictated by the position which G^2 is given by a rod connected to its small eyed outer end, the rod and position of G^2 being controlled by the driver of the car. An eccentric or cam path G is formed upon the big end of G^2 of a length that will permit a movement about the pin-formed end of M, through an arc of nearly 50°. A stop-pin H and the form of the ends of the path G, limit the extent of this travel. Behind the end of G^2, and upon the same pin is a roller G^1, between which and a roller F^1 carried at one end of the lever F pivoted as shown, the end D^1 of the eccentric rod D (see Fig. 280) slides.

A separate detail is given in Fig. 283 of the formation of the underside of the end D^1 of D. As D^1 slides to and fro between the rollers G^1 and F^1,

284

receiving its motion from the eccentric D², at a point in its travel depending upon the position of the roller G¹, it will ride up on the latter as soon as the thickened-up portion of D¹ arrives between the rollers. This repeated wedging apart action, occurring once every two revolutions of the engine gives motion to the rocking lever F, and through its suitably curved contact end, with the collar head of the inlet valve stem, gives movement to the inlet valve I. The moment of opening and closing of I, and the duration of the period of opening, and its amount are controlled by the position of the roller G¹. With G² in the extreme position shown, the stop-pin H not allowing of any further movement, G¹ has been pushed as far to the right as possible, and D¹ will ride up more on G¹, and give earlier and greater and longer maintained movement to the roller end of lever F, than when G² is in its other extreme position, and the end of the path G of least eccen-

Variable inlet valve opening

tricity in contact with H. G² is shown in this second position, when least movement would be given to the inlet valve, in Fig. 280. The small spring Z seen in the plan view keeps M moved rearwardly, and G² therefore in contact with H. The bracket casting O, which, in one piece, forms part of the carburettor, the inlet valve seat and inlet valve stem guide, and provides points of attachment for the inlet valve mechanism just described, is clearly shown in the drawing, and the attachment to the valve chamber casting by four studs seen in the plan view is clear.

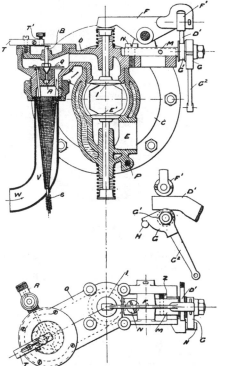

From what has been said it will be evident that control of the Cadillac engine is effected by varying the quantity of mixture admitted to the engine by change of the period and opening of the inlet valve, instead of governing in the usual way by throttle valve, and a constant valve opening.

The effect of this method of engine control upon the carburettor

Method of control

FIG. 283.—6½ BHP. CADILLAC CAR. ARRANGEMENT OF CARBURETTOR, VALVES AND OPERATING MECHANISM.

is peculiar. When the engine is running at its normal full load and speed the quantity of mixture and the suction inducing its flow to the cylinder are at a maximum, but as the power developed by the engine is decreased, the period, and not the fierceness, of suction is altered, resulting, therefore, in a series of sucks at the carburettor of about the same intensity, but getting shorter and shorter in duration with reduction in the load on the engine. This effect, compared with the constant period of suction with varying intensity obtained with the throttle valve, should result in a supply from the carburettor of more uniformly carburetted air, and is cheaply and ingeniously obtained, but in common with a few others who gain a like result in a less simple manner, these efforts to obtain a variable supply of a constant quality of mixture may ultimately prove to be misdirected.

Carburettor   Before leaving Fig. 283 some description may be given of the carburettor. Petrol is supplied at R (see plan view) through union and cock to the jet passage lettered R in the sectional view. A valve and deflector plate Q rest immediately on top of the jet passage, and normally close it, and the amount of lift that is permitted by suction effort is limited or adjusted by means of the setscrew T¹ and flat spring, the end of which is just above the valve spindle B. When starting the engine, flooding of the carburettor may be caused by lifting this valve and plate Q through the rod S, which headed at its upper end passes loosely downwards and through

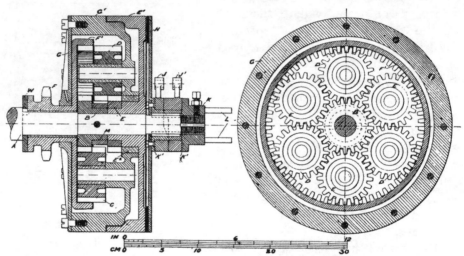

FIG. 284.—6½ BHP. CADILLAC. CHANGE SPEED GEAR.

a hole in the induction pipe at S where a spring normally keeps it down. The rod may be pushed up against this spring by finger pressure when it is found desirable to flood the carburettor and drench the gauze cone V through which all the air entering the carburettor passes. This gauze cone V is an interesting feature of the Cadillac carburettor and probably largely adds to its efficiency. When the demands upon the carburettor begin to increase some of the petrol flowing or spraying from the jet doubtless finds its way down to the cone, and is there greedily taken up by the air streaming through it. In the more recent cars the inlet to the induction pipe is not at W for the pipe is continued and upwardly curved, and the inlet is protected by a gauze covering from dust which would otherwise be carried in with the air, and be deposited upon and choke the cone V.

Fig. 284 illustrates the epicyclic change speed gear of more ingenious construction than usual. A is the engine crankshaft which at the part

286

occupied by the driving chain sprocket F is abruptly reduced in diameter and against the shoulder so formed is a hard steel collar W, caused to rotate, by the pin or key shown, with the crankshaft. The sprocket F is in one piece with the internally toothed ring F¹, and there is, as indicated, a bronze or gunmetal bearing bush at the position of the sleeve portion or bearing on the crankshaft. A little further along the shaft is again reduced in diameter, and to the first part of this reduced portion is pinned the toothed wheel B. B being fixed to the crankshaft is active when either the low speed or reverse gears are in operation, and always experiences the driving effort. The pin securing this wheel passes through its sleeve continuation, and is prevented from working out of position by the loose ring covering it. A wheel E having the same number of teeth as B is keyed upon the internal boss of the annularly formed casting E¹, which has a long bearing upon and is a running fit on the shaft. Upon a bush on the inner part of the internal boss of E¹ the deep casting G¹ is supported, and in bosses cast on the disc part of G¹ are 6 pins upon which planetary pinions are carried. The first, third and fifth of these pinions, lettered C in the cross section and E in the end view, gear with the centre or sun wheel B, and with the internally toothed ring F¹, while the second, fourth and sixth, D, gear with the other three planetary pinions, and with the centre or sun wheel E. A dust cover, or oil retaining plate G, necessarily in halves, is screwed to G¹, and completes the total enclosing of the gear. The outer face of E¹ forms part of the high speed clutch, the other part consisting of a plate H, which is leather faced where shown sectioned black, and is fixed to its centre sleeve boss K¹. K¹ is free to slide longitudinally on feathers sunk in the shaft and it, and therefore H to which it is connected, are driven by the shaft. J, J¹ are lever arms on ring pieces with cam formed edges. When one of these J is partly rotated and moved relatively to J¹ they separate by reason of the wedging apart action of the cam teeth partly shown in dotted lines, and end thrust occurs between the inner collar faces of the nut K and the boss K¹, and by movement of K¹ and H the clutch faces engage by face frictional contact. A rod connexion from the frame of the car to J¹ prevents its rotation, and similarly a connexion to a hand lever under the control of the driver enables J to be partially rotated in either direction to engage or disengage the high speed clutch.

During the greater part of the running of the car this high speed clutch is in engagement, and to reduce the frictional loss due to the end thrust which this entails, ball bearing thrust rings are used between the faces at the points lettered K¹ K¹. For the sake of clearness these have been omitted from the drawing.

Adjustment of the high speed clutch is simply carried out by means of the nut K. There are four longitudinal grooves at the screwed end of the shaft, and the adjustment having been made by screwing this nut more on or off the shaft by quarter turns, it may be there held by means of the setscrew shown, the point of which enters whichever of the four grooves it happens to have been brought opposite to. The gear is lubricated occa-

Construction of change-speed gear

287

sionally through the one oil hole provided in G, and shown in the left hand section where the boss of G rides on the sleeve of the sprocket wheel F.

Operation of change speed gear

Surrounding G[1] and E[1] are brake straps, that around the former controlling the reversing gear by means of connexion to the hand lever already referred to when describing the high speed clutch, and that around E[1] bringing the low speed gear into operation, when it is tightened, through connexion to a foot pedal.

When G[1] is held by its brake strap the gear reduction is that due to the proportion between the pitch diameters of B and F[1], the planetary pinions having only rotational movement on their pins, and the useful or active pinions being those lettered E, pinions D, and sun wheel E idling round at the same rates of revolution because they have the same number of teeth. The numbers of teeth are as shown, 63 in the internally toothed ring, and 21 in any of the others. The gear ratio for the reversing gear is, therefore, that due to the proportion between 63 and 21 or 3 to 1.

Gear Ratio

When G[1] is allowed to go free, and E[1] is prevented from rotating by the tightening of its brake strap the centre wheel E is also held, and any motion given to the pinions by rotation of the centre wheel B results in the movement of G[1] and F[1], and with it F around the shaft centre in the same direction as B. In a gear of this form in which the relation between the diameter of the pinions E and centre wheel B is the same as that between the pinions D and centre wheel E, the resulting gear ratio may be very simply expressed by the proportion between the diameters of the driver B and the driven internally toothed ring F[1]. This, as already expressed for the reversing gear, is 3 to 1, but by the introduction of the pinions D and direction dictating wheel E relative motion in agreement with instead of opposite to that of the crankshaft is obtained.

The weight of the car is 1,250 lbs., or considerably more than others of equal power and capacity. The 9½-HP. car of 1904 weighs about 1,300 lbs. Most of the bearing surfaces have been given ample area, and some of the moving parts are of such proportions that little or very slow wear will occur. The behaviour of one of these cars[1] in the 1903, 1,000 miles Reliability Trials, held by the Automobile Club of Great Britain and Ireland, was very good, and taking the Cadillac as one of the heavier examples of its type it appears that considerations of reliability and economy in running have largely controlled the design.

A 30-HP Cadillac[2] car is now made with a four-cylinder vertical engine. A planetary change-speed gear is used and the drive to the road wheels is through a propeller shaft and bevel gear.

The similarity of design of many of these light American cars, already mentioned, makes it unnecessary to do more than provide the two examples just dealt with.

[1] Parts of the car not here illustrated or discussed may be found in a description which appeared in the *Automotor Journal* of October 17, 24, and 31, and November 7 1903.

[2] *The Horseless Age*, May 24, 1905, p. 582.

AMERICAN PETROL MOTOR VEHICLES

The Oldsmobile and Cadillac have been chosen as more representative than the others, because they have been most used and seen in this country.

Among those other cars already mentioned the Pope Hartford,[1] has a single cylinder horizontal engine declared to give 10 HP. at 900 revolutions per minute. The cylinder bore is 5·25 in., and the stroke 6 in. The inlet valve is not of the mechanically operated type. The only brake provided works on both sides of the differential gear on the live axle, and is operated by a foot pedal through the usual equalizing connexions. The clutch or brake strap encircling the reversing drum of the change speed gear is depended upon as an emergency brake. The weight of the car fitted with detachable tonneau body is 1,600 lbs.

The Elmore [2] car has a two-cycle single cylinder horizontal engine, 4·5 in. bore, and 4 in. stroke, which is stated to give 8 HP. at 800 revolutions per minute. No valves are used in connexion with the engine.

The carburettor is combined with the throttle valve, and is of the float feed spraying type, with an adjustable spring-controlled supplementary air inlet. A screw-down needle valve, which may be partially rotated from the driver's seat, enables the rate of flow of petrol through the jet to be varied to suit the varying atmospheric conditions. These provisions for regulating the quality of the explosive mixture are of greater importance with the two-stroke valveless engine than with the usual four-stroke type. The weight of the car with detachable tonneau body is 1,200 lbs.

The Ford [3] car has a double cylinder horizontal engine, the cylinders are opposite each other, their diameter being 4¼ in., and the stroke of the engine 4 in. The power developed is said to be 9 BHP. at 900 revolutions per minute.

The change speed gear is, as with the other cars mentioned, of the planetary or epicyclic form, but it differs from the others in having no internally toothed wheels, and it will be remembered that the Oldsmobile has two such wheels in its change speed gear, and the Cadillac one. Full elliptic leaf springs are used between the road wheels and the body, but there are no radius bars or reaches to determine the position of the front axle. In addition to the elliptic springs at the rear axle, there is a transverse semi-elliptic spring which shares the load. The pivot connexion of these elliptic springs to the frame ensures their freedom from distortion when adjustment of the driving chain, by shortening or lengthening the radius rods, becomes necessary.

Planetary gear without internal teeth

The new Ford car which appeared during the autumn of 1904 departs in almost every essential part from the earlier design. It has a vertical four-cylinder engine occupying the now usual position and driving through planetary gear of the type already referred to, and by a propeller shaft the bevel driving gear on the rear live axle.

The Rambler [4] car with its detachable tonneau body weighs 1,650 lbs.,

---

[1] *Horseless Age*, April 27, 1904.   [2] *Horseless Age*, May 18, 1904.
[3] *Automotor Journal*, June 11 and 18, 1904.
[4] *Horseless Age*, April 20, September 14, 1904.

and has an engine rated at 16 HP., there being two horizontal opposed cylinders, 5 in. bore and 6 in. stroke. A single cylinder engine of the same primary dimensions and design is used for lighter cars. The arrangement of the valves and operating mechanism is similar to that in the Oldsmobile.

The method of relieving the compression to facilitate starting the engine consists in the use of a small cam fixed in such a position on the cam shaft that when the inlet valve lever is moved sideways from its normal position, the inlet valve is kept open during part of the compression stroke. The same method is employed with the Oldsmobile, but side movement of the exhaust valve lever results in the holding open of the exhaust instead of the inlet valve.

Thermo-syphon action is depended upon for the jacket water circulation, and the water capacity of the system is 6 gallons.

A cone clutch at the fly-wheel is used with the change speed gear instead of the face or internally expanding clutch employed with the cars already referred to. The objectionable end thrust which is present when the cars are running with the high speed clutch engaged, is absent, or only results in a balanced or self-contained effort in the direction of the crank-shaft axis, when the arrangement adopted by the manufacturers of the Rambler is employed. A further interesting feature is the automatic control by governor of the point of ignition.

Ignition timing by Governor

The brake equipment is good, there being in addition to a brake drum on the change speed gear, internally expanding brakes inside drum extensions of the road wheel hubs. The reversing gear could also be employed as an additional brake, but those provided and described as brakes constitute a preferable and safer arrangement.

Chapter XIX

SOME AMERICAN PETROL CARS (*Continued*)

THE PIERCE ARROW CAR.

AMONG those American manufacturers who have wisely profited by the experience of Continental and British car builders and designers are the George N. Pierce Co., of Buffalo, N.Y. In the design of their light touring car [1] they have embodied much with which we are familiar in this country, and the general external appearance as

FIG. 285.—THE PIERCE ARROW LIGHT CAR.

shown by Fig. 285 is more pleasing than, and is at the same time a departure from, that which has largely become standard American practice, namely the " Runabout " type with detachable tonneau or ready means of adding to or reducing the seating capacity of the car.

As a guide to the scale of the elevation and plan view drawings, shown by Figs. 286 and 287, it may be mentioned that the wheel base of the car is 6 ft. 9 in., and the wheel gauge 4 ft. 6 in.

A tubular under frame is used, the two long side members, the four

[1] *The Horseless Age*, July 1, 1903, p. 10.

suitably spaced cross connecting tubes, and the lugs for attachment being brazed together. To the first and second of the cross tubes are connected longitudinal steel angles, which carry the engine and form a frame for its support, and that of the change speed gear box Q. The ends of the third and fourth cross tubes or their extensions serve as points of attachment for the rear semi-elliptic springs or the hangers to which they are connected.

Attachment of rear springs

The plan view shows that the front springs are directly below the frame tubes, while those at the back axle are spaced rather wider or farther apart than the frame tubes, and so leave less length of axle between the road wheel centres and the points of application of the load at the spring seats. Truss rods or bars H stiffen the side frame tubes at their centre part between the springs. The engine employed is a two-cylinder 15 BHP. De Dion of the design shown by Figs. 95 to 98, and described in connexion with the De Dion Bouton 10 HP. light car. It will be remembered that among the distinctive features of this engine were the very complete system of gravity lubrication and distribution of oil to the main bearings by holes or passages through the crankshaft, and the method of control of the engine by varying the time amount and duration of lift of the exhaust valves. In the plan view, Fig. 287, P is the oil tank close up to the cylinder heads from which the oil flows by gravity to the main bearings. The oil is pumped by a worm and worm wheel driven gear pump from a well in the base of the crank chamber and returned to the tank P.

De Dion gravity lubrication with circulating pump

An abundant supply of oil to the bearings may be maintained by this system without the excessive lubrication of the cylinders which would result from too complete a use of the splash lubrication system. The positions of the carburettors are indicated at N N, and no description of the carburettor is here necessary since reference may be made to that described at page 24. A gilled tubular radiator is hung from the forward ends of the frame tubes, and may be seen in Fig. 286, while the flexible spring-driven water circulating pump may be seen in front of the engine in the plan view. The engine and change speed gear box are close together on the angle steel frame, and little relative movement between them is likely to take place as the result of springing or twisting movements of the main frame.

The driving effort from the cone clutch is received by the first gear transmission spindle E, Fig. 288, through the claw clutch shown. The clutch piece F fixed to E has two driving arms or fingers which may slide in slots in the extension boss of the centre part A of the cone clutch. Movement of the clutch towards the gear box when declutching does not affect F in any way, the length of the slots or the clearance being such as will allow the necessary movement to take place.

Adjustment of balanced clutch

As will be apparent from the arrangement of extension spindle G, clutch spring C, and ball thrust bearing D, the design of clutch is such that the spring pressure required to keep the clutch in driving engagement is self-contained, or does not result in continuous end pressure on any bearing. It may also be pointed out that, as is frequently the case with this arrange-

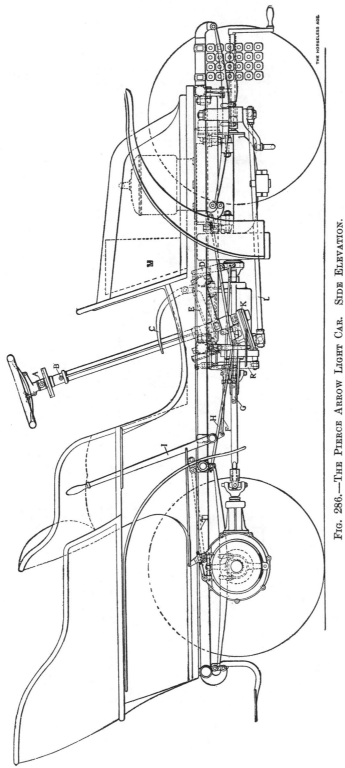

THE HORSELESS AGE.

FIG. 286.—THE PIERCE ARROW LIGHT CAR. SIDE ELEVATION.

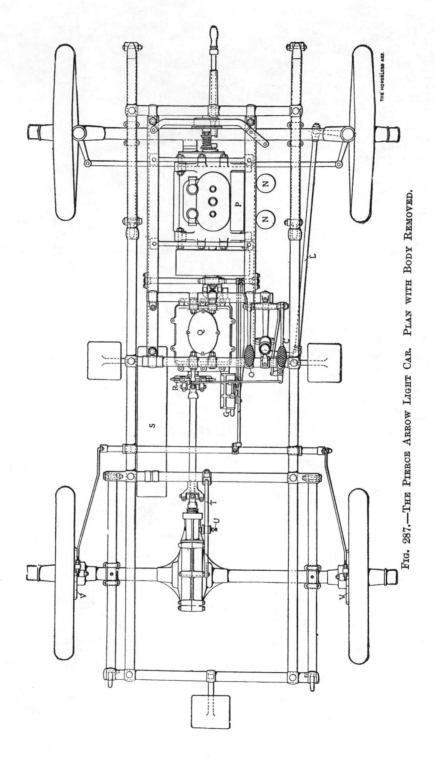

FIG. 287.—THE PIERCE ARROW LIGHT CAR. PLAN WITH BODY REMOVED.

294

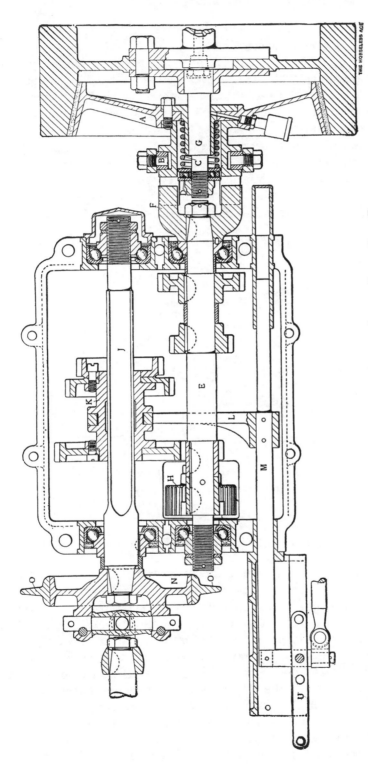

FIG. 288.—THE PIERCE ARROW LIGHT CAR. CLUTCH, CHANGE SPEED GEAR BOX, ETC.

295

ment, the threaded end of G, and the nut upon it, for adjustment of the spring pressure, is very inaccessible.

The arrangement of parts in the gear box is simple and may be readily understood from Fig. 288. On the squared spindle J is the sliding collar-piece K, with which the gear shifting fork L, carried on the spindle M, engages. The method of securing the three gear wheels upon K by screws without any lock nuts or pins is open to criticism. It has this advantage, however, that renewable or removable wheels may be used of smaller diameter than would be possible if lock nuts were employed, and provided that a sufficient number of thoroughly well fitted screws are used, and their projecting ends headed over the arrangement would be satisfactory in use, although the ultimate renewal of wheels would not be a simple matter.

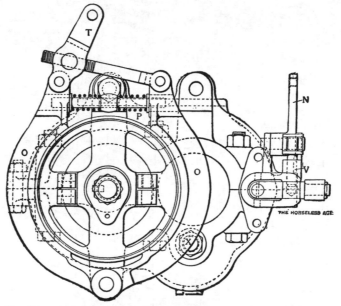

FIG. 289.—THE PIERCE ARROW LIGHT CAR. END VIEW OF CHANGE SPEED GEAR BOX.

In the position shown the wheels are in gear, for the low speed forward, a movement to the left brings the reversing wheels into gear while successive movements to the right put the second and third speeds forward into position. The wide reversing pinion H is carried on a short spindle below E, and is always in gear with its pinion on the extreme left of the spindle E, and therefore, runs idly when the reversing gear is not in action. The transmission shafts E and J both run in ball bearings. One half of the universal joint for attachment of the driving shaft to the bevel gear on the back live axle is formed on the brake drum N which is secured to the coned end of J by key, nut, and locking ring or plate. The construction of this brake may be understood by reference to the end view of the gear box, Fig. 289. The brake blocks O are pivoted on a fixed stud pin at their lower ends, and are normally kept out of contact with the drum N by the springs P. The connexion of T by the rods and bell crank to the foot operating pedal C may be seen in Figs. 286 and 287. Application of this pedal worked brake puts the operating rod connected to T in thrust.

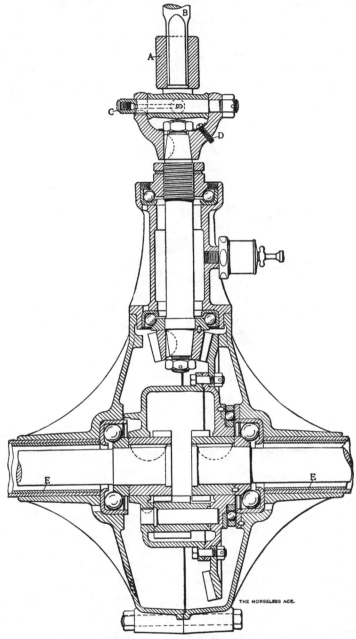

FIG. 290.—THE PIERCE ARROW LIGHT CAR. SECTION THROUGH BEVEL DRIVING GEAR AND DIFFERENTIAL GEAR.

An interlocking device, shown in Figs. 286 to 289, ensures that the cone clutch A is out of engagement when a gear change is about to be made,

Protection of gear by interlocking device

and prevents the return of the clutch until the gear is engaged and the gear wheels in correct position. The part U, Fig. 288, of the continuation piece at the end of M, to which the gear-shifting rod is shown connected and broken off, has holes drilled in it at intervals corresponding to the gear changes. A bell crank lever N, which may be seen in Fig. 289, and is shown, but not lettered, in Fig. 286, has a pin or bolt V in the end of its horizontal arm, which normally occupies one of the four holes mentioned according to which gear is at work. The bell crank lever N is connected by its vertical farm, and a rod to the clutch lever operated by the pedal O. When, therefore, the clutch is disengaged by depressing O, the locking pin V is withdrawn from a hole in U, and the rod M may then be moved in either direction

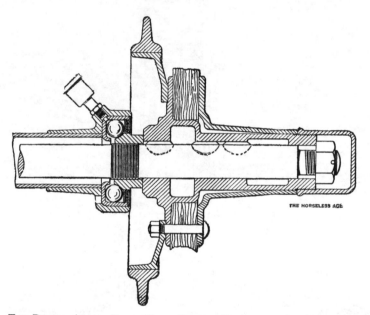

THE HORSELESS AGE

FIG. 291.—THE PIERCE ARROW LIGHT CAR. SECTION THROUGH OUTER END OF LIVE AXLE.

or change of gear. When the necessary gear change has been made and the gear changing lever left at a corresponding notch in its quadrant, the clutch may be engaged, the pin V at the same time being dropped into a hole in U which has been brought opposite to it. If the gear changing lever has not been moved into one of its speed notches in any of which positions the gear wheels are correctly in mesh, then V merely rests on the upper surface of U and prevents the engagement of the clutch. The interlocking device, therefore, prevents or largely reduces the possibility of damage to the gear wheels or their shafts which would otherwise occur as a result of improper use of the clutch pedal and gear changing lever.

Fig. 290 is a section through that part of the live axle occupied by the bevel driving and differential gears, and Fig. 291 a section of one of the outer ends of the axle. The bossed part A, Fig. 290, of one half of the universal joint has a square hole at its centre and receives the end of the driving shaft B, permitting a little sliding in and out, which will occur with movement of the car springs. A screw C covers a hole drilled down the centre of one of the pins of the universal joint through which oil may be fed to lubricate both pins, there being other holes at the centre at right angles, as shown forming passages for distribution to the second divided pin. These holes having some reservoir capacity, only occasional lubrication at this point would be necessary. At D is a locking-plate held in place by the screw shown, to prevent the nut securing the universal joint from working back, and similar to that shown in end view, Fig. 289, for a like purpose for the half universal joint at the gear box. The centre driving shafts are carried in ball bearings, adjustment being provided for at the outer ends where the cones are screwed on to the shafts.

Lubrication of universal joints

End thrust from the bevel gear is taken up by the ball thrust bearing shown, and within limits, adjustment of the centre shaft bearings does not interfere with the thrust bearing.

The bearing piece cast in the differential case immediately behind the bevel wheel at its point of engagement with its pinion supports the wheel laterally against springing resulting from momentary very heavy driving efforts.

The form of live axle construction adopted, involves the use of a screw thread at a critical part of the centre driving shafts, and is open to the criticism made at page 38. The side brake drums are suitably connected to the inner flanges of the road wheel hubs and their design is the same as that shown by Figs. 288 and 289.

The axle tubes, where they enter the bosses of the differential, are strengthened by internal liners or ferrules E.

The construction of the tubular steering column may be followed from Fig. 292, and the means of carrying and bracing it in position on the car frame is well shown by Figs 286 and 287. Referring to Fig. 292 short spindles are fixed in the upper and lower ends of the centre steering tube B, that at the upper end is keyed to the boss of the steering hand wheel A, and that at the lower end has the pinion C fixed upon its coned part by key, nut and split pin. The pinion C gears with the toothed quadrant D which is in turn keyed and pinned to the short spindle E carried in bearings formed in gear casing around C and D. By means of the arm F and connecting rod L (see Figs. 292 and 286) motion is given to the pivoted steering heads in the usual manner. Freedom of movement of L is provided for by its ball joint connexions with the steering arms. The outer tube of the steering column forms the supporting casing, and has attached to it, at the top, the rack plate or quadrant, over which the speed change lever H works, and which is held down normally in either of its gear notches by the bolt and spring S. The lever H is fixed to the middle tube which has an arm clipped to its lower end, shown broken away, and from which connexion by rod is made to the gear box.

Steering column and attached gear

The group of three small levers shown below the larger lever H are respectively for the carburettor control, the exhaust valve control already

299

referred to in connexion with the De Dion engine, and for advancing or retarding the time of ignition.

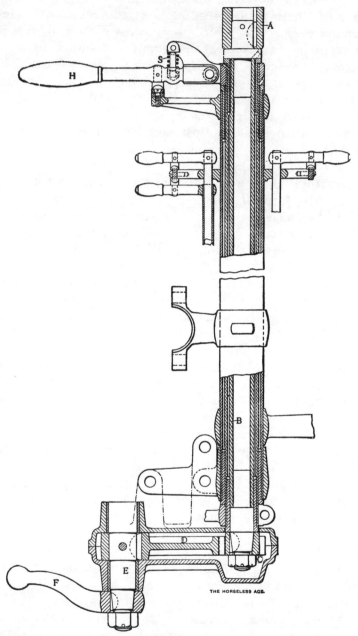

FIG. 292.—THE PIERCE ARROW LIGHT CAR. SECTION THROUGH STEERING COLUMN WITH PARTS OF CHANGE SPEED AND STEERING GEARS, ETC.

The weight of the car with the three-seated tonneau shown is given as 1,700 lbs., and provision is made for carrying lubricating oil and petrol sufficient for 350 and 150 miles running.

Among other letters not yet referred to M, Fig. 286, is the jacket water tank, I the side brake lever, the connexions of which may be readily followed, although the interconnexion with the clutch by rod F does not appear to be provided with the means of adjustment which it requires.

The rod T, Figs. 286 and 287, resists the tendency of the bevel pinion to rotate round the path of the bevel-wheel teeth, and to carry the fixed tubular part of the axle and gear casing with it around the axle centre. The position of this rod T or torsion brace, as it is sometimes called, does not always receive sufficient consideration; it has, however, been suitably connected with reference to the frame in this car. **Torsion brace**

The Geo. N. Pierce Co. have more recently undertaken the manufacture of a car[1] having a four-cylinder governed engine of 24 to 28 HP. of their own construction, but in which they retain the De Dion method of lubrication. The cylinder diameter is $3\frac{15}{16}$ in., and the engine stroke $4\frac{3}{4}$ in., and all valves are mechanically operated. The general design closely follows that of the lighter car, the main differences being in the engine and the pressed steel car frame.

THE WINTON CAR.

The four-cylinder 24 HP. car[2] of comparatively recent introduction, differs in arrangement from those of lesser horse-power and earlier design, but the changes made are mainly those necessitated by the use of the propeller shaft and bevel gear method of driving instead of the single driving chain, between gear box and live rear axle, hitherto used.

The engine is supported by suitable bearer brackets on the pressed steel side frame members in such a way that the countershaft occupies a longitudinal position in the car, and a little to the left of the centre line. The cylinders are $4\frac{3}{8}$ in. bore, and are on the right hand side of the crank chamber, and the crank chamber, gear cases, cover plates and other parts, wherever possible, are made of aluminium. The use of this metal for those parts which do not experience heavy working stresses has enabled the weight of the car to be reduced to 2,100 lbs.

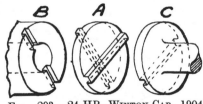

FIG. 293.—24 HP. WINTON CAR, 1904 TYPE. OLDHAM UNIVERSAL COUPLING.

The stroke of the engine is 5 in. The Oldham coupling, illustrated by Fig. 293, and mentioned with reference to Fig. 235, is between the engine shaft and first transmission gear shaft, and consists of a centre disc or floating piece A, having formed on its opposite faces a key or tongue. These key-formed projections are solid with the floating disc, and are at right angles to each other, and engage or register with corresponding grooves or slots, **Oldham universal coupling**

---

[1] A description and illustrations of this car appeared in the *Horseless Age* of February 3, 1904.

[2] *The Horseless Age*, July 20, 1904; *The Automotor Journal*, November 5, 12, 19, 1904.

one cut across the engine shaft and fly-wheel boss B, and the other across the suitably formed end of the first transmission gear shaft C. This arrangement satisfactorily provides for parallel and angular want of alignment of a driving and a driven shaft, when the two parts are intended to be normally in one straight line, and it further allows some end sliding movement to take place, and is, therefore, a true universal joint, suitable for purposes such as that here considered, where the movement at the joint or coupling would be slight and variable in amount and frequency of repetition.

The change speed gear[1] is designed to provide two speeds forward and one speed for reversing. The gear wheels are always in mesh, and three metal to metal cone clutches, with two shifting forks, collars and operating gear, in the gear box are required for the speed changes.

When the high speed clutch is engaged the drive from engine shaft to the bevel gear on the rear live axle is directly transmitted through the first gear shaft and double universal jointed propeller or driving shaft, the seven gear wheels used for the low speed forward and reverse speeds and their spindles running idle. The use of clutches in the gear box is reminiscent of the form of gear used by Messrs. de Dion et Bouton in their voiturettes as constructed three or four years ago (see vol. i. page 256), but the form of clutch was different and would transmit the driving effort with less liability to slip and consequently less wear of the clutch parts.

Gauze baffles in carburettor mixing chamber

A form of float feed spraying carburettor is employed in which the spray jet is inverted and delivers the petrol on to a series of gauze cones fitted in the induction pipe to the engine. The whole of the air and petrol received by the cylinders has to pass through the gauze baffles, and carburation is encouraged by the very fine subdivision of petrol or petrol mist formed during its passage along the induction pipe.

The speed and power of the engine is controlled by means peculiar to the Winton cars, and which have not hitherto been adopted for petrol motor car purposes. A small single cylinder single acting air-compressor driven from the engine cam shaft maintains a supply of compressed air in a receiver in communication with closed cylindrical chambers A, Fig. 294 adjacent to the inlet valve B. The inlet valve stems enter these air chambers as shown, through closely fitting guide bosses, and have pistons C fixed on their ends which are a working fit in the air chambers A. The compressed air pressure may be regulated by the driver by means of a controllable leak or exhaust valve, and the pressure in this way varied upon the inlet valve stem pistons. There are no inlet valve springs, and the variable air pressure upon the inlet valve stem pistons C not only fulfils the purpose of springs, but supplies the

Winton compressed air control

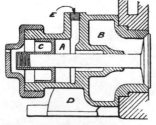

FIG. 294.—24 HP. WINTON CAR, 1904 TYPE. SECTION THROUGH INLET VALVE CAGE AND AIR PRESSURE PISTON AND CHAMBER.

[1] *The Autocar*, May 13, 1905, p. 655.

means of considerably varying the pressure, with which the valves are kept on their seats. Connexion between the air receiver and A is made at E, and a port, not shown, leads from the inlet bend D to the valve chamber B. No throttle valve for control of the quantity of mixture which the engine is allowed to take is used, and the engine always " does its best," or develops as much power as it can from the quantity of mixture it has been able to get by suction through the inlet valves. If the leak valve on the air receiver is opened, then the air pressure in the chambers A decreases, the engine obtains more mixture because the inlet valves open more readily and the amount and duration of opening of the valve is greater. Again, if the air receiver leak valve is throttled, the air pressure rises until the maximum pressure permitted in the receiver is reached and the inlet valve has little or no opening, while the horse-power output of the engine is proportionately reduced. This method of control by variable air pressure is ingenious, and is equivalent to the provision of very perfect automatically operated inlet valves, but the conditions of economical running are imperfectly met and must always be in an oil engine in which the supply of gaseous mixture is controlled by throttle valve or the equivalent of a throttle valve.

The live rear axle is of substantial design, is suitably reinforced with truss bars, and is generally of the type described with reference to Figs. 290 and 291. The bevel pinion spindle is well carried in one long plain bushed bearing, and the boss carrying this spindle is flanged at its point of attachment to the bevel gear casing, and is adjustable as to position by setscrews. Adjustment of bevel driving wheel and pinion as to depth of tooth engagement may by this means be made when the parts are put together, and subsequently when this adjustment has been affected by wear of the ball bearings in which the driving shafts run within the outer fixed tubular casings.

Advantage is taken of the available supply of compressed air in connexion with the lubricating arrangements. A pressure of one to four pounds per square inch carried in the oil tank is sufficient to lift the oil up from a lower level to the distributing sight feed drip fittings on the dashboard. A single tank below the bonnet in front of the car is divided into three separate compartments for petrol, lubricating oil, and cooling water.

The Winton [1] Cars for 1905 have four cylinder vertical engines. The control of the inlet valves by air pressure continues to be used, but the valves occupy a vertical position and springs are used to ensure the return of the valves to their seats or to assist their return when the air pressure is light.

[1] *The Horseless Age*, Dec. 14, 1904, p. 597.

Chapter XX

SOME AMERICAN PETROL CARS (*Continued*)

THE DURYEA CARS.

SOME description of these cars might have been given an earlier place in this section, partly because they may be looked upon as fore-runners in the development of the American motor car, and also because, although much still remains peculiar to the Duryea car, some of the

FIG. 295.—10 HP. DURYEA WAGGONETTE.

features of design have since been adopted by many who have not been as ready or as capable as Mr. Charles E. Duryea to persevere in design and experiment.

The general arrangement of the underframe, method of springing the car and supporting or carrying the engine on the frame, and the positions of the essential parts, are shown by the side elevation and plan, Figs. 296

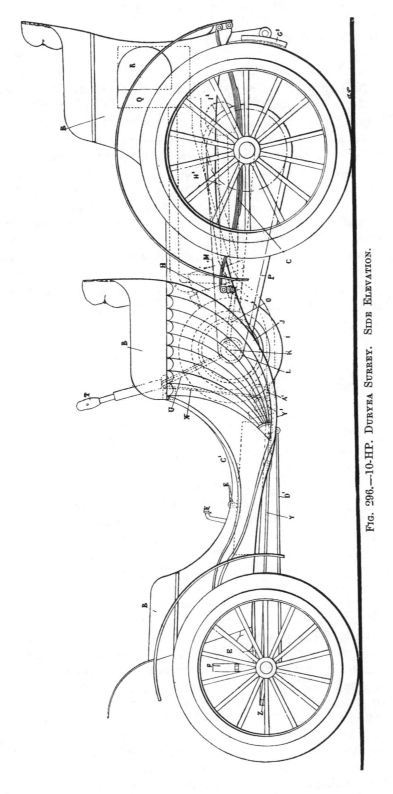

FIG. 296.—10-HP. DURYEA SURREY. SIDE ELEVATION.

305

20

and 297. The car fitted with the body shown in Fig. 296 is described as a 10-HP. Surrey, but with the back seats arranged a little differently, and with a higher back to the body it is sold as a tonneau car. The waggonette of the same power shown by Fig. 295, it will be noticed, only differs in the arrangement of the back part of the body, the front still retaining the characteristic outline by which the Duryea car has become very well known.

Although the seating capacity of the four-wheeled vehicle may be varied and the loads carried be sometimes considerably greater with one type of body than with another the general disposition of the parts and the design and arrangement of engine and transmission gear remain unaffected except in so far as alteration of the chain gear drive ratio from engine shaft to rear road axle may become necessary. Adherence to one design of running gear, combined with different arrangements of body work, to suit varying seating requirements, assists the economical production of cars, but it cannot satisfactorily meet the varying requirements as to economy, power and speed of heavy and light cars on easy and on difficult roads.

Referring to Figs. 296 and 297 it will be seen that the frame of the car in plan tapers regularly from back to front, and that the side members A, which are of wood suitably reinforced with steel flitches, are straight in plan view, but, as may be followed by the frame outline in Fig. 296, are curved behind the front seat, and passing beneath it, and at a lower level than the engine which is carried in an inclined position upon them, and two cross-connecting frame bars, inclines slightly upwards and finishes by connexion with the ends of a transverse scroll spring F upon the front axle G, through the dumb iron ends E. Reaches, or radius bars D, from the

General construction — front axle to a point of connexion on the frame visible in Fig. 296 prevent longitudinal displacement of the front axle. The rear wheels are larger than is now general, the front wheels are 30 in. diam., and the driving wheels 36 in. diam., and it is reasonably claimed that the use of larger wheels and tyres adds to the life of the latter and the smooth running of the car over poor road surfaces. Comparatively light leaf springs C attached to the rear axle-bearing shells, support the car at the rear, connexion from frame to springs being by shackles at both ends. The length and form of the springs are noteworthy, and as shown by Fig. 297 they follow in plan the gradual taper, and are spaced wider apart than the side frame members A, an arrangement which more favourably loads the rear axle than would be the case if the springs were immediately below the frame bars. Radius rods O maintain the engine shaft and rear axle centres K and D^1 approximately constant, relieve the rear springs from the duty of transmitting driving effort, and provide the means of adjustment of the single driving chain M. The live axle D^1 is not divided at the differential gear N, but is continuous from wheel to wheel, that on the off side being connected to D^1 and part of the differential gear direct, whilst the near side wheel receives its drive through a short tubular shaft connecting it to its side of the differential gear. D^1 is made of nickel steel and runs in long plain bearings.

The jacket water tank and cooler Q is combined, and is carried quite

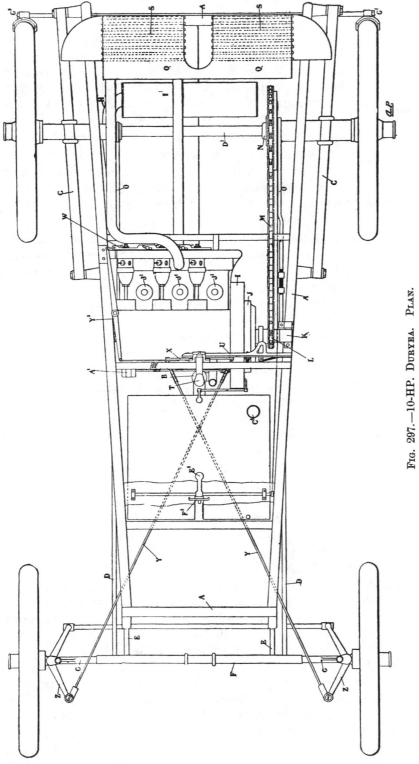

FIG. 297.—10-HP. DURYEA. PLAN.

at the back of the car. Two banks of tubes s connect the outer ends of the tank with an open centre space A. When the car is in motion the projecting side cars R collect and direct the air through the tubes inwardly to the space A from which the heated air leaves the cooler. The arrangement of the tubes in banks is similar to and is intended to follow engine condenser practice. Circulation by convection currents or thermo-syphon action is depended upon, flexible tubing of large diameter connecting the lower and upper sides of the engine jacket with the cooler tank. The petrol tank is below the footboards in front of the car, and is filled at a convenient corner c^1. The pipe leading from tank to carburettor float chamber B is short, and a needle valve in the passage between float chamber and spray jet is used to either regulate the flow of petrol or to shut off the supply from the tank when necessary.

Steering gear The steering gear connexions are plainly shown, but attention may be directed to one or two features which make it different to the majority of simple tiller and lever steering gears as distinguished from those using a wheel and some form of irreversible worm and sector or screw gear. There are double lever arms z on the front wheel axle ends, the rearwardly projecting levers being coupled together by a connecting rod in the usual manner, while light steering rods Y connect the front arms to the tiller bell crank c, Fig. 301, through the chains D D. The vertical arm of the bell crank c, lettered T in the plan view Fig. 297, receives motion in a plane across the car about the pin B to right or left of its vertical or mid position when it is desired to steer to left or right respectively, and the chains D D, passing round guide-pins or rollers, are required to convert the up and down motion given to them by the steering lever to a horizontal movement in the direction of the length of the tension rods Y. By crossing the rods Y their angle of connexion at z z is less acute, and as there are two of them they may be light rods, either of which according to the direction of steering transmits the pull from the steering lever. The axis of the steering head

Inclined steering pivots pivots are inclined to such an extent that the point of contact of the wheel with the road and the pivot axis produced are nearly coincident points. The effect of this is to reduce the intensity of road shocks experienced by the steering gear when the steering wheel or wheels receive blows or their equivalent at nearly road level from either bad road surface, stones or other obstacles. It is evident that if the axis of the steering head pivots were vertical, the actual distance between the centre line or plane of the wheel and the pivot would represent the effective leverage with which a road obstacle meeting the tyre at or about road level could attempt to overpower or reverse the steering gear. When, however, this virtual lever arm has been reduced by inclination of the steering pivot axis directly in proportion to the distance as already stated between the points, of wheel contact and pivot produced, with the ground, then at precisely the same rate has the power which road obstacles possess to reverse the steering gear been reduced. Further, when inclined steering pivots are used, angular movement of the wheels slightly lifts the car, the load on the front wheels resists the steering movement, and results in a tendency to always steer a straight line, or when the car is obliged to depart from the straight line to return to it. The use of inclined steering pivots is not, however, new with Mr. Duryea, and its importance is much reduced by getting the vertical pivot very close to the wheel nave.

Fig. 298 is a part sectional longitudinal view of the engine as fitted to the car illustrated, and Fig. 299 a transverse section.

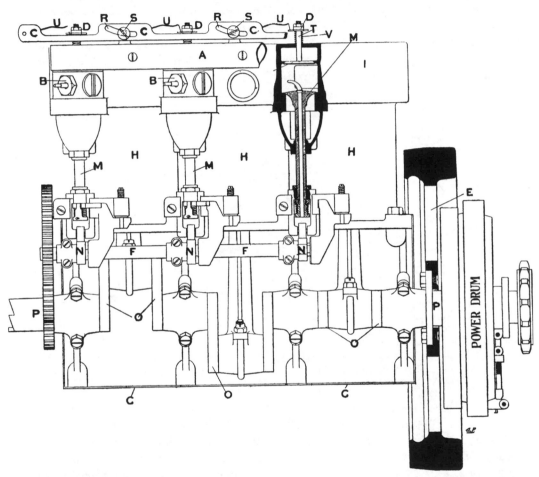

FIG. 298.—10-HP. DURYEA ENGINE. PART SECTIONAL PLAN.

The three cylinders H are cast in one piece, and with the water jacket I in the form of a belt round the upper part of the cylinder barrels or combustion chambers. From the sectional view, Fig. 299, it may be seen that no attempt is made to water jacket the top or head of the combustion chamber or the removable cover which in this engine forms combustion chamber head. That part of the cylinder barrel which is jacketed is exposed for the longest time to the fiercest heat, but the greater part of the

10-HP. three-cylinder engine

309

bore traversed by the piston rings depends for cooling upon conduction and radiation of heat. The cylinder covers are machined with a taper part to form a metal to metal gas-tight joint with a corresponding taper formed in the cylinder bore. This joint is made by screwing the cover into the cylinder, both being suitably screw-threaded. An approximation to a spherical form of combustion chamber has been obtained by casting the cylinder cover and piston top of the curved form shown.

The diameter of the cylinders is $4\frac{1}{2}$ in., and the stroke of the engine is the same. The crankshaft P is large in diameter, and provides large bearing surfaces without necessitating long journals, and although large in diameter is not heavy, the whole shaft having been lightened by boring out crankpins and main journal portions to such an extent that it might be described **Hollow** as a tubular rather than a hollow shaft. When the lightness of the tubular **crankshaft** parts of the crankshaft and the webs O is considered the satisfactory behaviour

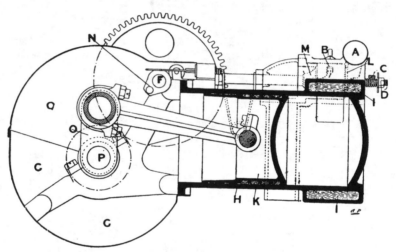

FIG. 299.—10-HP. DURYEA ENGINE. TRANSVERSE SECTION.

of these shafts is surprising even when the close support it receives from the bearings is taken into account. These shaft bearings, and those for the cam spindle F, are formed in the three-armed cast spider frames shown in the sectional view. There are four of them at the positions of the main shaft bearings, and they are screwed to the cylinder casting at the position of the bosses shown.

A light sheet metal casing around the spider frames prevents waste of oil thrown off the moving parts and protects the latter from dust and grit. The lower part of the casing G is fixed, and the upper part Q may be removed, when examination, cleaning or adjustment become necessary ; the lap joint of the casings and the line of the joint are shown in Fig. 299.

The fly-wheel E is fixed to the crankshaft by spigoted flange connexion, and the epicyclic change-speed gear or "power drum" is carried on an extension shaft which at one end enters the hollow crankshaft, and is driven by it, and is supported at the outer end in a bearing on the car frame. The magneto machine A¹, Fig. 297, which supplies current for ignition is driven by a torsion spring through friction wheel in contact with the tapered or coned portion of the fly-wheel rim. A governor on the magneto driving spindle gives longitudinal movement to the friction pulley, which by light pressure on the fly-wheel rim at its reduced diameter or heavier pressure at a path of greater diameter permits more or less slip to occur with increase or decrease of engine speed. By the use of this governor the magneto armature is not driven at injurious speed of rotation. The ignition system may be explained by reference to Figs. 298 to 300, the latter diagrammatic figure showing the essential parts conveniently assembled for explanation. A magneto B of the type having a rotating armature supplies low tension current to the primary winding of the induction coil C through the two-way switch N. The switch N, as shown, is open ; when moved to the right it closes the magneto circuit, and when moved to the left puts the dry cell battery or accumulators in circuit. The battery is only used when from some rare cause the magneto fails. High tension current from the secondary winding of the coil C passes by connexions D to the anvil plugs E which are insulated with talc or mica washers from the engine cylinders just as the live stem of an ordinary sparking plug is insulated from the body of the plug, though the insulation resistance in the latter case is ordinarily greater than at E. The position of the plugs E and the cap nuts B which secure them in position is shown in Fig. 298. The spindle of the little hammer or contact finger F passes down the hollow centre of the exhaust valve and spindle G, lettered M in Fig. 298, and is a free working fit therein. Normally the hammer F is in contact with the anvil E, and it is only at the instant of interruption of this contact that sparking occurs. As the strength or quality of the spark depends entirely upon the quickness with which contact is broken some form of spring-operated trip gear becomes necessary to ensure the required rapidity of interruption. On the lower end of the hammer spindle is a collar piece L¹ fixed in position by the small square-headed setscrew shown. Just above the setscrew is an upstanding part of this collar piece which by torsional effort through the spring L keeps the trip piece H in contact with the operating trip flap I and the hammer F in contact with its anvil E. The normal position of these parts is shown in the underneath plan given below the elevation view, and in which L¹ is the upstanding part of the collar piece, F is the central hammer spindle to which it is fixed, H is the trip piece which is a working fit on the reduced portion of L¹, and which may be seen in position in the part sectional view Fig. 298, while the operating trip flap I is shown on its pin, which works in an eccentric bush so formed for purposes of adjustment of the lap of the trip piece edges. One end of the spring L is fixed to L¹, the other end being sprung into one or other of the holes shown in H, to put more or less torsion on the spring. The flap I has a spring at the end of its pivot-pin to return it after each movement to the position shown. A stud-pin J in the side of the exhaust cam K, and lettered N in Fig. 299, at each revolution of

High tension magneto ignition

Ignition trip gear

311

the camshaft comes into contact with the lower part of I, twists it like a door upon its hinge-pin, and as long as the trip edges remain in contact also twists H upon L^1 putting more and more torsion upon the spring L. When finally the trip edges disengage H returns at high velocity under the influence of the spring L, strikes the upstanding part of L^1 a smart blow and so by slightly twisting the spindle F momentariiy separates the hammer F from the anvil E. At the instant of separation the spark occurs. The return circuit for both high and low tension currents is through earthed or frame connexions. After the release of the trip gear, H is left behind I, but presently the exhaust cam comes into operation, lifts the valve stem G, and with it the attached parts L^1 and H, the latter being moved above and out of contact with the flap I. When the exhaust valve drops and closes, H slides over and takes up its position in front of I ready for the succeeding firing stroke.

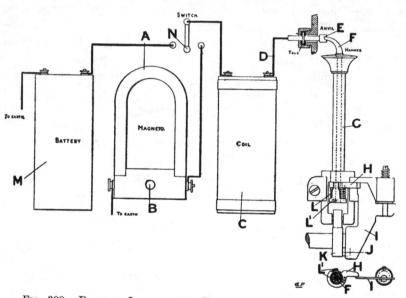

Fig. 300.—Diagram Illustrating Duryea System of Ignition.

Variable inlet valve lift

The power developed by the engine is controlled by varying the lift of the inlet valves, no throttle valve being used in the inlet pipe between the carburettor and the trunk pipe A through which the inlet valves receive their supply of carburetted air.

A bar c, Figs. 298 and 299, is supported and may slide longitudinally upon the headed screws S. Between those parts which have the curved slots R formed in them are parallel lengths at the position of the inlet valve spindles T, and by reference to Fig. 299 it will be seen that these parallel parts of c pass below the collar nuts D on the valve spindles. Movement of c to the left as far as the slots will permit, will raise it and bring its upper edge in contact with the projecting flanges of the nuts D ; the valves cannot then move from their seats. When c is moved as far as it will go to the right, it is lowered, and its upper edge is some distance from the lower faces of the collar nut flanges, and the valves are then free to open to their full extent. Intermediate positions of c more or less limit the opening of the valves, and so by a greater or less degree of throttling vary the quantity of mixture obtained, and the power developed by the engine. The illustrations relating to the Duryea car which have been described represent the car as constructed in America. During the past year, however, the English Duryea Co., of Coventry, have undertaken the manufacture of cars and have largely redesigned the Duryea.

Figs. 302 to 304 show modified change-speed gear, and two views of the new three-cylinder engine. Before describing these it may be mentioned **Control by one hand lever**

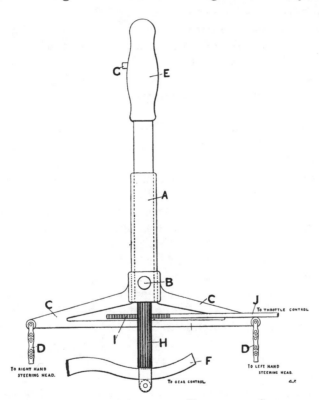

FIG. 301.—DURYEA CONTROLLING HANDLE AND CONNEXIONS.

that the methods of operating the old and the new change-speed gears differ, but as the original method might be applied to either it will not interfere with further description of the simple " one hand " car control with which most of the American cars are fitted.

Earlier reference has been made to Fig. 301 and the movements of the parts E A and c about the pin B for steering the car. The stem A is hollow, and through it slides a bar, at one end of which is the handle E and at the other the long pinion of small diameter, H, terminating in a forked end and roller engaging the curved end of the gear operating lever F.

When E is in its lower position the low-speed gear is in operation, midway the controlling friction bands and clutch are inoperative, and the engine runs freely without driving the car, whilst with E brought up to its top position the complete gear is locked and driven at the same speed as the engine crankshaft. The car is then driven at top speed, the only active gear being the chain drive from crankshaft to rear axle. By depressing a pedal, which, with its shaft and connexion to that part of the change-speed drum hidden by the fly-wheel, may be seen in Fig. 297, and lifting up E from the middle to the top position, the reversing gear is brought into action. A press button G in the handle E fixes the handle E at mid position when the car may be left stationary with the engine running.

When the handle E is twisted in either direction motion is given by the long pinion H to the rack rod I which, as indicated, leads to the throttle control bar C already described, and the connexions to which may be traced by reference to Fig. 297, and where the rack rod is lettered X, and the pivoted lever Y^1 connects X to the throttle bar, which has in this view been lettered W instead of C as in Figs. 298 and 299. This description of the method of control will be more intelligible when the change-speed gear, Fig. 302, has been described.

Planetary change-speed gear

The extension spindle P^1 is fixed by screw and taper fit in the end of the crankshaft P, and is covered by a long bearing sleeve K upon which the different parts of the gear are carried. The centre driving pinion D is fixed by screw thread on the crankshaft end and gears with three planet pinions H which in turn are always in mesh with the internally toothed wheel G. The pinions H run on stud-pins J formed with the concentric supporting ring N^1, and the clutch ring N is fixed on the reduced ends of the stud-pins by the nuts and pins shown. N^1 also provides points of support for the pivot-pins A which at their outer ends have the lever keys U pinned to them, and at their inner double armed ends the sliding bolts M which as may be seen by reference to the end view project a little way through the friction band or clutch surface of N. The springs W normally keep the bolts M pulled outwards, and the key ends of the levers U in engagement with teeth or notches cut in the edge of the part B. The parts N^1 and B are, therefore, generally locked together, and only have relative movement when the strap Y is tightened, and the bolts M pressed inwards so lifting the keys U out of engagement with B. B has two parts keyed upon it, the carrier ring for the internally expanding clutch shoes O, and the chain sprocket pinion L. The clutch ring F of the internally expanding clutch is in one piece with the internally toothed ring G, and they have bearing support, through their centre disc or web, upon the outer surface of the sleeve part of B.

By means of the shifting fork S, and sliding collars R, the three clutch shoes, one of which is shown, may be either pressed outwards or inwardly drawn through connexion at T with the toggle links C. With the toggle link in the position shown at right angles to the axis of the gear the clutch shoes are outwardly pressed and the clutch is engaged. A ball bearing on R reduces friction resulting from end thrust when engaging this clutch, but once engaged very little or no end pressure is necessary to keep the toggle

314

links in the position shown. The friction band x is required to hold the internally toothed ring G stationary when the low-speed forward gear is brought into operation.

When the friction bands x and y are freed, and the clutch F is disengaged no driving effort is transmitted to L, and the wheels G and H and their attached parts revolve idly at rates proportioned to the pitch diameters of D, H, and G.

If the clutch band x is tightened, G is held stationary, and power is transmitted to L through D, H, N[1] and B, and the gear ratio between P[1] and L depends upon the wheel diameters, and may be found from the expression $\frac{G}{D}+1$. With the relative gear-wheel diameters shown the ratio is $\frac{6 \cdot 75}{3 \cdot 375}+1$ or 3 to 1, that is the chain pinion L will rotate at one third of the speed of the engine shaft. **Gear ratio**

When the friction bands x and y are freed, and the clutch F is engaged

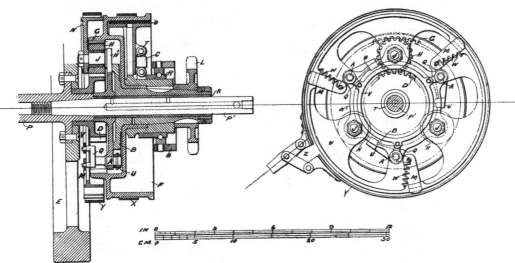

FIG. 302.—12-HP. DURYEA EPICYCLIC CHANGE SPEED GEAR.

both G and N[1], carrying the pinions H, are locked to B, and as there can then be no relative gear-wheel movement, the whole transmission gear is rotated by the crankshaft, and the speed of rotation of the chain pinion L is the same as that of the engine shaft. This provides the high speed or direct drive with no gearing active other than the chain transmission to the rear axle.

If when both friction bands and clutch are disengaged, the band y is first tightened, and then the clutch F engaged, the reversing gear is brought into play. The pinion carrier ring N[1] does not rotate, but G is held by the clutch F to L, and they are driven at a speed proportional to $\frac{D}{G} = \frac{3 \cdot 375}{6 \cdot 75}$ or at one half the speed of the engine.

A comparison of the new engine, Figs. 303 and 304, with the original American design, Figs. 298 and 299, shows that some of the latter has been incorporated in the design of the former, and that considerable changes **12-HP three-cylinder engine** have been made in the type and arrangement of valves and valve gear, and improvements in detail have been effected which remove sources of trouble, and indicate that by careful consideration of the design and performance of the earlier engine, alterations have been made which are not merely modifications. The inlet and exhaust valves v and v¹ are of the same size, are mechanically operated, and with their cages which form valve seat and valve stem guide, are interchangeable. The valve rods M, guided at the camshaft end, in easy-fitting gun-metal or bronze guides M¹, move the rocking levers about their pivots N, and so by contact with the projecting ends of the valve spindles operate them against pressure from the sugar-tongs shaped wire springs O.

The water jacket belt is wide, and that part of the bore traversed by the piston rings is cooled. The position of the valves has made it desirable

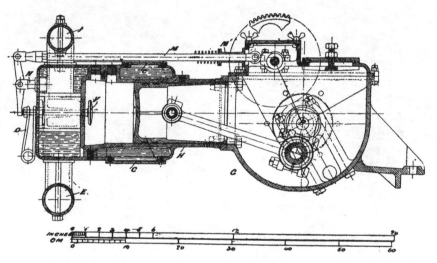

FIG. 303.—12-HP. DURYEA ENGINE. LONGITUDINAL SECTION.

to carry the water jacket round the head of the cylinder, and this unavoidably results in unnecessary waste and transfer of heat to and consequently higher temperature of the jacket water, and may under some circumstances necessitate the use of a larger quantity.

Trunk inlet and exhaust pipes I and E are simply held in position, between their branch connexions to the cylinder, by the long studs and nuts shown in Fig. 303. Foundry difficulties and subsequent trouble in removing the material of the cores from the water jacket spaces are lessened by casting the cylinders with large openings in their top and bottom jacket walls, which are afterwards closed with cover plates c. Those parts of the crank chamber which replace the bearing frames and sheet metal casing in Figs. 298 and 299 are cast in aluminium. The upper part of the crank chamber has been so arranged that the removal of one door allows the camshaft and attached and adjacent parts to be inspected, a second easily

316

removable oval door gives sufficient access to the connecting rods to enable an opinion to be formed as to the condition of their ends, and when adjustment or overhaul becomes necessary the main cover may be taken off with the attached covers referred to, and the camshaft complete in its bearings with the rods M in their guides M¹.   The piston is unusually long, and even when worn, or if purposely allowed to be a loose fit, there would be no tendency to cross-binding in the cylinder bore.   The crankshaft is hollow, but it has not been so much reduced in sectional area as the shaft of the American engine. The crankshafts in these engines are set at a lower level than the cylinder centre line, and as a result of the reduced angular movement of the connect-

Offset crankshaft

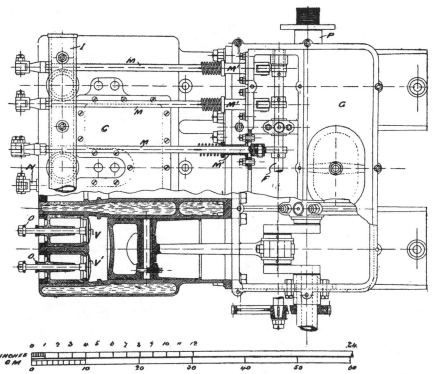

Fig. 304.—12-HP. Duryea Engine. Part Sectional Plan with Camshaft Cover Removed.

ing rod during the out stroke, the crosshead load upon the piston is less. The outer shaft bearings are in halves, and are spigoted in the circular openings in the crank chamber sides.   These circular openings are large enough in diameter to allow the shaft to be put in place or withdrawn lengthwise.

The cylinder diameters and the engine stroke are both 4½ in., and the normal speed 600 revolutions per minute.   At its normal speed the engine should develop its declared 12 HP. without difficulty.   High tension magneto ignition is employed with induction coil and sparking plugs, and means are provided for automatically advancing or retarding the ignition with rise or fall in engine speed.

The cars as made in America vary in weight from 8 to 10 cwt., according to the type of body with which the car is fitted. The waggonette shown by Fig. 295 weighs 10 cwt. The new cars constructed at Coventry are about 2 cwt. heavier, due to the additional weight of the new engine and other strengthened parts of the running gear.

The three-wheeled 10-HP. Duryea has attracted attention by its performance in racing events, and it weighs only 7½ cwt. It is probably the lightest 10-HP. car yet built.

Chapter XXI

RADIATORS AND WATER-COOLING REQUIRE-MENTS

MANY forms of radiator water coolers are in use, and they may be divided into four classes.

1. Those consisting of water tubes working by natural circulation caused by temperature difference in the water circuit, represented by the Renault, Argyll, and Brooke-Estcourt types, all having water-tubes with gills of some kind surrounded by air, and in some cases fixed in headers or small tanks.

2. Those formed of a continuous length of gilled water tubes as represented by the Loyal, Clarkson, and other forms, and generally worked in connexion with a tank as well as a circulating pump.

3. Those formed with round or flat vertical or horizontal water tubes fixed at their ends in headers forming reservoirs, such as the Albany, the recent Daimler, De Dion Bouton, and others, with a circulating pump.

4. The honeycomb type with the air passing through the tubes formed like the old steam condenser with horizontal tubes fixed in tube plates, or with the ends of the tubes swelled out to a square, rectangular or hexagonal form touching each other, and soldered into a compact honeycomb without the use of a tube plate. The use of this type for motor cars, though old in steam engineering, originated in the tube-plate condenser form by the Cannstatt Daimler Co., in 1898, and has since taken various forms in detail, as in the Mercedes, Albany, and others. In most cases the natural surface radiation of the tube surface increased by convection, more or less, in accordance with the speed of the car passing through the air is relied upon, but in many cases, not only with the honeycomb type, but with the continuous tube or tube and header type, a fan at the back of the radiator is relied upon for increasing the rate of passage of the air among or through the tubes.

The requirements of all or any of these radiator coolers depends almost entirely on the power of the engine for which jacket water has to be cooled, although the very great diversity in the practice of different car designers tends to show that they have not arrived at any conclusion on the subject. This is not very remarkable when it is remembered that, notwithstanding

319

the plentifulness of the literature of steam condensation, water heating and cooling and air heating, nothing has been published collating the information obtainable and making it directly applicable to the thermo-dynamic problem of the oil or gas engine.

Quantity of heat to be radiated We may estimate the radiator power required per horse-power of engine by reference to the heat efficiency of the average engine. The heat equivalent h in British thermal units of the mechanical work of 1 HP. is

$$h = \frac{33,000 \text{ ft. lb.}}{J} = 42.47$$

J being taken as 777.

This, however, assumes the engine to be of unit efficiency, and a large increase in this must be made to meet the thermo-dynamic requirements of the real engine. Some of the larger gas engines have shown by recent trials a very high heat efficiency as compared with the smaller gas and oil engines, but it will not be safe to adopt as the figure for present requirements more than 18.0 per cent., and even this will be high in many cases. We may, however, take this, referred to brake horse-power, as being approximately correct for our requirements.

The total heat units h^1 represented by 1 BHP. will then be

$$h^1 = \frac{HP \times 33,000}{J \times 0.180} = 236 \text{ BTU.}$$

Of this it may be assumed that from 27 to 33 per cent. will be carried off in the jacket water, and I have found by numerous experiments on oil and gas engines of various sizes that it varies through a larger range than this. Taking into consideration, however, the conditions of working of a motor-vehicle engine, it will be safe to assume that 30 per cent. of the actual heat equivalent of the work done will have to be carried off by the jacket water. The total heat H in BTU., then, to be carried off per BHP. of the engine when working at the power assumed, will be

$$H = \frac{33,000 \times .303}{777 \times 0.18} = 71.5 \text{ per minute.}$$

At whatever temperature, then, of jacket water it is decided to work with, 71.5 units of heat must be carried off by radiation from the surfaces of the jacket, the tank, if any, the pipe connexions and the radiator or cooler. The greater the temperature difference between these surfaces and the atmosphere, the smaller will be the necessary amount of those surfaces. This, however, will be modified by the rate at which the water is circulated within the cooling surfaces until the point at which the heat emissive power of the metal surfaces has been reached, either by the rate at which the water is represented to the surfaces by circulation or by the quantity of air passing them by the movement of the car or by fan. The usefulness of the surface is also affected by the quantity of water it envelopes, and is inversely proportional to it in the sizes of tubes usually adopted. In

RADIATORS AND WATER COOLING

some of the honeycomb coolers, however, the interstitial spaces separating the tubes is so small as to be almost capillary, with the result that circulation is so slow that the full value of the cooling surface is never reached, and the tubes would be better with slightly greater separation, and might without loss of effect be shortened considerably.

Practical considerations make very small tubes inadvisable, incrustation and sufficiently easy circulation dictating a limit to smallness of diameter. A year or so ago a form of Loyal cooler with tubes of only about a quarter-inch external diameter was fitted to a number of cars. A general return has, however, been made to tubes of about half an inch in diameter, and with or without gills or wire attachments for increasing radiating surface. This size seems to be found small enough when judged by practical considerations, and when the tube is in one continuous length even this size is small enough to throw a good deal of work on the pump.

Having found the **quantity** of heat which has to be carried off by the jacket water, we have now to find the quantity of surface necessary to keep the jacket water down to the temperature considered desirable as an ordinary maximum. Taking the temperature T of delivery from the jacket as 180 deg., and the temperature t of delivery of water from the cooler 130 deg., then the cooling water must be lowered in the cooler by a quantity $= T - t = 180 - 130 = 50$ deg., and the minimum quantity of cooling water Q required will be

$$Q = \frac{H}{T-t} = \frac{71 \cdot 5}{180-130} = 1 \cdot 43 \text{ lb. per HP.}$$

or for a 10 HP. engine about 1·43 gals. per minute at 130 F. must be circulated through a cooler, the surface of which is sufficient to lower the temperature from 180 to 130 deg., which are moderate limits. The larger the quantity of water circulated the lower the number of units to be taken up by each lb.

The question, then, is how much surface will be required for lowering the temperature 30 deg. of this 1·43 lb. of water per minute. If the area of surface be in such relation to the quantity of water passing over it that the radiation is proportional to speed of circulation, and if the cooling surface be in the best position, for instance, in the front of the car rather than at the back, then the results of various experimental inquiries on radiation losses and steam condensation, water cooling and air heating may be considered applicable to a solution of this problem.

My own experiments on long lengths of air cooled steam pipes,[1] and those of Anderson,[2] Audenet, Hudson,[3] Burnat,[4] Hagemann,[5] Craddock,[6] Clement,[7] Havrez,[8] Braussell, and Hudson, show that in free air 1 sq. ft.

[1] *The Engineer*, vol. xlix. p.199; [2] *Proc. Inst. C. E.* vol. xxxv. and vol. xlviii. p. 257; [3] *The Engineer*, vol. lxx.; [4] *Proc. Inst. C. E.* vol. xli.; [5] *Proc. Inst. C. E.* vol. lxxvii.; [6] "The Chemistry of the Steam Engine," "Condensation and Expansion of Steam," 1874 Reprint from *Mining Journal*; [7] Clark's *Rules, Tables and Data*; [8] Engineering, vol. vi.; *The Engineer*, 5 and 12 Dec., 1890.

21

of bare pipe surface will give up 0·0360 units of heat per deg. F. difference of temperature per minute.

More correctly it might be said that air will receive 0·036 units from 1 sq. ft. of heated surface per minute per 1 deg. F. difference in temperature of the steam or hot water on one side, and the air on the other side of the pipe.

Air is such a very slow or bad conductor, and the conductivity of almost all metals so high in comparison, that in heating still air the quantity of heat that the air will carry off is practically the same whether the heating surface be that of iron, copper, or brass in similar conditions. Even when the air is artificially circulated this remains true up to considerable speeds of passage of the air over the surface.

In consequence also of this slow receptiveness of air, the thickness of the metal of the tubes has little effect. The rate of transmission of heat under perfect conditions is inversely as the thickness of the metal tube or plate conducting it, but because it is so high as compared with the carrying off or receptive power of air, the supply from the hotter to the cooler surface even through a very thick plate is very much faster than the air will carry away. Hence a thin tube has in this respect no great advantage over a thicker one, and it is very questionable whether for the purposes under consideration a brass, or even an iron tube with the fine rust surface or the oxide formed on iron or steel in presence of steam under a high temperature, or coated with dull surface lamp-black, or white lead, would not be quite as good as copper, and if the copper be smooth and polished, much better. More depends with regard to the heat transference under consideration upon the velocity of the air passing over the surfaces than upon any other condition.

The following table gives some of the results of numerous experimental investigations for ascertaining the rate at which heat is conducted by iron, brass and copper tube walls, plates, and sheets from steam or water, on one side to water or air on the other.

RADIATORS AND WATER COOLING

TABLE No. 1.—CONDUCTION OF HEAT BY METAL SURFACES TO AIR AND TO WATER, IN UNITS PER SQUARE FOOT PER DEGREE F. OF TEMPERATURE DIFFERENCE PER MINUTE.

Metal.	Heating Medium.	Heated Medium.	Units per 1 deg., 1 ft., and 1 minute.	Authority.	Reference.
Galvanized piping	Hot water	Still air	0·045	Anderson	*Proc. Inst. C.E.*, vol. xlviii. p. 252.
Cast-iron, black painted . .	Hot water	Still air	0·0317 to 0·040	Tredgold	D. K. Clark, in *Rules, Tables, and Data.*
Cast iron pipes .	Hot water	Still air	0·0346	,,	,, ,, ,,
Sheet iron, smooth	Hot water	Still air	0·0280	,,	,, ,, ,,
Cast-iron pipes .	Steam	Still air	0·0362	Beaumont	*The Engineer*, vol. xlix. p. 199.
Cast-iron pipes .	Steam	Still air	0·0467	Burnat	*Proc. Inst. C.E.* vol. xli. p. 294.
Iron boiler plates	Steam	Open air	0·0960*	Fox, Head & Co.	Clark, *Rules*, etc., *Engineering*, vol. vi. pp. 317, 325.
⅛ in. copper plate	Hot water	Water	2·960	Peclet	*Traité du Chaleur.*
Copper boiler .	Steam	Water	5·600	,,	,, ,, ,,
Copper coils .	Steam	Water	5·200	Peclet-Laurens	Clark, *Rules*, etc.,
Copper pipe .	Steam	Water	4·660	Havrez	*Engineering*, vol. vi. pp. 293, 471.
Copper plate .	Steam ·	Water	2·880	Clark.	*Rules, Tables, etc.*
Steam Condenser	Steam	Water	6·100	Audenet	*Proc. Inst. C.E.* vol. xxxix. p.399.
,, ,, Brass tubes, 18 BWG.	Steam	Water	10·000	Nichol	Clark, *Rules, etc.*
Copper pan .	Steam	Water	3·50	Anderson	*Proc. Inst. C.E.* vol. xxxv.
Mean of Nos. 1 to 6			0·0377		
Mean of Nos. 8 to 15 omitting No. 14.			4·330		

* Experiment with a steam boiler outdoors with and without lagging.

Mr. J. S. V. Bickford made some experiments with a cooler formed of eight flat chambers or tubes connected to headers, and supplied with steam at about 210 deg. F. The flat tubes were 12 in. wide and ¼ in. thick with ½ in. space between them, giving a total surface of 28 sq. ft. The rates of condensation were taken with still air and in moving air with results as follows—

In still air Steam condensed per hour	8 lb.	units per deg. per sq. ft. per min.								0·0356
In gentle breeze ,, ,, ,, ,,	10 lb.	,,	,,	,,	,,					0·0397
In fresh breeze ,, ,, ,, ,,	18 lb.	,,	,,	,,	,,					0·0750
In fan blast ,, ,, ,, ,,	22·75 lb.	,,	,,	,,	,,					0·090

In all these experiments the smaller tubes gave the higher results, and still higher results would be given by the much smaller and very thin tubes used for the coolers of motor vehicles. The surfaces were in almost all cases the slightly rough finish of drawn tubes and cast-iron pipes. The temperatures in many cases were similar to those which prevail with car-coolers, namely, from 120 deg. F. in hot water heater pipes, and air at from 42 to 62 F. ; and with steam at slightly above and below atmospheric pressure, or from 208 to 220 F. All had plain surfaces without any gills. As the latter give a large increase to the heat transmitting power of any length of tube, it may be useful to inquire into the probable value of the increase of emissive surface by these additions.

The conductivity of copper being greater than brass and much greater than iron or steel, the relative figures being 0·89, 0·75 and 0·5, gold being 1·0; copper will be referred to in that which follows.

Attachment of gills to tubes In almost all cases the gills are fixed to the copper tubes by solder, which has a conductivity of only about 0·31 of that of copper, and where they are only partly soldered, or, as in some coolers, the thin gill disc is only slipped on the tube, the conduction from the tube is almost lost. The metallic connexion must be perfect, and even then the heat transmission resistance offered by the solder is considerable, and lowers the value of the gills as emissive surfaces.

The relation between the receptive and the emissive capacity of the inside and outside of a copper tube respectively has not been definitely ascertained, but is probably much the same as that of iron tubes and heating surfaces. The heating power of the Serve boiler tubes with six interior radiating ribs was found to be about 15 per cent. greater than that of the plain tube, indicating that the emissive power to water is greater than the receptive power from the heated products of combustion passing through the tube. The interior surface, however, of such a tube may be quickly though thinly coated with non-conducting material while the outer plain surface in the water may remain clean or suffer very little diminution of its emissivity. Hence there is sufficient reason for expecting considerable increase in heat transmission from the increase in receptive surface.

Mr. C. Wye Williams found by experiment that the heat transmission of a given tube to water was largely augmented by increase in the receptive surface without any addition to the emissive surface,[1] while an increase by the same means, namely, projecting pins fixed in the tube, of the emissive surface made comparatively little difference. Circulation of the water being heated would however probably have increased the rate of transmission.

Experiments on steam condensation by water and air condensers, and on cooling water by water and by air, by Craddock,[2] are still of much interest in connexion with our subject. His results were confirmed in 1875 by experiments made by Mr. B. G. Nichol,[3] who used brass tubes 0·75 in. outside diameter, 18 wire gauge thick.

---

[1] *On Increasing the Evaporative Power of Boilers*, by C. Wye Williams. London, 1842 ; and *Mechanics' Magazine*, 1841-2.
[2] *The Chemistry of the Steam Engine*. Thos. Craddock. London, 1846.
[3] *Clark's Rules, Tables, and Data for Mechanical Engineers and Engineering*. December 10, 1875.

RADIATORS AND WATER COOLING

The highest rates of heat transmission obtained by Nichol in condensing steam at 245 deg. outside the tube with water circulated at high speed through it, was 10 units per sq. ft. per deg. difference of temperature per minute. This may be taken as the maximum receptive and emissive power of the plain brass tube, the conductivity of which was about 85 of that of copper.

A copper tube would, therefore, presumably have transmitted $\frac{10 \cdot 0}{0 \cdot 85} = 11 \cdot 75$ units per sq. ft. per deg. difference of temperature per minute.

Relative rate of radiation to air and water

Taking as the cooler co-efficient C the best mean rate of transmission to free air as given in the table, page 323, namely 0·0377 units, it is clear that the emissive surface might be increased very largely before it exceeded in emissive capacity the receptive capacity. The relation between the receiving power of still air, and the conducting power of the copper pipe, will, according to the experiments quoted, be as one to $\frac{11 \cdot 75}{0 \cdot 0377} = 311$. The experiments of Craddock would put this at approximately 250, while other experimenters arrived at a similar value. The mean results of Audenet's experiments would give $\frac{6 \cdot 10 \times 0 \cdot 89}{0 \cdot 0377 \times 0 \cdot 75} = 192$.

Taking these three values we have a mean of $\frac{311 + 192 + 250}{3}$, or 1 to 251 as the relation that might exist between gill emissive surface and the receptive surface of the tube. Taking then the ordinary circumstances of a car standing in the open with the engine working, or running slowly up hill with a stern wind equal to the speed of the car, and the engine working at from three-quarters to full power, it would appear that the exterior surface of the plain cooler tubes might be effectively increased in area by gills metallically united to the tube, and having certainly over twenty times the area of the part of the tube to which they are attached, before the conducting power of the gills exceeded that of the tubes. These conditions do not, however, obtain with sufficient frequency to make this area of gills from thermal reasons necessary, and for practical mechanical reasons such large gill surfaces are impracticable. With the car moving there is almost always some encouragement of air movement at the cooler which would increase the co-efficient of 0·0377 materially. Moreover gills giving the area above mentioned would be large and easily bent, or if smaller and nearer together they would be so close together that the air would be held between them until moved by considerable speed of the car or by exhaust fan. Then the large area would become unnecessary, for the specific heat of air being 0·238 it would only require $\frac{1 \times 10 \times 13}{0 \cdot 238} = 547$ cubic feet of air per unit of time over each square foot of surface to equal in effect the cooling power of 1 gallon of water passed over the tubes in the same time, air at a temperature of 60 deg. F. occupying about 13 cubic feet per lb.

325

To return, then, to the cooler requirements dictated by the work done by the engine, namely 1·43 lb. of water to be lowered 50 deg. per minute. Assuming the cooler to be capable of giving off to the air 0·0377 units per minute per degree difference of temperature T of the hot water in the tubes and the temperature t^1 of the outer air per square foot of surface, we then have as the surface S theoretically required per BHP. assuming a mean air temperature of 60 deg.

$$S = \frac{Q \times T - t}{C \times T - t^1} = \frac{1.43 \times 50}{0.0377 \times 120} = 15.80 \text{ sq. ft.}$$

This is assuming that the car is standing or is running under such conditions that there is no movement of the air through or among the tubes and gills of the cooler except the very slight current produced by convection. It also assumes that the whole of the cooling is done by the cooler tube and gill surfaces and neglects that done by any other surfaces. Allowances must, therefore, be made with respect to these points.

Firstly, when the car is standing the engine is doing very little work, seldom more than one-fourth full power, but the quantity of fuel consumed and the temperature of combustion remain large in proportion to the work done.

Effect of speed on radiator capacity The radiation and convection from the cylinder and pipe connexions and from parts of those coolers which have surfaces other than those of the tubes and gills must also be taken into account, and finally the increased efficiency of the cooling surface with the movement of the car through the air. The latter must here be considered.

Craddock found that hot water in tubes moved at 40 miles per hour lost heat at 12 times the rate of tubes in still air.

Taking Craddock's results and the mean difference in the rates of cooling in still and in moving air as obtained by the foregoing calculations, it will be seen that the rate of cooling for any other speed below Craddock's would be as follows—

M=Craddock's experimental speed=40 miles per hour.
    t=time of cooling at M speed=12 minutes.
m=speed in miles per hour for time of cooling t^1.
t^1=time of cooling at m speed in miles per hour.
    Then

$$m = \frac{M\, t^1}{t} \text{ and } t^1 = \frac{t\, m}{M}$$

At a speed then of 7 miles per hour the rate of cooling would be

$$t^1 = \frac{t\, m}{M} = \frac{12 \times 7}{40} = 2.1$$

times the rate in still air, or the rate would be twice that in still air when

$$m = \frac{M\, t^1}{t} = \frac{40 \times 2}{12} = 6.65 \text{ miles per hour.}$$

RADIATORS AND WATER COOLING

At 6·65 miles per hour then the required combined tube and gill surface S would be reduced to $\frac{15\cdot80}{2}=7\cdot90$ sq. ft. per BHP., or it may be safely taken that $S=8\cdot0$ sq. ft. per BHP.

Surface required per horse-power

If, however, it can be assumed that less work will be done by the engine when the car is standing, or less fuel consumed, than has here been taken into account; and if further it be assumed or found that the engine of a given type of car is developing a smaller fraction of its full power than is here assumed when travelling at the above speed, then it is possible that S may fall as low as 5 to 6 ft,. but the conditions must be all properly and fully taken into account.

This area would, of course, be insufficient with the engine working at anything like full power, and the car either standing, unless a fan were used, or running under conditions of stern wind which produced a dead calm at the radiator. Such conditions, however, seldom or never exist, and it may be safely assumed that a car, without a fan, running up a stiff hill with engine working at full power, would have the benefit of an air current of at least 5 to 8 miles per hour; $S=8\cdot0$ sq. ft. would therefore suffice.

This would, perhaps, allow the temperature to rise somewhat at lower speeds than 6·65 miles per hour, or on a long hill, but the full power of the engine is seldom exerted for more than a very few minutes at under 7 miles per hour, and this amount of surface would without doubt be sufficient, especially as the greater temperature difference between the air and the water would increase the effectiveness of the cooling surfaces.

Additions would, no doubt, in some cases be necessary to allow for imperfect metallic connexion between tube surface and gills; for increase in incrustation, though this would generally be small, as the receptivity of the tube is so much greater than the emissivity of it, and the gills in and to air; and for loss of emissivity due to smooth or polished surfaces of any part of the cooler.

On the other hand improvement of the emissivity may be effected by a thin coating of dead lamp black paint or dead white lead paint coating, the emissivity of these being 100 as compared with 17 for very smooth or polished steel or only 7 for polished copper.

Character of surface

It may for practical purposes be assumed that the improvement in emissivity by a proper coating balances the loss by solder connexion of gill to copper tube surface when the gills fit, and the solder is merely a thin sweated connexion. Fastening by galvanizing is better in this respect than with solder, as zinc is of higher conductivity, and galvanized surfaces are good as heat emissive surfaces, the superiority being in proportion of about 1·5 to 1 with relation to ordinary bare iron surfaces.

The quantity of surface required, moreover, depends very much upon the arrangement of the tubes. They should not be packed closely and not placed directly behind each other in rows. The back rows only receive the air heated by those in front of them, and their effectiveness is seriously reduced, and in some cases their utility nominal. Freedom of air movement must be provided.

Considering the conductivity relations involved it may be assumed that sufficiently rapid water circulation is generally obtained, but care must be taken to secure rapid water movement.

Assuming then the 8·0 sq. ft. of tube and gill surface to be the surface required per BHP. it is necessary to consider the relative amounts of tube surface and of gill surface. This resolves itself into a question of practical expediency, as it is impossible to provide gills having an area proportional to the relation between the heat receptivity and conductivity of the material of a small hot water tube, and the power of the gill surface to transmit the heat to the air at low speeds.

The maximum rate of transference from hot water and steam tubes to water as shown by, for instance, Burnat's experiments among those already quoted was as 132 to 1 to still air. Peclet arrived at a higher result, and the rate deduced from the figures already given $\dfrac{4\cdot33}{0\cdot0377} = 115$.

The rate, however, of transfer of heat from the gill surface would be a decreasing rate from a maximum at the tube to a minimum at the periphery of the gill. A curve representing it would be asymptotic for a large part of the outer area of the gill, and this would reduce the useful ratio of gill surface to tube surface by at least one-half. Even this, however, would give much larger gills than are practicable.

Gills placed much nearer together than a quarter inch are inadvisable ; firstly, because they would hold the air too much with the car standing or at very slow speeds ; and secondly, the means of metallically fixing them would not be satisfactory.

Gills of large size are also from practical considerations inexpedient, and the ratio of gill to tube surface must be much less than thermal considerations alone would dictate.

Relation between tube and gill surface

The question may be considered with reference to two of the several practical forms of tube, namely the Loyal and the Clarkson.

A tube of $\frac{1}{2}$ in. external diameter will carry gills at about $\frac{1}{4}$ in. centres. They vary in size, but $1\frac{1}{2}$ in. may be taken as a size which is convenient as to distance apart of tubes, and which though thin will withstand ordinary usage.

The $\frac{1}{2}$ in. tube will have an 18·85 sq. in. per 12 in. of length, and in this length it will have, say, 48 gills. If of $1\frac{1}{2}$ in. square they will, allowing for central part displaced by the tube, and for the slightly rounded off corners, have an area of 2 sq. in. each, or 4 sq. in. taking both sides. Their total area per ft. of length will then be 48 × 4=192 sq. ins. The tube area is then to the gill area as—

$$\frac{192}{18\cdot85} = 10\cdot2 \text{ to } 1,$$

and the total surface is 192 + 18·85=1·50 sq. ft. per ft. length of tube.

At this same rate a cooler would require $\dfrac{8 \cdot 0}{1 \cdot 5} = 5 \cdot 33$ ft. length of gilled tube of the above size per BHP. of car engine.

The Clarkson spiral wire-covered cooler-tube of $\frac{1}{2}$ in. diameter, the wire being No. 20 gauge, the spirals 0·33 diameter, gives a ratio of gill surface to tube surface of approximately 6·1 to 1. The total surface of this tube then $= \dfrac{18 \cdot 85 + (18 \cdot 85 \times 6 \cdot 1)}{144} = 0 \cdot 93$ sq. ft. per ft. of pipe, and a cooler would require $\dfrac{8 \cdot 0}{0 \cdot 93} = 8 \cdot 6$ ft. length of the tube per BHP., assuming the spiral wire gills to have the same heat transmitting efficiency as flat surfaces.

On the cooling power of these spiral wire-covered tubes Mr. Clarkson made some experiments which should be quoted here. They were made in 1900 with steam passed into the tubes at about 20 lb. per square inch, and with an air blast estimated to be about 8 miles per hour, and at a temperature of about 60 deg. F. A form of Loyal disc gill cooler with $\frac{5}{8}$ in. tube as then used by the Daimler Co. was tested at the same time for comparison.

The spiral wire-covered tubes tested were of $\frac{1}{2}$ in., $\frac{5}{8}$ in., and $\frac{3}{4}$ in. outside diam. The following are the results obtained as given by Mr. Clarkson, the Loyal being taken as unity in the comparison for relative condensing power.

TABLE 2.—COOLING POWER OF LOYAL AND OF CLARKSON TUBES.

Description of Tube.	Diam. of Tube.	Thermal Units per lb. of Metal in Coolers per Hour.	Efficiency per lb. of Metal in Cooler.	Ounces of Water Condensed per ft. Length of Tube per Hour.	Relative Duty of Tubes as Condensers.
Loyal Corrugated disc gill . . .	0·625	866·32	1·00	15·00	1·00
Clarkson tube . .	0·50	2558·80	2·95	18·75	1·25
,, ,,	0·625	1916·40	2·21	19·90	1·32
,, ,,	0·75	1274·00	1·47	21·70	1·44

From the results of these experiments numerous other values may be calculated, and they lead to the figures given in Table 3.

329

TABLE 3.—ANALYSIS OF EXPERIMENTS WITH LOYAL AND WITH CLARKSON COOLER TUBES.

Name of Tube.	Diam. of Tube. Inch.	Surface of 1 ft. Length of Tube.		Length of Tube to 1 sq. ft. of Surface.		Steam Condensed.				Thermal Units per 1°F. difference of Temperature per Minute.		
		Plain Tube. sq. inches.	Gilled Tube sq. feet.	Plain Tube. Inches.	Gilled Tube. Feet.	Per Hour per 1 ft. Length of Gilled Tube, lbs.	Per Hour per sq. ft. of Surface of Tube reckoned as Plain Tube, lbs.[2]	Per Minute per sq. ft. of Surface of Tube reckoned as Plain Tube, lbs.	Per Hour per sq. ft. of Surface of Gilled Tube, lbs.	Per sq. ft. of Plain Tube.	Per sq. ft. of Gilled Tube.	Per ft. Length of Gilled Tube.
1	2	3	4	5	6	7	8	9	10	11	12	13
Loyal ·	0·625	23·56	1·310[1]	6·10	0·770	0·94	5·73	0·095	0·71	0·575	0·071	0·950
Clarkson ·	0·500	18·85	0·785	7·62	0·275	1·17	8·90	0·147	1·49	0·885	0·149	1·176
,,	0·625	23·56	0·980	6·10	1·020	1·24	7·57	0·126	1·26	0·763	0·126	1·247
,,	0·750	28·30	1·170	5·10	0·85	1·35	6·90	0·115	1·15	0·685	0·116	1·357

[1] The gills assumed to be 8 times the area of the tube surface.
[2] Assumed for the purpose of calculation, see p. 331.

RADIATORS AND WATER COOLING

It will be seen that the Loyal tube of $\frac{5}{8}$ in. diam., and 1 ft. in length condensed 0·94 lb. of steam with air moving at about 8 miles per hour. Taking the gill surface of this tube as 8 times the surface of the 1 ft. of $\frac{5}{8}$ in. pipe, the total surface of the gilled tube would be $\dfrac{23\cdot56 \text{ sq. in.} \times 8}{144} = 1\cdot31$ sq. ft. and the steam condensed per sq. ft. of such tube surface $= \dfrac{0\cdot94}{1\cdot31} = 0\cdot715$ lb. per hour.

Supposing for the moment the tube be without gills, or that the gills be ignored, it would require $\dfrac{144}{23\cdot56} = 6\cdot1$ ft. run of the tube to give 1 sq. ft. of surface, and on this assumption 1 sq. ft. would condense
$$0\cdot94 \times 6\cdot1 = 5\cdot73 \text{ lb. per hour.}$$
From this it would seem that the gills multiply the radiating power of the plain tube, under the conditions, by from 6 to 8 times, and that the Clarkson radiating wire surface has the full value of the amount of its surface.

The 8 miles per hour would increase the cooling power of the tubes as compared with still air by the value $t^1 = \dfrac{t\,m}{M} = \dfrac{12 \times 8}{40} = 2\cdot4$

Taking then the mean of the results obtained with $\frac{5}{8}$ in. Loyal tube, and $\frac{5}{8}$ Clarkson tube, namely 0·07 units and 0·126 units respectively per deg. difference in temperature per sq. ft. of tube and gill surface per minute, we find that it accords with the co-efficient deduced from experiments tabulated in Table 1 namely 0·0377, as 0·0377 × 2·4 = 0·090. The mean of the values given above is 0·098, the difference between the calculated co-efficient, and that obtained from these experiments being only about 1 per cent.

A series of experiments were carried out two years ago by Col. R. E. Crompton, R.E., with a single row of plain bright brass tubes 0·75 in. diam., and 18 in. long, and weighing 0·174 lb. per ft. run, or 0·26 lb. for the 18 in. length; with a similar row of Clarkson tubes, Rowe indented tubes and a Loyal cooler with three rows of tubes with corrugated gills. Steam was admitted at a little over atmospheric pressure and regulated so that a little vapour came away with the condensed water at the outlet. An air current of about 1,000 ft. per minute, or 11·3 miles per hour, was maintained by a fan, the atmospheric temperature being not less than about 62 deg.

Some of the results of these experiments are given by Col. Crompton in the annexed Table No. 4. It may be noted that the figures given as the quantity of steam condensed per square foot refer to a square foot of front of cooler, not of tube surface. It should also be mentioned that Col. Crompton found that freshly blacked tubes condensed about 15 per cent. more than the bright tubes, but that this fell to about 10 per cent. in some cases, some time after the tubes had been blacked. The tubes were tested in the horizontal and vertical positions with very little difference in result, but it was necessary to place them so that drainage took place freely.

TABLE 4.—COOLING POWER OF PLAIN TUBES AND LOYAL TUBES.

Name of Tube.	No. 13 Blacked.	No. 9 Blacked.	Clarkson ⅜″ Bright.	Clarkson ¾″ Bright.	Rowe Tube Bright.	Plain Tube Bright.
Weight of Tube used in Experiment.	lb. ·543	lb. ·906	lb. 1·024	(a) ·595.[1] (b) ·75.	·55.	·260.
Tubes changed in position

Changed 2nd time | 10·37 10·20 10·40 10·0 10·35 10·6 10·12 10·50 | 9·6 9·20 9·37 8·45 8·50 9·18 9·30 9·25 | 9·0 9·0 9·25 12·10 14·10 9·2 9·50 9·35 | 11·5 10·55 11·0 10·57 11·4 10·40 11·6 11·0 | 25·56 25·46 26·30 25·40 25·55 26·30 26·40 26·30 | 33 33 33 32·30 33·30 33 32·15 33·55 |
Mean water condensed in lb. per hour	lbs. 2·9	lbs. 3·26	lbs. 2·94	lbs. 2·75	lbs. 1·14	lbs. ·91
Lbs. of water per hour per lb. of condenser tube	lbs. 5·35	lbs. 3·6	lbs. 2·86	lbs. 3·7 or 4·6	lbs. 2·07	lbs. 3·5
Lbs. water per hour per sq. foot condenser surface 3 row tube	34·5	38·7	35	32·5	13·4	10·8
Fig of merit	7·3	5·5	4·0	4·8 & 6·0	1·42	1·48

[1] (a) Lightest tube.   (b) Heaviest tube.

It may be concluded then that practical considerations indicate that the radiating gill or wire surfaces may usefully be from 6 to 8 times the surface of the tubes to which they are attached.

The practice of different makers differs almost as widely as it is possible in this matter.

Difference of amount of surface used Measuring the tube surface only, and neglecting the gills, of a large number of the small cars up to 8 BHP., the tube surface varies from 0·23 sq. ft. with gills to 2·1 sq. ft., without gills per HP. In many cases the gill surface reaches 10 to 1 of tube surface, and the cooler has tank or header surfaces which receive the full benefit of the air current, so that it may be said the practice varies from a total of 2·3 sq. ft. to 9 sq. ft. per HP. No wonder that some of them get very hot and want a lot of water on trying roads.

The following Table No. 5 gives the approximate measurement of the tubular part of the coolers of a large number of cars, all the smaller of which ran in the Hereford Trials last August and September.

RADIATORS AND WATER COOLING

TABLE 5.—RADIATOR WATER COOLERS OF VARIOUS CARS.

Name.	H.P.	Outside Dia. of Tube Inch.	Length of Tube Feet.	Area of Tube Surface Sq. ft.	Approximate Gill Surface to 1 of Tube Surface.	Total Surface Estimated.	Tube Surface per HP. Feet.	Total Surface per HP. Feet.	Description.
Rover	8	0·5	14·0	1·84	7	12·88	0·230	1·61	Round disc gills —Fan.
Prosper	8	0·55	13·0	1·87	8	14·96	0·234	1·87	Flat gills 1·5 in. sq.
C. Service	6·5	0·62	10·0	1·63	8	13·04	0·251	2·01	Laeis gills
Jackson	6	0·56	11·0	1·62	8	12·96	0·270	2·16	Square gills.
Croxted	9	0·56	20·8	3·06	?	?	0·331	?	?
Alldays	8	0·44	24·3	2·80	6	16·80	0·351	2·10	3 tubes in 1 gill.
Star	7	0·56	23·0	3·42	7	23·94	0·490	3·42	Corrugated gill.
Swift 1 Cyl.	7	0·50	27·0	3·54	8	28·32	0·507	4·08	Gills 1·5 in. sq.
Swift	7	0·50	28·6	3·75	8	30·00	0·535	4·28	Gills 1·5 in. sq.
Speedwell	6	0·62	21·7	3·55	8	28·40	0·594	4·73	Square gills
,,	9	0·56	36·8	5·41	8	43·28	0·602	4·81	,, ,,
Chambers	7	0·50	36·0	4·72	0	4·72	0·675	0·675	Plain tube, dull black.
Enfield	8	0·56	37·4	5·48	8	43·84	0·685	5·48	Square gills.
Siddeley	6	0·56	30·16	4·80	6	28·80	0·800	4·80	Clarkson tube.
De Dion	6	—	10·0	4·90	0	4·90	0·816	0·816	Flat tubes 2·25 in. wide.
Humber	7·5	0·50	50·0	5·50	6	33·00	0·737	4·42	Gills 1·25 in. sq.
Wolseley	6	0·62	30·0	4·91	6	29·46	0·820	4·91	Slit gills.
Mobile	6	0·50	39·3	5·16	9	46·44	0·862	7·73	Gills 1·75 in. sq.
Brown	8	—	32·0	16·80	0	16·80	2·100	2·10	Flat tubes 3 wide.
Wolseley	10	0·75	45·5	9·00	9	81·00	0·900	8·10	Gills 2·0 in. sq.
Belsize	14	0·56	36·5	5·38	9	48·42	0·385	3·44	Round corrugated gills.
"HONEYCOMB" TYPE OF COOLER.									
Martini	18	—	—	—	—	—	1·75	—	—
Talbot-Clement	16	0·42	411·0	61·0	0	61·0	3·80	3·80	1,234 sq. tubes 4 in. long.
Prunel	24	0·407	830·0	90·0	0	90·0	3·50	3·50	2,222 round tubes 4·5 inches long Hexagon ends.
Humber	30	0·25	1428·0	119·0	0	119·0	3·97	3·97	4,284 sq. tubes 4 in. long.
Rochet-Schneider	8	0·393	528·0	54·5	0	54·5	6·80	6·80	1,269 round tubes 5 inches long in tube plates.

This Table gives the cooling surface of the tube coolers, the surface of the tube alone being given, and the estimated surface of tubes and gills combined.

Among these coolers are several with plain tubes without gills. These are the Chambers (Downshire) Car, the De Dion-Bouton, and the Brown. These have respectively 0·675 sq. ft.; 0·816 sq. ft., and 2·1 sq. ft.

of surface per HP. declared, and it may be assumed that the power actually used was not at any time in excess of this.

The Chambers cooler consists simply of two vertical rows of staggered thin brass tubes $\frac{1}{2}$ in. diam., and 33 in number, connecting a small bottom tank header, and at their upper ends a small tank forming the connexion between two side water columns. The tubes are sufficiently far apart **Horse-power** for the air to move freely past and among them, and the back row is almost **and radiator** or quite as well placed as the front row. The surface of this tank would, **capacity** of course, add materially to the tube surface, and probably make the total effective surface 5·5 sq. ft., or nearly 0·8 sq. ft. per HP. This, at 8 miles per hour, taking the figures given on page 326, and the constant for heat transmission to air as 0·0377, would represent

$$0·8 \times 0·0377 \times 2·4 \times (180-60) = 8·64 \text{ units.}$$

Theoretically 71·5 units must be carried off, and, therefore, this surface would appear to be only sufficient with the car running at 8 miles per hour to allow the engine to work at $\dfrac{8·64}{71·5} = 0·12$, or about one-eighth of its power. which in this case would be $7 \times 0·12 = 0·85$ HP.

The engine might, of course, be run with a higher jacket-water temperature, and the water in the cooler might be, say, 205 deg. instead of 180, which would give a temperature difference of 145 deg., instead of 120 deg., and this would raise the units transmitted under the above conditions to 10·44 units, and the power to 0·146 of the engine power or to 1·02 HP. of the 7. If standing still in quite still air the engine power would have to be cut down to $\dfrac{1·02}{2·4} = 0·42$ HP.

It remains to be seen whether this surface is sufficient even with the surfaces coated with lampblack or dull black paint, or white lead. Even if this surface were very much increased it would seem to be insufficient, judging by experience with other forms of coolers and the practice of different makers of high repute, but no doubt the spacing of the tubes adds to their comparative effectiveness.[1] The same may be said of the De Dion and the Brown smooth tube cooler, for the whole of the $2\frac{1}{4}$ in. width of the De Dion tubes cannot quite be as efficient as the small $\frac{1}{2}$ in. tubes, because the air temperature after its passage over the front part of the tube would be raised. Assuming, however, that the efficiency of the wide tube is not thus lessened, then the power of the cooler, if it may be so called, would only be $\dfrac{0·12 \times ·816}{0·675}$ or 0·145 of the engine power, or under one-seventh of the whole, with a cooler temperature of 180 deg. With a cooler temperature of 205 F. this would be increased to 0·175, or over one-sixth of the whole,

---

[1] Since the above was written Messrs. Chambers and Co. have increased the number of the tubes to 47, although with judicious driving the 33 were sufficient. The surface and the above-mentioned results will be increased by about $47 \div 33 = 1·42$ times.

which would in this case allow the engine to work at $0.175 \times 6 = 1.05$ HP., the engine being 6 HP., or if in still air it could only be worked at 0.44 HP.

The Brown Cooler has thin vertical tubes or chambers 3 in. wide, and $\frac{1}{4}$ in. thick at the centre, giving about 6.25 in. of surface per inch length. The tubes are closely spaced, and the air passing through would be considerably warmed before it reached the back part of the tube surface; but neglecting this, as with the De Dion, this cooler would give off in still air

$$2.1 \times 0.0377 \times (180-60) = 9.5 \text{ units,}$$

or sufficient to allow the engine to work at $\dfrac{9.5}{71.5} = 0.133$ of its power, or

between one-seventh and one-eighth of its power. At 8 miles per hour it permits $(0.133 \times 2.4)8 = 2.56$ HP. to be used without a fan, and with the assumed temperature difference of 120 degs. between maximum temperature of water entering the cooler, and of the outer air.

If this proves sufficient with these smooth tubes it is clear that they are considerably aided by the surrounding tank part of the cooler, and by radiation from pipes and engine.

Further, it would appear questionable whether several of the various forms of gills add anything like as much as is usually supposed to the plain tube power of air heating. There can be no doubt that several of them soon get coated with non-conducting dirt, and prevent the intimate contact of the air with the hot surface.

Turning now to the honeycomb form of cooler it will be observed that they have all much larger surface per HP. For calculation purposes this **Honeycomb radiators** may be taken as 3.75 sq. ft. per HP. This in still air gives 17.7 units, or allows the engine to work at 0.25 of its power. At 8 miles per hour the engine could be worked to 0.6 of its power, or in the case of the Talbot-Clement car of 14–18 HP., or, say, 16 HP. to 9.6 HP. at that speed; or to full power at about 14 miles per hour. In this calculation no allowance is made for adverse conditions of interior or of exterior surfaces of the tubes.

These coolers would, therefore, appear to be sufficient except when travelling up a long, stiff hill with a stern wind, but for this and to allow for slight coating of the inside of the tubes some allowance must be made. Although, however, all the surface is tube surface, and none of it gill surface, which has such a very much smaller value, these coolers are not effective without the assistance of an exhaust fan even at considerable speed. The causes of this want of efficiency are several. Firstly, nearly all of them have tubes too small in diameter, and much too small in proportion to their length. The air without considerable help by the exhausting fan packs in these small tubes, and most of the temperature difference is gone before the air has passed more than about five diameters of the length of the tube, the remaining length of the tubes having little effect. Secondly, the water space between the tubes is so small that without very much more powerful pressure circuation than is used, the very thin film of water gives up all the heat that can be carried off under the conditions to the first portion of the great batch of tubes commencing at the inlet of the hot water. Only a part of the surface thus does anything like full duty. It would seem that these coolers should have larger and shorter tubes, and the water space between them should not be a mere film. These may be called air tube coolers, and are now much used. With some improvements they might be made as efficient as properly made simple water-tube coolers, but the advantage of the air-tube over the water-tube cooler is not obvious and the necessa y employment of an exhaust fan is a disadvantage.

In some of the most recent arrangements the water-tube cooler with gilled tubes is employed because of its superiority over the close tube honeycomb form, and a fan is employed. In the Renault cars the radiator is fitted against the dashboard and the fly-wheel acts as fan. In some cases the fly-wheel fan may make very small, if any, addition to the lost work in the engine, as the speed of the air through the fan-blade spokes may approximate to that of the car through the air. Radiators of all kinds and of the best make can now be obtained from British manufacturers.

It may be mentioned that among the many others, including Craddock, who experimented with air condensers and hot-water coolers, was E. Perrett, who in 1886 patented air condensers for use with locomotives, like the Cannstatt coolers with an exhausting fan at the back, and W. Turnbull, who in 1877, patented somewhat similar apparatus for cooling condensing water on locomotives, the tubes being only separated by the thickness of a wire mesh which determined and regulated their distance apart.

Chapter XXII

CRANKSHAFTS AND AXLES

ONLY the use of steel of very high mechanical properties has enabled the constructors of motor vehicle engines to produce crankshafts which can withstand the heavy and ambiguous stresses which are visited upon them chiefly by work and abuse, but partly as a consequence of the form given to some of the cranks.

My object in this chapter is not to enter at length into the theory of the means of determining the strength of crankshafts generally, as these can be found in several text-books[1] on applied mechanics, so far as the methods of first principles have permitted those who have written these books to give them. To the indications obtained by these methods the practical engineer has to make his own additions or modifications, so that the theory is made complete by taking into consideration the practical conditions which are only realized by that type of practical man who can through experience realize the feelings of a piece of metal under given circumstances, think as it would think if it could think, and put the thought into words or dimensions. An estimate of the torsional force and the torsional resistance of a revolving shaft transmitting a given load may be made with a useful approximation to accuracy for a simple shaft or spindle, or for one element of a crankshaft. The stresses in and strength of a crankshaft for a double or multi-cylinder engine are, however, difficult to determine, and appeal must be made to experience with reference to the very high speed and high grade steel of the modern motor vehicle crankshaft. The estimates which may be made by reference to cylinder pressures, strength of the steel of which the cranks are made and the stroke, form a guide or check which is useful, but experience is necessary to enable the designer to determine the dimensions which shall include his estimate of the resistance to torsion of the particular material he is using, and his

[1] Cotterill, *Applied Mechanics*, p. 362 ; Rankine, *Machinery and Mill Work*, p. 544 et seq. ; Jamieson, *Applied Mechanics*, p. 256 ; Unwin, *Elements of Machine Design*, part i. pp. 90, 208–25 ; part ii. p. 85 ; Rankine, *Applied Mechanics*, p. 358 ; Unwin, *Testing Materials of Construction*, p. 333 ; Davies, *Solutions to Board of Trade Questions*, pp. 85, 252. *Proc. Inst. Civ. Eng.*, vol. lxxiv. p. 258 ; vol. xc. p. 382 ; vol. cxxxviii. p. 1.

estimate of the necessary factor of safety, when the effect of shocks due to sudden application of force as in an internal combustion engine, and equally sudden variations of load, are taken into consideration.

The text-books referred to all deal more or less with the keys to a solution to the problem here presented, but they none of them sufficiently separate mathematical playfulness from the serious and definite questions which present themselves to the designer.

As it is troublesome to all, and exceedingly tiresome, and even difficult, to most to thresh out the practically applicable essence from the elaborations and incomplete examples in most books, it may be useful here to present some simple tools of calculations which will perform definite practical work.

It will simplify a perusal of these expressions if it is stated that some experiments for determining the resistance of materials to torsion were made with fixed bars of iron and steel 1 in. in diam., to which a lever 12 in. **Determination of strength by experiment** long was fixed. It was found that a load of 1,200 lb. was sufficient to twist a 1 in. round bar of steel, and it is known that the torsional resistance of a shaft is proportional to the cube of its diameter. From these experiments, and from others set out at length in the *Proc. Inst. C.E.*, already referred to, the following expressions have been derived.

Let K = The torsional resistance of a 1 in. bar loaded at 12 in. radius.

TR = The twisting moment or product of load W and lever length, or crank radius L in inches.

f = Factor of safety.

D = Diameter of shaft.

Then the safe load $W = \dfrac{12\,K\,D^3}{L\,f}$ lb.

$$W\,L\,f = 12\,K\,D^3$$

$$D^3 = \frac{W\,L\,f}{12\,K} \text{ or } \frac{TR\,f}{12\,K}$$

$$D = \sqrt[3]{\frac{TR\,f}{12\,K}} = \text{diam. of shaft.}$$

The first requisite is an estimate of the force likely to be brought to bear by the application of the force upon a piston transmitted to a crank (1) when explosive combustion occurs ; (2) when a back ignition occurs ; (3) when an engine is abruptly brought to rest by sudden application of the clutch.

For the purpose of determining the suitable diameter of a shaft it will be advisable to take the maximum pressure on the piston, and assume the whole of it transmitted to the crank, although in practice the maximum pressure occurs when the crank, is in such a position that the twisting or torsion moment is not the maximum. A pressure, equivalent to the mean effective pressure occurs after the crank has reached a position of right angles to the produced piston centre line, and it may afford some guide if this be determined with reference to the HP. of the engine under consideration.

The mean twisting moment T M in statical inch pounds will be—

$$T M = \frac{12 \times HP. \times 33000}{2 \times 3\cdot1416 \times N}$$

$$= 63024 \frac{HP}{N} \text{ inch lbs.}$$

N being the number of revolutions per minute, and $2\pi N$ the angular velocity.

For a single-cylinder engine then of say 10 HP. running at 800 revs. per minute, the mean twisting moment in inch lbs. is,

$$T M = \frac{63024 \times 10}{800} = 788 \text{ inch lbs.}$$

Taking L=the radius of the crank, or the length of the arm at which the twisting force acts, the twisting moment is $\frac{788}{L}$, or if the stroke is 6 in.

then the twisting moment is $\frac{788}{3} = 262\cdot6$ for the mean acting pressure or

load W on the crank-pin. For steam engines this quantity is useful as the ratio of the maximum and the mean twisting moment may be approximately estimated. For the steam engines used in motor vehicles it may be taken as 2·25 to 1, and the diameter of a shaft calculated with reference to the mean acting pressure would have to be multiplied by 1·3 to increase its strength in agreement with the ratio 2·25 to 1. For internal combustion engines, however, it is simpler to take a maximum pressure on the piston at the period of the stroke, when the angle a the crank makes with the centre line multiplied by the then pressure P is a maximum, or when the load W=A P sin a, and the torsional moment is approximately

$$T M = A L P \sin a.$$

A being the area of the piston.

It will, however, generally be sufficient to take the twisting moment as

$$T M = W L \text{ and } W = A P.$$

With a 5 in. cylinder and an assumed maximum pressure of 300 lb. per sq. in.

$$W = 5^2 \times \cdot7854 \times 300 = 5,889 \text{ lb.,}$$

and if the stroke be 6 in.

$$T M = 5889 \times \frac{6}{2} = 17,667.$$

This has to be met by a diameter of crank and crank-pin sufficient to have a considerable factor of safety in view of the conditions 1, 2 and 3 above mentioned.

Taking the twisting moment then as 17,667, and assuming a factor of safety f of 10, and the ultimate torsional strength K of the steel to be used as represented by a force of 1,800 lb. acting with a leverage of 12 ins., or $1,800 \times 12 = 21,600$ in. lbs., we have—

$$T R = T M = 17,667 \text{ and } T R f = 176,670.$$

To obtain this torsional resistance with steel of the strength named we shall require a diameter D of crankshaft—

$$D = \sqrt[3]{\frac{T R f}{12 . K}} = \sqrt[3]{\frac{176670}{21600}} = \sqrt[3]{8 \cdot 17}$$

$$= 2 \cdot 0 \text{ in. for the case taken.}$$

For a given diameter of crankshaft

$$T R . f = 12 \, K \, D^3,$$

and the greatest piston load with the assumed factor of safety would be—

$$W = \frac{12 \, K . D^3}{L . f}.$$

Strength of solid and hollow shafts

For hollow shafts the torsional resistance would be found to be equal to that of a solid shaft, when

$$D^3 = \frac{D_1{}^4 - D_2{}^4}{D_1} \text{ and}$$

if the internal diameter D_2 be selected, and be represented by the relation $D_2 = x \, D_1$, then D_1 being the external diameter,

$$D_1 = \sqrt[3]{\left(\frac{D^3}{1 - x^4}\right)}$$

Thus the $2 \cdot 0$ in. shaft of the preceding example would if a hollow shaft of equal strength with a hole through it of say 1 in. diam., or $0 \cdot 5$ of the 2 in. have an external diameter of—

$$D_1 = \sqrt[3]{\frac{2^3}{1 - 0 \cdot 5^4}} = \sqrt[3]{\frac{8}{0 \cdot 9375}} = 2 \cdot 043 \text{ in.}$$

On the other hand, if the 2 in. crankshaft had a 1 in. hole through it, it would instead of carrying the load W of 5,889 lb. carry the lessened load of

$$W_1 = 12 \, K \frac{(D_1{}^3 - D_2{}^3)}{L f} = \frac{21600 \times (8 - 1)}{3 \times 10} = \frac{151200}{30} = 5,040 \text{ lb.}$$

or $14 \cdot 4$ per cent. less than the solid shaft, while it would be 25 per cent. lighter.

The application of the load W on a crank-pin is attended with a twisting effort in the crank-web, and a bending effort on crank-pin. These forces may also be estimated, but it is not necessary to give examples here because the web will as an accident of construction always be of sufficient strength if the thickness be one-half of the breadth. The breadth settles itself, as the rounded shoulder at the journal or at either end of the crank-pin makes it necessary that the width of the web shall be greater than the diameter D already found by from 20 to 25 per cent.

Thus the stresses on, and dimensions of, a crank which is a single over-hang crank or a bent crank with a bearing close to each, Fig. 305, are easily determined. This is not, however, the case with one form of crank, namely Fig. 307, the double crank without central bearing, and with one long arm connecting the two crank-pins, a form to which I have alluded unfavourably in earlier pages. With the double-bearing crank there is comparatively small torsional stress on the crank-pin, but in the double cranks without centre bearing a very heavy torsional effort is thrown on the crank-pins, and the bending stress is enormous, and is complicated by a very heavy twisting stress.

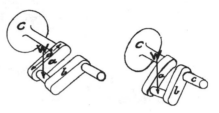

FIGS. 305 AND 306.—CRANK WITH BEAR-ING ON EACH SIDE OF CRANK ARMS.

Turning to Fig. 305 the crank a b has visited upon it a load w tending to produce motion which is resisted by say a clutch c. If w is in excess of the resistance of the web a to bending and twisting, it will tend to take the form shown in Fig. 306. The pin may not bend, but it will go with the twist of the web a, and probably break at its junction with one of the webs, although the web b is free to move, even though somewhat confined in direction by the bearing at c. The twisted arm has a length d, and the torsional stress on the crank-pin is negligible. Cranks built up of crank-shaft, discs and pin occasionally loosen at the disc centres, and then throw very heavy stresses on the crankshaft bearings.

Nature of crankshaft stresses

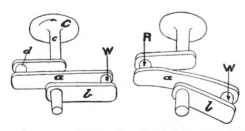

FIGS. 307 AND 308.—CRANK WITH LONG CEN-TRAL ARM AND NO CENTRAL BEARING.

Turning now to diagram, Fig. 307, it will be seen that the turning effort of the load w acts through the long web a, and rotates the crankshaft c by a very heavy torsional stress on the crank-pin d, this stress converting the downward movement at w into a lifting move-ment at d. The result is a bending and twisting stress on the web a, a wrenching torsional stress at d, a twisting stress on the web b, as indicated by Fig. 308.

The bending stress in the web a is twice that in the crank with a bearing between the cranks, and the torsional stress in the crank-pins is many times greater, and this means that the crank-pins must break away at their junctions with the webs or the webs must break off close up to the crank-pins.

The determination of the actual stresses resulting from these combined bending, twisting and torsional efforts is a matter of much difficulty. The twisting of one web may be easily determined, but the effect of the combined twisting effort on the three webs renders any attempt to determine dimensions by a resolution of these stresses unsatisfactory.

It may, however, be concluded that the strength of the shaft may be satisfactorily determined, and that the webs may easily be of sufficient strength in the cranks with bearings on either side of each crank-pin; but that those cranks that have one long dip between two crank-pins need much greater strength in the webs and crank-pins than the other form.[1] It may also be said that generally crank-pins should be larger than in common practice, even though this may increase the weight of the big end of the connecting rod, which might, however, be rather the narrower for the increased diameter.

AXLES

Fixed axles   The fixed axles of chain-driven cars may be considered as subject to bending stresses only, although sometimes they are subject to the simultaneous orthogonal stresses, due first to the stress resulting from the load carried, and, secondly, to the stress resulting from contact of the wheels with obstructions on the road. Numerous minor elements affecting these stresses might be taken into account in an estimate of the total, but the main stresses may be calculated as follows—

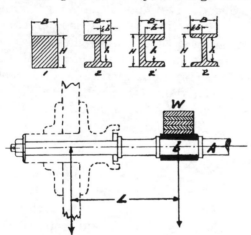

FIG. 309.—DIAGRAM OF DRIVING-WHEEL AXLE.

Suppose A, Fig. 309, to represent a fixed axle of the round type held firmly in a clamp bearing under the springs seen in section at w, as in the Wolseley **Calculation of stresses** cars. The length of the bending arm may be taken as the distance from the centre of the spring bearing b to the centre of the wheel. In practice it would often be less than this, as the point of fixture might be taken as near the outside edge of the spring bearing, and inside the wheel centre, but on the other hand it might occasionally exceed this length through looseness of the spring clip, or of the wheel on the axle, or through the varying incidence of the load as affected by road conditions.

Let L=length of the bending arm.
    D=required diameter of the axle in inches.
    K=Strength in lbs. per sq. in. in tension within the elastic
        limit of the material of which the axle is made.
    W=load on axle in lbs. at w, Fig. 309.

[1] See paper by Author—*Automobile Club Journal*, 1 Dec., 1904; *The Horseless Age*, July 26, 1905. A study in four-cylinder crankshafts.—P. M. Heldt.

CRANKSHAFTS AND AXLES

f = a factor of safety.

BM = Bending moment or corresponding moment of resistance.

Z = Modulus of the section, or value of the section for purposes of calculation of the moment of resistance B M.

For circular axles of diameter D, $Z = \dfrac{D^3}{10 \cdot 2}$.

For square axles where the size is s, $Z = \dfrac{S^3}{6}$.

For other axles (see 1 Fig. 309) of rectangular section, where B is the breadth, and H the depth, $Z = \dfrac{BH^2}{6}$.

For girder sections (see 2, Fig. 309) where B = breadth, H = depth, and b = width of flange on both sides of the web, or in the case of channel sections on the side of the web, and h = depth between flanges.

$$Z = \frac{BH^3 - bh^3}{6H}.$$

The modulus of the section multiplied by K must have a value equal to B M.

That is $Z K = B M$, and $K = \dfrac{BM}{Z}$.

The axle is considered the equivalent of a beam fixed at one end and loaded at the other, when

$$B M = W L$$

Then for circular shafts $WL = \dfrac{K D^3}{10 \cdot 2 f}$, from which

$$D = \sqrt[3]{\frac{10 \cdot 2\, WL f}{K}}.$$

For square shafts $WL = \dfrac{K f S^3}{6}$, from which

$$S = \sqrt[3]{\frac{6\, WL f}{K}}.$$

For rectangular shafts $WL = \dfrac{K f BH^2}{6}$, from which

$$BH^2 = \frac{6\, WL f}{K}.$$

For axles of girder section $WL = ZK$ or $Z = \dfrac{WL}{K}$.

Manufacturers of steel joists and other sections habitually give the value of z for all descriptions of sections which they usually supply. When

Strength of different types of axles

343

the value of z has, therefore, been ascertained from the above expression an axle section of suitable dimensions may be selected from the manufacturer's catalogue. If, however, unusual dimensions are contemplated then a section must be designed in which $Z = \dfrac{BH^3 - bh^3}{6H}$. As exemplifying the application of these expressions, calculations for a circular, for a square, and for a rectangular section will be given. Let W=1,000 lb. L=9 in. K=34,000 lb. and $f=4$.

$$\text{Then } D = \sqrt[3]{\frac{10 \cdot 2 \times 1000 \times 9 \times 4}{34000}} = 2 \cdot 21 \text{ in. diameter,}$$

$$\text{and } S = \sqrt[3]{\frac{6 \times 1000 \times 9 \times 4}{34000}} = 1 \cdot 85 \text{ in. square,}$$

$$\text{and } BH^2 = \frac{6 \times 1000 \times 9 \times 4}{34000} = 1\frac{9}{16} \text{ in.} \times 2 \text{ in. deep.}$$

The assumed material is high quality fagoted iron of great toughness, having an assumed elastic limit of about 11 tons per sq. in., and a very high factor of safety is assumed because of the uncertain stresses to which axles of high speed cars are subjected.

Strength of material

For the high strength tough steels, as used in the axles turned down from the rolled bar for shoulders out of the solid in the Wolseley cars, a very much higher elastic strength would be taken, probably as high as 50,000 lb., or say $22\frac{1}{2}$ tons per sq. in., and this would reduce the necessary diameter in the example given for a circular section to 1·94 ins.

If some of the very high strength recent nickel alloys be used an elastic strength of from 70 to over 100,000 lb. may be assumed. See chap. xl.

Live axles

The stresses in live axles in which the rotating part acts as propeller and as load-carrying axle exceed those in the fixed type, but these axles should be much larger in diameter than is apparently necessary to meet the combined bending and torsional stresses, because the stress at every part of the bending locality of the axle is changed from tension to compression, and vice versâ at every revolution. Although the resultant loss of life following the fatigue brought about by this rapid alternation of stress cannot be accurately estimated, it must be allowed for in assigning a diameter.

The combined bending and twisting moments may be provided for by addition, to the bending moment imposed by the load on the axle, of the equivalent bending moment represented by resolution of the twisting moment.

Let it be assumed, for purposes of illustration, that the diameter of the driving wheel is 30 in., and that at a speed of 10 miles per hour 15 horse-power is exerted at the driving-wheels, or 7·5 horse-power at each wheel. The effort at the periphery of the wheel is then

$$\frac{7 \cdot 5 \times 33000}{5280 \times 0 \cdot 166 \text{ miles per min.}} = 283 \text{ lb.}$$

344

Taking $K = 50,000$ lb., $L = 9$ in., and $W = 1,000$ lb. as before. $W_1 = 283$ lb.; the effort at the driving-wheel periphery, and $l = 15$ in., the radius of the driving-wheel.[1]

Then using the expression $D = \sqrt[3]{\dfrac{10 \cdot 2\,WL\,f}{K}}$ already employed for ascertaining the shaft diameter, W L is in this instance made up of a bending moment W L due to the load of 1,000 lb. on the axle, and a further bending moment the equivalent of the twisting moment due to the torsional driving effort. The second of these bending moments may be found from the expression

$$\frac{W_1}{2}\left(L + \sqrt{L^2 + l^2}\right).$$

The shaft diameter is, therefore,

$$D = \sqrt[3]{\frac{10 \cdot 2 \times f}{K}\left[W.L. + \frac{W_1}{2}\left(L + \sqrt{L^2 + l^2}\right)\right]}.$$

Putting values to these letters

$$D = \sqrt[3]{\frac{10 \cdot 2 \times 4}{50000}\left[9000 + \frac{283}{2}\left(9 + \sqrt{81 + 225}\right)\right]},$$

$$D = \sqrt[3]{\frac{40 \cdot 8 \times 12,749}{50000}} = \sqrt[3]{10 \cdot 4},$$

$$D = 2 \cdot 185, \text{ or say } 2\frac{3}{16} \text{ in. diameter.}$$

The designer must use his own judgment with regard to the factor f, and must be guided by the nature and rigidity of the axle fastening at b, by the permissible elastic flexure as affected by the distance or length L, by the size and flexibility or inflexibility of the driving-clutch and propeller shaft, by the character of the drive, i.e. whether by gear throughout or by one chain, two chains or three chains. He must be guided not only by the ultimate strength and elastic limit of the material he proposes to use for his axle, but, by the quality of the material as indicated by the co-efficient of toughness shown by the relation of the values for elastic resistance and elastic extension,[2] as well as the similar relations for maximum resistance and total extension before rupture. It is important for axles and steering gear above all things that the quality of toughness should be superior to all other mechanical properties ; while on the other hand a material with less toughness, but with higher elastic resistance may be adopted for crankshafts and transmission spindles. The factor f should, therefore, be taken not only with reference to the particular part to which it is to be applied, the results or dangers of failure being the determining consideration, but with reference to the characteristics of the material adopted for that part.

[1] The length L would be taken from the point of entry of the axle into the hub of the wheel to the centre of the axle bearing for those live axles which are rigidly secured in the hub or boss.

[2] See *Steel as a Structural Material*, by the Author. *Trans. Soc. Eng.*, 1880.

Chapter XXIII

POWER SPEED AND TRACTIVE EFFORT

FOR some time makers have gone on adding to the power of the engines specially designed for speed-trial cars. In doing this they have done great service in aid of motor vehicle development. They have added enormously to the horse power obtainable per lb. weight of engine and of car, and particularly their labours have been valuable in respect of increase of power of engine in proportion to its weight.[1]

Adhesion of rubber tyres

The limit to the useful or utilizable power, however, is given by the tractive adhesion of the wheels on the ground. Under the best possible conditions the co-efficient of adhesion of rubber tyres is not more than 0·7, and it often drops as low as 0·5.

With this as basis we may calculate the maximum useful power for a car with a given weight on the driving-wheels, and it will be found that the available adhesion is always greater than any of the powers hitherto applied to a motor vehicle except when starting or running at a very low speed.

Taking the mechanical efficiency E of the transmission mechanism between engine and road-wheel periphery as $0·70 = C$, and the adhesion co-efficient of rubber tyres on average surface as $0·65 = k$, and if W = the weight on the driving-wheels, and $v =$ speed in miles per hour, we then have for the maximum power which can at any speed be applied without quite slipping the wheels—

$$HP = \frac{W \, k \, 88 \, v}{33000 \, E}.$$

Possible power utilisation at various speeds

Thus, with a car having a load of 1,250 lb. on the drivers the greatest power that can be applied on the level at a speed of 1 mile per hour

$$= \frac{W \, k \, 88 \, v}{33000 \, E} = \frac{1250 \times ·65 \times 88 \times 1}{33000 \times ·7} = 3·10 \text{ HP.}$$

and the maximum power that can be utilized at starting when a very slow movement is required is a mere fraction of the power available on any car. If the car moves at the rate of say 40 ft. per minute only 1·4 HP. can be utilized once the inertia of rest has been overcome. This shows the great

---

[1] The engines used with the large racing cars have been constructed to weigh 13¼ lb. per HP. The weight of water-cooled engines of the types in general use is from 20 to 22 lb. per HP.

importance of an easy gradually operating clutch. The maximum power that can be employed is thus very small at low speeds, but it rises with great rapidity with increase in speed, so that in hill climbing even at, say, 5 miles per hour, the power which can be applied with the load mentioned of 1,250 lb. on the drivers reaches 15·5 HP.

At 30 miles per hour 93 HP. could be used, but, of course, this power would rapidly accelerate the car and drive it at a very much higher speed than this until the gradient or wind or other resistances raised the necessary effort at the periphery of the road wheels to greater than the 1250 × 0·65 or 813 lb., or to what would be a draw-bar pull of 813 lb., if the car were hauled by some form of tractor.

The proportion, however, of this available tractive adhesion which could be utilized rapidly decreases as the speed with a given power increases, and hence the tractive adhesion which is balanced by the effort of a 15·5 HP. engine at 5 miles per hour is more than double that which can be utilized by a 90 HP. engine at say 55 miles per hour.

For instance the 93 HP. would, were it not for indeterminate increase of resistances, decrease in mechanical efficiency, and of tractive adhesion at very high speeds, propel a car having with its load a total weight of say 23 cwt. up a hill of say 1 in 15, the effective area exposed to air resistance being 15 sq. ft., and the road resistance 50 lb. per ton, at a speed which can be thus found.

Speed calculation

Let W =the weight of the car and load.

P =brake horse-power.

R =resistance to traction of the vehicle on the road in lbs. per ton.

G =resistance due to gravity.

E =mechanical efficiency of transmission from crankshaft to road =0·85.

a =resistance due to air in lbs. per sq. ft.

A =effective area exposed to air resistance.

v =speed in miles per hour ; then

$$v = \frac{P \times E \times 375}{[(R + G) \times W] + A \times a}.$$

The relations between speed, power, resistances and tractive adhesion can be found by means of the formulae given at p. 58, vol. i.

In the case assumed : W =1·15 tons, R =50 lb., and G =149 lb. per ton, and the value A × a has to be found. Taking the expression given at p. 59, vol. i., for air resistance where the total resistance =V²A × 0·0017, and where V =speed in feet per second, v × 1·466 may be substituted for V, since 1·466 feet per second =1 mile per hour. The expression then becomes $(v \times 1·466)^2 \times A \times 0·0017$, and

$$v = \frac{93 \times 0·85 \times 375}{[(50 + 149) \times 1·15] + (v \times 1·466)^2 \times A \times 0·0017},$$

$$v = \frac{29625}{230 + 0 \cdot 054 v^2},$$

$$0 \cdot 054\, v^3 + 230\, v = 29625.$$

$$v^3 + 4259\, v = 548611$$

$$v = 64 \cdot 5$$

Air resistance   Since from the second stage of this reduction the total air resistance $= 0 \cdot 054\, v^2$ we may now find the value of this quantity.

$$A \times a = 0 \cdot 054 \times 64 \cdot 5 \times 64 \cdot 5$$
$$A \times a = 224$$

$$a = \frac{224}{15} = 14 \cdot 93 \text{ lb. per sq. ft.}$$

This is in agreement with the result given by the formula on p. 59, vol. i., and confirms the results of the above means of finding the speed $v = 64 \cdot 5$ miles per hour, part of the calculation, namely air resistance not being previously ascertained.

Tractive effort at high speed   But in propelling the car at this speed of 64·5 miles per hour it would only have called for a tractive adhesion of 454 lb., or little more than half that available under the conditions of lower speeds.

At the higher speeds the actual pressure upon the treads would vary so much from moment to moment owing to irregularities of surface, and rebounding of wheels jumping from point to point, that the tractive hold would no doubt fall very much below the 813 lb., available at low speeds, but not so low a proportion of the whole as is found to be necessary for the 64·5 miles of the example.

We thus see the necessity for a maximum weight on the driving-wheels for starting purposes, and for slow speeds, and the desirability of keeping **Conflicting require- ments** that weight down or of being able to lower it as the speed increases in high speed or racing cars. That which is necessary for " getting away " quickly is a burden, and cause of otherwise unnecessary wear of tyres of such cars for the greater part of their time of running. We also see that a racing car which might be the best in design for short races with standing starts or for races in which controls were frequent, might be beaten easily by a car designed for a long distance race with few controls.

It will be noticed that the tractive effort T, of a 1 HP. engine at 1 mile per hour is the same as that of a 10 HP. engine at 10 miles per hour, or 100 HP. at 100 miles the tractive effort being $T = \dfrac{33000 \times HP \times E}{88 \times v}$, or per 1 HP., and per mile per hour $T = \dfrac{33,000 \times 1 \times 0 \cdot 7}{88 \times 1} = 262$ lb.

A higher value of $E = 0 \cdot 85$ has been selected for the second example as more closely representing the efficiency to be expected with a racing car running at high speed with a direct drive and reduced transmission losses.

POWER SPEED AND TRACTIVE EFFORT

HILL CLIMBING

The ability of passenger motor vehicles to surmount most hills at speeds not greatly lower than those desirable on level roads necessitates the provision of engine power much in excess of that normally required. The use of powerful and heavy engines is, however, attended with disadvantages and compromise limits the speeds at which hills may be ascended.

The many hill-climbing trials that have been carried out from time to time have provided opportunities for making observations from which the performances of the cars may be known and from which much information of a comparative nature may be obtained. An inquiry into the question of the engine powers likely to be necessary, for the ascent of hills at required speeds, shows that calculations must be made embodying the whole of the available data affecting the work to be done.

The results of consideration of the subject are here given, as they may have useful application. In the formulae presented the letters have the following significance.

W=Weight of car and load in pounds.

w=Weight of car and load in tons.

H=Total ascent in feet.

L=Length of hill in feet.

R=Road resistance in pounds per ton weight of car and load.

r=Air resistance in pounds=V^2A 0·0017.

V=Velocity of car in feet per second.

A=Area of car in square feet presented to air resistance.

T=time occupied in ascent in seconds.

E=Efficiency of transmission of power from engine to road wheels.

HP=Effective horse power of engine.

p=Power given off at road wheels or work done in ascending hill.

S=speed of ascent in miles per hour.

Then if P=mean effective cylinder pressure.

L=length of stroke in feet.

A=area of piston in square inches.

N=number of useful or firing strokes per minute,

the brake horse-power of petrol or steam engines, having a mechanical efficiency of E is

$$\frac{EPLAN}{33000}$$

For compound steam engines the number of useful working strokes are those occurring in the low-pressure cylinder, and the mean effective pressure in the high-pressure cylinder or cylinders is referred to the area of the low-pressure cylinder.

The power given off at the road wheels during the ascent of the hill is

Power at road wheel

$$p=\frac{HW+(R\,w+r)\,L}{T\,550}$$

349

and the brake horse-power required at the engine with a selected value of E is

HP. of engine

$$HP = \frac{HW + (R\,w + r)\,L}{TE\,550}$$

With the other conditions known the efficiency may be found from

$$E = \frac{HW + (R\,w + r)\,L}{HP.\,T\,550}$$

the speed of the ascent in miles per hour from

Possible speed

$$S = \frac{L}{\dfrac{\overline{HWL + (R\,w + r)}}{HP.\,E\,550}\,L} \times 1 \cdot 466$$

and the time occupied in the ascent from

Time occupied

$$T = \frac{HW + (R\,w + r)\,L}{HP.\,E\,550}$$

Unless the probable speed of the car is high, or the wind velocity such that the pressure upon the surface of the car presented to air resistance opposing motion approaches that due to the speed of the car through still air, then the quantity r may be omitted from the above expressions without introducing more than slight error.

Calculation of air resistance

The following table of air resistances at speeds of from 10 to 100 miles per hour calculated from the formula hitherto used $V^2 A\ 0 \cdot 0017$ will make it easy to judge the speeds which may or may not sensibly affect results.

Comparison may be made between the horse-power of the engine of the car, and the fraction of its power represented by column 4, multiplied by the front area of the car or the surface exposed to wind pressure in square feet.

TABLE 6.—AIR RESISTANCE AT VARIOUS SPEEDS.

Velocity.		Resistance.	
Miles per Hour.	Feet per Second.	Pounds per sq. ft.	Horse-Power per sq. ft.
10	14·66	0·36	0·009
15	21·99	0·82	0·032
20	29·32	1·46	0·078
25	36·55	2·27	0·151
30	43·98	3·28	0·262
35	51·31	4·47	0·417
40	58·64	5·84	0·623
50	73·30	9·13	1·219
60	87·96	13·15	2·103
70	102·62	17·90	3·342
80	117·28	23·38	4·980
90	131·94	29·59	7·090
100	146·60	36·53	9·740

From the above table it would appear that at very high speeds, the constant 0·0017, obtained experimentally, and correct up to high speeds, is too high, although it is much lower than the constant deduced theoretically from the laws of falling bodies, namely 0·0025. Up to 60 miles per hour at least the air resistance as given in the above table seems to be confirmed by the practical results of running with racing cars, with fronts made of easy entrance, and tails of easy leaving form.

For instance assuming the surface presented by an 80 HP. racing car to be equivalent in area to 10 square feet of flat surface, the horse-power absorbed will be 49 at 80 miles per hour, leaving 31 HP. available for road resistance and friction of transmission gear. Taking the road resistance as not more than 45 lb. per ton at this speed, and the mechanical efficiency of transmission of a direct drive racing car in good condition as say 85 per cent., the power required for these two items would be 11·3 assuming the car to weigh one ton. There would thus be 20 horse-power remaining available of the 80 horse-power considered.

At the lower speeds there is a still larger margin, but at 85 miles per hour, and above it, the air resistance appears to be too high as found by the expressions which are perfectly correct for the lower speeds.

Necessity of further experiment

It would thus appear that the suggestion of the late Sir Frederick Bramwell [1] as to an " extensive envelope of air travelling if not at the speed of the train yet with high velocity " would apply to motor cars, and that the envelope so travelling provides a resistance lessening intermediary between the high-speed parts of the car and the stationary air at some distance outside it. Actual experiment at the very high speeds is still required to settle this point.

Formulae of simple form, derived from those already given, may be used to determine the relative performance of cars run in hill-climbing trials and speed contests. With such formulae a constant or factor may with advantage be employed, in order that the high-powered car with relatively small seating capacity may be discouraged, and, further, that the car with greater seating capacity but of unduly light construction may not be encouraged. That is to say a factor should be used to allow for the very variable relation of $\frac{w}{W}$, where W = the weight of the complete car, and w = weight of passengers or load carried.

Meritorious performance and handicapping

With a formula of the form $\left(\frac{W}{HP \times T} \times 100\right) - K$, the factor K may for instance have a value equal to $\left(\frac{W}{HP \times T} \times 100\right)\left(0.25 - \frac{w}{W}\right)$, and as long as a positive quantity is obtained it should be subtracted from the figure of merit given by the formula.

[1] *Proc. Inst. C.E.* vol. cxlvii., paper read by J. A. F. Aspinall; vol. clvi., paper by T. E. Stanton.

Chapter XXIV

VIBRATION AND TURNING EFFORT

The Influence of Combination and Sequence of Cyclic Efforts

A PROBLEM of considerable interest and importance to the constructor of motor vehicles is that of reduction of vibration, with consequent improvement in smoothness of running, and freedom from noise.

Principal cause of vibration
In vol. i. chapter xxxiv. it was pointed out that efforts were being very generally directed to internal balancing of the moving parts of the engine, regardless of the fact that irregularity of impulse and turning effort was, except in a few cases, more productive of vibration than the smaller and more rapidly fluctuating acceleration forces.

It is now more generally recognized that the problem differs from, and requires different treatment to that relating, for instance, to the control or balance of forces, and elimination of vibration, with fixed or stationary engines upon foundations of relatively great weight, or even the railway locomotive. The latter, although it has more in common with the motor vehicle, presents conditions differing widely from or not appertaining to it.

As the result of experiments, by many makers of motor vehicles, during the past four years, very marked improvement has been achieved by a few, and results have been obtained which, considered solely with regard to the requirements of smooth running and freedom from vibration, leave little to be desired.

These qualities have, however, been obtained in some cases by considerable addition to the cost of production, and by multiplication of parts to an extent that suggests an incomplete study of the possible means of effecting improvement. The subject will, therefore, be considered a little more in detail than in vol. i.

Uniformity of turning effort with steam engine
A comparison of the steam engine, as applied to motor vehicles, with the light oil engines of equal power is unfavourable to the latter when their relative equality of turning effort is considered. Not only may the frequency of repetition of effort be four times greater in the steam than in the oil engine, but the variation of pressure during each effort is usually considerably less with the former than the latter. The quiet running of a

352

motor car has, by the behaviour of most steam cars, become associated with the steam engine, and is undoubtedly in part due to the greater continuity or uniformity of driving effort. It may be urged that the construction of the steam engine and the use of the condenser account for apparent smoothness of working, but it must be remembered that in some cases the design of the engine and particularly of the valve gear is almost identical with the petrol motor, that the condenser is not invariably used, and that in some cases also the transmission gear between engine and road wheels is of precisely the design adopted in connexion with the petrol motor, but without the advantage sometimes conferred by freedom or relief at the clutch.

Some consideration may, therefore, be given to the possible means of improvement in regularity of turning effort with the petrol motor as having an important bearing upon the quietness of running of the car. While there has been recognition in design of the improvement to be effected in the directions here indicated, undue attention has not infrequently been given to some points with corresponding neglect of others of equal importance.

An estimate of the relative influence upon design of the following leading conditions may be expected to affect beneficially the whole arrangement.

1. Increase in speed of revolution of engine.
2. Increase of number of cylinders.
3. Increase of fly-wheel energy in relation to engine power.
4. Sequence of cycles in relation to sequence of cranks.
5. Method of governing output or horse-power of engine.
6. Position of the engine in the car.

Conditions 1 and 2 are interdependent, and from what follows, are, it will be seen, controlled by 4. An increase of one or other of these, or infrequently of all of them, has proved advantageous.

Before discussing the effect of sequence of cycles and efforts with various numbers and combinations of cranks, an example will be given showing among other things the relative values of the inertia forces and the extent of the variation, during one complete cycle, of the pressure in the cylinder, and the turning effort at the crank-pin.

Diagram, Fig. 310, is based upon the following data, selected as representing the average and not exceptional conditions.

The indicator diagram,[1] curve A, was taken with a Crosby indicator from a Wolseley petrol engine running at 380 revs. per minute. The cylinder diameter and stroke were $8\frac{1}{2}$ in. and 10 in. respectively. At the speed given a reliable diagram can without doubt be obtained with the type of indicator used. At much higher speeds, diagrams have been obtained, free from apparent distortion, with mirror reflecting indicators

---

[1] Information relating to indicator diagrams, indicators, etc., may be found in *The Steam Engine Indicator and Indicator Diagrams*, by the Author.

of the type introduced by Professor Perry and MM. Hospitalier and Carpentier, but they have also exhibited peculiarities which threw some doubt on their accuracy. The Wolseley diagram has, therefore, been employed, and in order that normal figures may be employed for the following calculations a cylinder diameter of 4·5 in. and a stroke of 5·75 in., with a speed of revolution of 900 per minute have been adopted. The weight of the reciprocating parts, namely, the complete piston and connecting rod, has been taken as 12 lb.; and the length of the connecting rod equal to twice the stroke of the engine. The mean effective pressure throughout the four strokes is 20 lb. per sq. in., or 80 lb. per sq. in. if considered during the firing stroke only. The negative work during exhaust, suction, and part of the compression stroke, was allowed for by measurement of the diagram by planimeter. The maximum pressure during the explosion stroke is 270 lb. above, and the minimum pressure occurs during the suction stroke, and is nearly 10 lb. below atmospheric pressure. Diagrams from the same engine show maximum pressures of 330 to 340 lb. per sq. in. when ignition has occurred a little earlier. It is, however, preferable that the pressure rise should continue a little after the commencement of the stroke, and as shown in diagram, Fig. 310.

Inertia forces   The curve of inertia of the piston and connecting rod has been plotted, from calculations of the amount of the acceleration force expressed in its equivalent value in lbs. per sq. in. of piston area, at the commencement and termination of the stroke, the curve connecting these points crossing the zero line at the point of no acceleration, or where the moving parts have attained their maximum velocity. With a ratio of connecting rod length to crank radius of 4 to 1 the point of zero acceleration occurs as indicated by vertical dotted line on the diagram at 43 per cent. of the out stroke.

The formula used for calculation of the acceleration force is

$$\frac{W\,v^2}{gr}\left(1\pm\frac{1}{n}\right),$$

where W = weight of the moving parts in lbs.

    v = circumferential velocity at crank-pin in feet per second.

    g = 32 feet per second gravity, acceleration.

    r = radius of crank. 2·875 in. or 0·24 ft.

    n = ratio of length of connecting rod to crank radius

    a = area of piston 4·5 in. diam. = 15·9 sq. in.

The total amount of the force is then

$$\frac{12\times\left(\dfrac{2n\times0\cdot24\times900}{60}\right)^2}{32\times0\cdot24}\left(1\pm\frac{1}{4}\right)=990 \text{ and } 594 \text{ lb.}$$

or $\dfrac{990}{15\cdot9}=62\cdot25$ lb. and $\dfrac{594}{15\cdot9}=37\cdot35$ lb. per sq. in. of piston area. The dotted curve[1] on the diagram is the indicator diagram curve corrected for the positive and negative inertia forces represented respectively above and below the zero line, by the areas B, C, D, and E.

The turning effort at the crank-pin has been obtained from the dotted curve by means of the expression $P_1 = aP \sin \theta$, where θ is the varying angle which the crank makes with the centre line of the engine at different points of the stroke, P the pressure in lbs. per sq. in. of piston area, shown by the dotted curve, and P_1 the tangential effort at the crank-pin, a being the piston area in square inches.

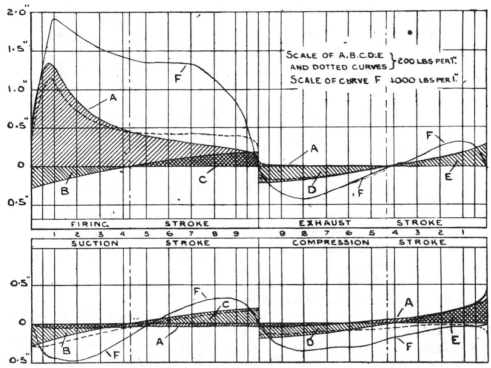

FIG. 310.—CURVES OF PRESSURE, INERTIA, AND TURNING EFFORT FOR A SINGLE-CYLINDER, FOUR-STROKE CYCLE, PETROL MOTOR.

The following table gives the results of the calculations made to find the value of the turning effort during the firing stroke. The quantities in column 4, plotted to a scale of 1,000 lb. per inch of height, are shown by the turning effort curve F. Calculations for the exhaust, suction, and compression strokes are made in the same way, and need not be repeated here as examples—

Value of turning effort

[1] See paper by the Author on " Development of the Light Car," in the *Automobile Club Journal*, Dec. 1, 1904, p. 420.

Point of Stroke.	Pressure shown by Indicator Diagram Corrected for Inertia. Dotted Curve, lbs. per sq. inch.	Value of Sin θ.	Value of $a\,P$ Sin θ lbs.
0·0	24	0·000	0·00
0·05	194	0·393	1210
0·1	220	0·55	1925
0·2	146	0·749	1740
0·3	112	0·872	1550
0·4	94	0·951	1420
0·5	86	0·993	1360
0·6	86	0·999	1368
0·7	88	0·957	1340
0·8	82	0·857	1115
0·9	80	0·662	843
0·95	76	0·47	568
1·00	52	0·00	0·00

Having given the data employed, and the selected method of representing the results obtained, it may prove interesting to refer to some of the information which an inspection affords.

To those who have not considered the question it will be surprising to note that at about 0·1 of the firing stroke, when the explosion pressure reaches 270 lb. per sq. in. there is a load on the piston of 4,300 lbs., or an amount approaching two tons, and that during the suction stroke the maximum suction effort opposing movement of the piston is 159 lb. From these figures it will be gathered how variable the load upon the piston is, and, as mentioned already, the mean effective pressure throughout the cycle is only 20 lb. per in., equivalent to a load on the piston of 318 lb., and an amount that would be represented by a horizontal line only 0·1 of an inch above the zero line which with this diagram is also the atmospheric pressure line.

Variation of turning effort

The turning effort curve is indicated by the letter F. The maximum effort occurs at 0·1 of the firing stroke and is equal to 1,925 lb. tangential effort at the crank-pin, whilst the minimum effort is at 0·2 of the suction stroke, and has a negative value of 480 lb.

It will further be noted that the turning effort has a positive value during the firing stroke, a negative value during the compression stroke, and a value changing from negative to positive during the exhaust and suction strokes, due to the value of the inertia forces in relation to the pressure in the cylinder. A tendency to knock, if there is any looseness at connecting rod and crankshaft bearings, would occur during the exhaust and suction strokes at the instants when the curve F crosses the zero line. As, however, the curve crosses slowly, and at a comparatively small angle with the base line at these points, the knock would not be severe, unless there was heavy wear at the bearings. The most serious knock would occur at the end of the compression stroke at light loads, when the compression is less, and the speed frequently high. Under these conditions the curve F would cross the zero line towards the end of the compression stroke, there would be a positive turn:ng effort on the crank-pin and separation, to the extent of the freedom in the bearings, of crank and gudgeon pins, from their bearing surfaces. The increase of pressure upon the piston accompanied by change of direction of movement of piston would then result in compression instead of tension, at the connecting rod, and the production of a knock by impact of the separated bearing surfaces.

Pounding or thumping, due to badly timed period of ignition, is not an uncommon cause of so-called knocking, but is distinct from the knocking to be expected under normal conditions of running.

The equivalent value of the inertia force is drawn to the same scale as the indicator diagram A, and an idea of the extent to which inertia modifies the load on the connecting rod may be formed by direct comparison of the negative areas B and D, and the positive areas C and E with that of the indicator diagram during the four strokes. The areas enclosed by the curve F and the base line are, 3.75 sq. in. for the firing, 0.3 in. for the exhaust, 0.25 for the suction, and 0.6 for the compression stroke.

Referring to Figs. 312 and 313 relative areas have been taken of 12 positive, and, therefore, above the base line, for the firing stroke, and 1 and 2, negative, and below the base line, for respectively the exhaust and compression strokes. No value has been assigned to the suction stroke, it is less than the negative value of the exhaust stroke and of small value.

These areas have been selected to represent the relative values of the turning effort during each stroke of the cycle, and there is, it will be seen, practically the same relation between 12, 1, and 2, as there is between 3.75, 0.3, and 0.6.

Before dealing with Figs. 312 and 313 some explanation may be given of diagrams numbers 1 to 12, shown by Fig. 311. These figures have been prepared to represent diagrammatically the frequency of repetition of the working strokes and to enable an estimate to be made of the comparative regularity and comparative amount of the turning effort. Each black area in these figures represents a firing stroke, and the three following open areas give the time spacing for the exhaust, suction, and compression strokes. Diagram 1 shows the frequency of effort for a single crank engine and for this simple case there are only 3 firing strokes for the length of diagram used, and which in each of the twelve cases represents 3 cycles or 6 revolutions.

Comparative uniformity of turning effort

Diagrams 2 and 3 are both for two-crank two-cylinder engines, and it may be seen at a glance that whereas in diagram 3 with cranks arranged opposite, or at 180 deg., two idle strokes separate pairs of firing strokes, the one following the other, there is in diagram 2 a better distribution of strokes, alternately a firing and an idle stroke.

Diagrams 4 and 5 present the case for three-crank engines, and it should be mentioned that the figures and letters on the crank circle diagrams indicate the sequence of the cranks, and the part of the cycle each crank is at the moment engaged upon, and not the order of arrangement of cranks side by side. F E, S, and C denote respectively firing, exhaust, suction and compression.   The distribution of efforts with the three-crank, shown in No. 4, is bad, and the sequence of firing there shown should be avoided. It will be seen that the cylinders fire in the order 1, 2, and 3, given, that the efforts overlap each other, and that there is an idle interval between each series of firing strokes of $1\frac{2}{3}$ strokes.   By changing the sequence of firing to 1, 3, 2, a uniform spacing of firing strokes occurs with a short idle period between each firing stroke of only $\frac{1}{3}$ of a stroke as seen in No. 5.

Diagrams 6 and 7 relate to the two usual arrangements of four-cylinder

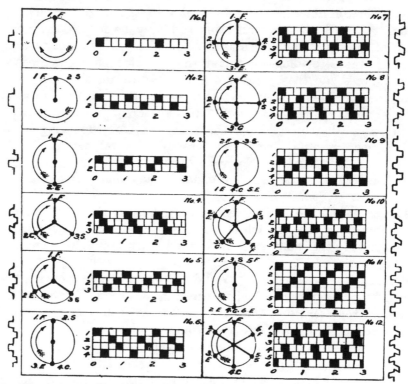

FIG. 311.—DIAGRAMMATIC REPRESENTATION OF FREQUENCY OF REPETITION OF FIRING STROKES.

engines.   In the first of these, with pairs of cranks at 180 deg., there are two cranks up and two down, and the arrangement is comparable to the coupling together of two engines, of the class represented by diagram 2, with cranks opposite. Firing occurs at every stroke, but there is no overlapping of efforts, and there remain points, at the dead centres with this engine, where there is no turning effort, and where but for fly-wheel energy the running of the engine would be spasmodic.   The occurrence of these points of no turning effort helps to explain why some engine con-structors prefer to set the cranks at 90°, and why in practice the engine so arranged appears to be livelier and capable of more rapidly accelerating the car. The occurrence of efforts with four cranks at 90° is represented by diagrams 7 and 8. In the first of these the sequence of firing in the cylinders is as indicated 1, 2, 4, 3, and in the second 1, 3, 4, 2. It will be apparent that this alteration in sequence does not affect the period of recurrence of efforts. The idle interval is only half a stroke in duration, and during each cycle there is an overlapping of two firing efforts which periodically gives an increase of turning effort. If the cylinders fire in their numerical order 1, 2, 3, 4, a very irregular turning effort results from the overlapping of the four firing strokes, with an idle period intervening equal to one and a half strokes, or three-quarters of a revolution.

Diagrams 9 and 10 show arrangements of cranks at 180 deg. with five-cylinder engines. No. 9 is the equivalent of coupling a single-cylinder engine to a four-cylinder engine of the type shown by diagram 6. Two firing strokes occur simultaneously, followed by three single successive efforts. As with the four-cylinder engine there is no idle period, but at the commencement of each stroke there is a dead point where fly-wheel energy is called upon to sustain the turning effort. When the five cranks are equally spaced at 72 deg., as in diagram 10, and the sequence of firing is as there shown 1, 3, 5, 2, 4, there is regularity and continuity of effort, all firing strokes uniformly overlap during one-fifth of their stroke, and the turning effort from such an engine would be more uniform than from any of the others.

Sequence of firing in the order of cranks 1, 2, 3, 4, 5, would destroy this regularity, and result in a period, between five overlapping strokes, of one and a half idle strokes.

Diagrams 11 and 12 show the effect of arranging the cranks of a six-cylinder engine, firstly, with three cranks opposite three, and secondly with the cranks spaced equally or at 60°. As might be expected, diagram 11 is the equivalent of combinations of three No. 3 diagrams, and the effect of such a combination is to produce, during a cycle, two succeeding double-firing strokes, with two following single strokes, and a resulting irregular turning effort. Diagram 12 may be looked upon as the combination of two No. 5 diagrams, the two three-crank engines being coupled together with an angular lag or lead of 60° between sets of cranks. A peculiarity of this combination is the variation, from stroke to stroke, of two-thirds lap of firing strokes, to no lap at all. This would lead to irregularity of turning effort and inferiority to the five-crank engine, diagram 10.

A brief consideration of the subject suffices to show the importance of attention to the sequence of working strokes as well as to the question of number of cylinders.

Sequence of working strokes

The comparative uniformity of turning, of such internally balanced engines as the Lanchester and Gobron-Brillie, is represented by the diagrams for two-cylinder engines. If each black area be assumed for purposes of comparison to have a double value, or if a diagram 3 be added below diagram 3 or diagram 2 below 2 then the cases of the Lanchester and Gobron-Brillie respectively are illustrated.

Before continuing the subject a little further, attention may be called

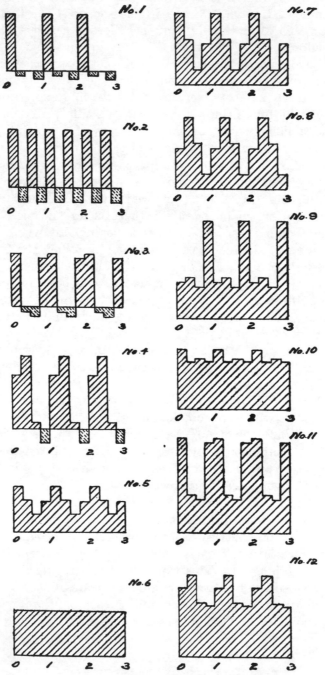

FIG. 312.—DIAGRAMMATIC REPRESENTATION OF COMPARATIVE REGULARITY OF TURNING EFFORT.

to the evil effect of incorrectly making electric ignition connexions.   When
the designer of the car has satisfied himself as to the order to be adopted,

360

there are occasions in the works and at the repair shops when, inadvertently, wrong connexions may be made, which in the case of four, five, or six-cylinder engines would require some time and patience to detect and rectify. Instances, such as those represented by diagrams Nos. 4, 5, 7 and 8, of the effect of possible errors in sequence of efforts occasioned by, for instance, badly placed cams are interesting in this connexion.

It has already been mentioned that in Figs. 312 and 313 relative areas of 12, 1, 0, and 6 have been adopted to represent the amounts of positive and negative energy delivered to the crank-pin during the four strokes of the cycle. By addition of these positive and negative amounts for the multi-cylindered engine a fairly correct opinion as to relative uniformity of turning effort may be formed.

Comparative representation of turning effort

There are, it will be observed, 12 diagrams, given to show the total energy per revolution of the engine, and the variation, during the cycle, in the rate of delivery of that energy.

Thus diagram 1, Fig. 312, shows, for a single-cylinder engine, the variation in the amounts of energy per stroke during three repeated cycles. It is an elementary representation of the turning effort curve F, Fig. 310. The diagram commences the first cycle with a firing stroke, shown by a vertical area equivalent to twelve arbitrary units above the atmospheric line. The second or exhaust stroke is shown by a small proportional area below the atmospheric line and equal to 1 unit. The suction stroke is then shown by a space between the exhaust pressure area, and the fourth, or compression stroke, shown by the larger negative area equal to 2 units, which completes the first cycle. The two following cycles are repetitions of the first.

In Fig. 313, diagrams 6 and 7 of Fig. 312, namely, those of the four-cylinder engines with cranks at 180° and 90° respectively, are reproduced to illustrate the method of obtaining the resultant areas given in Fig. 312.

Taking the first column of No. 6, the value of the work done during this space or half revolution is the maximum of the arbitrary 12 units for No. 1 cylinder, 0 for No. 2, −1·0 for No. 3, and −2·0 for No. 4, or the net result is 12−3, or 9 for the first column of the resultant shaded area below the separate cylinder diagrams. Treating the other columns in the same way it will be found that their sum is always the same, and the resulting figure rectangular. (As mentioned in connexion with diagram 6, Fig. 311, there is an instant of no turning effort at the end of each stroke, but this is of less duration than can be shown by a line, and, therefore, is not shown by the method here employed. If, however, more precise results are required, a resultant turning effort curve can be plotted from the combination of curves F Fig. 310.)

When the cranks are separated by any other angle than 180°, these equivalent figures have been moved to left or right by an amount, in terms of one stroke, proportional to the angular spacing of the cranks. Thus diagrams representing cylinders 2 and 4, in diagram No. 7, have been moved respectively to left and right half a stroke because they are 90°, or half a stroke behind and ahead of the operation of cylinder one.

Additions of the positive and negative efforts in the first column then gives for cylinder 1 a value of 12, for cylinder 2, $6-1=5$, for cylinder 3 a negative value of 1, and for cylinder 4 a negative value of 1, or a resultant figure of 15. Proceeding in the same manner with the other columns, the resultant figure shown is obtained. As with the diagrams given in Fig. 311 the figures are plotted for three complete cycles. An examination of Fig. 311 in conjunction with Fig. 310 will convey more exactly for any particular case an idea of the work performed, and the variation of effort for the twelve sets of conditions represented.

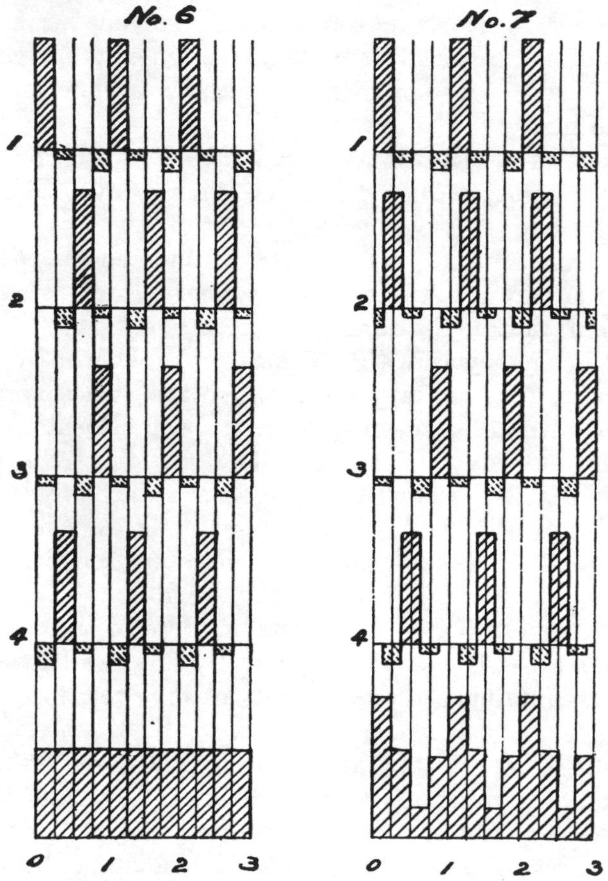

The characteristics of these diagrams need not be discussed here, reference being made to them in connexion with particular engines described in this volume. It will be noticed that the difference between the maximum and minimum efforts in Nos. 4 and 5, 6 and 7, 9 and 10, and 11 and 12, confirms the deductions made from the firing diagrams in Fig. 310.

FIG. 313.—EXAMPLE OF METHOD OF OBTAINING DIAGRAMS OF COMPARATIVE REGULARITY OF TURNING EFFORT.

Increase of speed of revolution Returning to the first of the six stated conditions relating to the speed of revolutions of the engine, it is obvious that an increase of speed up to the maximum number of revolutions per minute practicable, would for a given required horse-power result in a greater number of lighter impulses in unit time, and therefore a more uniform driving effort. The improvement would be directly proportional to the increase of speed of rotation, and as engine weight is at the same time reduced such an increase may, on these grounds, be decided upon. The limit to profitable increase of speed with some of the designs of engine now in general use is reached by either insufficient port and valve capacity, or of the pipes and passages

362

communicating with the latter, or weight of reciprocating parts, and sometimes it is questionable if the carburettor is equal to the duty imposed upon it.[1]

Conditions 1, 2 and 4 are interdependent, and are subject to compromise; the useful limit to the number of cylinders is partly determinable by practical and mechanical objections, and to multiplication of parts and too small dimensions having regard to thermodynamic efficiency. **Many small cylinders objectionable**

Greater fly-wheel energy in relation to horse-power results in less variation of angular velocity and exchange of energy between engine and fly-wheel, or in other words smoother and more continuous driving effort. The capability of the engine to accelerate the car would also be increased. The energy for a given speed of rotation can only be increased by additional weight or by as favourable a disposition of the mass as possible so that the radius of gyration is relatively great. The disadvantage of addition to weight must, however, be justified by improved behaviour of engine and car. **Fly-wheel energy**

The method of governing, or control of the power developed by the engine, may considerably affect both the quietness and smoothness of running. The almost universal use of throttle governing or its equivalent by variable opening of inlet valves, shows that there is little difference of opinion as to the necessity for a method of control, not by varying the number, but by varying the value or power of the impulses. The power of the impulse may be controlled by variation of quantity and quality of explosive mixture per stroke and this may involve compound governing of the quantities of fuel and air. The latter method is seldom employed, and throttle control with the more recent forms of carburettor does not, or is not intended to, change the richness of the mixture. **Control of power**

The position of the engine in the car need not affect the uniformity of the driving effort from the engine, but it may considerably affect the nature of the vibration of the car. In vol i. chapter xxxiv., reference was made to the tendency of the engine to rotate around as well as to rotate its crankshaft. This tendency has to be opposed by the car or that part of the car to which the engine is held. The design of the engine, and of the frame and springing of the car must be considered together, and the engine should be arranged with reference to movement resulting from this tendency. The improvement in uniformity of turning effort will not reduce this tendency, but reducing its variation will also reduce the vibratory movement, which is only set up by such variation. The horizontal engine arranged transversely in the car is invariably in a better position, as regards reduction of vibration set up by the engine, than the vertical type arranged longitudinally. **Effect of position**

These questions, as so far discussed, open up others which cannot now be minutely entered into, but they have perhaps been sufficiently elaborated to confirm the author's views with regard to vibration, and its causes as set forth in that part of vol. i. already alluded to.

---

[1] "Valves and Valve Mechanism of Internal Combustion Engines," by R. E. Phillips, M.I.Mech.E., *Automobile Club Journal*, March 10, 1904.

Chapter XXV

STEAM AND PETROLEUM ENGINE LORRIES AND WAGONS

THE past three years have seen a large increase in the number of makers in this class, several new designs have been made and tested, and a great deal of experimental work has been followed by the accumulation and recording of useful experience.

Unsatisfactory record We have had to lament that so little was recorded of the actual work and practical experience and details of Hancock, Gurney, and others. Our successors will lament the trouble of sifting the enormous quantity of useless and untrue publication which has marked the revival of the mechanical road vehicle.

Since 1900 almost everything in a motor wagon has been strengthened, and the heating surface and evaporative efficiency of some of the boilers has been increased. Water-tube boilers have been in most cases given up and improvements have been made in the design and in the tube plate **General** staying of vertical boilers. The general tendency has been to increase dead weight, and from the point of view of the public carrier still heavier tare weights are necessary in order that a net load of from 8 to 10 tons may be carried. In the interests of those who hold this view, and who appeal to experience with heavy wagons in public service to support it, a modification has been made in the tare weight regulations promulgated under the 1896 Light Locomotives Act. This view is based chiefly on the fact that with smaller loads and ordinary haulage distances, and frequency of stoppages, it is cheaper, or as cheap, to haul by horses.

Local Government Board Regulations Hence a very high tare weight has been asked for, and the Departmental Committee of the Local Government Board have recommended that 5 tons be now the regulation weight, and $6\frac{1}{2}$ tons with trailer, with a maximum total load of 8 tons on any one axle. Various other weight and dimension regulations have been recommended, and, although not confirmed by the L.G.B., it may be assumed that they will be adopted, at least so far as they relate to the diameter, width, and load on wheels.

It is assumed that the objections to the very heavy load per axle are

364

removed by increasing the width of the wheels, but road surveyors and others concerned in road upkeep will not be likely to agree entirely to this.

The following are the existing regulations—

Maximum tare weight of vehicle (instead of three under the
    1896 Act) . . . . 5 tons.
Maximum tare with trailer . . . . 6½ ,,
Maximum load on any one axle of motor wagon . . . 8·0 ,,
Maximum load on any axle of trailer . . . . 4·0 ,,

The chief recommendation regarding wheels and axle weights is that if the wheel is 3 ft. in diameter the tyre shall be not less than ½ in. wide for every 7½ cwt., or fraction of 7½ cwt., of the certified gross weight to be carried on the axle. If the wheel exceeds 3 ft. in diameter the foregoing rule shall apply with a proportionate increase above the 7½ cwt. in the ratio of 1 cwt. for every 12 in. increase in the diameter of the wheel. If the wheel be less than 3 ft. diameter the foregoing rule shall apply with a proportionate decrease below the 7½ cwt., of 2 cwt. for every 12 in. decrease in the diameter of the wheel.

From this I assume that a 3 ft. 6 in. wheel will be allowed 8 cwt. per ½ in. width and a 4 ft. wheel 8½ cwt., and correspondingly a 2 ft. 6 in. wheel will be allowed 6½ cwt.

On this basis the following Table has been made for ready reference :—

RELATIVE DIAMETER, WIDTH AND WEIGHT PERMITTED ON WHEELS OF MOTOR VEHICLES WITH METAL TYRES, AND EXCEEDING 2 TONS IN WEIGHT UNLADEN.

	in.	in.	in.	in.	in.	in.	in.	in.	in.	in.	in.
Diam. of wheel . . .	24	24	24	24	30	30	30	30	30	36	36
Width of Tyre . . .	5	6	7	8	5	6	7	8	10	5	6
Gross weight per Axle, cwts..	55	66	77	88	65	78	117	104	130	75	90

	in.	in.	in.	in.	in.	in.	in.	in.	in.	in.	in.
Diam. of wheel . . .	36	36	36	36	36	48	48	48	48	48	48
Width of tyre . . .	7	8	9	10	11	5	6	7	8	9	10
Gross Weight per axle, cwts..	105	120	135	150	160	85	102	119	136	153	160

A development which has been noteworthy in the heavy wagon haulage by steam is in the construction and use of very small traction engines, weighing only about 3 tons empty. Numbers of these have been made and are in use hauling nightly large quantities of market garden produce to London. They have the advantage of a large diameter of driving-wheel of considerable width, can travel over roads or ground which would not carry a motor wagon of the same capacity, and by working with spare trailers wagons need not wait for unloading or loading. They will only haul about four and a half tons, but this may prove a paying load if the

Small traction engines

cost of repairs and maintenance does not exceed that which appears likely, and when the length of haul is such that the day or the working hours are not split up into lengths causing waste of time.

I think that there can be little doubt that as a steam tractor this type might be developed into one which might find very wide application provided the speed be kept moderate, that they be not overloaded, and that they are run by men who can be trusted to keep within a moderate speed limit, to run at speeds according with the character of the road, and to take an interest in their engine. To secure these things the men must be considered with reference to their working hours, and exceptional circumstances must not be appealed to to excuse long and late hours nearly every day the tractor goes out, as they are with some traction engine owners. Moreover, these tractors must be owned and worked in numbers under one organization so that not less than 15 per cent. of their number can be looked upon as stand by stock under inspection, repair, or adjustment, and so that no tractor shall be worked that needs even the smallest repair.

Importance of good organization

The general character of these young traction engines may in many respects be adhered to, but there should be little difficulty in removing some of the objectionable " rattly-steam-winch-broke-away-from-its-moorings " character in appearance and working.

Much of the argument which has been urged in favour of very heavy motor wagons will, I think, in a very few years be shown. to be, like many of the roads they have to run on, without any good foundation.

Commercial success with them depends very much on constant working, and this almost of necessity means continuous working on the same roads. To run wagons with 10 tons on an axle constantly over the same route will impose very heavy demands on the road constructor, and almost equally on the wagon maker and user, even if what are considered moderate speeds, namely, about double the speed of traction engines of the same total load, are observed. The wagons cannot call for very strong roads without calling for great strength in themselves, and to ask to run them on return empty journeys at 8 miles and more per hour, as is proposed, is to postulate as a necessity that which must be denied them on various grounds. Only moderate loads and moderate speeds will ever be permanently successful on the average good roads as now made or likely to be made. If the motor wagon cannot be made to pay on average good roads, it is clear that its environments are more fatal to its survival than to other forms of transport. For exceptional purposes the very heavy wagon may have its uses, just as traction engines have, but an enormous proportion of all the heavy road transport is occasional, casual, and between points unfavourable to them for loading up and getting away, though less so to the traction engine. As I wrote in vol i., and as I still must think, the conditions as to quantities of materials or goods to be moved, which are requisite for the continuous profitable use of very heavy wagons on any route, are those which are best met by the construction of some form of tramway or railway.

Limitation to successful use

366

HEAVY MOTOR VEHICLES

An interesting paper on " Motor Transport for Goods " [1] was read before the Society of Engineers last year. This paper very fairly puts the advantages and disadvantages and difficulties yet remaining to be overcome very fairly before its readers, and I am generally in accord with the views and facts of the author, Mr. Douglas Mackenzie. With one of his conclusions I do not agree, namely, that " for small traders or retailers, or for those whose goods are sold in comparatively small parcels, the motor vehicle offers no economy, but is occasionally used as an advertisement." The words " motor vehicle " used here make this conclusion much too general. If the author had said motor wagon I could agree with him. The light motor vehicle can and will, I believe, be made which will meet the requirements of many traders and retailers, but the vehicle must be suited to the trade, and to the districts within which it is to work. There can, however, be little doubt that even this form of delivery work will be most economically done by large delivery vehicle concerns, by whom all the working arrangements and details are carried out with the completeness of the organizations of some of the existing omnibus and delivery companies.

Light commercial vehicles

The relative fitness of the steam engine, the internal combustion engine, and the electric motor, for the vehicles which will ultimately be most used is a subject of some conjecture. There can, however, be no doubt that the second of these will ultimately be the form of motor used for the greater part of the whole of the motor vehicle work for both light and moderate weight vehicles.

Nothing but praise can be offered the two or three makers of the light steam vehicles and omnibuses now at work, and admiration of the ingenuity and high class of the design and workmanship of their engines and gear must be sincere on the part of every one who has any knowledge of them. A considerable measure of commercial working success, moreover, provides a backing of reality for this sincerity.

Steam vehicles

The design and construction of the engines, gear, boiler and connexions of several of the heavier vehicles also command a measure of praise that cannot be withheld, but must be spontaneously offered.

For much of the work these designs will continue to be used.

But—in the steam vehicle there is a boiler, and it and everything between it and the engine, the steam pipes, the feed-water pipes, the pump, the regulator pipes, the gauge, and other fittings, all are under the maximum pressure at any time available on the engine pistons and most of them at the temperature due to that pressure.

In the internal combustion motor vehicle there is no boiler, and there is nothing under pressure except the cylinders at the moment of combustion.

This statement of fact carries with it an indication of some of the most important reasons for the preponderating use of the internal combustion motor. The steam engine does not now, in most cases, materially reduce the quantity or weight of the gear required, and its comparative superiority

[1] *Transactions of the Society of Engineers,* 1903.

367

in flexibility and handiness of operation is lessened by the improvements in the other motor. The steam engine permits the use of cheaper fuels, either liquid or solid, but the growing thermodynamic efficiency of the internal combustion motor and probability of its successful use of kerosene for fuel, is reducing this advantage. There is, moreover, the distinction which theory confers on the newer motor, namely, that its possible efficiency is greater.

Ultimate success of internal combustion motor

Convenience is the name given to numerous otherwise undescribed grounds for a preference of one thing rather than another. The balance of advantages and disadvantages in this respect will, in some degree, remain a matter of opinion dictated by circumstances. Ordinary circumstances and conditions will, I think, ultimately award that balance to the internal com-bustion motor.

Of the electrical motor there is not much to say. The changes, or improvements in the construction of the vehicles, or the motors, or the secondary batteries, have not in the past three years been such as to lead to any noteworthy extension of the use of the electro-mobile. As a *voiture de luxe*, a brougham, private carriage, almost entirely for town use, it is perhaps the " nicest " vehicle one can have, but the cost of running it, or of running its storage of electricity, does not encourage great expectations, or any expectations at all with regard to heavy vehicles.

The London General Omnibus Company, the London Road Car Company, the London and District Motor Bus Co., the London Motor Omnibus Co., Great Eastern Omnibus and Tramway Co., the London and Suburban Omnibus Co., Messrs. Tillings, and Messrs. Birch and Co., are all now running motor omnibuses. The three first-mentioned com-panies have the large Milnes-Daimler omnibuses, and Clarkson steam omnibuses carrying 34 passengers, and the Road Car Co. have some Straker Durkopp and Germain omnibuses, carrying 32 passengers. Messrs. Tilling and Co., and Messrs. Birch and Co. are using Milnes-Daimler omnibuses. The Cannstatt omnibuses are of large size, have 24 HP.Daimler engines, and well designed and well made transmission gear, as will be seen on page 434.

The tyres have hitherto given more trouble than any other part of these public service vehicles, especially on the heavier double-deck omnibuses carrying the larger number of passengers. In spite of the cost of tyres, however, they have proved profitable at Eastbourne and elsewhere, and in spite of the tyre troubles the Great Western, and other railway companies are finding it to their advantage to adopt the motor omnibus.

The lighter vehicles have the advantage as to tyres, and it is question-able whether even with coming improvements in tyres, the moderate size omnibus will not ultimately prove the more suitable as street vehicles, and as profitable as circumstances governing the use of the streets will permit any vehicles to be. The largest omnibuses do not approach in size the enormous gilded, bedizened and electric-lighted monster and monstrous tramcars that are at present permitted to thunder along some of our streets and suburban roads, and it is good to think that there are physical difficulties that will probably prevent any such size ever being approached.

Tyre manufacturers, and probably users, have been somewhat to blame, and have not done all that might have been expected to provide a suitable tyre for vehicles which with their load weigh 5¾ to 6½ tons, with approaching 2 tons on each driving-wheel.   In dealing with tyres, and referring to solid tyres for heavier work in vol. i. p. 603, I gave reasons for stating that these tyres should be broad and thin as compared with the prevailing fashion in sections ; that opinion still holds good, and many of the solid tyres of very deep section put on the wheels of motor vans and omnibuses of to-day would last longer if about 30 per cent. of the radial depth of the tyre were cut off the tread before being put to work.

Since 1900, trials have been carried out with heavy vehicles by the Liverpool Self-Propelled Traffic Association, and numerous figures were deduced from them as to the cost per ton mile of transport by different vehicles.

The use of many vehicles, however, in the sterner, though less academic, trials of daily commercial use, has materially qualified the arithmetical deductions on this subject.   The makers of the best of the vehicles have got their meed of praise and patronage, and, in some cases, profit, but the enthusiasm that breathed loudly in some reports in expectation of vast economies, and a " revolution in mechanical locomotion " has softened down to the milder and more tenacious enthusiasm and expectation of those to whom experience has given judgment in excess of their acquisitions in what is called knowledge.

To some of the results of the trials made under auspices of different bodies, reference will be made in the following descriptions of vehicles.

Reference may, however, first be made to the effect of varying road resistance on the carrying and hauling power of heavy motor wagons. The advocates of the motor wagon sometimes forget its limitations as imposed by steep hills, by weak, or by bad roads, or by roads in bad condition after continued bad weather, conditions which rapidly reduce the carrying power, though in some respects they less rapidly reduce the hauling power.

A motor wagon with boiler and engine capable of maintaining 30 HP. will carry and haul a very big load on a good hard level road, but its capacity falls off to a surprising degree on bad roads.

Great disappointment has, in many cases, resulted from the under estimation of this fall of carrying power or from neglect to consider it. As an example of this we may take the case of a motor wagon weighing 5 tons ready for the road, and with 7 tons load, or a total of 12 tons gross, and with engine and boiler capable of maintaining 30 BHP.   Assume mechanical efficiency of 0·70.   For such vehicles and loads the road resistance on the level can seldom be less than 70 lb. per ton, except on first-class roads in good condition, and it will often exceed this by from 20 to 100 per cent. and more.

Solid rubber tyres

Reduction of carrying power with increased road resistance

Assume in the first instance that a motor wagon only is used without a trailer.

Then the load W which the wagon will carry at 6 miles per hour, using the formulae given at p. 58, vol. i. will be

$$=\frac{PE\ 375}{(R+G)\,v}=\frac{30\times0\cdot70\times375}{(70+0)\,6}=18\cdot75 \text{ tons.}$$

This wagon could then haul a trailer weighing $1\frac{1}{2}$ tons, and carrying the $18\cdot75$ tons—$13\frac{1}{2}$ tons, or $5\cdot25$ tons.

On the other hand with a soft winter country road, and a resistance of say 120 lb. per ton, the 30 HP. will only move a total of

$$W=\frac{30\times0\cdot70\times375}{120\times6}=11 \text{ tons,}$$

so that the 7 tons load must be lowered to $11-5=6$ tons.

If the wagon has to mount a hill of say 1 in 15 (see Table 12, p. 57, vol. i.), and the road resistance = say 100 lb. per ton, then the load the 30 HP. will move is

$$W=\frac{30\times0\cdot70\times375}{(100+149)6}=5\cdot27,$$

and hence the load the wagon can carry falls to $5\cdot27-5=0\cdot27$ tons.

Calculated on the same basis the following table relating to a running tare of 5 tons may sometimes be of service.

TABLE No. 7.—GROSS AND NET LOADS CARRIED BY MOTOR WAGONS AND TRAILERS.

	HP.	Road Resistance.	Gradient.	At 6 Miles per Hour.		At 3 Miles per Hour.	
				Gross Load.	Net Load.	Gross Load.	Net Load.
1	30	70 lb. per ton	level.	18·75 tons	12·25[1] t's	37·5 tons	28·0 tons[1]
2	30	70 ,, ,,	1 in 30	9·00 ,,	4·00 ,,	18·0 ,,	11·5[1] ,,
3	30	70 ,, ,,	1 in 20	7·20 ,,	2·20 ,,	14·4 ,,	7·9[1] ,,
4	30	70 ,, ,,	1 in 15	6·00 ,,	1·00 ,,	12·0 ,,	7·0 ,,
5	30	80 ,, ,,	level	16·50 ,,	10·00[1] ,,	33·0 ,,	23·5[1] ,,
6	30	80 ,, ,,	1 in 15	5·75 ,,	0·75 ,,	11·5 ,,	6·5 ,,
7	30	100 ,, ,,	level	13·00 ,,	6·50[1] ,,	26·0 ,,	18·0[1] ,,
8	30	100 ,, ,,	1 in 30	7·50 ,,	2·50 ,,	15·0 ,,	8·5[1] ,,
9	30	100 ,, ,,	1 in 15	5·27 ,,	0·27 ,,	10·54 ,,	5·5 ,,
10	30	120 ,, ,,	level	11·00 ,,	6·0 ,,	22·0 ,,	14·0 ,,

The use of a trailer will often considerably increase the paying load, and will sometimes make it possible to do road contractors, and other work,

[1] Under these circumstances a trailer, or trailers, must be used not greater than the proposed regulation weight of $1\frac{1}{2}$ tons, and carrying respectively at the most 8 tons.

when hauling out of a pit or from a low-lying wharf, when the load could not be worked on a motor wagon at all, not only because of the gradient, but because of the weakness or softness of the road.

An inspection of the table shows the enormous value of speed reduction for weight carrying. Reduction from 6 miles to 3 miles only doubles the gross load, but, as in instance No 4 it increases the net load that can be carried 7-fold, and in instance No. 9 it increases the net load from 0·27 ton to 5·5 tons or over 10 times. In some cases the full possible load could not be taken as three trailers would be required, and this will probably not be allowed.

The Thornycroft Heavy Vehicles.

Since 1900 little change has been found necessary in the design and general arrangement of these steam wagons, other than those modifications mentioned in the description which was given in vol. i., and any further reference may be considered to supplement rather than supersede the earlier description.

During the last four years, further experience as to the continued performance and durability of these wagons has proved both the fitness of design, and the nature of the material and workmanship employed.

Figs. 314 and 315, showing respectively top and under views of a standard 4-ton lorry with the platform removed, are interesting views and show clearly the complete arrangement of the running gear and the location of parts which, dealt with separately, would require much needless description. In Fig. 315 the double helical toothed wheels for the final gear reduction to the road wheels may be seen, and the distinctive triangular bar frame which maintains the centres or spacing of the helical wheels, and while taking its bearing upon the road-wheel axle, on each side of the differential gear, provides at its apex the bearings carrying the helical pinion. The tension bar or torsion brace, connected to this bar frame at one end, occupies as it should a nearly horizontal position, and may be seen connected at its other end to the cross-frame member above the engine crank chamber. No radius bars are used in the wagon shown, but they have been employed in some of the heavier design, and in them the springs between frame and rear axle are relieved of the driving effort, which must otherwise be transmitted through them. The range of movement of the universally jointed or articulated driving-shaft, connecting the helical pinion and intermediate gear-wheel shafts, is about 7 in., an amount which sufficiently provides for the relative positions of rear axle and wagon frame for the extreme conditions of running light or fully loaded over bad road surface.

Some further information of the construction of the engine may be gained from Fig. 317. The section is taken through the low-pressure cylinder and that end of the crank chamber through which only the reversing shaft projects, and, as may be seen, at this end of the crankshaft there is a crank

Transmission parts

Fig. 314.—Thornycroft Standard 4-Ton Steam Wagon.—Plan with Platform Removed.

Fig. 315.—Thornycroft Standard 4-Ton Steam Wagon.—Underneath View with Platform Removed.

Easy examination of engine

disc and overhung crank-pin upon which the low-pressure rod of the engine works. The large crank-chamber cover, carrying a smaller inspection door, may be removed without disconnecting any running gear or disturbing parts which examination may show to be in perfect order and requiring no adjustment.

A very compact form of valve gear is employed requiring the use of only two eccentrics ; constant lead at varying points of cut-off is obtained, and reversing through link motions is effected through connexion with the lay shaft which may be seen in its bearings below the crankshaft.

Slide valves

Balanced D slide valves are used, and are preferable to small piston valves, more particularly at the low-pressure cylinder where steam leakage would be extremely wasteful.

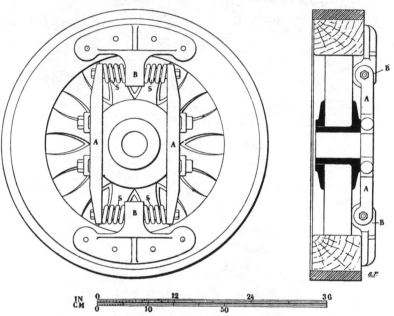

FIG. 316.—THE THORNYCROFT SPRING DRIVE.

Importance of lagging

The economical running of these small steam engines when well constructed is more adversely affected by heat radiation losses than direct loss by steam leakage, and careful and complete lagging of. pipe connexions and cylinder surfaces is desirable. Reference to Fig. 317 will show that the exposed cylinder covers may be well lagged, and that they are fitted with thin sheet metal false covers having little more than line contact with the cylinder covers.

The boiler is usually fed by the worm and worm gear-driven pump below the engine, and fixed to a bracket cast with the crank chamber, seen in Fig. 317.

A modification of the form of spring drive seen in Fig. 315 is given in Fig. 316. In this form the laminated plate springs are replaced by equivalent driver arms A which are fixed to the driving parts of the rear axle, and through the coil springs s deliver their effort to the lug plates B. The plates B are each held by four plates to two felloe sections of the wheel rim. The provision of this spring drive very considerably relieves both transmission gear and road wheels from the heavy shocks which occur in various ways.

Spring drive

Messrs. Thornycroft have wisely decided that the carriage of heavy loads may be more economically effected by the use of either the steam wagon as carrier and tractor, or more and lighter steam wagons. While the 7-ton lorry shown by Figs. 318 and 319 has proved its capability of dealing with the greater loads, experience suggests that any advantage to be obtained by the use of such high capacity wagons may be nullified by their destructive effect upon themselves and the roads. These considerations have led Messrs. Thornycroft to discontinue the construction of the 7-ton wagon, and to replace it by one of 5 tons carrying capacity. In all these wagons the same design of transmission gearing is retained, and any departures in detail from the standard 4-ton lorry are mainly those necessitated by change in dimension and nature of load. A locomotive type boiler, similar to that shown by Fig. 320, p. 382, is used with the 5-ton lorry, and an engine having cylinders $4\frac{1}{2}$ and 7 in. diam., and a stroke of 7 in., and capable by comparison with the smaller engine of developing a maximum of 50 BHP. The vertical water-tube boiler (illustrated and described in vol. i.) is used for the 4-ton wagon, and has been improved by addition to the depth of the upper annular steam and water chamber. The improvement in the dryness of the steam supplied to the engine means also additional economy in fuel and water consumption, and is equivalent to an increase of boiler evaporating power. A lorry of the type illustrated by Figs. 318, 319 was awarded a gold medal as a result of its performance in the heavy vehicle trials held at Liverpool in June, 1901, and some description will be given of it.

Very heavy wagons abandoned

Dry steam important

The water-tube boiler has a total heating surface of 132 sq. ft., and a grate area of 4·25 sq. ft., the working pressure being up to 225 lb. per sq. in.

Boiler dimensions

According to the recorded trials figures the boiler evaporated 6 lb. of water per pound of gas coke, a satisfactory result from a small boiler even if some allowance is made for occasional priming or wetness of steam.

The boiler is fired from the top at the central opening and regulation of fire or draught may be done by means of the hinged door R, and by the cover lid M. Of those boiler mountings which are lettered, H is a steam dome, to which the pressure gauge is connected, and steam is taken from the top of the dome through an internal pipe connected with the starting stop valve Q. O are the water gauge, and P the injector fittings, the injector being the alternative means of feeding the boiler. A feed-water heater J is used, the exhaust steam passing through it on its way to the exhaust nozzle at the base of K. The driver is seated at N with the steering-wheel opposite to him, and the reversing lever s on his right. The screw down

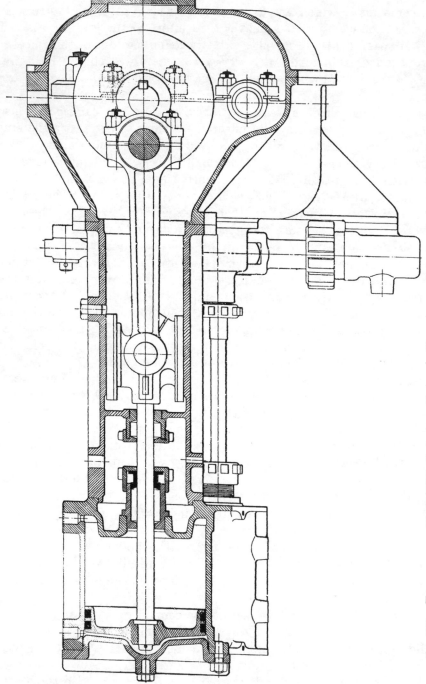

FIG. 317.—THORNYCROFT 35-40 BHP. COMPOUND ENGINE.—TRANSVERSE SECTION THROUGH LOW PRESSURE CYLINDER.

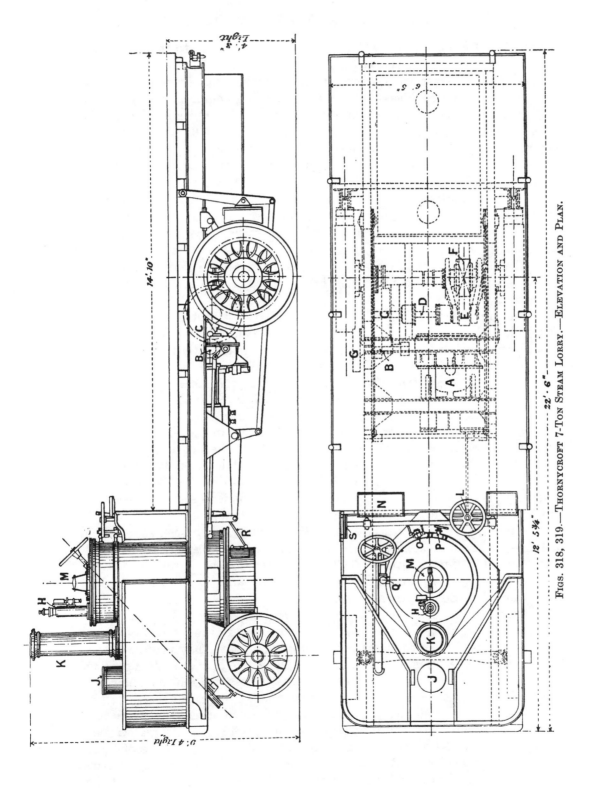

FIGS. 318, 319.—THORNYCROFT 7-TON STEAM LORRY.—ELEVATION AND PLAN.

377

brake wheel L is connected by the system of levers shown, to the shoe-brake blocks at the rear road wheels.

As already stated the arrangement of engine and transmission gear is like that shown by Figs. 314 and 315. A is the engine having high and low pressure cylinders 4 and 7 in. diam. respectively, and a stroke of 5 in. There are cut-off notches for normal running at 62 and 87 per cent. of the stroke, and at 500 revs. per minute 30 HP. is developed. On the squared extension of the crankshaft are the high and low gear pinions B, and at the extreme end the fly-wheel G.

General arrangement

The universally jointed shaft D is driven by C through whichever of its two gear-wheels is in mesh with one of the pinions B, and in its turn drives the double helical pinion E in mesh with the wheel F upon the live axle.

The boiler in this lorry occupies a position to the rear of the front axle, and where it may be carried lower in the wagon frame than when immediately over the axle.

An improvement has been effected by discontinuing the use of a firing door or clinkering hole through the lower annular water chamber of the boiler, and by arranging the grate and ash pan so that they may be lowered, and the whole grate exposed, for convenience in lighting up at the start of the day's run, or for clearing the fire occasionally. Apart from the question of evaporative efficiency the facility with which this boiler may be opened up for complete internal examination is to be commended.

Internal examination of boiler

The leading dimensions of the wagon are given in the plan and elevation views, and those show the clear platform area to be 144 sq. ft.

Platform area

During the Liverpool trials an average useful load of 6·33 tons was carried, consisting of general merchandise, of large and small bulk for weight, and the platform area was apparently about sufficient for such loads or up to the seven tons for which the lorry was designed.

A number of this type, and of the lighter class, have been sent to the colonies, and among these latter was the successful 3-ton lorry which obtained the War Office first award of £500 for its performance during the trials which took place in December, 1901. It was afterwards sent by the Government to South Africa.

Among other modifications necessary to make these vehicles suitable for Colonial use is an increase in road-wheel diameter, and width of tyres, and the use of a three-point method of support, so that they may be better able to travel over rough and soft roads. The whole of the toothed gearing is cased in and dust excluded, a protection that is essential in those countries where dust storms occur, or where roads are loose and sandy.

Protection of gearing

Steel wheels, with cut away plate spoke discs, and connexion by circumferential angles to the rim and cross slats, are used for the colonial wagons in preference to the usual wood artillery type. The increased use of these steam wagons has resulted in their employment for a great variety of purposes. Municipal authorities have not been slow to take advantage of the economy and convenience resulting from their use for street watering,

378

and for the transport of materials required in large quantities for various works of public importance, but they are not extending their use of them for the collection of town refuse purposes. They have so far been little used for public service vehicles.

It is significant that although the steam wagon has met with much success Messrs. Thornycroft are paying great attention to the design of heavy vehicles propelled by the ordinary petrol motor and engines using the heavier classes of oil, and it is reasonable to infer that the 20 HP. petrol motor wagon now introduced by them is the forerunner of a type that will ultimately survive the stage to stage method of improvement by trial and error. **Petrol engined wagon**

A four-cylinder motor having cylinders 4 in. diam., and a stroke of $4\frac{3}{8}$ in. is supported longitudinally on the frame, and through a cone clutch and change-speed gear drives by bevel wheels a shaft carrying an overhung chain sprocket wheel. The final drive to the road wheels is by a single chain to a large chain wheel fixed to the differential gear on the live rear axle. The standard form of spring drive used with the steam vehicles is retained. The engine is governed by throttle valve, and the inlet valves are not mechanically operated. The sliding gear wheels in the gear box are arranged for forward speeds of 3, 6, 9, and 12 miles per hour, and reversing. Used as a passenger vehicle it has been designed to carry seventeen passengers, and about half a ton of luggage. A similar arrangement of engine and transmission gear is being used in the construction of the several small public service vehicles which Messrs. Thornycroft are making for the Lake District.

Figs. 319A and 319B show the general arrangement of the chassis for a double-deck omnibus seating thirty-four passengers.[1] From the leading dimensions given and the outlines of the principal parts the design may be sufficiently understood. The 24 BHP. four-cylinder engine runs at 900 revs. per minute and it has cylinders 4·25 in. diameter and a stroke of 5 in. It drives through an enclosed clutch of the multiple disc type and through a jointed propeller shaft to the change-speed gear box situated centrally in the frame. The engine and gear box are supported on an underframe and the longitudinal members of the latter are fixed to the cross members of the main frame. From an overhung pinion at one end of the transverse shaft in the gear box the power is transmitted by a single heavy roller chain to the chain wheel surrounding the differential gear on the live rear axle. Four speeds are provided for in the gear box, of 3, $5\frac{1}{2}$, $8\frac{1}{4}$ and 12 miles per hour and the reverse speed at 3 miles per hour, and the speed changes are effected through the hand lever shown to the left of the driver's seat. Through the hand lever at the right-hand side of the driver the band brake acting upon a drum on the chain wheel at the rear axle is applied. A drag bar in a suitable position prevents rotation of the brake strap and the latter is prevented from continually dragging or rubbing upon the brake drum. A second brake operated by a pedal acts upon a **General design of omnibus chassis**

[1] *The Tramway World*, March 1905.

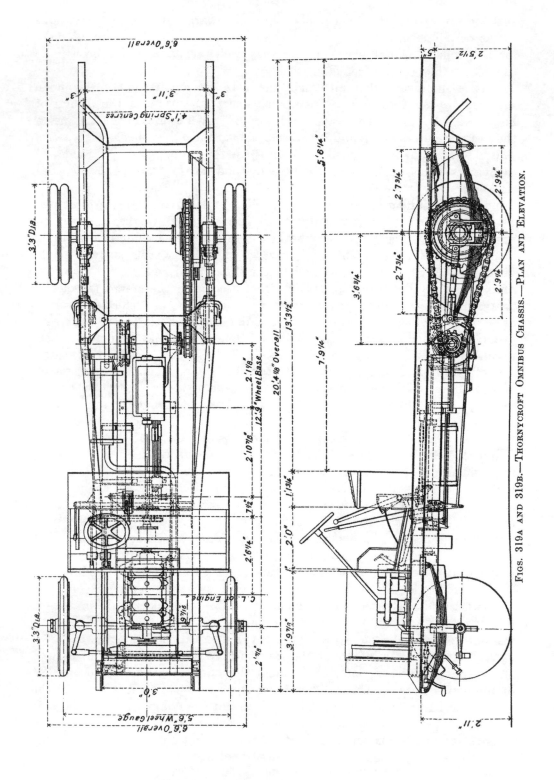

6'6" Overall

3'11" 3"

4'1" Spring Centres.

3'3" Dia

2'1⅛"

12'9" Wheel Base

2'10⅞"

7½"

2'6¼"

9⅞16"

C. L. of Engine.

3'3" Dia.

2'4⅛"

3'0"

5'6" Wheel Gauge

6'6" Overall

5"

2'5½"

5'6½"

2'7¾" 2'9½"

3'6¾"

2'7¾" 2'9¼"

7'9¼"

13'3½"

20'4⅛" Overall

1'1¾"

2'0"

1'

3'9½"

2'11"

Figs. 319a and 319b.—Thornycroft Omnibus Chassis.—Plan and Elevation.

380

drum at the right-hand end of the transverse shaft of the gear box. Attention may be called to the design of the rear springs and their attachment below the axle bearings.

An omnibus with a chassis of the type here shown is running successfully at Hastings, and a number of them are under construction for the London Motor Omnibus Company, and others for use as lorries suitable for carrying 5-ton loads.

A modified form of transmission gear is employed in which the radial frame and horizontal torque bar adopted with the steam wagons is retained, but the power is transmitted from the overhung chain pinion through a Hans Renold chain to a chain wheel alongside and fixed to the double helical pinion of the final drive.

MANN'S PATENT STEAM WAGONS.

The outline of the lorry illustrated by Figs. 320 to 323 suggests the combination of light traction engine and wagon which characterized the earlier design of steam tipping cart. It may be looked upon as an intermediate type, preceded by the tipping car, and followed by a design which has been adopted for a heavy load 7-ton lorry for colonial use.

The continued use of the locomotive type of boiler with rearward extension of the side fire-box plates for formation of a box girder frame, remains a distinctive feature of the Mann vehicles, and a feature which by retention suggests proved advantages of form and cost of production sufficient to neutralize the disadvantage of heavy construction.

Referring to the longitudinal and transverse sections of the boiler, Figs. 320 and 321, it will be obvious to those familiar with locomotive boiler construction that there is little in the design that calls for comment, and that in general the methods employed are those common to small well made boilers of this type. The connexion of the fire box D at the bottom to the outer shell is peculiar, for in addition to the outward flanging of the fire box plates a light spacing or foundation ring is also used. No girder or bridge stays are used, both the top and sides being supported by screw stays, the top or crown plate stays having nuts on the ends exposed to the fire.

Boiler design

The smoke tubes B are not more than 3 ft. long, but their whole surface is very efficient as heating surface, and any further addition to their length would not add, in proportion to such increase, to the evaporative power of the boiler.

The tubes are given a slight upward inclination towards the smoke-box end, and as may be seen in the transverse section the bottom tube of the centre row has been left out, and its place is filled by a plug at the smoke-box end. Through this plug hole the sludge or deposit along the barrel bottom may be dislodged and washed out through the fire-box hand-holes. The accumulation of deposit is largely prevented in this type of boiler by leaving plenty of room between the tubes and the barrel towards the bottom.

The effective heating surface is 48, and the grate area 2·5 sq. ft. The working steam pressure is 160 lb. per sq. in.

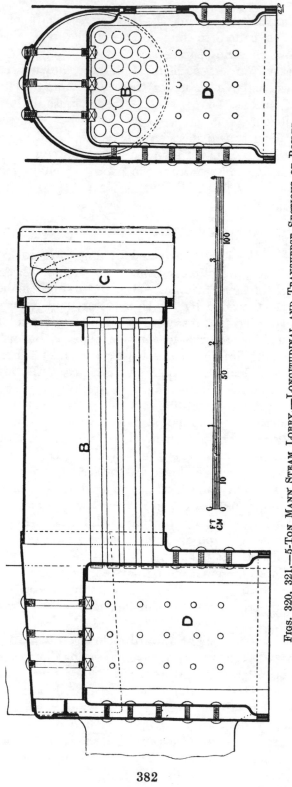

Figs. 320, 321.—5-Ton Mann Steam Lorry.—Longitudinal and Transverse Sections of Boiler.

382

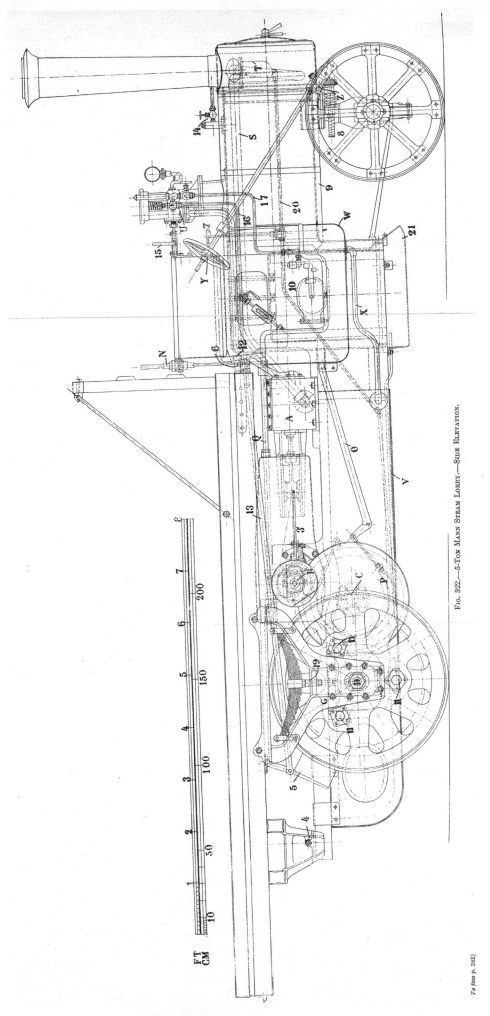

Fig. 322.—5-Ton Mann Steam Lorry.—Side Elevation.

[To face p. 382]

MANN'S STEAM WAGON

The fire hole is necessarily arranged at 10, Fig. 322, on one side of the boiler, instead of as usual at the end, for convenience in firing the boiler in its rather restricted position. It is evident from Figs. 322 and 323 that accommodation for the driver in the side box w is limited, and that attention to the fire is not very easily given.

A mounting about central in the length of the boiler carries the starting valve or regulator u of slide valve type operated through the lever handle 15, the duplex safety valve with spring in compression is above the starting valve casting, and to the front of it is attached the pressure gauge as shown. On one side of the same casting are two steam branches 16 and 17 with their stop valves. Through the first of these live steam may be supplied direct to the low-pressure cylinder steam chest, either momentarily for starting, or sometimes for longer periods when load, road, and gradient are heavy and some increase of horse-power at the engine becomes necessary.

Boiler fittings and mountings

It should be remarked that unless this method of bye passing steam has been originally limited by the makers, by restricting the capacity of the steam pipe, it is within the power of the wagon driver to very heavily overload the low pressure side of the engine, while at the same time the useful work done in the high-pressure cylinder will be reduced, and the total power developed by the engine not much increased. The pipe 17 supplies steam to the feed-water injector, and the check valve connected with it may be seen on the fire-box side in Fig. 322. The supply from the feed-water tank v is by pipe x. Exhaust steam from the engine passes by way of pipe s to the separator box T, and thence to the coil c, Fig. 321, becoming superheated on its way to the blast nozzle, and less visible as it leaves the funnel. A branch pipe 20 is led from the bottom of T, by way of the plug cock 7, and as indicated by dotted lines enters the feed-water tank v near the bottom, takes the form of a long U in the tank, and leaving the tank at the forward end is finally taken to the ashpan 21. Water is in this way drained from the exhaust steam which is then superheated to a higher degree during its passage through c. By regulating the cock 7, more or less exhaust steam as required will find its way to the coil or U bend in the tank, and so be utilized for heating the feed water.

When there is steam pressure in the boiler the water tank may be filled by the ejector or water lifter R, Fig. 323. The tank holds 86 gallons, and the quantity in the tank at any time may be judged from the level in the gauge glass on the side of the tank which may be seen in Fig. 326.

Tank capacity

The forward end of the boiler is supported below its barrel 9 by a saddle connected with the locking plate or turntable above the worm-wheel quadrant 8. Connected with and below 8 is a horn-plate casting in the fork of which the front wheel axle may rise and fall with deflection of the transverse spring below it. A distance bar or stiffener between the fire-box and horn-plate casting assists the latter to resist road shocks tending to displace it. The method of steering by hand wheel Y and shaft connected through bevel wheels to the worm spindle z is clearly shown in both Figs. 322

and 323. Coal or coke fuel is carried in the open tip box above the fire-box end of the boiler.

Reference has been made to the rearward extension of the boiler side shell-plates in box girder form, and in Fig. 321 the cut away part is shown in dotted lines where the rear axle passes through. The pocket so formed is three-quarters of the depth of the girder frame, but it will be seen that sufficient strength is ensured by continuity of the corner angles across the pocket.

Engine The engine cylinders A, A¹ cast in one piece are bolted to the top of the tank girder, and the upwardly projecting side plates, cut away for the width of the cylinder, serve as snugs or check blocks for definite fixing of the cylinders independently of the fit of studs or bolts in their holes. The bored trunk crosshead guides are cast in one piece with the valve spindle guide and the front cylinder cover, and with an additional disc and flange having a gland at the piston rod to prevent the loss of lubricating oil from the casing, surrounding the moving parts of the engine, one end of which is attached to the disc flanges on the crosshead guides. It will be noticed that there is a distance of about 6 ins. between this flange and the cylinder cover, and the packing of the glands may be much more easily done than when this clearance is less.

Dimensions The cylinder diameters are 4 and 6·5 ins., with an engine stroke of 8 ins.

The two-throw crankshaft, with dips at 90°, is carried in bearings at B and C fixed to upstanding parts of the tank side plates.

A form of reversing gear requiring only one eccentric to each cylinder s used, and with it constant lead is obtained for intermediate points of cut off. It is known as Mann & Charlesworth's Patent. The eccentric is not moved round but across the shaft and follows a straight-line movement from side to side from the full forward to the full backward positions. The outline of this gear is indicated in the plan, Fig. 323, and parts of it in Fig. 322.

Single ex-centric reversing gear

Between each eccentric and the centre fork collar 18, is a disc keyed to the crankshaft. Bell-crank levers are pivoted in these discs and their free ends are connected at one end to the eccentric sheave ring, and at the other to the fork collar. Sliding movement of the collar 18 along the crankshaft moves the bell cranks about their fixed pivots when a change of eccentric position becomes necessary for alteration of the point of cut off or direction of running.

Movement of 18 in the direction of the axis of the crankshaft results through the bell cranks in movement of the eccentric sheaves in their own planes, and in a direction normal to the crankshaft axis.

The collar 18, it will be understood, is connected with and controls the position of the two eccentrics and the shifting fork which may be seen in plan, Fig. 323, and in dotted outline in Fig. 322, is keyed to the shaft Q having the reversing lever N at its other end.

Transmis-sion gear The gear reduction from engine to road wheels is through gear wheels

384

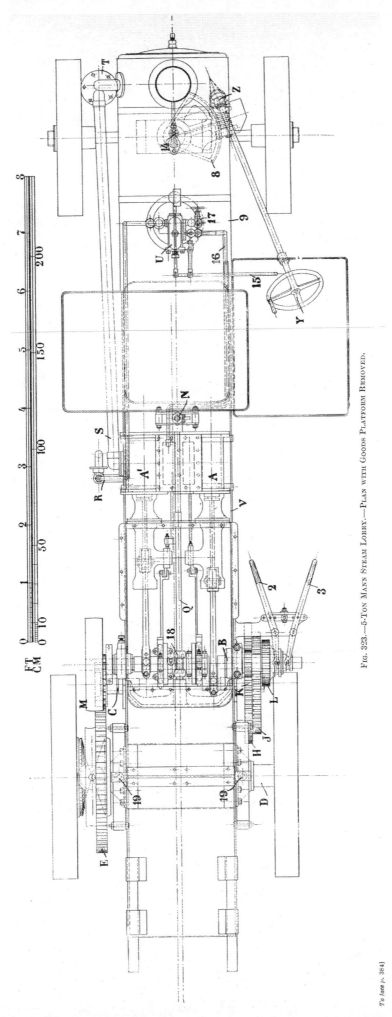

FIG. 323.—5-TON MANN'S STEAM LORRY.—PLAN WITH GOODS PLATFORM REMOVED.

and by two stages, from the engine shaft B to the intermediate C and from the intermediate C to the live axle D.

The method of changing gear may be followed by inspection of Fig. 323. When the hand levers 2 and 3 are in the positions shown, pinion K is in gear with the spur-wheel H, and there is a speed reduction of 5 to 1 in the rates of revolution of the shafts B and C. When 2 and 3 are approached to each other, K is taken out of gear with H, and L put into gear with J. The relative diameters of L and J, and, therefore, the ratio of speed reduction is in the proportion of 1 to 2·33. The size of the pinion on C which is always in gear with the large spur-wheel E is such that a second reduction of 3·8 to 1 **Gear ratios** is obtained. The total gear reductions for the high and low road speeds are, therefore, 8·85 and 19 to 1. The larger pinion L slides on feathers on the tubular stem or extension of K which in its turn slides on feathers sunk in the crankshaft. When L is in gear with J, K is moved outward and into the recessed or dished part of L. If when the pinions are in the position in Fig. 323, lever 3 only is moved inwards, then neither of the pinions is in gear, and the engine may be run without driving the wagon, or conversely the wagon may be moved without driving the engine. On the crankshaft, but at the end away from K and L is the fly-wheel M.

Fig. 324 serves to show the arrangement of live rear axle, and the method **Removable platform or body** adopted by Messrs. Mann of coupling up a lorry platform, or other form of body, carried on its own road wheels, to the driving-wheels and frame of what might be correctly described as a steam tractor, since, without the added platform or body, it has no goods carrying capacity.

The large spur-wheel E is cast with the crown ring, and drives the small pinions of the differential gear, and through them the large bevel wheels in the usual way. Of the last-mentioned wheels that on the left or near side is connected directly to the hub of the road wheel F, and it takes its bearing on the central solid axle, and is kept there by the collar nuts shown on the axle ends. The large right hand bevel wheel is fixed on the squared part of the axle and so drives the offside road wheel F, the hub of which is on the square formed end of the axle.

There is a casing over the central part of the axle, with bearings at its ends, and above these bearings are pillars 19 which convey the load from **Position of springs** the springs to the bearing casing. A transverse semi-elliptic spring shares the load with the side springs, and is connected at its centre with a cross T-iron, with web of deep section at the centre, tapered off towards the ends.

The road wheels F^1 of the wagon platform are of similar construction to the driving-wheels F. Reference to the different views shows them to be of the all steel built up type, with pressed plate centres flanged at the outer diameter, and having steel tyres riveted to them. The hubs are made of cast steel and have three equally spaced arms bossed at their ends. Fitted coupling bolts 11 hold the wheels together, and so oblige the outer wheels F^1 to serve as drivers and load bearers. Short stud axles D are reduced in diameter at their outer ends as shown, and are held in blocks or axle boxes which may slide up and down in the guide castings G. These guides or bearing frames are connected directly to the frame longitudinals of the platform, and have the semi-elliptic springs shackled at each end to

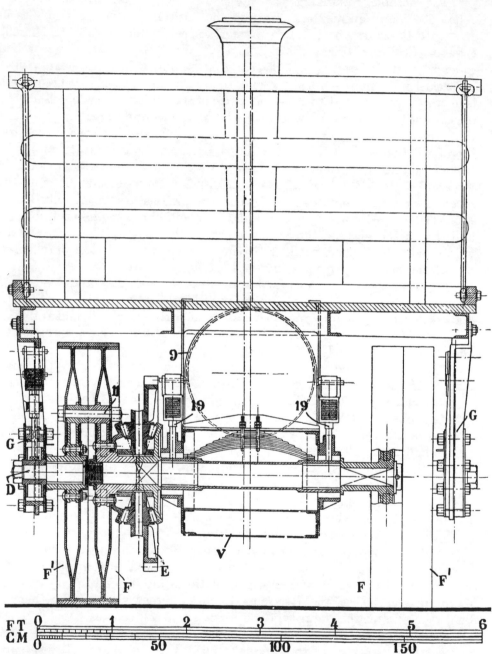

FIG. 324.—5-TON MANN STEAM LORRY.—TRANSVERSE SECTION AT REAR AXLE, LOOKING FORWARD

their Y arms.   The platform load borne by these springs is supported through pillars 19, Fig. 322, upon the sliding blocks in which the stud axles D are held.

MANN'S STEAM WAGON

Referring again to Fig. 322, when the platform has been run into position, and the coupling bolts 11 inserted, the fore-part of the platform is pin-coupled to the frame lugs 6, whilst the rear part is supported by the little carriage 4 on rearwardly projecting bearers on the top of the tank frame. The connexions at 4 and 6, it may be pointed out, transfer a part of the load carried to the centre pair of springs, and to a certain extent act as load equalizers. **Coupling goods platform to tractor**

The bearers are not held to the tank top but slide in clips or plate straps, and when pushed into position are there held by the bolt shown connecting the carriage 4 to the bearers. The latter are, therefore, removable, and when they and the pin 6 are taken out the lorry top may be tilted rearwards about its axles D. When this is done the engine and more of the gearing is left quite accessible, and, being so, is more likely to receive the attention periodically required.

FIG. 325.—5-TON MANN STEAM LORRY.—GOODS PLATFORM DISCONNECTED FROM TRACTOR.

Fig. 326 shows a steam tipping wagon with the steel tank body tipped, and from which it may be seen that the engine and top gearing are comparatively easy of access. The arrangement of the tractor is similar to, and has already been described with reference to Figs. 322 to 326. An arrangement of gearing has been provided which may be connected to the engine for tipping the body and returning it to the normal position. The separate goods platform already referred to is shown uncoupled and away from the tractor in Fig. 325, and it is provided with coil springs instead of the leaf springs shown by Fig. 322.

It is claimed that this distinctive combination of lorry and tractor makes it possible to comply with the provisions of the Locomotives on Highways Act, and Local Government Board Regulations, and to reduce to some extent the difficulty attendant upon the use of, and the operation of, brakes in connexion with a trailer. The advantage, considered from the standpoint of road endurance, of the wide tyres provided by the coupled wheels is questionable ; it is, moreover, not an inseparable feature of the Mann wagons, and the method is not adopted in the construction of the heavy colonial type of wagon. Now that an increase of tare weight has been permitted the necessity no longer exists, and no advantage is obtained by the combination of truck and tractor in such a way that, while for compliance with regulations the tare weights could be separated, they were merely the equivalent of the more general design of steam wagon which, when the 3 tons tare limit was in force, could only by the exercise of considerable skill, and the use of special means and materials, be made to satisfy the requirements as to weight.

Brakes The arrangement of brakes may be understood by reference to Figs. 322 and 323. The strap of the bandbrake, lettered P, and shown in dotted

FIG. 326.—5-TON MANN STEAM WAGON WITH TIPPING GEAR.

outline behind the gear-wheel casing, is anchored at the upper end, and connected at about the position P by an adjustable link to the short arm of the pivoted bell crank O. The brake-wheel or drum may be seen in Fig. 323 on the shaft C, and beside the spur-wheel K. The brake is foot-operated through the long arm O of the bell crank, which works in a slot in the driver's side box W, and enters it far enough for convenient operation by the driver. A suitably cut ratchet plate holds the brake on or off. There are also brake blocks 5 acting upon the driving-wheel tyres, and operated by means of the hand-wheel 12, on the screwed end of the connecting rod 13.

The brake equipment is, therefore, liberal, for in addition to the brakes provided, some braking effect may be obtained by judicious use of the engine reversing gear.

Chapter XXVI

THE ROBERTSON LORRY

THE Robertson lorry is one of those which have recently appeared and in the arrangement of which it may be supposed the designers have been ready to profit by the experience gained by makers earlier in the field. In outline and external appearance it resembles a number of wagons by other makers, and may be said in some respects to conform to practice that has gradually become general. Considered in detail there are certain distinctive features, chief among which is the boiler shown by Fig. 332, and others to which reference may be made in connexion with the general arrangement drawings, Figs. 328 to 331. From these drawings it will be gathered that the frame is of channel and angle steel construction, the main Frame side channels running parallel in plan for the greater part of their length, converging at about the position of the driver's seat, and continuing forward parallel, but at a reduced distance apart in order that there may be clearance between the frame and the steering wheels when the latter are turned to their greatest steering angle. Little appears to have been done to strengthen the frame against racking stresses by the use of either diagonal members or plate gussets, some reliance presumably being placed upon the added strength of attached parts, particularly the goods platform. A short longitudinal angle secured to two transverse angles, together with the frame channel A forms a rectangular bed up to which the engine is rigidly held, at the positions of a bracket on each side of the crank chamber and suitable facings on the upper side of the cylinders. The latter are 4 and 7 in. diam., Engine the stroke being 5 in. and the speed of revolution 435 per minute when the wagon is travelling at 5 miles per hour. The engine is rated to develop 25 BHP., and at the stated speed would do this with an initial pressure of 195 lb. per sq. in., and 7 expansions, or with cut off in the high pressure at 45 per cent. of the stroke, and at a suitable point in the low-pressure cylinder for equalization of work. With link reversing motion with which this engine is fitted, controlled by a single reversing lever, an arrangement Advantage of separate expansion gear which is commonly used by locomotive engineers, and also by most other steam wagon builders, it is possible to effect this equalization with a compound engine through a limited range only, and for the sake of simplicity

389

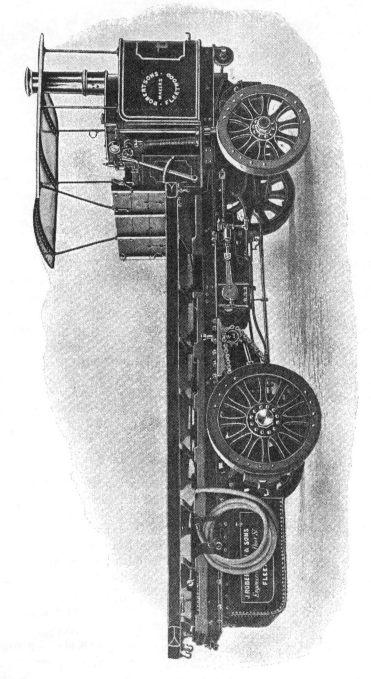

FIG. 327.—ROBERTSON 5-TON STEAM LORRY.

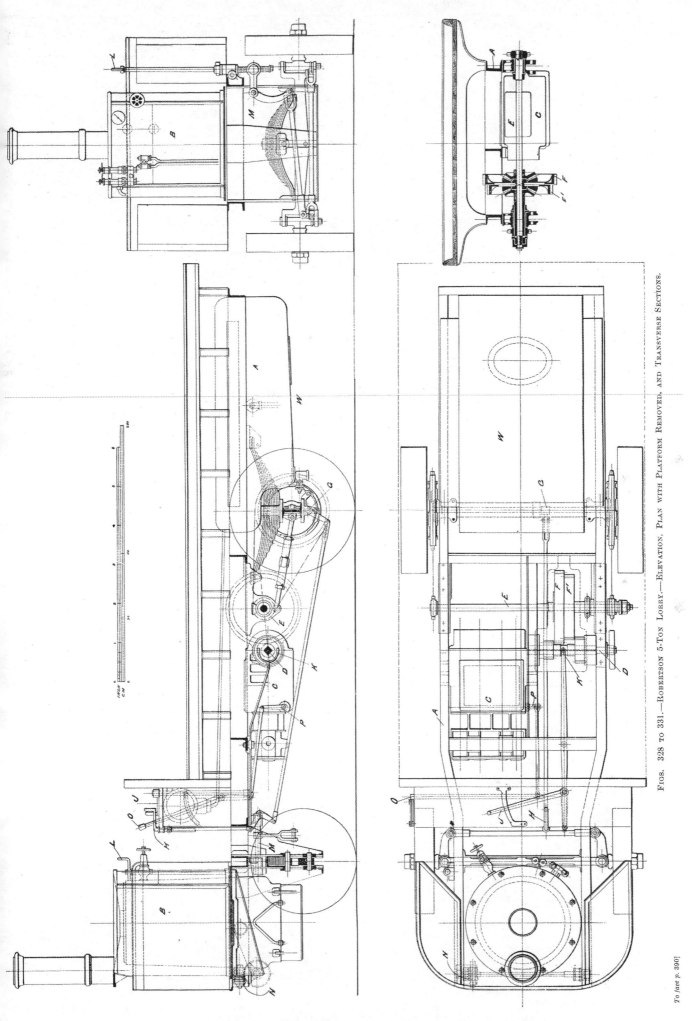

Figs. 328 to 331.—Robertson 5-Ton Lorry.—Elevation, Plan with Platform Removed, and Transverse Sections.

To face p. 390]

of control it is a convenient approximation. Separate control of the expansion gears of compound locomotives has been tried with success, and a method of rendering such duplicate control easy of operation would make it possible to obtain better results from the road locomotive. The lay shaft of the reversing gear is in this engine at P, and it is connected as shown to the side lever O. An extension of the engine crankshaft squared for part of its length is supported in a bearing at D on the main frame, and carries a fly-wheel at its extreme end. Two pinions connected by a collar sleeve K, may be moved, by the system of levers controlled by that lettered J, along the squared portion of the shaft to mesh with spur-wheels F or F^1, for the high and low gears. The spur-wheels F F^1, Fig. 331, form the outer or crown ring of the differential gear, and drive through the right and left hand differential bevel wheels the chain sprockets fixed upon one end of the shaft E, and on the sleeve on the left side of E, and which while running normally at the same speed may, when the differential gear operates, revolve upon E. The outer end face of this sleeve is toothed, and a similarly toothed piece, capable of sliding on the squared end of E may be moved into engagement with the sleeve when it is required, to lock the differential gear by means of a small steam cylinder controlled from the driver's seat. Bearings for E are provided in brackets to which the radius rods are also connected, and extensions backward of these brackets form guide shoes in which the forward ends of the rear springs may slide. The other ends of the springs slide in similar shoe castings.

Transmission gear

Differential locking gear

The total speed reductions between the engine shaft and the road wheels are 18·5 and 10·15 to 1, the side chain sprockets and pinions having a ratio of diameters of 3·7 to 1. The rear axle ends are steel forgings suitably formed and machined for connexion by riveting to the centre part, which is rolled steel girder of I section.

The method of support of the front part of the lorry is indicated in Figs. 328 and 329. The load is taken at the centre of the transverse spring, and delivered through its ends to points on the front axle where guided by the side horn plates. The horn plates are stiffened at their lower ends, and cross connected by bent angles, and by attachment to the centre column. The transverse spring buckle carries a pin on its under side upon which the centre column takes its bearing through a seating block in position at the top of the fork in the centre column. The front axle can thus rise and fall or cant, and the spring may also turn about its pin in agreement with such movements, and a little side play of the axle is permitted according to the amount of freedom of the pin at the centre of the axle in the slot formed in the centre column. The front axle ends are forked to take the pivoted stub axles, and the wide spacing of the fork jaws is a good feature. Arms and jointed rod connexions for steering on the Ackerman principle are shown, and their control by screw column, L pivoted nut, and bell crank M may be followed from Fig. 329, which is a section across the lorry just behind the steering gear, looking forward.

Front carriage

The engine reversing gear is considered equivalent to a brake, and a screw-down brake is fitted of the type shown by Figs. 328 and 330, or like that shown on the general external view Fig. 327, where the brake blocks act upon the road-wheel tyres.

Boiler   The fire tube boiler, Fig. 332, is of considerable interest, and well suited to the requirements.   It has in common with other fire tube boilers relatively large water content, and water surface of sufficient area, from which steam may be liberated without promoting priming or necessitating large steam drum or separator space capacity.   The main parts of the boiler are of steel and of cylindrical formation, and, therefore, naturally strong, and the flat surfaces at the junction of the fire box c and the centre uptake B as well as that of the top cover plate L are of small extent, and require no staying other than that already afforded by connexion to adjacent parts.   The outer shell A is in one piece, and is connected by inward flanging, and riveting to the fire box.   It is inwardly flanged at the top, whilst B is outwardly flanged, and the annular cover plate L is jointed to the surfaces so formed with concentric rings of well fitted studs.

Internal examination   The removal of L makes examination of some of the internal surface possible, particularly that part about the water line.   The design does not permit of such complete opening up for internal examination as for instance the type of boiler illustrated by Fig. 357, but this disadvantage is to some extent nullified by freedom from liability to deposit, or the probability of almost the whole of the solid matter in the water finding its way to the bottom of the annular space around the fire box, where it may be readily raked and washed out.

A large number of radially arranged steel tubes connect the inner chamber B, and the outer shell A, and the hot products of combustion pass from the fire box through these tubes, and so to the outer smoke box casing, surrounding the greater part of the outer boiler surface.   The tubes provide most of the heating surface, but that in the fire box, and as much of the surface of B as is water covered, is valuable, and contributes to the evapora-**Heating surface** tive power of the boiler.   The water-covered or effective heating surface is 80 sq. ft., and the grate area 2·6 sq. ft.   The working pressure is 200 lb.

The arrangement of smoke box casing around the boiler is advantageous when it can be done in such a way that occasional examination of the external surface of the boiler may be made without disconnecting fittings or mountings and pipe connexions to them.   The casing in this boiler is made in three sections, and connexions to the boiler are made through that lettered K, which is above the level of the tubes.   The taking down of the movable sections permits practically the whole of the outside boiler surface to be seen.   The whole grate with the ashpan below it is slung by wire ropes or **Fire grate suspended** chains from the side handles H passing over the guide pulleys J and connected to drums on the spindle of the worm and worm-wheel gear N, see plan, Fig. 330.   By means of this gear the grate may be held up to the boiler bottom or lowered for lighting up at the start or for cleaning the grate. When the grate is lowered at the conclusion of the day's run, the tubes may be cleaned from their inner ends by blowing through with a steam jet pipe

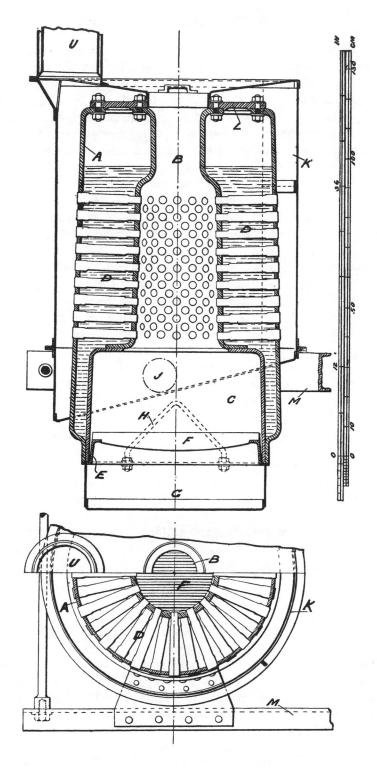

FIG. 332.—ROBERTSON 5-TON LORRY.—SECTIONAL ELEVATION AND PLAN OF BOILER.

provided for this purpose. The boiler is conveniently fired through the centre opening at the top of B. It is supported on the lorry side frame members M, by plate and angle brackets bolted to it, and riveted to the shell A. The funnel U is double or air jacketed, and is fixed to the front of the casing as shown. The question of lagging in this design has been avoided altogether, the smoke box casing and the hot gases from the smoke tubes constituting a very effective protection from heat losses by radiation.

Feed-water heater
Some of the exhaust steam is profitably used in a tubular feed-water heater, and in addition to the fuel economy to be expected as a result of supplying feed water at a temperature of 180 to 190°F. to the boiler, such an economiser is equivalent to an addition to the boiler evaporative capacity. Rowe tubes are used in this heater, and the condensed steam is filtered and returned to the feed-water tank.

Difficulty of feed-water purification
Purification of feed water is, however, a difficult matter, particularly when oil has to be separated from it sufficiently completely to render the feed suitable for use in a steam boiler. If the treatment is by filtration, vessels of relatively large size are required to contain enough medium, and means must be provided for readily removing it, and the entrained oil and other impurities. It is unlikely that any economy will be effected in this way when the costs of filter material and extra attention are considered, and when neglect of the filter would sooner or later result in the presence of oil in, and some damage to, the boiler.

While the engine is running the boiler is fed by the main feed pump, which is worked by gearing from the crankshaft enclosed in the crank-chamber. The pump and its plunger rod and pipe connexions may be seen at the side of the engine in Fig. 327.

An independent steam pump is carried with separate connexions between feed tank and boiler, and the boiler feeding arrangements are thus in duplicate.

The water tank W contains 190 gallons, is provided with water lifter and armoured suction hose pipe, and has necessarily provision for filling by hand. Three try cocks are fitted to the tank instead of a gauge glass, so that the quantity of water in the tank may at any time be known approximately. The lorry is designed to carry 4 to 5 tons on a platform 13 ft. 4 in. long, and 6 ft. 5 in. wide.

Hydraulically operated gear inadvisable
Convenience in driving the lorry has been borne in mind, and it will have been observed that gear changing may be done from the driver's seat. The method of locking the differential gear through a steam relay cylinder is ingenious, but unless the lorry is to work in a district where roads are very bad or gradients exceptionally heavy, such a convenience cannot be regarded as a necessity. There would also be difficulty with such an arrangement in frosty weather.

THE NAYLER STEAM WAGON

Messrs. Nayler, of Hereford, are among the older engineering firms who have more recently undertaken the construction of steam wagons,

The type now built by them, and introduced two years ago, is shown in side elevation by Fig. 333.

The vertical fire-tube boiler is mounted directly over the front axle, and it has 4·9 sq. ft. of grate area, and 120 sq. ft. of heating surface. The upper part of the boiler shell is removable, so that occasional complete examination may be made, and the less accessible places, which are frequently those which scale or accumulate deposit most, may be exposed. The tubes are arranged in straight, through rows, and are not staggered; cleaning or scaling could, therefore, be attempted, which would, with other arrangements, necessitate removal of some of the tubes. **Arrangement of tubes**

The boiler is fired from the top through a large centre tube, and removal of ashes and lighting up, cleaning or clinkering the grate may be done through a door in the ashpan at the front of the wagon, the fire grate being hinged. The working pressure is 200 lb. per sq. in. The feed-water pump is driven

Fig. 333.—Nayler 5-Ton Steam Lorry.

from the engine at one-fourth its speed through gearing, and a Penberthy injector is used as alternative method of feeding the boiler.

The high and low pressure cylinders of the compound engine are 4 and 7·5 in. diam. respectively, the stroke is 7·5 in., and the normal speed 400 revs. per minute. Ordinary link motion provides for occasional running at very heavy load, for hill climbing, and for reversing. **Size of engine**

The system of transmission gearing is noticeable for its simplicity, and is representative of that used by some other heavy wagon builders. That used in the Straker vehicle is almost identically the same, and that in the Foden of the same type though differing in structural detail.

The engine-shaft extension is of square section, and has upon it two pinions of suitable diameters relatively to the large spur-wheels upon the intermediate shaft with which they gear, for road speeds of 2 or 6 miles per hour. The second speed reduction is by a single heavy chain from the sprocket on the intermediate shaft to a large chain-wheel on the live axle,

Two-stage gear reduction

and connected with the differential gear. When the wagon is travelling at either of its two road speeds there are only two gear-wheels, two chain-wheels, and a chain in active operation, but the simplicity of a two-stage speed reduction is generally obtained at the expense of either objectionably large or small gear-wheels, and a reduction of engine speed.

The support of the front part of the lorry and the method of steering follows traction engine practice to some extent. There is a ball bearing between a cradle surrounding the front axle and the throat or bearer plate

Ball bearing to front carriage

connected to the wagon frame, and by the use of this form of bearing the steering effort is considerably reduced. Two transverse springs are fixed at their centres to the cradle, one on each side of it, and the front axle which it surrounds, and their outer ends are connected by shackles to points on the front axle not far from the axle ends.

The only positive connexion between the front axle and the wagon is by the spring shackles, and the front axle is free to ride up and down or cant in the cradle under spring deflection. The worm-wheel of the steering gear is fixed to the cradle, and as the latter surrounds or slides over the front axle, it gives the angular motion it receives from the worm-wheel to the front axle for steering. This method of steering by locking plate or fifth wheel may be more satisfactorily employed for heavy than for light vehicles. It necessarily limits the diameter of the front wheels for a given height of frame or platform from the ground, and further, any pipe or rod connexions, or other parts beneath the frame, must be arranged clear of the area swept by the wheels through their steering angle.

The front wheels are 2 ft. 6 in. diam., with 5 in. tyres, and the driving wheels 3 ft. diam., with 9 in. tyres.

Those shown in Fig. 333 are of very light construction, and have been superseded by heavier all steel wheels having cut away pressed disc centres.

The goods platform is 12 ft. long, and 6 ft. 6 in. wide, and 4 ft. from the ground when the lorry is unloaded. The wagon is designed to carry 5 tons, and to haul 2 tons in a trailer.

It will be seen that the goods platform is 8 or 9 in. above the frame

Access to engine

channels, and that there is sufficient room to remove the cover to the engine crank chamber. It would thus be possible to examine and, under difficulties, to adjust some of the engine bearings without removing the platform.

The Garrett Steam Wagon

The accumulated experience of many years in the construction of heavy road engines, and familiarity with the problems of transport on common roads have enabled Messrs. Richard Garrett & Sons, Ltd., to produce a steam wagon, which, while possessing some original and good features of design, bears evidence of appreciation of the good work done by other firms who have associated themselves earlier with the construction of heavy motor vehicles. They have in fact wisely profited by the lessons which others have gained by particular experience and have added the results of their own intimate knowledge of engineering matters.

From the external view of the first wagon constructed by them, shown by Fig. 333A it will be seen that the vertical boiler is carried at the front of the wagon a little ahead of the steering-wheel axle, low down upon the longitudinal frame channels, that a combined engine and gear box is hung from side brackets at the rear, and a centre suspension bolt at the forward end, at about the middle of the wagon frame, and that the feed-water tank is held up by strap bolts to the back end of the frame.

The boiler is of the top-fired fire-tube type with the funnel set forward on the smoke box to leave the large centre tube free for the introduction of coal or coke fuel. As with other similar types of boiler already referred to, the fuel having been dropped on to the grate, distributes itself

FIG. 333A.—GARRETT 4–5-TON STEAM WAGON, 1905.

by the shoggling action set up by vibration as the vehicle travels over the road.

The dimensions of the boiler are liberal and its heating surface and grate area are respectively 65 and 3·75 square feet. The heating surface mentioned is that of the fire box, and as much of the surface of the fire tubes as is effective or water covered. If the whole surface of the tubes is called heating surface, as it is often incorrectly described, then the total heating surface of the boiler becomes 112 square feet. The diameter of the shell is large compared with other boilers used for wagons of similar capacity, being 2 ft. 7 in., but this permits a large number of relatively short fire tubes to be employed. The tubes, numbering 216, are 1 inch internal diameter and their length between the fire-box crown plate and the top tube plate is 18 inches, and it is claimed that by using short tubes, trouble due to differential expansion and leakage is reduced. There are six bar stays which relieve the tubes of the duty of staying the tube plates and it is

Dimensions of boiler

quite possible that in this instance being short, they will prove as satisfactory a method of staying as the bar stays of great length employed under less advantageous conditions in large locomotive type boilers. The working steam pressure is 225 lb. per square inch.

Combined engine and gear box The arrangement of the combined engine and gear box shown by Fig. 330B is interesting. The whole of the engine running parts, the intermediate and differential gear shafts 2 and 3, and their wheels run in an oil bath, and the driver is not required to, and should not be expected to, separately lubricate these parts. Oil ways are cut in favourable positions to distribute the oil and in the case of the built-up tubular differential shaft openings are left to ensure that, the differential gear, inside the drum G and of the spur-wheel type, and the driving shafts, shall receive copious lubrication.

The high-pressure cylinder A and the low-pressure cyclinder B are both

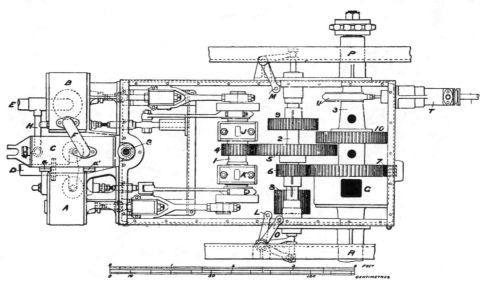

FIG. 333B.—GARRETT 4–5-TON STEAM WAGON.—ARRANGEMENT OF ENGINE AND GEARING.

carried upon stools fixed to the cast-steel end plates of the gear box, and these stools, it will be observed, also form cylinder covers and have the piston rod and valve spindle stuffing boxes cast with them. More space has been allowed than is usual for access to the glands for repacking. With these small high-pressure steam engines soft packing is almost invariably used, and the glands require a good deal of attention from the driver. The application of metallic packing to small rods is not easy, but for these steam wagons it should, wherever possible, be adopted.

Difficulty in starting The cylinder dimensions adopted are $3\frac{3}{4}$ and 7 inches for the high and low pressure cylinders and the stroke is $6\frac{1}{2}$ inches. Originally the cylinder dimensions were $3\frac{1}{2}$ and $6\frac{1}{2}$ with a stroke of $5\frac{1}{2}$ inches, and it is worth recording that while these smaller dimensions enabled the engine to develop ample power, difficulty was sometimes experienced in starting the lorry smartly, particularly on hills where a heavy torque or starting effort is demanded. A reasonable attempt is made to combine the advantages of both piston and slide valves, by adopting the former type for the high-pressure cylinder where steam temperature and pressure are high, and the latter type for the low-pressure cylinder where it is essential that there should be no loss by steam leakage past the valve.

Use of piston and slide valves

A feature of some interest in connexion with steam motor wagons is the use here made, for the first time, of an engine arranged to run always in one direction, single fixed excentrics being used with the simplest form of valve gear possible. Reversing is effected as in petrol motor vehicles through the transmission gearing, the reversing pinion 8 being visible below the intermediate shaft 2, and carried upon a fixed hollow shaft.

The steam pipe D is connected, as shown in dotted lines, to a junction piece, one branch leading to the high-pressure cylinder and the other to the receiver vessel or controlling cylinder C. The steam having done its work in the high-pressure cylinder, exhausts at the ends of the piston valve and passes by the branches a or a^1 to C. When the piston valve, of peculiar form inside C, is in its normal position the steam may continue its course through the diagonal pipe F to the low-pressure cylinder, which it leaves by an outlet shown dotted below it, and the exhaust pipe E. Before the steam is finally exhausted up the boiler funnel, it passes, first through a tubular feed-water heater and then to a drying coil in the smoke box, the latter by slightly superheating the steam rendering it less visible as it passes away to the atmosphere.

Controlling cylinder

If the controller valve is pulled forward the engine becomes a simple or two-cylinder high-pressure engine, capable of developing for short periods a high horse-power or of exerting a heavy starting effort. It is also necessary sometimes to momentarily move the valve to this position, by means of the hand lever at the driver's seat, when the engine has been stopped with the high-pressure crank-pin on either dead centre. With the valves cutting off at about ¾ stroke the low-pressure crank-pin would then be in a favourable position for starting the engine.

When the controller valve is in this position exhaust steam from the high pressure passes direct to the exhaust pipe, through C and by the branch pipe H, which, it will be seen, joins E. High-pressure steam is supplied to the receiver chamber C through the branch from the main steam pipe already referred to, and goes by F to the low-pressure cylinder, which it leaves, as before, by the main exhaust pipe E. If the controller valve is set to its middle position the exhaust passages a a^1, from the high-pressure cylinder are closed and the low-pressure cylinder receives no steam. By compression of steam on what would ordinarily be the exhaust strokes at the high-pressure cylinder, braking effect through the engine may be obtained, just as with link motion reversing gear the same effect is obtained by reversing the engine, while it is still running. Locomotive type flat single guide bars are used and the excentric rods are both inside or towards the centre of the engine. The balanced crank discs are of cast steel and in one piece with the crank-pin and excentric. They are forced tightly on to the end of the shaft I and keyed in position. The driving pinion 4 is fixed upon the centre of the shaft and is well supported between the bearings J and K.

Transmission gear The total gear reduction from engine shaft to road wheels is 9·5 to 1 for the 6 miles per hour or high speed, and 18·5 to 1 for the low speed or about 3 miles per hour. At these road speeds the engine runs at 475 revs. per minute. The gear-wheels are shown in position for running at the low speed, the drive being through wheels numbers 4, 5, 6 and 7. If 6 is **Reversing** moved out of gear with 7 and the reversing pinion 8 moved into its running position, the drive is then through 4, 5, 6, 8 and 7, the direction of rotation of 7 and hence of the chain pinions being changed. The intermediate shaft 2, like the fixed shaft upon which the reversing pinion runs, is hollow except at the centre part, and the wheels 6 and 9 may be moved along the squared parts of the shaft they occupy by means of rods connected with cotters through the bosses of the gear-wheels. The gear-shifting rods slide in the hollow shaft 2, in which cotter ways are cut, and are given the sliding motion for putting the wheels into and out of gear by means of bell cranks L, M, with forked ends engaging with the outer collared ends of the shifting rods. The bell crank O and the pinion 8 which it moves have double the movement of L and M. The gear changes are made from the driver's seat by three levers and connecting rods to the bell cranks.

Means of feeding boiler The complete engine and gear box is supported by hanger brackets on the longitudinal frame girders at P and R, and at the forward end from the centre spring buffered suspension bolt S. The boiler feed pump is shown at T and it is worked by an excentric U cast with one of the tubular parts forming the shaft 3. When the lorry is standing the boiler is fed by an injector with independent connexions from the feed tank and boiler. The whole gear box is of light construction, the bottom being formed of a ½-in. flat steel plate, the sides are of ¼-in. steel plate, and the ends are flanged steel castings. The gear shafts are correctly spaced and have their white metalled bearings in two steel castings bolted to the box side plates. The top cover with suitable inspection doors is shown removed.

The wagon is substantially built and with the cast-steel wheels shown is intended to come just within the new tare regulations. The front wheels are 2 ft. 9 in. diameter and the rear wheels 3 ft. 3 in. diameter. In addition to the braking effect at the engine, there are internally expanding compensated brakes acting on the inner surface of the chain-wheel drums on the road wheels and operated by a foot lever, as well as screw down brake blocks acting on the peripheries of the rear wheels.

The wagon is shown with a lorry platform, 12 ft. 6 in. long, and 6 ft. wide, and 3 ft. 10 in. from ground level when unloaded, and it is designed to carry a load of from four to five tons and to draw a trailer wagon.

The overall length is 18 ft. 9 in. and the width over the rear-wheel hub caps 6 ft. 8 in. Messrs. Garrett also construct small high-speed traction engines of the type earlier referred to.

HEAVY MOTOR VEHICLES

The Simpson and Bibby Steam Lorry

The vehicle made by Messrs. Simpson and Bibby, of Manchester, and illustrated and described here with reference, Figs. 334 to 350, differs very materially in design and construction from the original productions of the firm of Messrs. Simpson and Bodman, described in vol. i. of this book.[1]

The boiler is practically the same, but nearly everything else has been changed in design and position on the vehicle The original three-cylinder engines were entirely discarded, a different arrangement of three-cylinder engines was also displaced by the two sets of two-cylinder engines coupled together illustrated, a differential countershaft being introduced to replace the separate engine drive to each wheel, and the engine centrally placed instead of at the rear of the lorry, only one fixed speed reduction being employed. Hydraulic steering gear is also shown in place of the chain gear formerly used.

Changes in design

A reproduction from a photograph, Fig. 334, shows the general appearance of this lorry, the engines on which are of about 45 horse-power, and capable of propelling the vehicle with a load of about 7 tons, and a trailer with 3 tons, at a speed of 6 miles an hour, on average give-and-take roads. The speed reduction by gearing between engine and road wheels is 11·7 to 1 for the low speed.

Capacity

Figs. 335 and 336 show elevation and plan of this vehicle. The main frame A is of teak reinforced by top and bottom steel plate, as seen in section in Fig. 343. It has an overall length of 18 ft., and a platform area of 12 ft. × 6 ft. 4 in.

The boiler B is of the flash type with " Rowe " tubes, and is similar to that described in vol. i., a modification being the substitution of ordinary unions for the " Haythorn " joints then employed. Fuel is fed through the door 23, Figs. 335, 336, in the rear side of fire box, a hinged ashpan 24 being provided.

For starting and occasional purposes the boiler is fed with water by means of the hand pump v, see plan Fig. 336, worked by the handle 9.

The engines C and countershaft G are placed centrally in the vehicle with crank and second motion shafts across the frame, the spherical hanger bearings for the latter being held in brackets bolted to the underside of the framing A, as seen in Fig. 336, and in detail in Figs. 337, 338, and 340. Power is transmitted from the engine crankshaft by spur gearing to the sprocket countershaft G on which is mounted the differential gear contained in casing M, pinions E on the ends of the crankshafts, gearing with the wheels F on the differential gear.

General arrangement of gearing

On the ends of shaft G are the sprocket pinions H, H, which drive by chains the large wheels J, J, fixed to the rear road wheels on the fixed back axle.

The feed pump P (see Figs. 336 and 342), is driven direct from countershaft G, though not shown in Fig. 340. It is a three-throw pump worked

Three-throw feed pump

[1] Vol. i. pp. 543–558.

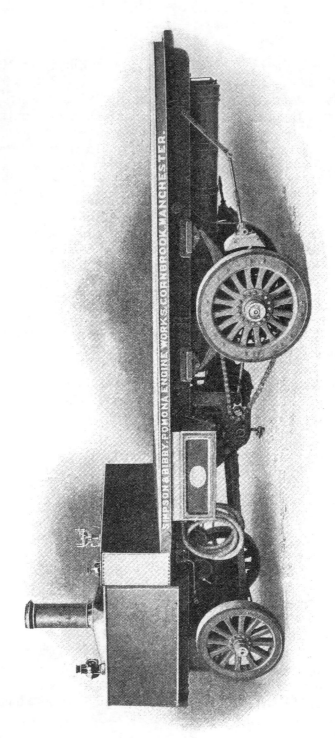

Fig. 334.—Simpson and Bibby 7-ton Steam Lorry.

402

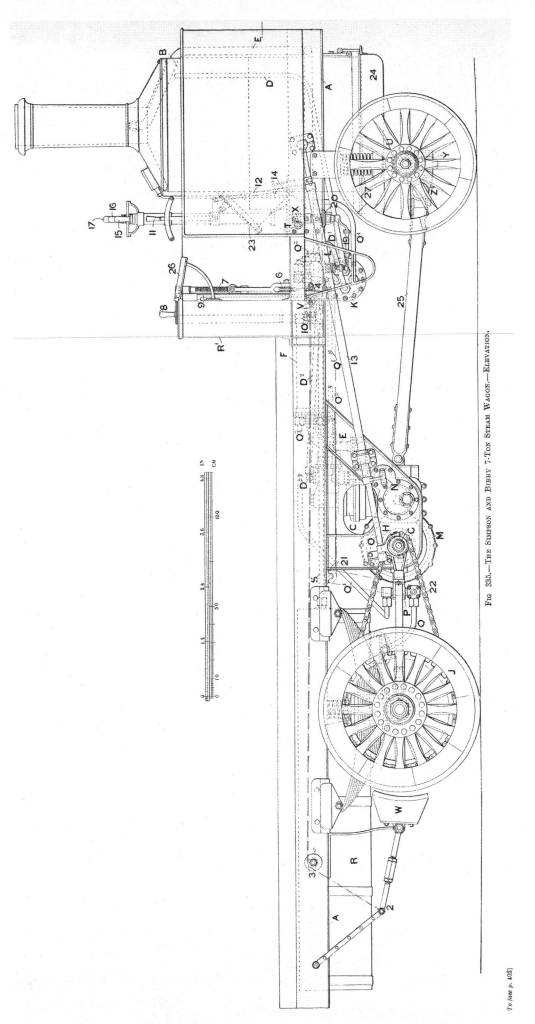

Fig. 335.—The Simpson and Bibby 7-Ton Steam Wagon.—Elevation.

by one wide eccentric F, the plungers P¹ P² P³ being of different diameters, the one or the other being put into use as hereinafter described by cams on a shaft C, which lift and hold up one or other of the three valves A¹ A² or A³, according to the quantity of water required for regular feed. R (Figs. 335, 336) is the main water tank, and R¹ containing an auxiliary supply is connected to it by the pipe S.

The main feed pump P sucks water through the pipe Q and forces it through pipe Q¹ to a valve box (to which the hand-pump feed-water pipe is also connected), through pipe Q², thence through the heater F, heated by exhaust steam from the engines, through pipe Q³, drum K, and pipe Q⁴ to boiler.

The highly superheated steam generated in the boiler passes from the upper superheating tubes through pipe D to the receiver drum K, where it gives up some of its heat and also raises the temperature of the already warmed feed water, passing out of K through pipe D¹ to the stop valve L; from L it is allowed to flow through pipe D² to branch pipe D³, which is connected to the four cylinders C of the engines. The stop valve L is operated by means of the handle, 17, which partially rotates a spindle passing through the steering pillar, an arm on the lower end of the spindle actuating the valve by a lever on the valve stem and connecting link 19.

The exhaust steam from the engines enters at one end of the heater F through pipes E, circulating round and heating the feed-water pipes, and leaves it at the other end and passes through pipe E, which leads it to the base of the chimney, where it escapes in a more or less superheated condition.

The main triple-feed water pump P is shown by Figs. 336 and 342. It is entirely cased in.

It is driven by the shaft G, Figs. 340, 342, on which an eccentric is formed. The eccentric strap F surrounding this is formed integrally with its forked connecting rod, holding the pin of crosshead S. The crosshead S is formed in one with the largest of the three different sized plungers P¹, P², P³. Stop pieces for the suction valves A¹, A², A³, are held in place by screws in the bridge K. The water entering at Q, and passing through suction valves A¹, A², A³, is forced through the valves B¹, B², B³, to a common chamber above them, to which the delivery pipe is connected at Q¹. When either of the suction valves is held up by the cams on the spindle C the plunger to which that valve belongs is of course inoperative, the water it takes in is only returned to the suction pipe.

The spindle C, which may be partly rotated from the foot plate, has one cam on it beneath each of the suction valves A¹, A², A³, which are thus lifted the required amount.

Referring to Fig. 336, it will be seen that there is a chain pinion 22 on the projecting end of the valve cam spindle of this pump P connected by chain to one of two pinions which are fixed to one another, and are mounted on a stud spindle 21 fixed to the frame A. The other pinion is connected by chain to a pinion on a stud spindle fixed to the frame immediately below the steering column. On this spindle, and fixed to the chain

Live steam heated feed water

Control of feed to boiler

pinion, is also mounted a small bevel wheel gearing with wheel 20, he bevel wheel 20 being fixed to the end of a tubular spindle passing up through the steering column and rotated by means of the handle 16. By these means the cam spindle of the pump may be set so that the pump will go on feeding at any one of the three rates of feed, or by setting the cams so that A^1 and A^2 or A^2 and A^3 are simultaneously operative, two more different rates of feed may be obtained making five in all. A Marsh direct-acting steam pump is also provided.

The steering of this vehicle is effected by means of the hydraulic gear shown by Figs. 343 to 346.

Figs. 343 and 344 show the front axle Y supporting the main frame A, A, by round steel spiral springs S, S, held in cup brackets bolted to the main

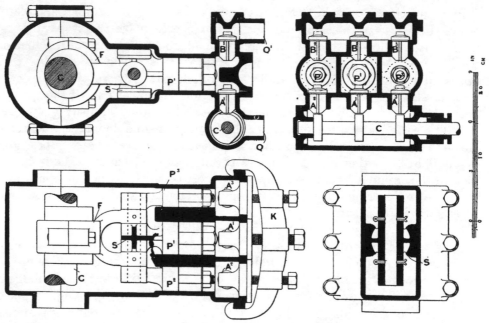

FIG. 342.—THREE-THROW DIFFERENTIAL FEED PUMP.

Hydraulic steering gear

frame, and in cups held by the steering arm pivots U; the pivoted road wheel axles D, D, have the usual arms formed on them connected by the rod F. A weight-bearing ball race is used inside the upper jaw of the fixed axle on each pivot. It has been found by experience that these balls stand the work providing the race is not filled. To the web of the axle Y cylinders Z, Z^1 are bolted end to end with a cover plate between them; the pistons B are fitted with Nixon's hot water hydraulic leather cups, and a connecting rod C pivots on a gudgeon-pin in each piston, and is connected at its outer end to the pivoted axle arm.

Water is always present inside the cylinders Z, Z^1, Water from the lower tubes of the boiler and therefore under steam pressure is forced through ports E, E^1 into either one or the other cylinder pushing piston

404

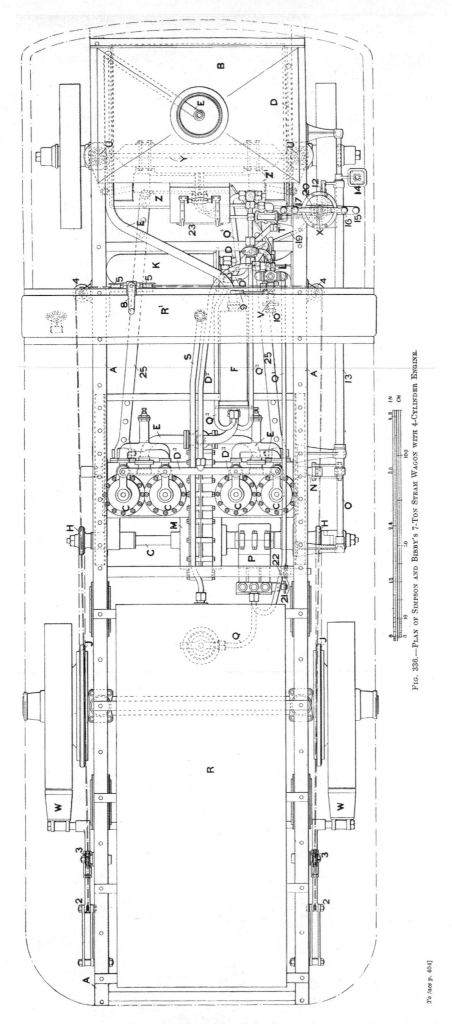

FIG. 336.—PLAN OF SIMPSON AND BIBBY'S 7-TON STEAM WAGON WITH 4-CYLINDER ENGINE.

To face p. 404]

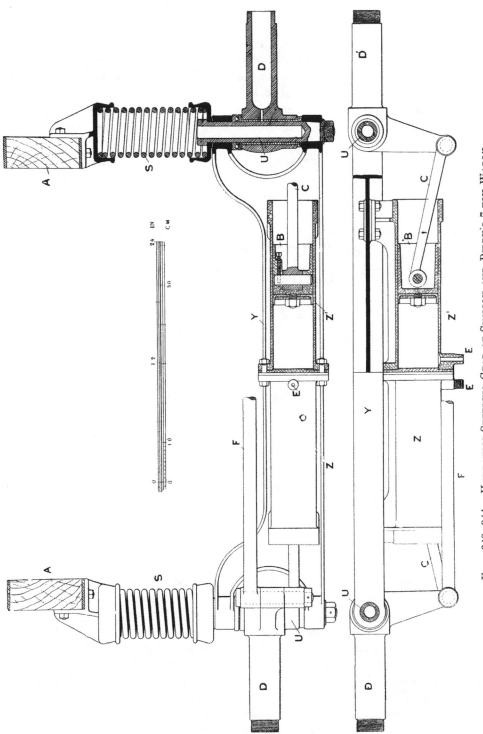

FIGS. 343, 344.—HYDRAULIC STEERING GEAR OF SIMPSON AND BIBBY'S 7-TON WAGON.

405

B outwards, and thus moving by rod C the pivoted arm and axle D of either one or the other road wheel.

Figs. 345 and 346 show the distributing valve for this gear by vertical and horizontal sections through the valve chamber T, the position of which is seen in Figs. 335, 336. Water under steam pressure is always present in the valve chamber T. It is admitted at S, the valve U sliding over the small portholes W W, allowing this pressure water to flow either through W or W^1 to the back of one or other of the pistons connected to the steering axle arms.

The passages W and W^1 are connected up by pipes to nipples E, E^1, formed on the cylinder castings Z, Z^1, Figs. 343, 344. The position of these cylinders and pipe connexions is seen in Fig. 336, from which it will also be

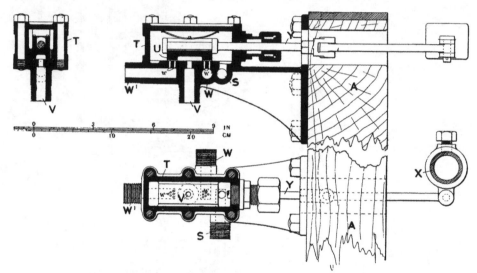

FIGS. 345, 346.—HYDRAULIC STEERING VALVE.—SECTIONS.

seen that stout stiffening pieces of teak are bolted on either side of the web of the steering axle.

The distributor valve is bracketed to the inner side of the main frame A through a hole in which works the forked end of the valve rod Y and rod connecting it to the lug or the collar on the tube X; this tube X passes up through the fixed steering column, and is partially rotated by means of a handle 15 at its upper projecting end.

Very little water is used for working this gear, as that which is forced out of one cylinder Z while the other is being filled goes back to the water tank, and very little leakage occurs at the distributing valve. The water discharged from either of the steering cylinders is returned through one or other of the ports W, and so to the centre exhaust port V and back, by a pipe thereto connected, to the feed-water tank.

Single acting engines The engines are illustrated by Figs. 337 to 341, which show them in part-sectional elevation, part-sectional plan, vertical section, showing

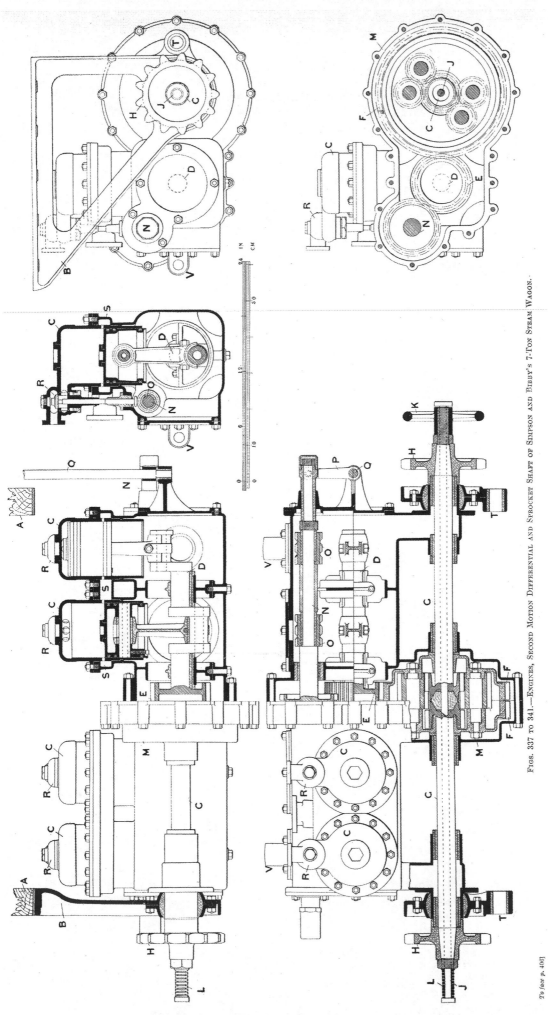

FIGS. 337 TO 341.—ENGINES, SECOND MOTION DIFFERENTIAL AND SPROCKET SHAFT OF SIMPSON AND BIBBY'S 7-TON STEAM WAGON.

To face p. 406]

valve, end view from outside, and central end view. The engines are single acting, and comprise two sets of two cylinders c c, c c. Gudgeon-pins in the open-ended pistons receive the small ends of connecting rods, the big ends of which have a bearing surface only on the upper half of the crank, only a U-strap passing underneath the crank, and held in place by a single bolt through the rod to prevent the big end leaving the crank.

Each pair of engines has a separate crankshaft on the end of which is a spur-gear pinion E; these pinions already referred to mesh with the larger gear-wheels F, F, mounted round and containing the differential spur gear.

Mushroom inlet valves R, Fig. 338, are used for steam admission to the cylinders. These valves are made of cast iron, and were found to work perfectly under steaming conditions with the engine running at about 500 revs. per minute, but sometimes broke when the engine was worked at high speed, light, on down hill runs. They broke off at about where a shoulder is shown in the engraving. This shoulder is incorrect, and the stem was after experiments carried up to the valve without reduction in diameter, and the shoulder was never a square one.

Mitre admission valves

The valve seat is, it will be seen, inserted complete with valve from below, and held in place by a gland nut which forces the seat tightly upon a copper joint ring.

No exhaust valves are used, the exhaust steam passing through holes round the cylinder at the end of the stroke of the piston into an annular chamber s which is connected to the exhaust pipe E, Fig. 335, 336. These admission valves are lifted by cams o on the shaft N, raising the pusher rods as in petrol engines, and pushing the valve stems. The cam pieces o which are fixed to shaft N have three different rubbing faces for the valve-lifting rod to bear upon. They give two points of cut off and admission at the proper period, and reverse running of the engine. The shaft N has on one end a gear-wheel driven by the pinion E, the end sliding motion necessary for shifting position of the cams o being obtained by means of a forked lever P on the pivoted rod Q. A modified form of this arrangement of shifting levers is shown by Figs. 335 and 336, the tubular shaft 13 held in bearing brackets has on one end a lever and links connecting it to the end of the sliding bearing of shaft N, to the other end is fixed an arm with rod 12 connecting it to the short arm of hand lever 11, Fig. 335, which pivots on a pin fixed to the steering column, a notched quadrant being provided.

Valve-operating gear

It is possible that with a four-cylinder engine this form of valve gear and no exhaust valves may be practicable, but it would not do for a two-cylinder engine, the want of freedom of complete exhaust making it difficult to start under load or even sometimes without it.

Exhaust valves preferable

The cams o are held tight on the spindle by the insertion of split taper brass bushes or cylindrical keys, a distance tube being placed between them and both kept in place by the nut shown.

A six-cylinder engine, or rather a pair of three-cylinder independent engines as designed to be used with the form of wagon illustrated in vol. i.

is shown by Figs. 347 to 350. In main features these engines are similar to those of the four-cylinder engine just described. These engines are not now however made, and Messrs. Simpson and Bibby's rights have been acquired by Messrs. Alley and McLellan, who, however, are making from other designs.

The differential countershaft is in two parts G G, Fig. 340, a chain pinion H being mounted on the outer end of each. The chains used are Renolds' 2-inch pitch, having a breaking strength of 53,500 lb. It is provided with a differential gear lock for greasy roads and bad road hill climbing.

Means of locking differential

A rod J passes through G G, one projecting end having a compression spring L on it.

The hand-wheel K is screwed, and runs on the end of the rod seen passing through both parts of the shaft G G. By turning the hand-wheel the rod J is pulled toward K and a pull put on it subject to the resistance of the spring L, and this forces the two inner ends of shafts G G on to the coned piece between them, thus locking their differential gear and preventing differential action.

A different method of locking the compensating gear is shown by Figs. 335 and 336, a claw-clutch being used on one end of shaft G, a collar with forked lever moves this clutch into or out of engagement ; the forked lever is fixed to one end of the shaft O, which is held in bearing brackets and passes through the tubular shaft 13. By depressing the pedal lever 14 the shaft O, to which it is fixed, is partly rotated, forcing the claw clutch into engagement and locking the gear.

The whole of the crankshafts, camshafts, gear-wheels, and differential gear, are enclosed in an oil-retaining case M, the piston, etc., being splash lubricated. The engines are hung by brackets B bolted to the underside of main frame A. Lugs V, V, bolted to crank case, and lugs T, T, formed on brackets B, B, Figs. 337, 339, receive pins for connexion of distance bars to the front axle and back axle respectively. Figs. 335 and 336 show forward reach bars 25 on the crank case lugs connected to the steering cylinders bracketed to the front axle, and radius rods embracing the countershaft G and pivoted to lugs on the back axle. Brake blocks W, W, are applied to the tyres of the rear road wheels. These blocks have spring hangers from the frame, and two arms of a toggle couple jointed at 2 and pivoted to the

Brakes on road wheels

frame and to block W. These brake blocks are applied equally by pulling at 2, by means of a single rope passing over pulleys 3, 3, 4, 4, 5, 5, and 6, as seen in Fig. 336. The two sets of pulleys 3, 4 and 5, are held in brackets fixed to the frame, the rope passing under pulleys 5 and over pulley 6. The spindle of pulley 6 is held by a forked link, to the upper part of which is attached a collar screw 7 with handle 8, and a bearing fixed to the front of tank R. By turning this screw the forked link and pulley 6 are lifted, tightening the rope and forcing the brake blocks on to the tyres. The back axle supports the frame by means of laminated springs to the ends of which shoes are fastened which slide in guides bolted to the frame, central spiral buffer springs between frame and top of laminated springs being also used.

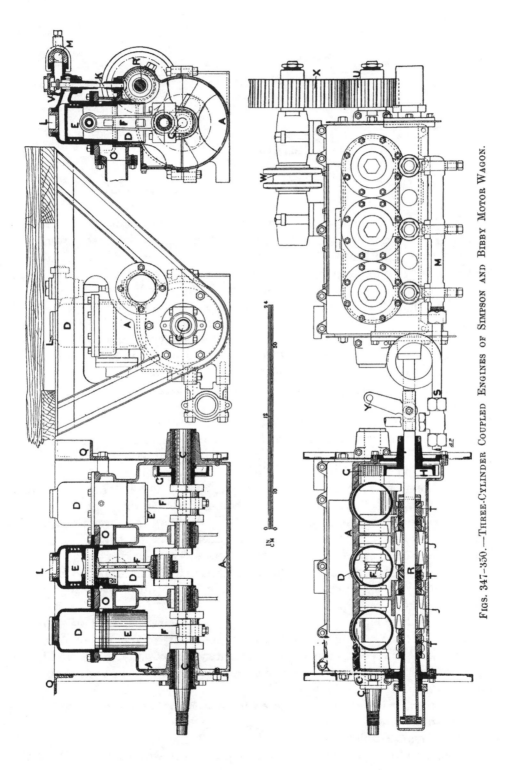

FIGS. 347-350.—THREE-CYLINDER COUPLED ENGINES OF SIMPSON AND BIBBY MOTOR WAGON.

409

Chapter XXVII

COULTHARD FIVE-TON STEAM LORRY

THIS vehicle, illustrated by Figs. 351 to 357, is manufactured by Messrs. T. Coulthard and Co., Ltd., of Preston, and is designed to carry a load of from 4 to 5 tons. The tare weight fully provisioned is 5 tons. Gearing is provided for two speeds of 5½ miles and 2½ miles per hour with engine running at its normal speed. The boiler supplies steam to a compound engine, on one end of the crankshaft of which is a spur pinion gearing with a pinion on an intermediate shaft, which drives, through either of two sliding pinions, gear wheels on the differential shaft, on the ends of which are pinions driving the rear road wheels by means of Renold's chains.

The general arrangement, Figs. 351 to 353, shows the vertical boiler B well down in the channel steel frame, the angle bearers supporting it being bolted thereto.

Fig. 357 is a vertical section of this boiler with steel fire tubes D and screwed stay tubes E, which with the fire box B give a heating surface of 77 sq. ft. The fire bars F afford a grate area of 2·75 sq. ft. It will be noticed that the upper half of the boiler shell is bolted to the lower half and the crown plate, and may therefore be removed. The lagging frame and sheets around the upper part of the shell are shown in Fig. 357. The boiler is fired through a side door with gas coke, the bunker capacity being 10·8 cub. ft. There are 180 fire tubes, three of them being stay tubes. A boiler arranged with a large centre tube and fired from the top has been used, but the design illustrated shows the essential features of a type used by Messrs. Coulthard, and by the Lancashire Steam Wagon Co. for some years, and since adopted by other makers

Accessibility and internal examination For a given size of boiler a very large heating surface may be obtained, the tubes may be easily got at for repair or renewal or cleaning, and by jointing the shell as shown the internal examination of the boiler is rendered possible and its condition may be definitely known, a matter of great importance.

The boiler shown is constructed for a working pressure of up to 225 lb. per square inch.

The steam regulator at V is a balanced valve used for regulating quickly the amount of steam admitted to the engines; it is worked by means of the

410

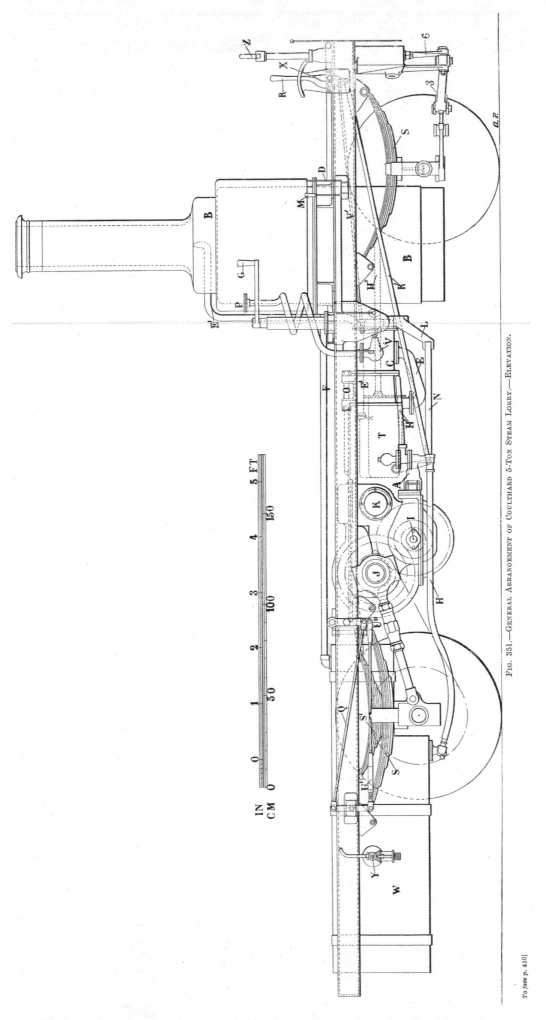

Fig. 351.—General Arrangement of Coulthard 5-Ton Steam Lorry.—Elevation.

[To face p. 410]

handle x, Figs. 351, 353, rod v[1], Fig. 351, and lever pressing on the projecting
end of the valve spindle.   A byepass valve M[1], Fig. 353, and M, Fig. 355,
attached to the combined cylinder and piston valve cover and receiver C,

**By-pass
valve**

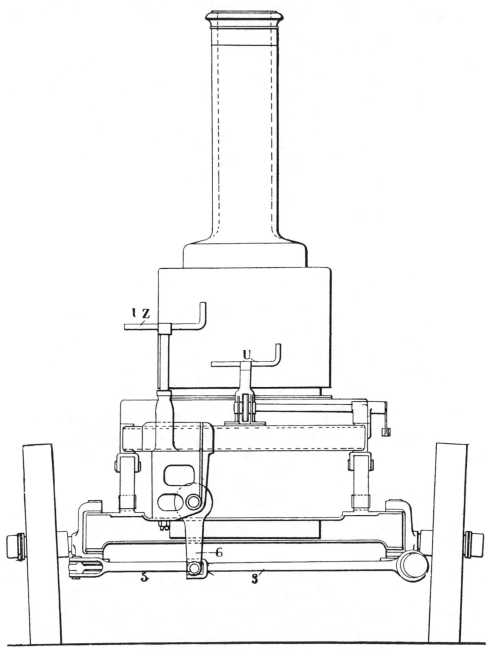

FIG. 352.—FRONT VIEW OF COULTHARD 5-TON STEAM LORRY.

allows high pressure steam to be admitted directly to the low pressure
cylinder.   This valve is actuated by the spindle f, which shifts the double-

411

seated valve from its normal position shown on one seat to its other seat, thus closing the opening for exhaust from high-pressure cylinder to receiver *c*, and allowing the exhaust steam to pass out direct to the exhaust. At the same time the smaller valve *n* is lifted off its seat where it is held normally by high-pressure steam, which then flows through it into receiver *c* and so

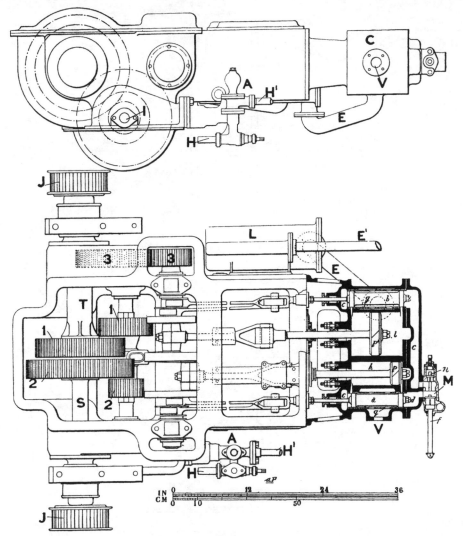

FIGS. 354, 355.—ELEVATION AND SECTIONAL PLAN OF 25-BHP. ENGINE FOR COULTHARD 5-TON STEAM LORRY.

to the low-pressure cylinder. This valve M¹ also does duty as a relief valve, allowing any accumulated water to be blown out. It is operated by the hand lever M, Figs. 351, 353, and suitable connexions.

**Arrange-
ment of
gearing**   The crank case and speed-changing gear box, Figs. 354, 355, and 356, are combined, power being transmitted from a pinion 3 on one end of the crankshaft to a gear-wheel 3 on an intermediate shaft, on a squared portion

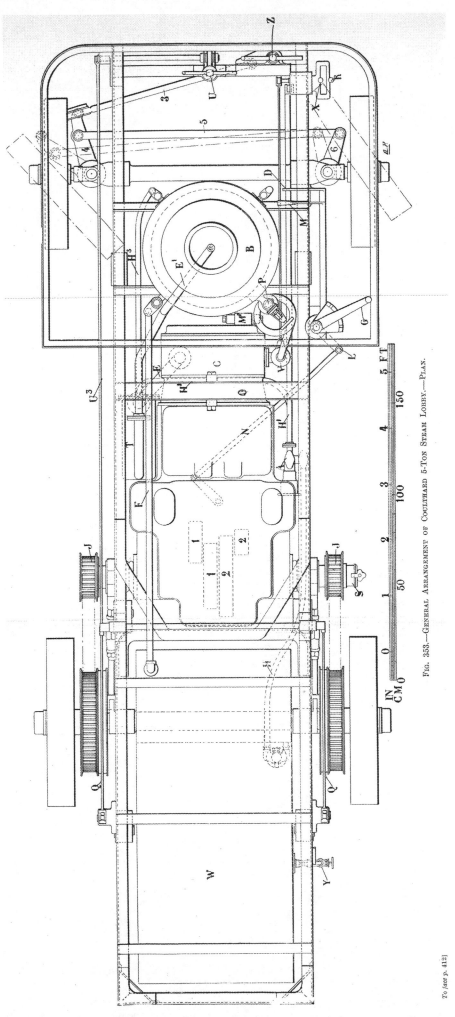

Fig. 353.—General Arrangement of Coulthard 5-Ton Steam Lorry.—Plan.

To face p. 412]

of which slide the pinions 1 and 2, these driving the larger gear-wheels

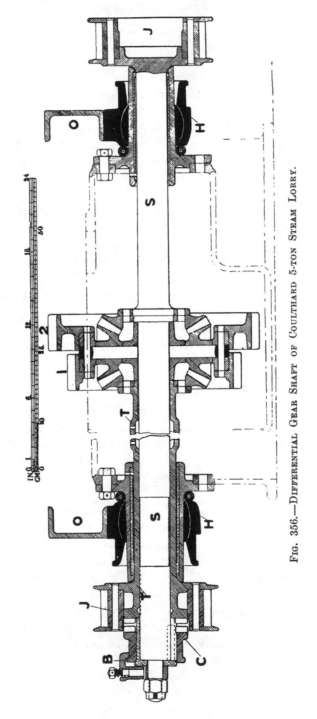

FIG. 356.—DIFFERENTIAL GEAR SHAFT OF COULTHARD 5-TON STEAM LORRY.

1 or 2 respectively, mounted on the differential gear shafts. The sliding gears are put in and out of mesh by means of a forked arm and lever,

413

connected by the rod N, Fig. 351, to the arm L, which is actuated by the handle G at the top of the pillar, and working on a quadrant behind the driver.

In another form [1] of the speed gear used in the Coulthard wagons the intermediate spindle carrying the pinion 1 and wheel 3 is dispensed with.

The ends of the crankshaft are squared, and a sliding pinion mounted upon it at each end ; either of these pinions is made to mesh with wheels mounted upon the differential gear shaft on the ends of which are the chain-driving pinions. By this arrangement the combined oil-containing crank case and gear box is retained, the intermediate shaft and gear wheels are dispensed with, and the differential and sprocket shaft brought closer to the engine.

Differential shaft. Fig. 356 shows a section of this shaft which is carried in long bushed bearings bolted to the casing ; these bearings are held in pieces with spherical bearings in brackets H, H, rigidly bolted to the main channel frame O, O, Fig. 356 ; the shaft carries at each end a pinion J for a Renold's chain. A locking device is provided, the clawclutch c feather-keyed to shaft s engaging with a similarly formed clutch face on end of sleeve T, the finger piece B, which slides upwards on pin shown against the action of the spring, keeping the clutch either in or out of lock.

The arrangement and design of this shaft is in many ways good. The use of a divided shaft at s is avoided or the equivalent of a solid straight through shaft is obtained, one of the chain pinions J being driven, it will be seen, through the sleeve T, and by flange coupling to the differential bevel wheel at that side. The gear-box is hung from the bearing sleeves, and the shaft s is relieved from any weight-carrying duty and from any binding or straining that would occur from relative movement of the frame channels O O by flexure of the frame. By use of the spherical bushed brackets H H and the single point of suspension at O, Fig. 351, the whole engine and gear box is hung from three points in a mechanically satisfactory way.

Boiler feed pumps. Feed water is drawn from the tank W, which holds 180 gallons, through suction pipe H, and delivered through pipe H^1, heater T, pipe H^3, and check valve to boiler. An auxiliary steam pump is provided with independent suction from tank and delivery and check valve on boiler. These check valves may be examined while the boiler is under steam, being so constructed that should the driver forget to close the suction, the pump's delivery would be discharged through a byepass in front of him. The usual water and steam gauges are fitted on front of boiler. A throttle valve P behind the driver regulates the steam supply, which passes through a spiral flexible pipe connected to regulator valve v on the engine case.

To one side of the case is bolted the feed-water heater L, the feed-water pump A on the other side being driven off the end of the shaft I by means of an eccentric. The horizontal double-acting compound engine has cylinders of $3\frac{3}{4}$ in. and 7 in. diam., with 6 in. stroke and develops 25 BHP.

[1] The *Automotor Journal*, vol. vii. p. 649.

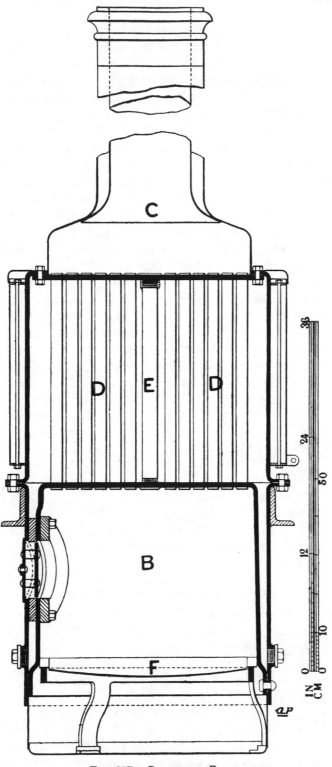

Fig. 357.—Coulthard Boiler.

415

at the normal speed of 450 revs. per minute. High-pressure steam is admitted at v, and passes to an annular space round and through ports in the liner q, in which the piston valve a slides, to either end of piston p, and exhausts through receiver c to either end of piston valve b, which admits it to the piston p^1 of the low-pressure cylinder, the exhaust steam then passing through pipe E, feed-water heater L and pipe E^1 to the boiler chimney.

The Renold's chains on the pinions J J mesh with teeth of gear drums on the hubs of the rear road wheels, these drums being formed in part to receive brake bands ; radius bars pivoted on the projecting ends of the brackets H H are attached by pivots to the fixed back axle and the springs S, S^1, supporting the frame, the brackets H H taking the thrust of the driving-wheel effort in propelling the vehicle.

Band brakes on driving wheels The band brakes are applied to the drums on the road wheel hubs by means of rods U^1 U^{11} pulling on either end of the band ; rod U^1 is connected to one end of a lever arm pivoted on a bracket on the frame, the other end of which arm is connected by a tension rod Q to arm fixed on ends of transverse shaft which actuates both sets of brake rods and levers. The tension rod U^3 on the near side of the vehicle is attached to the arm on transverse shaft and to an arm on the end of a short shaft operated by the hand lever U through a worm and wheel.

The steering handle Z moves the pivoted axles through worm and sector lever arm 6 and connecting rods 3 and 5 and arms 4 and 6.

In the Trials for Motor Vehicles for Heavy Traffic, carried out in 1901 by the Liverpool Self-Propelled Traffic Association, this vehicle, with a mean tare weight of 4·65 tons, carried an average load of 4·48 tons during four days of Trials, a total distance of 148·7 miles at an average speed of 5·68 miles per hour, the consumption of gas-coke fuel averaging 3·77 lb., and water 2·29 gallons per ton of load carried for every mile run. A gold medal award was made for this vehicle.

In the Automobile Club of America Commercial Vehicle Trials, 1903, a Coulthard lorry was awarded a gold medal, its performance throughout the trial being very favourably commented upon.

The Yorkshire Steam Wagon

We have now one example of a type of steam wagon construction differing in many respects from those already described.

The vehicle, illustrated by Figs. 358 to 364, is the standard type of 4-ton lorry, constructed as originally by the Yorkshire Steam Motor Company in 1901–2. Similar vehicles are now being built by the Yorkshire Patent Steam Wagon Company of Leeds.

Fig. 361 shows the general appearance of this wagon, and shows one of the pair of engines and the position of the boiler, which is distinctive in design, placed across the frame. This arrangement of the boiler makes it easily possible to keep the ashpan above the front axle, and thus avoid overhang of the frame in front of the vehicle. The wagon is fitted with two

Fig 361.—Yorkshire Steam Wagon Company's 4-Ton Lorry.

417

27

General arrangement separate engines, one fixed on each side on the main frame where they are readily accessible. This will be gathered from Fig. 361, but in Figs. 358 to 360 it is more clearly shown. The feed pump D on the water tank, Fig. 360, is also in a very accessible position. Spur transmission and driving gear is used throughout, providing for two different reductions of engine speed, and giving 5½ and 2½ miles per hour at normal speed of the engines.

Another distinctive feature is the method of carrying the intermediate or countershaft in bearings on pivoted brackets, which also carry the main axle bearings and the main springs supporting the frame. With this arrangement the gearing is kept in proper radial relation, and the play of the springs provided for.

The main frame is constructed of channel steel longitudinal members, with cast steel transverse connexions and diagonal stay bars. It

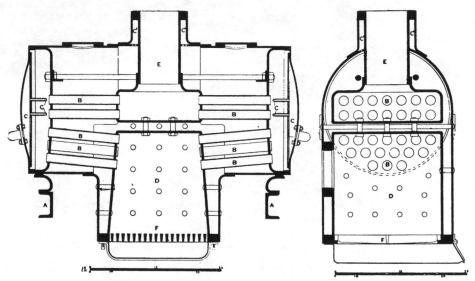

FIG. 362.—YORKSHIRE WAGON COMPANY'S STEAM BOILER.

is supported on the front axle by one strong transverse spring bearing on the forked ends of the axle, which is guided in cast-steel horn plates. Cheeks are provided on the axle, as seen in Fig. 359, which act as guides on the inside surfaces of the forked or horn plates to prevent end motion of the axle. The steering is of the Ackerman form, operated by hand-wheel S, Fig. 358. Drag rods are fitted to the rear road wheels to transmit the driving stress direct to the wheel felloes.

A screw-down brake, not shown, applies a block to the tyre of each driving road wheel.

Figs. 358, 359, and 360 show in elevation, front view and plan, the general arrangement of the lorry machinery and connexions.

Locomotive type boiler The boiler B is shown in section by Fig. 362. It is a form of double locomotive boiler with central fire, the fire-box D and two short barrels, and with a secondary distinct chamber over the fire-box, and surmounted

418

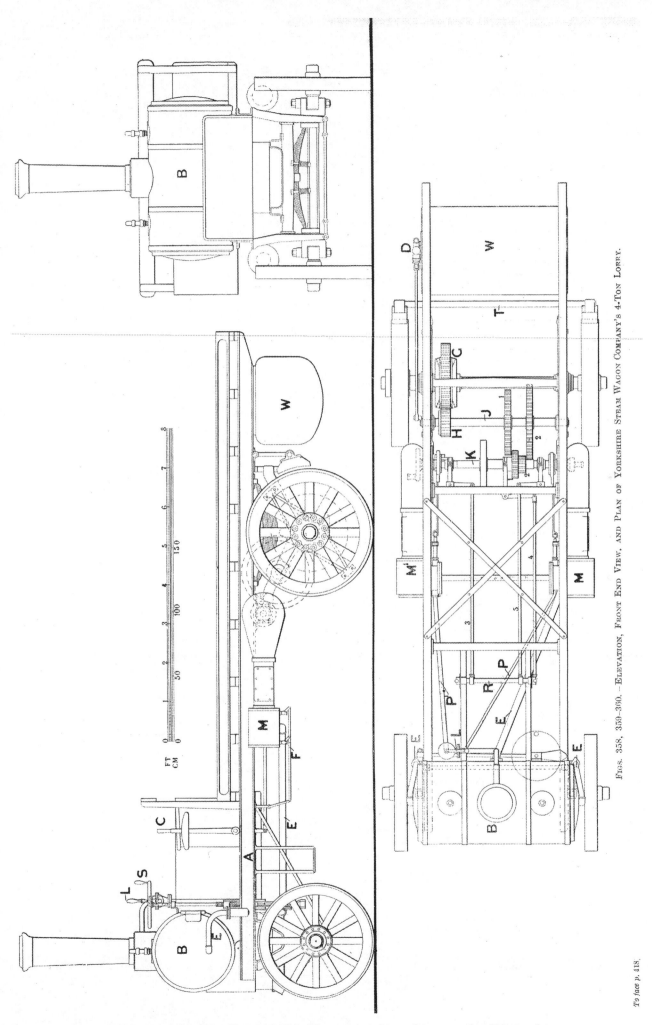

Figs. 358, 359, 360.—Elevation, Front End View, and Plan of Yorkshire Steam Wagon Company's 4-Ton Lorry.

To face p. 418.

by a central uptake E. The fire-box is connected to the end plates of the barrels by steel smoke tubes B B, the lower tubes delivering into the two smoke boxes, and thence through the upper tubes into the upper chamber and uptake. Hollow smoke-box doors are provided, having spaces c into which the exhaust steam passes, and issues by two rows of jets G opposite through the upper rows of tubes.

From these jets it passes through the tubes into the secondary chamber, is superheated by mixing with the products of combustion, and passes away up the chimney.

The exhaust steam on its way from the low-pressure cylinder to the hollow smoke-box door spaces c c passes through a separator and thence to the pipes E, Figs. 358, 360, which are pivoted gland fashion, in axial alignment with the hinges of the smoke box-doors.

Steam is raised to a working pressure of 160 lb. per sq. in., and led from the dome c¹ to the main stop valve, the fuel being coke fed in at the side door, Fig. 362.

The grate area of this boiler is 2·25 sq. ft., and the total heating surface about 51 sq. ft. In many respects this is an excellent boiler and a particularly suitable design for steam wagon purposes although of rather expensive and complicated construction.

The smoke tubes may be very readily brushed through and the greater part of the heating surface kept in clean and efficient condition. Some exception may be taken to the use of the secondary or internal smoke-box and to the difficulty in detecting and stopping leakage at the inner ends of the tubes. The tubes are however short, and the trouble due to tube leakage, common to locomotive type boilers, should be absent in this design if the boiler is reasonably well treated. The flat surfaces are well and sufficiently stayed.

The pump D, Fig. 360 feeds the boiler with water from the tank W, which has a capacity sufficient for a ten-mile run.

An injector is also provided.

Mr. G. H. Mann, of this Company, has recently taken out a Patent, No. 25,125, of 1903, for a vertical fire-tube boiler. Fig. 363 shows a section of this taken from the Patent drawing, the special feature being the enlargement of the upper part c of the shell so as to give a large water surface and steam space, the tubes B being below water level, or drowned.

The hollow smoke-box cover A over the sunk space above the tube plate receives the exhaust steam from the engines through pipe E. The steam thence passes down the tubes F to the hanging superheating

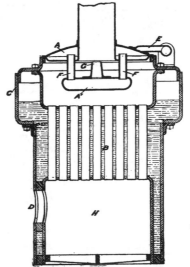

FIG. 363.—MANN'S VERTICAL STEAM BOILER.

Condition of heating surface

Design for a vertical fire-tube boiler

chamber A^1, in which it is highly heated without contact with any heating part of the boiler, and from which it escapes through the nozzle G up the chimney.

The chamber A^1 forms also a baffle to the hot gases, causing them to distribute themselves more than they would to the outer circles of and larger numbers of tubes rather than concentrate their flow through the central tubes.

Fuel is fed through the fire door D.

Cylinder dimensions The engines are compound, the high-pressure cylinder M^1, with a bore of 4 in., having an air jacket which makes its external diameter the same as that of the low-pressure cylinder M, the bore of which is 6 in., the stroke of both engines being $7\frac{1}{2}$ in. High-pressure steam is admitted to cylinder M^1 through the pipe P^1, exhausting by pipe connexion into cylinder M and passing to exhaust pipe E. High-pressure steam may on occasion be admitted direct to low-pressure cylinder, the byepass being actuated by depressing a pedal.

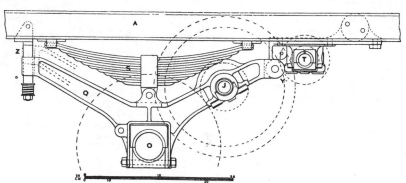

FIG. 364.—YORKSHIRE STEAM WAGON COMPANY'S DRIVING AXLE SPRING SUSPENSION AND COUNTERSHAFT.

Single excentric reversing gear The crankshaft K is fitted across the main frame A, and has on each end a balanced crank arm and pin, a central fly-wheel being used. A simple form of single excentric reversing gear of the Mann and Charlesworth design is used and operated through suitable bell-crank levers and connecting rods 3 and 5, and by a lever, on the driver's right-hand side, pivoted on the shaft R. The engine cranks, connecting rods and crossheads are enclosed in oil-tight casings between which are the excentrics and driving pinions as shown in the plan Fig. 360.

Arrangment of spur gearing The spur pinions 1 and 2 slide on a squared portion of crankshaft K, and are moved by forked bell-crank arm and connecting rod 4 actuated by hand lever C pivoted on R. These pinions are made to mesh with the large gear-wheels 1 and 2 respectively on the intermediate shaft J, providing the two different reductions of speed, the engine, as shown in Fig. 360, running free. The gear pinion H is fixed to shaft J, and is in gear always with wheel G, which is mounted on the differential gear on the live rear road wheel axle.

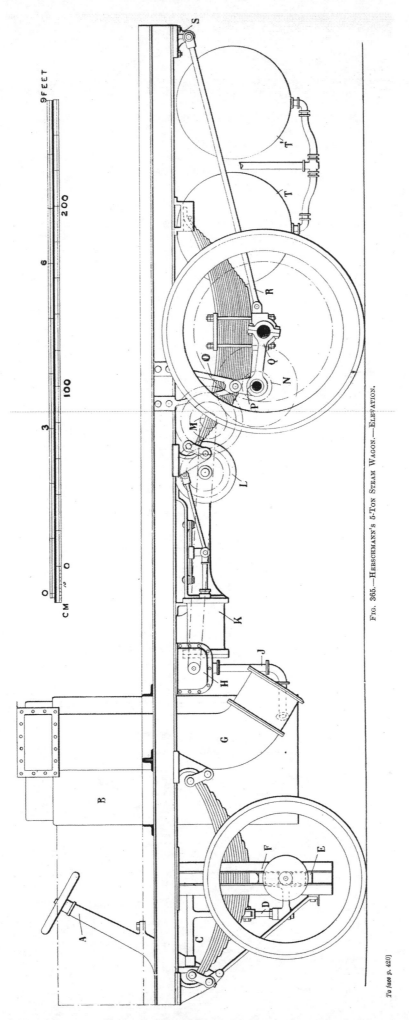

FIG. 365.—HERSCHMANN'S 5-TON STEAM WAGON.—ELEVATION.

[To face p. 420]

HERSCHMANN STEAM WAGON

The feed-water supply pump D is driven off the end of shaft J by crank-pin and connecting rod.

Fig. 363 shows the arrangement of countershaft and axle-bearing brackets employed. The frame A is supported by spring S pivoted to the bracket which holds the bearings of the rear road wheel axle O and intermediate shaft J, the axle bearings being formed to allow a limited swivel movement. The arm Q of this bracket has a forked end which is free to slide vertically on the piece Z, as shown by dotted lines, a spring buffer being provided at the lower end. The other arm of the bracket is pivoted at Y to the crankshaft bearing bracket T. The centres of the rear axle O and intermediate shaft J are thus fixed relatively to one another, and the difference in the lengths of the radii J Y and J T being small, and total movement at O being small, the effect on the pitch engagement of the crankshaft pinions 1 and 2 in their wheels is negligible. It will also be seen that differential play of the main axle and springs on the two sides of the lorry will not affect the cross contact of the teeth of the pinion H and wheel G, Fig. 360.

THE HERSCHMANN STEAM WAGON

The problems involved in the design and construction of heavy steam vehicles are no easier of solution than similar questions concerning the lighter and higher speed motor cars; but while there is in America great activity in the production of motor cars, little appears to have been achieved in the production of wagons for the transport of heavy loads. Some interest, therefore, attaches to the steam wagon, the subject of this description, not only because it originated in America, but also on account of its construction and improvement in England by the English Steam Wagon Co. During the six years that have elapsed since Mr. Arthur Herschmann designed the first mechanically propelled wagon for the Adams Express Co., of New York, various alterations or changes have proved to be necessary, experience gained with successive designs pointing out, as is usual, the directions in which improvements might be effected. The most important of these modifications are referred to in the following description of the final form taken by some of the parts. The elevation and plan line drawings, Figs. 365 and 366, show the general arrangement of the frame and machinery as made in America in 1903, and until recently in England. The frame longitudinals are of girder or H-section, and following general practice converge at about the middle of their length, continuing parallel to each other forwards, and at a suitable distance apart to obtain clearance at the front wheels when the steering angle is acute, and to facilitate the hanging or carriage of the boiler B. The latter is supported in a cradle made up of three curved angles bearing upon the side frame members, and upon the transverse pieces immediately in front of and behind the boiler.

The type of boiler indicated is like that now adopted by the English company, and is generally of the form already described, namely, a vertical firetube multitubular boiler. The peculiar side downtake G which discharged the hot products of combustion from the smoke box at nearly ground level

American origin

Vertical fire-tube boiler

421

has been abandoned, and a single ordinary vertical funnel is used situated toward the back of the boiler, and on the near side, and connected to the smoke box by a rather short right-angle bend. A large central tube is also used, and the boiler is top fired, that is to say, the fuel is dropped in at the top of this centre tube and becomes distributed over the grate by the continuous tremor or shoggling that results from road vibration.

Differential expansion The length of the tubes in this boiler is rather greater than is usual, a feature which, although tending to increase of both power and fuel economy, would also tend to aggravate the stresses set up by differential expansion. When the rates of evaporation per square foot of tubular — that is to say, smoke-tube—heating surface are moderate, and the whole external surface of the tube is water covered, excellent results may be obtained. If, however, either the evaporative rate at which the boiler works is high, or the tubes are not completely water covered or drowned, then more or less pronounced trouble inevitably results by deterioration of the tubes and by leakage at their ends.

Interesting boiler design The boiler originally used with these wagons is shown by Fig. 367. It possesses distinctive features, and though no longer used in this country is worthy of description. It may be seen that about 120 curved water tubes are connected between the fire box crown plate and side plate. The internal diameter of these tubes is approximately 1 in., and it is evident from the half plan of the tube plate that the tubes are not too crowded, and that they may, therefore, be more readily kept clean and in an efficient condition as heating surface. The fire box and combustion space is relatively large, whilst the water content of the boiler is, if anything, small. It would appear, for instance, that unless the steam space is unduly encroached upon that the depth of water over the fire-box tube plate is such that a temporary cessation of feed from any cause for a short period would have harmful results.

Live and exhaust steam super-heaters There are two superheating coils, the first of which, D, dries and super-heats the steam on its way from the pipe F to the engine. The steam enters the coil by the stop valve E, so that when the lorry is standing or when the steam demands are light, there is some danger of burning the coil. The life of the coil may be, and probably is, prolonged by the use of a second stop valve at F to ensure that there shall be steam in the superheater.

The conditions of working of the exhaust steam superheating coil to and from which the inlet and outlet pipes G conduct the steam are less severe, because, although exposed to high temperature from the fire, they are under only low internal steam pressure. The grate is provided with rocking fire bars H capable of being moved to break up or detach clinker, by means of an external lever arm connected with the shaft upon which the arm J is held. There are two uptakes K, which without doubt help to make the draught more uniform than would be possible with this construction if a single uptake were employed without the complication and cost of exterior connecting flue ways. The boiler is fired through the centre tube L, whilst access at grate level is obtained by the door M. Complete opening out

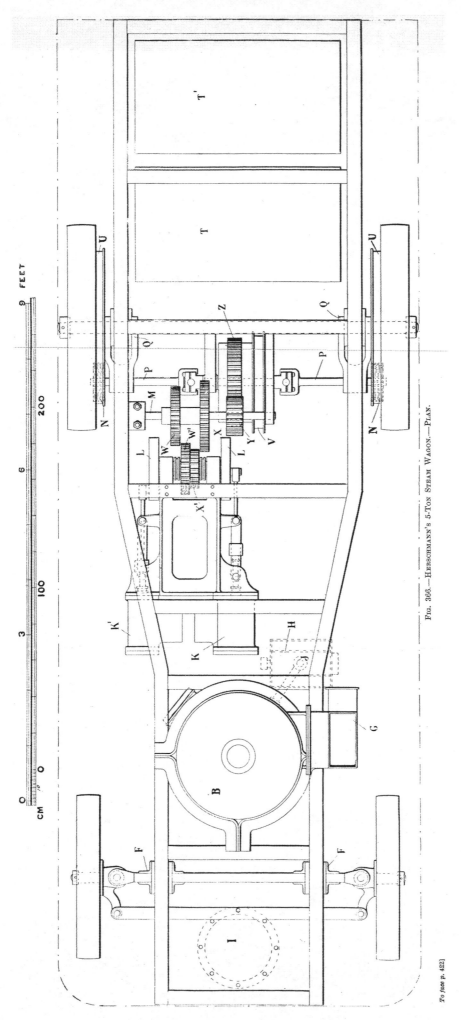

Fig. 366.—Herschmann's 5-Ton Steam Wagon.—Plan.

[To face p. 422]

The material originally positioned here is too large for reproduction in this reissue. A PDF can be downloaded from the web address given on page iv of this book, by clicking on 'Resources Available'.

of the boiler for internal examination and cleaning is possible by breaking joints at the heavy seating blocks shown connected together at the top of the boiler by long screws, at the angle rings connecting, through bolts

Internal
examination
difficult

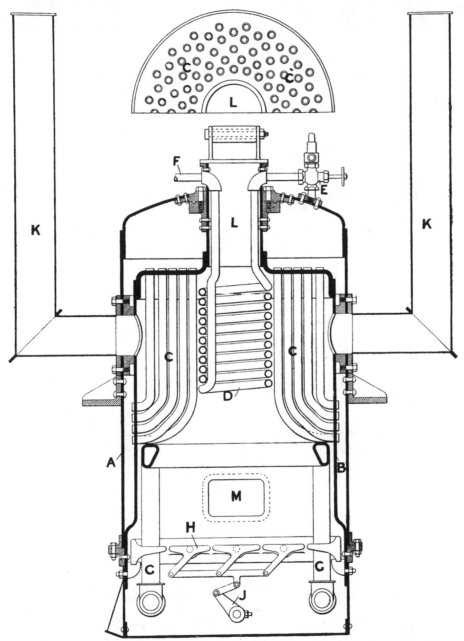

FIG. 367.—HERSCHMANN'S 5-TON STEAM WAGON.—SECTIONAL VIEW OF BOILER.

and nuts, the outer shell A and the fire-box B, and by removing the set-screws which maintain the joints, at the branches, leading to the uptakes K and between the shell A and the fire-box B.

These last-mentioned joints would probably be unsatisfactory in practice and difficult and troublesome to make.

Design of front carriage

Attention may be directed to the method of supporting the front part of the wagon through its springs upon the front axle. From Fig. 365 it may be gathered that the axle is guided in angle horn plates, and that

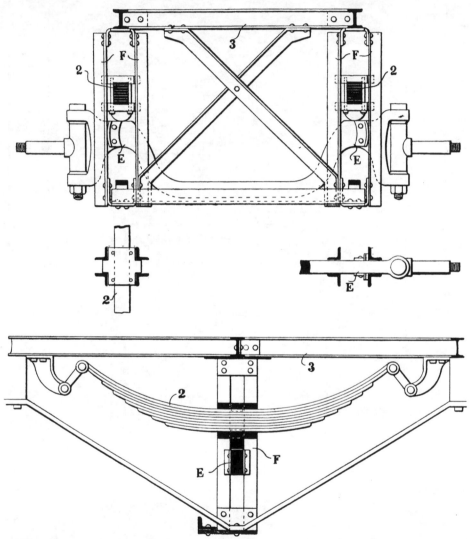

FIG. 368.—HERSCHMANN'S 5-TON STEAM WAGON.—ARRANGEMENT OF CARRIAGE AT FRONT AXLE.

the springs are inside the guide plates, and the method of bracing is partly indicated. Fig. 368 has been prepared to clearly show the form which the very strong guide frame takes.

The groups of four-angle guide bars F form double guides, for the front axle in one direction, and for the springs 2 in another direction at right angles to the former. From the small detail of the springs 2 in plan

424

it will be understood that the spring centre clamp plates are formed to fit and slide within the guide bars F, and that the springs are only permitted to have motion in a direction normal to the horizontal plane of the wagon frame, and are relieved from lateral stresses. The spring bearing upon the front axle is through a semi-circular pad, it is not held or anchored to the front axle, and the latter may rise and fall or cant in its guides with comparative freedom. Stop plates E, shown in the elevation and plan views and by a separate detail, necessarily have the curved outer edges shown in order that side play of the axle may be restricted, while at the same time there is no restriction of angular movement of the axle, or tendency to bind in the guides when such movement occurs under spring deflection. With this form of construction of front truck or guide frame double-shackled spring attachments must be used, and further, in considering the complete guide frame, though capable of resisting heavy stresses and allowing easy riding, it is without doubt more costly in construction, and possesses no greater advantages as regards the particular requirements of free and restricted movements than other designs which have in addition the advantages of a central support for the front part of the wagon by the use of a transverse spring, and the relief of the frame to some extent from twisting stresses. For exceptionally heavy work, and over extremely poor road surfaces, the additional weight and cost would not perhaps be prohibitive, but it may be interesting to here state that the English Steam Wagon Co. have modified and lightened this construction, though still retaining the same general design.

The gear reduction from engine to road wheels is in three stages, and by toothed gearing.

Referring to Figs. 365, 366 and 369, the engine may be seen held up **Engine and** rigidly to the two transverse frame members at the position of the cylinders **gearing** K and K^1, and near the crankshaft end of the engine, and from the outline drawing it may be seen that the gear-wheels X^1, W^1, are on the straight shaft between the crank discs. The high and low pressure cylinder rods work on to overhung crank-pins, and the steam chests being between the cylinders, the single reversing excentrics operating the valves are also inside the main bearings, and each side of the wheels X^1, W^1. With these latter wheels others of larger diameter—X, W—may be moved into mesh sideways, X for the low-speed or W for the high-speed gear. In the positions shown neither of them is in gear, and the engine can be run freely or without moving the wagon. On the same shaft M upon which X and W are arranged to slide upon their key feathers is a fixed pinion Y, which is always in gear with Z. Z is mounted upon the outer casing of the differential gear, to be described hereafter, and through the arms P, P, drives bronze pinions N, N gearing into internally toothed rings U, U, fixed to the road wheels.

The intermediate shaft M runs at one end in a bearing attached to the main frame, and at its other end in a bearing connected with one of the hanger brackets which may be seen in Fig. 369, and which provide bearings for the differential gear box. The shafts P run in bushed bearings formed in the outer ends of the forked radius links Q, and on either side of Q are short upstanding links, pin-connected to the brackets O bolted or rivetted to the main frame (see Fig. 365). The extremely heavy laminated springs are free to move at either end ; they slide in housings at the rear, and are shackled at their forward ends.

Transmission of driving effort
The driving effort from the road wheels is transmitted through drag rods R attached at one end to the spring pillow blocks, and at the other to brackets S bolted to the extreme end of the frame girders. When the wagon has to be run backwards the driving effort is less directly taken through the links Q and P to the brackets O, and the drag rods R have to some extent to serve as thrust bars. This system of links and bars is not as shown entirely satisfactory. Deflection of the springs must result in

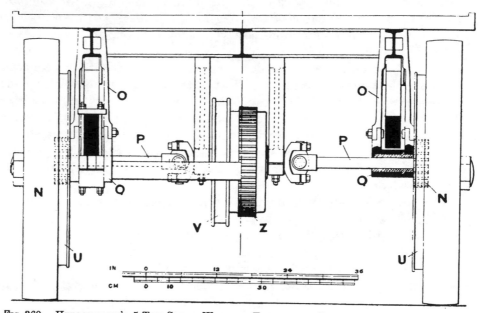

FIG. 369.—HERSCHMANN'S 5-TON STEAM WAGON.—TRANSVERSE SECTIONS AT POSITIONS OF REAR AXLE AND DIFFERENTIAL GEAR.

movements of the short links about their pin connexions with the brackets O, and in bending stresses in the shafts P, which will be more or less severe, according to the range of movement of the back axle under spring deflection, and the relief afforded by looseness at the universal joints on the inner ends of the shafts P, and in the bearings Q.

Improved design
In the more recent designs the English company have effected the improvement required in this respect by providing a second universal joint at the outer ends of the shafts P. These may be seen by reference to Fig. 371, which is a plan view of the rear end of the wagon, and which shows other modifications in the arrangement of the gearing, and the more complete protection and lubrication afforded by casing in the engine and most of the gearing.

The sliding gear wheels x^1, w^1, for the high and low speeds, are now carried on an extension shaft connected by a universal joint to the engine crankshaft. All the other wheels are fixed to their shafts and remain in gear.

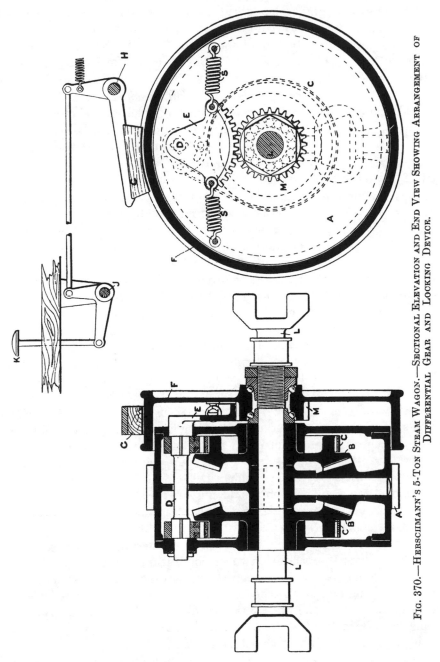

FIG. 370.—HERSCHMANN'S 5-TON STEAM WAGON.—SECTIONAL ELEVATION AND END VIEW SHOWING ARRANGEMENT OF DIFFERENTIAL GEAR AND LOCKING DEVICE.

Fig. 370 illustrates the arrangement of the differential gear and the means provided for preventing or hindering differential action when required.

Differential gear

427

The outer toothed wheel A formed with the casing drives in the usual manner the bevel pinions B, which, being in gear with the bevel wheels B, drive the divided centre shaft L. The bevel wheels B have brake rings formed upon them, and are encircled by brake straps as indicated by dotted lines in the end view. On the cross spindle D are fixed toggle plates providing the points of attachment for the ends of the brake straps. A part rotation of D in either direction will tighten the brake bands, and when the toothed sector E is in mid position, or in the position shown in the elevation, the tension in the two springs S is equal, and the brake straps are free upon the brake rings. The transverse section shows that E is fixed upon D, that the springs S are attached at one end to E, and at the other end, to pins in the disc, part of the large brake drum F, and further that the teeth on the sector E mesh with those of the centre pinion M, which is solid with, or part of, F. It will be noticed that the rim of F is of rather heavy section, and that it is mounted on ball bearings on the divided shaft L. On the outer surface of the drum F a brake block G may be brought into action by the wagon driver by applying pressure upon the pedal K, and by transmission of this effort through the bell-crank levers pivoted at J and H.

Peculiar method of locking differential When the differential gear and its casing is being driven in either direction, application of the brake G reduces the rate of rotation or hinders movement of F and M. As the rate of rotation of the differential gear and casing undergoes no change, the sector E advances or gains upon F and M, and during this advance around M is obliged to undergo a partial rotation with its pin D. As a result of this movement of D, and with it the toggle plates, the brake straps are applied to the bevel wheel brake rings and differential action is retarded or prevented according to the pressure upon the brake ring F. It may be pointed out that although braking action between F and G is undesirable, and may absorb power at a time when the propulsion of the vehicle requires all the power available, yet the amount of this braking action is not often considerable, and as a rule it would be of short duration. Adjustment of the internal brake bands is possible, and would necessitate dismantling the gear box, but with original good adjustment and construction would only require to be done occasionally. Referring to the general elevation and plan, Figs. 365 and 366, letters H and J refer to an enclosed blower, and its delivery air pipe to the ashpan for increasing the draught. It was used in the American wagon, and, as indicated, was driven by belt from the second gearshaft M.

The arrangement of the steering gear presents no novel feature. The inclined steering column A through two sets of bevel gear wheels gives the necessary motion to the horizontal spindle C, and to the downwardly projecting lever arm fixed to it. A suitably jointed rod connects the end of this lever arm with one of the steering heads at D. A more recent form of the gear avoids the use of bevel wheels, and employs instead a vertical steering column, with its lower threaded end working in a nut pivoted at the end of an arm on the spindle C.

The brackets O are not now used, the short upstanding links P (see

Fig. 365) having given place to longer links pin-connected to blocks bolted to the underside of the frame girders.

The engine in the more recent wagons, of the completely enclosed double-acting compound type, has cylinders 4 and 7 in. diam., the stroke being 6 in. The road wheels are smaller than originally used, those in front being 3 ft. diam. with tyres 5 in. wide, and the driving wheels 3 ft. 9 in., with tyres 6 in.[1] wide. The total gear ratios used are about 17 and 8·5 to 1, so that at road speeds of 3 and 6 miles per hour, for which the wagon is designed, the engine speed is moderate. The boiler-working pressure is 200 lb. per square inch.

The brake equipment is not illustrated, but when occasion arises the engine reversing gear and the brake described in connexion with the differen-

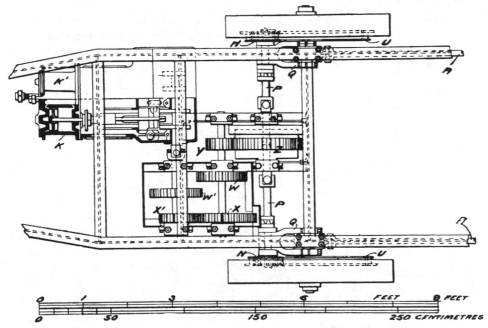

FIG. 371.—ENGLISH STEAM WAGON CO.'S 5-TON STEAM LORRY.—PART PLAN SHOWING ARRANGE-MENT OF GEARING.

tial gear may be used to assist the large diameter band brakes arranged to work on the outside of the internally toothed rings U.

There are numerous makers of steam lorries, wagons, and omnibuses **Jesse Ellis** other than those mentioned. Among them, and early in the field, are Messrs. Jesse Ellis & Co., Maidstone, whose long experience in the use of

[1] Tyres of greater width than these are now required to comply with the recent Local Government Board Regulations referred to at the commencement of this section.

traction engines has enabled them to appreciate many of the difficulties arising from heavy wear and tear of steam motor vehicles carrying loads on the common roads. Their lorries and wagons conform in general to what may be called the prevailing type of construction.

They use a long rectangular channel steel frame, vertical steam boiler with slightly bent tubes secured in the top of a convex fire-box top, and to a concave-top tube plate, the forms given to these tube plates facilitating the escape of steam globules as formed. The boiler is placed immediately over the front axle and behind it, and suspended below the main frame is a small double-cylinder compound engine working at a maximum pressure of 200 lb., fitted with ordinary link motion, and driving, through double countershaft two-speed gearing, a heavy spur-wheel on differential gear on the live driving axle.

Savage steam wagon Messrs. Savage Bros., King's Lynn, also experienced in the traction engine work, are makers of a steam wagon with water-tube vertical boiler, and with chain-driven main-road wheels on a fixed main axle, the chains being driven by a differential gear countershaft, and a double-cylinder compound engine working at a normal speed of 450 revs. per minute, and a pressure of 200 lb. ; the cylinders are 4 in. and 7 in. diam., both having a stroke of 5 in., and both having piston valves operated by single excentric reversing gear motion.

Messrs. Hindley & Sons are also making steam lorries with gear-driven live axle and a vertical boiler directly over the front axle, and constructed for a working pressure of 225 lb., and with a compound horizontal engine running normally at 400 revs. per minute, and having cylinders 3¾ in. and 7 in. diam., and 6 in. stroke. The boiler is of the smoke-tube vertical type,

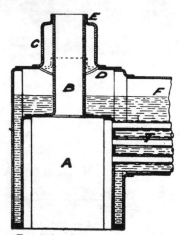

Hindley

FIG. 372.—BRETHERTON'S TOP-FIRED LOCOMOTIVE TYPE BOILER.

but recently Messrs. Hindley have been making a boiler of the type patented by Bretherton, which is a top-fed locomotive fire-box boiler with the barrels cut off to a diminutive length. The Bretherton boiler is illustrated in vertical section by Fig. 372, in which A is the fire-box, B vertical uptake in the centre thereof, surrounded at its upper part by the steam dome, and drying chamber riveted to the outer shell of the fire-box at D, and to the upper part of the uptake at E. A boiler similar to this is also being used by Messrs. Thornycroft & Company, but the uptake is not surrounded by a steam dome as shown.

Bomford and Evershed Messrs. Bomford & Evershed, of Pershore, are making a steam wagon with a channel steel main frame outside all the road wheels, and with a small Belpaire locomotive type boiler placed transversely on the frame behind the front axle, the driver's footplate being quite in front of the vehicle under a canopy through which the boiler chimney passes on one side. The engine employed is double-cylinder compound, with radial valve gear and slide valves. The road wheels are driven by an internal toothed ring fixed to them to which motion is given by pinions on the ends of a differential gear countershaft running in bearings carried in a steel casting which also carries the main axle. The power is transmitted from the engine by a high-speed silent chain to the first motion shaft.

The Wantage Engineering Company is also making a steam lorry with vertical water-tube boiler in front of the leading axle, or with smoke-tube boiler having a total of 80 sq. ft. of heating surface, and made for a working pressure of 225 lb. The driving-wheels are chain-driven on a fixed axle **Wantage** driven by a differential countershaft and Renold roller chain. The engine is horizontal compound with link motion reversing gear, and having cylinders $3\frac{1}{2}$ and $6\frac{1}{4}$ in. diam., with 6 in. stroke, and running at 500 revs. per minute, the lorry being made to carry 4 tons. As in each of the preceding cases the differential gear can be locked when necessary.

Messrs. Howard, of Bedford, are also making a steam wagon to carry 5 tons. It has a vertical boiler immediately over the front axle, and a compound engine with cylinders 4 and 7 in. diam. and 5 in. stroke, with **Howard** link motion reversing gear. The rear wheels, 3 ft. 6 in. in diam., run on a fixed axle $3\frac{1}{2}$ in. square, and are driven by a heavy roller chain.

Fodens, Limited, have made a considerable number of their steam wagons, which consist of a locomotive boiler and chimney arranged, in the ordinary traction-engine form, with horizontal engine over the boiler driving a second motion shaft traction-engine fashion, from which motion is taken by a pair of roller chains running close together side by side on sprocket **Foden** wheels on the outside on the main axles traction-engine fashion, close to the off driving-wheel. The channel steel frame surrounds the boiler from near the smoke box, and widens out so as to enclose the main or driving wheels, the axle of which runs in a form of horn plate, and the springs bear upon the axle bearings and underside of the frame as in railway practice. The vehicle is a very heavy one, but is a strong one, and the machinery is of known kind.

The Hercules Motor Wagon Co., Levenshulme, are also making a 5-ton steam lorry with vertical boiler behind the front axle, the driver sitting in front. The engine is compound and drives the main wheels by heavy **Hercules** roller chains through reduction gear and a differential sprocket counter-shaft.

Clarksons, Ltd., Chelmsford, have now supplied a large number of their steam omnibuses, carrying 14 passengers, to the Torquay, Eastbourne, and other provincial omnibus companies, and the experience gained with **Clarkson** them is satisfactory to makers and users, and the London Road Car Company has been running two of them continuously, but are now using the double-deck bus. These vehicles are referred to on p. 474.

If the talented mechanical engineers were as perspicacious regarding the utility of an object as they are capable and painstaking in devising a means of attaining it, they would be the leaders of the world. But they seldom unite the originating and executive capacity of the inventor or designer with the business sagacity of those who care not for the means, but look only to the commercial utility of the end. The man of commerce must wait on the man of mechanics, and he sometimes finds—too often, in fact—that mechanical ingenuity has disregarded fitness in the selection of means.

The originator of a new type of machine, or thing which does work quicker, better, or at lower cost than hitherto, is a necessary innovator, though sometimes perhaps a nuisance : but once his new type has been proved practicable and useful, though susceptible of improvement in details, it is nearly always the wiser course for his followers to make an inventory of all its good features and make the best of all, rather than to magnify the few bad features and set to work on some radical changes which may remove these defects, but which ruin all the rest of the design that was good.

This course would have saved an enormous expenditure of money and time if it had been followed in the heavy as well as in the light motor vehicle world.

When, for instance, Levassor's Paris-Bordeau Car was improved by Panhard and Levassor, and the British manufacturers were free to make cars of the type with improvements, most of them hunted round for the defects, magnified them, started, without half Levassor's ability, on an entirely new tack, with an engine that would not work or would have been no better if it had worked, and would not suit the car or the gear ; or they started on new gear that did not work or which would have involved other radical changes in the car if it had. After this they began to count up the good parts that had led to success so far, and then to do what the follower generally should do, namely, make better in detail the thing that the originating mind has devised in main features, and shown to be feasible. Some works in England would have saved their shareholders many thousands of pounds sterling if they had made proper use of their opportunities in this way, but here the commercial man went wrong because he did not insist for a time on the best possible copy of the best thing which they had as models, with acquired rights to all designs.

Laudable as is the endeavour to make improvements, it must be remembered that mere changes of form of parts are not necessarily improvements at all. See carefully the good points of, for instance, a motor carriage, and unless you can see your way to make the thing that has done well either cheaper or of better material, let it alone or worship it, and usefully apply the lesson it should convey ; don't change the whole arrangement of engine and gear to get in some totally different kind of boiler that you have not tried for such purposes, or change all the gear for the sake of some kind of multiple-everything engine that is to save a farthing a mile in fuel, while the other costs per mile reach forty farthings. Don't, in fact, do as men did when they rejected the Panhard and Levassor model in 1897

432

only to come gradually out of a scrap heap to recognize and accept its good points, and improve the means of carrying them out. The De Dion-Bouton heavy vehicle, and the Lifu [1] provided models which have been much better utilized abroad than at home, although more relative advance has been made in this country in the construction of heavy steam vehicles than on the Continent.

[1] Vol. i., pp. 481, 486, Figs. 360–365, and pp. 494–511, and Figs. 374 to 382.

Chapter XXVIII

THE MILNES-DAIMLER AND OTHER HEAVY PETROL MOTOR VEHICLES

Early vehicles

TWO of the petrol lorries of the Daimler Motoren Gesellschaft, of Cannstatt, went through the Richmond Trials of the Automobile Club in 1899. This was the first time that vehicles of the kind carrying considerable loads were tested in this country, and their performance was looked upon as a revelation. The successful use of the petrol engine for propelling light carriages seemed natural enough, but a new view of their possibilities was accepted when these two lorries, the one with an engine of 11·8 BHP carrying 6 tons, and weighing empty about 2 tons 19 cwt., and the other, with an engine of 7·8 BHP., carrying 2 tons 4 cwt., and weighing about 2 tons 1 cwt., showed their capabilities, and worked without a hitch. The trials were easy both for these and the competing steam vehicles, but they were sufficient to show that the petrol motor vehicle for this work was a most promising innovation, and we consequently awarded them a gold medal. The expectations of their future development have been fully realized, and the whole of the main features of their design are those of the Daimler vehicles of to-day, but numerous modifications and improvements in detail have been made, and for most purposes the power has been increased, more especially in the omnibuses made by the same builders, and known in the country as the Milnes-Daimler vehicles. Of these omnibuses there is now a large number in use in this country in the hands of the London and some provincial omnibus companies, and of several of the railway companies. The success in London is leading to their rapid extension in numbers Although the rubber tyre difficulties are not entirely removed, the use of the double tyres and of less radial depth than those formerly used, has reduced the practical difficulties considerably, and the commercial objections correspondingly, or at least sufficiently so to put commerical success within sight where the slight excess of speed and handiness as compared with the horse omnibus, enables each omnibus to attract and carry more and more of a ready traffic, and run constantly with full or nearly full load.

It is found that some of the single-deck 14-passenger omnibuses

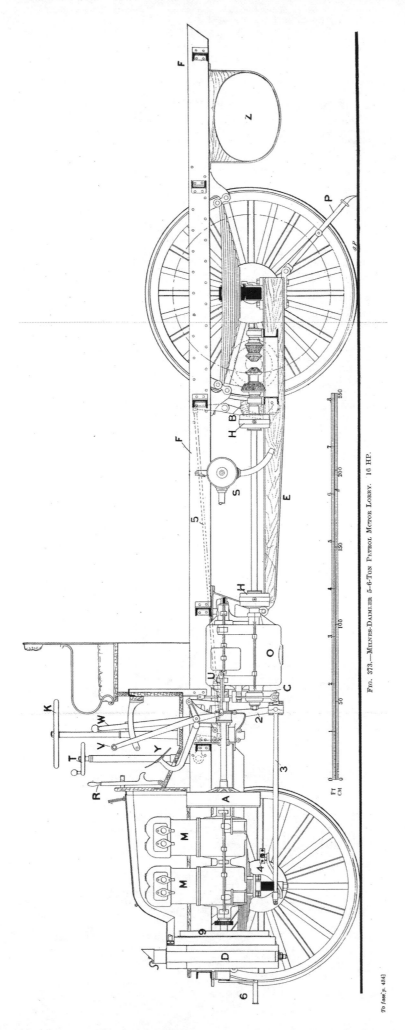

FIG. 373.—MILNES-DAIMLER 5–6-TON PETROL MOTOR LORRY. 16 H.P.

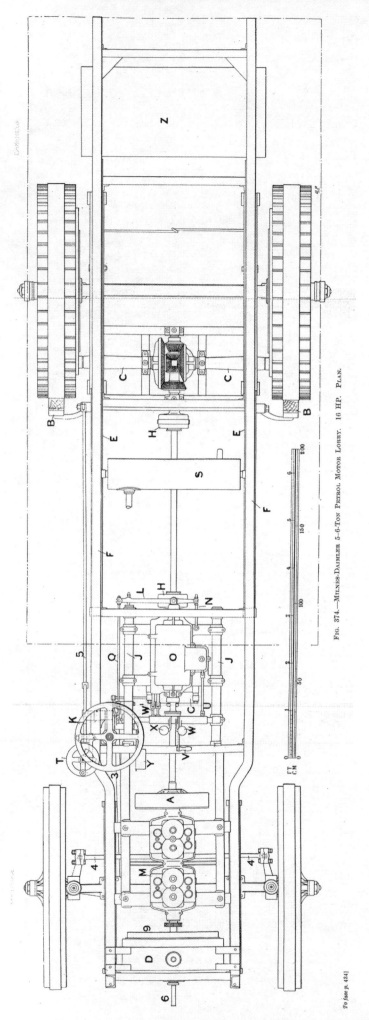

Fig. 374.—Milnes-Daimler 5-6-Ton Petrol Motor Lorry. 16 H.P. Plan.

[To face p. 484]

in London carry about as many passengers per day as the 26-seat horse omnibus, owing to the higher speed and other attractions. There is, however, a tendency on the part of purchasers to call for the large double-decked 32- and 34-passenger omnibuses. This means great total weight, and it may be questioned whether the omnibus of more moderate seating capacity, smaller size, lower weight, greater handiness, and frequency of service, will not in the end prove to be the greater commercial success. **Single and double deck omnibuses**

The omnibus has developed from the lorry, and in most cases is the lorry frame and machinery with small modifications fitted with an omnibus body.

The illustrations, Figs. 373 to 380, illustrate the Cannstatt Daimler, or Milnes-Daimler 6-ton lorry as made three years after the Richmond Trials,

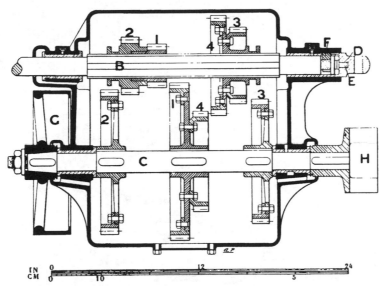

FIG. 375.—TRANSMISSION AND CHANGE-SPEED GEAR BOX, VERTICAL SECTION. MILNES-DAIMLER LORRY.

and Figs. 381 to 385 show some of the recent detail modifications as used in the omnibus machinery.

From the first of these, which are given to a large scale, it will be seen that the frame is formed of parallel longitudinal channel steels, cramped near the fore end to reduce the width to give room for greater steering angle for the front wheels. These longitudinals are cross-connected by smaller channel steels, to which are connected lower longitudinal angle bars carrying the engine and tubes, J carrying the gear box. As at present made the angle bars are carried backward to carry the gear box. **Frame**

The engine is of about 23·5 BHP It has four cylinders with exhaust and inlet valves in chambers on opposite sides of the cylinders, and all mechanically operated. The cylinders are 4·15 in., 105 mm. diam., and 5·12 in., 130 mm. stroke, the normal speed being 800 revs. per minute. The engine governor acts on a throttle valve, not by controlling the exhaust valve

Control of engine

as formerly, and as described in vol. i., and the action of the governor may be restricted, and the position of the throttle valve controlled, so that the speed of the engine may be accelerated. By the lever R, Fig. 373, the governor is assisted, the throttle valve being by it shifted to the position of minimum supply to the engine, when the hook half way up the lever is used to hold the clutch pedal down and the engine running free.

The carburettor is of the float-feed jet-spray Daimler-Maybach type, and is fed from a tank under the driver's seat.

The carburettor is sometimes duplicated so far as the float-feed vessels are concerned, the carburettor being placed above the engine and arranged to receive a pressure supply of kerosene as well as petrol for starting. The

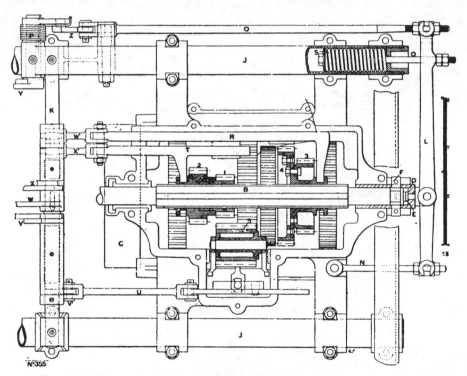

FIG. 376.—MILNES-DAIMLER 5-6-TON LORRY CHANGE-SPEED GEAR.—PLAN.

Kerosene as fuel

kerosene pipe, and the carburetted air pipe from this carburettor, are jacketed and heated by a shunt pipe from the exhaust. By using the petrol in this carburettor for a few minutes when starting, and occasionally when the load varies or falls very low, kerosene can with a little care be used in this way, but for the very variable work of an omnibus it is better to use petrol always.

Ignition is effected by low tension rotary magneto and apparatus not shown, but similar to that used on the Mercedes cars, which are made by the same builders.

The exhaust gases pass to the silencer at s.

The jacket water is cooled by a honeycomb radiator D, through which a current of air is drawn by a fan 9, now replaced by exhaust-fan blade spokes in the fly-wheel. The honeycomb cooler now used is carried by brackets projecting from its sides and bearing on rubber cushions on the forward extension of the frame. The water is circulated by a gear-driven rotary pump not shown.

The power is transmitted from the cone clutch in the fly-wheel A to the upper and first motion spindle B in the gear box O, shown in detail by **Trans-mission gear** Figs. 375, 376, 377. This spindle carries the sliding change-speed pinions 2, 1, 4, 3, which gear alternatively with the corresponding four gear wheels

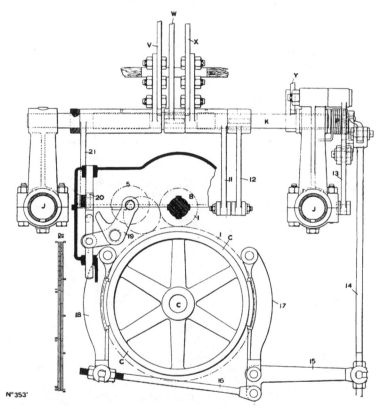

FIG. 377.—CHANGE-SPEED LEVERS, COUNTERSHAFT BRAKE, AND REVERSING GEAR.

on the second motion shaft C, Fig. 375, one end of which carries the brake drum G, and the other end the universal joint H. Connected to this joint and a corresponding joint H is the propeller shaft, which drives the differential gear and transverse countershaft C, Fig. 374, and K, Fig. 379.

The sliding gear wheels are of the usual form, providing four speeds forward, viz. 1½, 3½, 6, 8 miles per hour; the wheels 1, 2, 3 and 4 sliding on the milled feathered shaft B, which is driven by the motor, meshing with those of same numbers respectively keyed to the shaft C. A small pinion 5, Figs. 376 and 377, is interposed to mesh with the two wheels 1, 1, giving a reverse motion to the shaft C. The reverse speed is 4 miles per hour.

The sliding gears are operated by forked rods R and T, Fig. 376, moved by means of the hand levers w and x and arms 11 and 12; lever w and arm 11 being keyed to shaft K, and lever x and arm 12 being on a sleeve pivoted on shaft K. The reversing pinion is brought into play by the hand lever v and arm 21 pulling the rod U, Figs. 373, 376, 377 and 378, to which a bar sliding in the casing is connected, having on it a pin (see Fig. 378), which gives a vertical motion, through an angular slot, to rod 20, and so to the bell-crank forked lever 19, Figs. 377, 378. The bell-crank lever has on it

Reversing gear

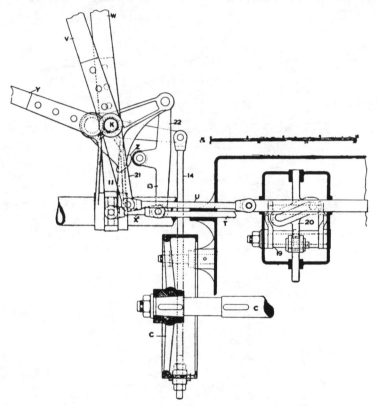

Fig. 378.—Sections of Gear Box. Reversing Gear.

two projecting arms which engage with stops on the inside of the casing, and fix position of the pinion.

The clutchshaft is connected to one end of the shaft B, Figs. 375 to 377, at the other end of which is a thrust-sleeve bearing D to which is pivotally attached the arm L, Figs. 374, 376. This bearing is free to slide longitudinally through a limited range, and is prevented from rotating by the forked piece F which fits into it and projects on either side into slots in the casing. This forked piece fits over the shaft B which is turned down to receive it, and is moved with it. At E are hardened steel thrust-pins with cheese heads. Pressure on the clutch faces is maintained by the compression spring s in the tube J held in place by a screwed cap. This spring is adjustable by the

Arrangement and operation of clutch

438

locknuts on the rod O which rest on lever L. The pull of the spring S on the arm L forces the bearing D, fork F, and shaft B, and thus the clutch cone inwards. A tension rod N pivoted to the casing holds the opposite end of the lever L. This, as well as the connecting rod Q, is adjustable as to length, so that clutch-pressure adjustment is readily made.

By depressing the pedal Y, Figs. 373, 374, and 378, which is pivoted to the bracket bolted to the tube J, the arm Z is caused to push the roller on the lever 22, one end of which is pivoted to the bracket on tube J, Fig. 376, and forces the other end with rod Q to move the arm L outwards, and by pulling on the rod O further compressing the spring S, thus relieving its pressure on the end of shaft B and on the clutch-cone surfaces.

A further depression of the pedal Y causes the arm on the outside of

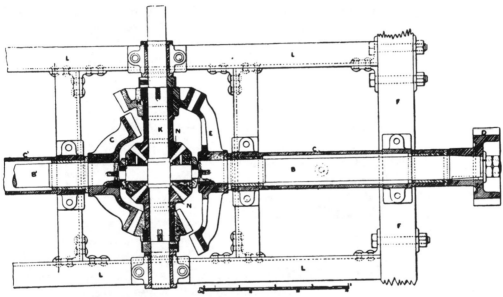

FIG. 379.—DIFFERENTIAL GEAR. MILNES-DAIMLER LORRY WITH DOUBLE INCLINED COUNTERSHAFT.

spring P, Fig. 376, to lift the rod 14, Fig. 377, and apply the shoes to the **Foot brake** brake drum G on the shaft C ; these shoes are held by arms 17, 18, pivoted at one end to the casing, the other ends being pivoted to the bent lever 15 and rod 16, the tension on which forces the shoes inwards equally upon the drum G.

The brake is normally kept out of action by means of the spring P which holds the two arms connected to the pedal Y downwards.

Powerful block brakes B B are applied to the tyres of the rear-wheels, **Tyre brakes** a screw down hand-wheel T being employed, pulling on the rod 5, and brake-block arms which are connected by a shaft across the frame.

Bolted to the undersides of the frame F are brackets connected by a tie rod, to which are pivoted the ends of radius beams or reach rods E E, the other ends of these beams being bolted rigidly to the rear axle. Two transverse steel channels connecting these reach beams carry the differential counter-shaft and bevel gear wheels. Figs. 373, 374, 379, and 380.

The wooden beams E E are reinforced on the top and bottom by steel bars, a sprag P being bolted on beneath the back axle. The springs on the rear axle have shackles on either end giving them free play equally from the centre, the whole of the rear axle, springs and gear moving as one piece with the beams E E. The arrangement is an excellent one, keeping all the rear driving spindles and gear in their true relative positions, but they are, of course, subjected to the severe stresses resulting from running on bad granite pavements as they are not spring supported.

Differential and bevel driving gears The differential gear has been specially designed to permit the use of dished driving-wheels on axles pointing downwards as in ordinary road vehicles.

In Fig. 379 the motor-driven shaft K drives the bevel pinions N N¹ by the usual smaller differential pinions, the small bevel pinion J being rigid with N¹, and the larger pinion H with N. The bevel pinion H drives wheel E, which is shrunk on and setscrew pinned to one end of shaft B,

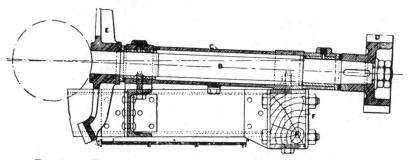

FIG. 380.—DIFFERENTIAL GEAR-DRIVEN INCLINED COUNTERSHAFT.

to the other tapered end of which is keyed and lock-nutted the spur pinion D which meshes with the internal toothed ring on the rear road wheels. In the same way the bevel pinion J drives the shaft B¹; the ratio of teeth of bevel wheels J and C and H and E being equal, the driving pinions on the ends of transverse shafts B B¹ rotate at the same speed. The advantage of the arrangement is, as will be seen from an inspection of Figs. 379 and 380, that the two halves of the countershaft B B¹, though they must, with the wheel proportions shown, remain at 90 deg. with the driving shaft K, may within limits be placed at any relative angle in their vertical position. This is shown by Fig. 380, from which it will be seen that the shaft B and its tubular lubricating bearing shaft C point downwards from the centre, and the spur pinion D will gear properly with the internal spur gear on the driving-wheel on a similarly inclined main axle.

The steering gear is of the well known hand-wheel and worm and sector form, with an angle of lock of about 40 degrees.

As with every other form of motor vehicle, the design of the details as last described has been modified in many particulars in the past two years, more particularly in the machinery for the lighter vans, lorries and omnibuses. Fig. 381 shows the exterior of one of the smaller omnibuses fitted with a four-cylinder 18-HP. engine on a 1½ ton underframe.

The general arrangement of the change-speed and transmission gear **Modification** for this and the larger and heavier omnibuses and wagons is the same **of design** as that already illustrated, but ball bearings are now used throughout except in the road-wheel axles, and for reversing a wide sliding pinion is now used instead of that having an angular movement as described with reference to Figs. 376 to 378. Fig. 382 is a vertical section of the gear box through the centre of the bearings of the first and second motion spindles, and Fig. 383 is a plan of the same partly in horizontal section along the line of the first motion or clutchshaft bearings. From these it will be seen that the clutch cone is no longer forced in by pressure at the back end of

FIG. 381.—MILNES-DAIMLER SINGLE-DECK OMNIBUS 18-HP.

the first motion shaft, but by pressure on a ball thrust bearing G^4 upon which pressure is maintained by a spring G^9 acting through a lever G^5 which embraces a double ball-thrust collar as seen at G^5, Fig. 382.

The reversing is effected by the pinion J^5, which runs on two ball bearings on the spindle J^8, which is, when required, slid inwards into gear with the pinion H^4 and wheel J^1, when they are in the position shown in Fig. 382. The spindle J^8 is moved by the hand lever J^6 and link J^7. The clutch G^3 is released as before by relieving the pressure on the thrust lever by compressing the spring G^9 by the right foot pedal G, pivoted at N^8, and acting on the lever G^7 continued depression of the pedal causing the pedal lever to lift the pedal brake lever N^7 and rod N^5 connecting it to the lever N^5

on the brake-shoe levers, as seen in Fig. 384. The various parts shown by

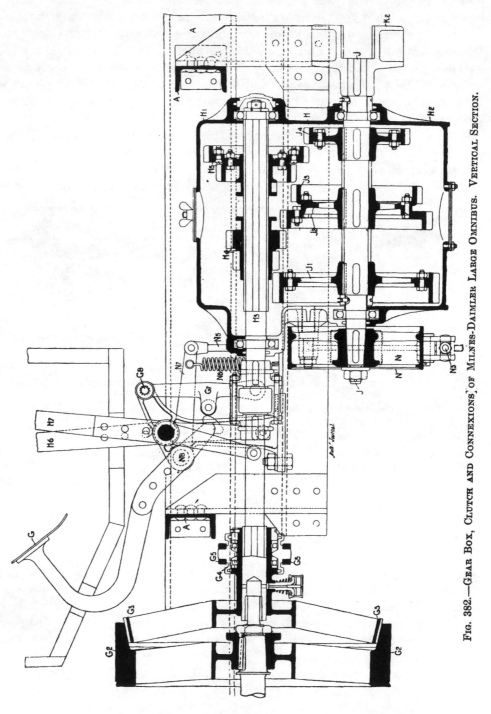

FIG. 382.—GEAR BOX, CLUTCH AND CONNEXIONS, OF MILNES-DAIMLER LARGE OMNIBUS. VERTICAL SECTION.

these engravings are similarly lettered, and thus make further description unnecessary.

The differential gear is of the same design as that already described, but instead of the open shafts and plain bearings, the gear and shafts are

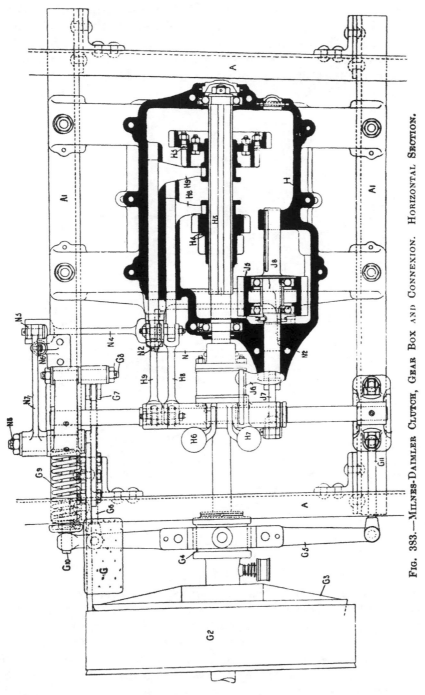

FIG. 383.—MILNES-DAIMLER CLUTCH, GEAR BOX AND CONNEXION. HORIZONTAL SECTION.

enclosed, as shown by Fig. 385, by an oil-tight case extending to the ball bearings at the ends close up to the pinions L^2 which drive the road wheels, the ends of this casing being fixed by lugs to the wood reach bars A⁴ by vertical and by horizontal lugs.

The hand-applied brakes act upon the drums F to which the renewable pinions L² are fixed by screws, as shown. More recently these brakes have **Differential** been in some of the vehicles of the internal expanding type instead of the **shaft** exterior form, and they are close in. Ball-thrust bearings have also been **brakes** provided at the back of the large bevel wheels of the differential gear, those shown not providing for the horizontal component of the thrust of the bevel pinions M³ and M .

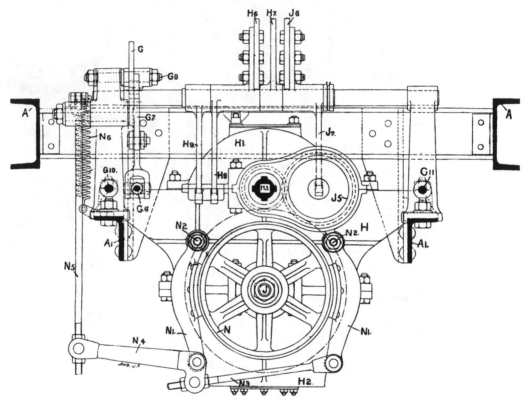

FIG. 384.—MILNES-DAIMLER GEAR BOX, FRONT END VIEW. PEDAL AND HAND LEVER
CONNEXIONS.

The balls used in most of these bearings are ⅝ in. and ¼ in. diam., and **Size of balls** are kept about ¼ in. apart by small spiral springs, this arrangement being adopted partly to keep the balls from rubbing together and partly as a means of getting all the required balls into place between the internal and outer rings by putting the outer ring excentric to the inner ring, putting the balls into the index crescent thus made, and afterwards separating them by the springs, and thus forcing the rings into concentric relationship.

It may be mentioned that the naves of the road-wheels are all bored **Loose axle** out to about ½ in. larger diam. than the axles, and fitted with a phosphor **bushes** bronze bush ¼ in. thick. This bush may turn on the axle or in the wheel

444

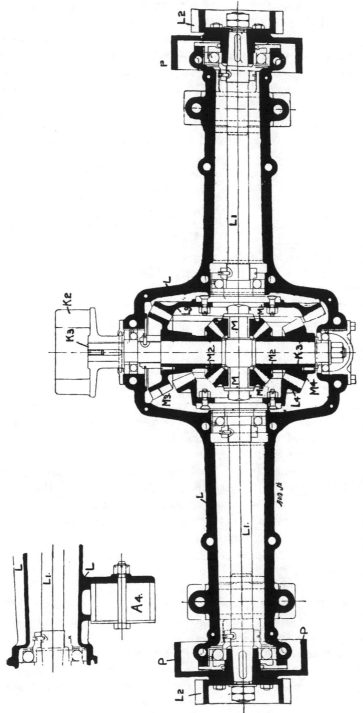

FIG. 385.—MILNES-DAIMLER DIFFERENTIAL COUNTERSHAFT. HORIZONTAL SECTION.

445

whichever it finds easiest, and as it is perforated with several holes it helps to keep and distribute the lubrication given to the axle.

The large omnibuses now running in London and elsewhere are built upon the chassis that is used for a 3-ton lorry. These are geared for $2\frac{1}{2}$, $4\frac{1}{2}$, 6 and 12 miles per hour, and have engines with four cylinders of 100 mm. diam. 3·94 in., 130 mm. stroke, 5·11 in., and run at a normal speed of 800 revs. per minute. The 20 and 25-HP. engines used for some of the vehicles have cylinders 105 in. diam., 4·13 in. and 130 mm. stroke, 5·11 in. ; and 115 mm. diam., 4·52 in., and 140 mm., or 5·51 in. stroke respectively, all running normally at 800 revs.

The wheel base of the large double-decked omnibuses is 11 ft. 3 in., the gauge centre to centre of tyres 4 ft. 11½ in. ; overall length 24 ft., and overall width 7 ft. 2 in. The driving-wheels are fitted with double solid Turner or Shrewsbury-Challiner tyres of 40 in. diam., and each 4 in. wide at the base ; the front wheels have a single tyre of 32 in. diam.

All these omnibuses are fitted with the Cannstatt honeycomb cooler, gear-driven circulating pump, and the fly-wheel and clutch constitute an exhaust fan for increasing the air circulation through the cooler tubes.

Low tension magneto apparatus is used, the ignition tappet rods being worked by cams on the inlet valve camshaft.

For the illustrations, Figs. 381 to 385, I am indebted to the *Automotor Journal*.

In the introduction to this section dealing with heavy vehicles reference was made to other manufacturers engaged in the construction of petrol-engined wagons and omnibuses, but no mention was made of the vehicles made by the Wolseley Company and those known in this country as the Straker-Squire. A brief description and illustrations of a Straker-Squire omnibus chassis follow,' from which the leading features may be noted. **Wolseley** With regard to the excellently designed and constructed Wolseley vehicles it can now only be remarked that they closely follow, in the type and arrangement of the engine and transmission gear, the passenger vehicles of different powers and similar design, made by them in such large numbers and described earlier in this volume.

Wherever necessary changes have been made in the nature of the details to fit the vehicles for the different work and the heavy loads to be borne, and it may be mentioned that the great mistake made by some makers of attempting to utilize gearing and parts hitherto used for the lighter passenger vehicles, with engines of similar powers to those required for omnibus work, has not been made. The design of the Wolseley underframe and running gear is advantageous in that it admits of a reduction in the length of the vehicle compared with those using vertical engines, but difference in general arrangement from the majority of makers and from what is becoming more and more a standard arrangement, is unfortunately disadvantageous.

Straker-Squire Figs. 385A and 385B show a part sectional elevation and a plan of the underframe and running gear of the Straker-Squire [1] omnibus. The first

[1] The *Autocar*, Jan. 21, 1905; The *Tramway and Railway World*, March 1905.

omnibuses of this type supplied to the London companies were designed by Herr Roth and constructed by Herr Bussing of Brunswick and some of the similarity in detail between these vehicles and those of the Canstatt Daimler Co. may be attributed to Herr Roth's earlier connexion with the latter firm.

Referring to the illustrations it may be pointed out that a four-cylinder engine of 24 HP. is used for the doubled-deck 34-seated omnibuses instead of the two-cylinder engine there indicated. A distinguishing feature of its design is the arrangement of the horizontal camshaft and pivoted valve-operating levers on the top of the cylinders, driven by a vertical shaft and skew gearing in very much the same way that the Maudslay engine valve gear described at pages 194–6 is arranged, but in the Maudslay engine the valves are in the cylinder head and in the other case the inlet and exhaust valves are in chambers B B and C C at both sides of the cylinders. *Valve operating gear*

The arrangement of transmission gearing is peculiar in that two gear boxes J and N are employed, the drive from the first to the second being through the universally jointed propeller shaft M. *Gearing arrangements*

In the first gear box J, which receives the drive from the cone clutch at the engine fly-wheel through the coupling H^1, provision is made for three forward speeds and one reverse. By means of two movable gear-wheels, on the first motion shaft in the second gear box N, gearing with wheels fixed to the differential gear, two rates of reduction of speed in this gear box may be obtained, the relation between them being 2 to 1. The ratios of the gearing are such that including the transmission by roller chains to the rear-wheels, speeds of $1\frac{1}{2}$, $4\frac{1}{2}$ and 7 miles per hour are obtained with the engine running at normal speed and with the low-speed gear in use in the gear box N. If the high-speed gear in the box N is put into operation, then a second series of three speeds of 3, 9 and 14 miles per hour are obtained. If the series of two reverse speeds are included, this combination of gears provides for eight different speeds, six being for forward running. The speed changes are effected by two levers L, and the third or side lever Q is shown connected with compensated brake gear and straps acting upon drums fixed to the chain-wheels on the road-wheels. The chain-wheels, it will be observed, are connected to the road-wheels by a series of radial arms and driving-shoes fixed directly upon the wood felloe and embracing alternate spoke ends.

The provision of a type of radius bar with a slotted end at the point of connexion with the rear axle, and the use of the horizontal drag bar thereto connected and anchored to the vehicle frame through the volute spring R^1 and bracket shown, gives some spring relief to the transmission gear during the starting effort or when heavy shocks are experienced. The adjustment of the driving-chains would require to be carefully done and with recognition of the range of movement or change of the distance between the centres of the pinion and sprocket wheels. A^3, D^3, and G^1 are gear-wheels upon the end of the crankshaft, on the armature spindle of the magneto D^2 and on the water-circulating pump spindle. With the four-cylinder *Spring drag bars*

447

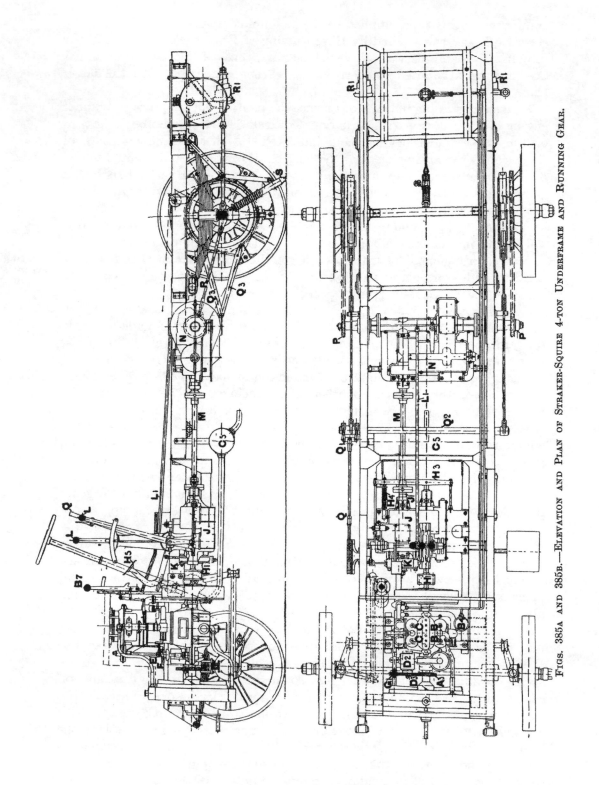

FIGS. 385A AND 385B.—ELEVATION AND PLAN OF STRAKER-SQUIRE 4-TON UNDERFRAME AND RUNNING GEAR.

448

24-HP. engine the underframe shown is suitable for use for omnibus purposes or as a goods wagon to carry 4 tons.

Petrol motor vehicles of the heavier class are also being made by the Durkopp Company. They have supplied the London Road Car Company with 32-passenger double-decked omnibuses with 4-cylinder engines of 24 HP. **Durkopp**

The Germain Motor Company have also supplied the same Company with a petrol motor omnibus having an engine of 20 HP. nominal. The gearing of these was of the same pattern as that of a light car of the same power, and was of course insufficient for the heavy vehicle. They are being altered. **Germain**

Messrs. Moss and Wood have supplied the London General Omnibus Company with " Orion " two-cylinder horizontal opposed cylinder petrol motor chassis of 20 nominal horse-power fitted with the body of the Company's ordinary horse-drawn omnibus. The chassis weighs under two tons. The rear wheels are driven by roller chains from a differential countershaft, and are fitted with the double or twin solid tyres. **Orion**

Messrs. Glover and Co. are also the makers of petrol motor wagons fitted with a 12-HP. Astor engine, three-speed gear and side chains driven by differential countershaft connected to the gear box by propeller shaft. The frame is of wood with steel flitch plates. The engine has two cylinders 105 mm. diam., and 130 mm. stroke. **Glover**

An omnibus chassis constructed by Scheibler of Aachen is now in service with the London Motor Omnibus Company. It has a four-cylinder vertical engine, and the road wheels are driven by side chains. The design of the transmission gear is good and suitable proportions have been adopted for the change speed gear and bevel driving-wheels. **Scheibler**

With regard to the commercial aspect of the working of these omnibuses, it is sufficient now to say that where sufficient numbers are employed to give proper time for overhaul and adjustments, where the moderate speeds are observed, and where the heaviest loads are held more in respect than in reverence, the motor omnibus is a proved practical profitable success, and must now, as rapidly as the best of these omnibuses can be made, replace the horse-hauled vehicle.

The large number of motor omnibuses now in use in London—over one hundred—will rapidly accumulate that experience necessary to show whether a somewhat lighter vehicle carrying fewer upper-deck passengers and with a somewhat wider wheel gauge will not be the more desirable for several reasons. In the recent (July, August, 1905) French trials of industrial vehicles there were several vehicles worthy of special notice in the heavy class. Among these were the Brillie omnibus, made by Mr. M. Schneider et Cie., Le Creusot, the De Dion et Bouton omnibus, the heavy lorry of the Neuer Allgemeine Gesellschaft, which is a particularly good job, much on the lines of those of the Daimler Co. of Cannstatt, which also competed with great success. The Turgan lorry, a strong vehicle, did well, and the Latil front drive vehicles were a surprise. They ran well, withstood high speed with heavy load on the dreadful *pavé*, and appeared to be in excellent order at the finish. **French industrial vehicle trials**

Chapter XXIX

THE WHITE STEAM CARS

THE makers of these cars are to be congratulated upon the possession and development of a type of steam generator, which, in combination with a cleverly designed engine and well devised and constructed auxiliary gear and parts, has made it possible to produce a car capable of continuing to compare favourably with those propelled by the petrol engine.

Survival of satisfactory types

The efforts of the White Company in America, and of Messrs. Clarkson of Chelmsford, Turner of Wolverhampton and others in England, and the continued development of the Serpollet and Miesse cars, have done much to disprove the unnecessarily severe criticism and condemnation of the use of steam as a power medium, which resulted from the unsatisfactory performance, or perhaps mainly from the troublesome and short life of the large number of light steam cars, chiefly of American origin, which were until recently in use. Although there have been detail modifications and changes of form and number and position of elements, it may be said

6-HP. Stanhope

that the 6-HP. car of 1902 had embodied in its design the type of steam generator and the necessary auxiliaries for fuel and water supply and control which are retained in use, and to the use of which the success of the White cars is largely due.

10-HP. Tonneau

The 10-HP. tonneau car, an external view of which is given by Fig. 386, was introduced into this country in 1903. It differs considerably from its predecessor in external appearance, and closely resembles in general outline the form taken by many petrol-engined cars. A vertical compound engine is used instead of the two-cylinder simple type previously employed, and it is carried at the front of the car, longitudinally arranged, instead of transversely as before. The single chain transmission direct from engine to live rear axle has been replaced by a bevel gear drive of the type in which tubular extensions from the casing containing the bevel and differential gears house the driving-shafts upon which the road wheels are fixed, a universally jointed shaft transmitting the power from the engine to the bevel driving-gear.

An armoured wood frame is now also employed instead of the arrangement of tubular underframe directly connected to the car axles with the body spring mounted above it.

450

WHITE STEAM CARS

These were the most noticeable departures in design from the 1902 6-HP. Stanhope car. Quite recently a larger car of 15 HP. has been completed, and as it contains some improvements dictated by experience and not by fashion it may be looked upon largely as representative of the 10-HP. car, but with these further modifications to which reference will be made. An external view of the 15-HP. car is given by Fig. 387. The design of body there shown is one of the standard types adopted, and is of pleasing proportions. The leading dimensions are as follows: The wheel base is 7 ft. 9 in., the tread or gauge is 4 ft. 8 in., and the highest point of the body

15-HP. car

FIG. 386.—WHITE STEAM CAR—10-HP. TONNEAU.

is 5 ft. 6 in. from the ground; the road wheels are 34 in. diameter and the weight of the car is 2,000 lb.

Before proceeding to consider some parts of the car in detail it may be advisable to indicate the position and purpose of those most essential to the running of the system.

Figs. 388 and 389 show respectively top and under views of the 15-HP. car with the body removed. From these views a good idea may be formed of the arrangement of the mechanism and the suitability of the positions chosen with regard to accessibility, to the dual use of some parts, and with a view to the reduction of the total number and weight of parts and

451

Frame   connexions between them. The side and back end frame members are of wood reinforced on their inside faces with steel flitch plates. At the rear corners, these plates leave the wood frame and, serving as diagonal connexions or braces, add considerably to the resistance which the whole frame can offer to racking stresses. There are four cross-connecting frame bars, one of channel section a little behind the condenser c, one also of channel section below the dashboard and feed-water tank D, and the other two of L-section in front of and behind the steam generator A. The latter is here shown complete with its casing; the rectangular flue openings to be seen on each side of and near the top of the casing are connected with longitudinal flue tubes of oval section, and one of the latter is visible in the car elevation, Fig. 387, and is still more clearly shown by Fig. 386 of the 10-HP. car.

Flue baffles   There is inside each of the oval side flues a pivoted baffle plate to

FIG. 387.—WHITE STEAM CAR.—15-HP. TONNEAU.

prevent downdraughts and interference with the steady burning of the fire and the updraught from it. The baffles take the form of flap plates arranged vertically like doors, and their positions are chosen so that whether the wind draught through the oval flues is in one direction or the other the right-angle flue connexion from the boiler is shielded.

The steam generator is supported on the cross angles in front and behind, and it is side-stayed to the longitudinal frame bars by tie rods which are visible in the underneath view (Fig. 389).

The engine B is supported by four brackets on two parallel steel channels. The shorter channel on the left side connects the transverse frame channels behind the condenser c and below the water tank D, and that on the right side is longer and runs through to its point of connexion with the transverse angle bar in front of the steam generator. This longer

FIG. 388.—WHITE STEAM CAR.—PLAN OF 15-HP. CAR, WITH BODY REMOVED.

channel, it will be seen, supports one side of the engine at its forward end, provides a point of attachment for the steering gear, and stiffens the frame by interconnexion of the three transverse frame bars.

The cylindrical fuel tank E contains twelve gallons of petrol or benzolene, and it may be conveniently filled through the screw plug at the top **Fuel feed by air pressure** towards the back. An air pressure of about 30 lb. per sq. inch is maintained in the tank, and the supply to the burner being under pressure and also not influenced by gravity or difference of fuel level, is not liable to interruption by sluggish flow. There is a stopcock close up to the tank in the supply pipe leading to the burner, which should be closed when the car is not in service, or when for any reason it becomes necessary to uncouple any pipe connexions. The hand-wheel belonging to this stopcock may be seen in Fig. 387 standing up above the floor-board level and below the lubricator fitting on the dashboard.

The lubricator is double-chambered, and a single worm and worm-wheel drives a small plunger pump in each chamber. The lubricator **Mechanical lubricator** chambers are separately used for the oil supply to the engine crank chamber, and for the forced feed of special cylinder oil to the high-pressure cylinder steam chest through the pipe H. The positive and continuous supply of oil to the engine cylinders and crank chamber may be supplemented by the hand-plungers above the lubricator chambers, or at the conclusion of a run copious lubrication of the cylinder and valve surfaces may be ensured. Above the cylinders of the engine there is a light shaft J (Fig. 388), driven by a round belt from the coupling disc between engine crankshaft and fly-wheel. This drives, at the rear end, the worm-spindle operating the mechanical lubricator, and at the forward end by pulley and round belt, a fan behind the condenser. The very complete system of lubrication of the engine is necessary and cannot be looked upon as a complication. Efficient lubrication was one of those questions which did not receive sufficient attention in most of the early steam cars.

On the dashboard, and to the right of the lubricator, are two pressure gauges, the one indicating the pressure in the steam generator and the other the air pressure in the fuel tank.

Above the steering-wheel and concentric with it is a smaller hand-wheel. The spindle operated by it passes down the centre of the steering column ; it has a short lever arm at its end, and, by inspection of Fig. 389, the rod connexion between it and the throttle or stop valve for control of the steam supply to the engine may be seen.

The throttle valve fitted to the 10-HP. cars is of the disc or circular grid type having similarly spaced holes in a pair of disc faces in contact **Throttle valves** with each other. A movement of the throttle-controlling hand-wheel through about 45 degrees is sufficient to bring these holes opposite each other for the free passage of steam, or blind and closed when the steam supply to the engine is to be stopped.

A type of valve more suitable for use with high temperature and high pressure steam is used with the 15-HP. cars. It is of the screw-down

Fig. 389.—White Steam Car.—Underneath View of 15-H.P. Car with Body Removed.

455

needle-valve type, with a coarse pitch screw which allows of sufficient opening to steam with limited rotary motion.

Pumps

There are six pumps, four of them operated by the engine, and two hand-worked. The position and purpose of these will be indicated, but they will be more particularly referred to in connexion with following diagrams and illustrations.

Of those worked by the engine two are operated from an excentric on the crankshaft, through a short excentric rod and a rocking lever, the two plungers being in line and connected by a common rod. These are both generator feed pumps, and they are visible in the side elevation of the engine (Fig. 397), from which the arrangement of pump rod with slotted boss or cross head at mid-length for the reception of the outer forked end of the rocking lever may be seen. Being positively driven, the pumps work

Power worked pumps

continuously with the engine, but their delivery is variable and is controlled primarily by an automatic water regulator, and the delivery from the upper pump may be entirely or partially returned to the water tank by a hand-operated bye-pass valve, for a purpose which will be described. At the forward end of the engine there is a water pump worked from an extension of the low-pressure cylinder crosshead pin, which projects through the crank chamber and works in a slot or race way in the crank-chamber wall. The duty of this pump is to return the condensed steam from the condenser to the water tank. The fourth engine-worked pump is driven by a connecting rod from an extension of the low-pressure cylinder crank-pin. It may be allowed to pump air into the fuel tank when the pressure gauge on the dashboard shows this to be necessary, or it may be caused to run idle by holding the inlet valve to the pump cylinder open.

Auxiliary pumps

The position of the two hand-worked auxiliary pumps, which are partly hidden from view in Fig. 388, may be seen by reference to the elevation of the car (Fig. 387), where they stand up through the footboards just in front of the driver's seat; one of them is a feed pump to the generator, and the other an air pump. The feed pump is seldom required, but at the commencement of a run in frosty weather or at any other time when the steam generator has been completely emptied, it would be used to pump water into the generator to form steam enough to get the engine and its power-driven pumps running. Similarly after refilling the fuel tank the hand-worked air pump would be required to obtain an air pressure in it of about 10 or 12 lb., at which pressure fuel would be supplied to the generator burner, and when the engine-worked air pump would again be available, as soon as steam was supplied to the engine.

Condenser

The frame of the condenser c (Fig. 388) is cast with side lugs, and it is supported upon them at the front of the frame and braced in position by the attached pipes and brackets; the bonnet over the engine also fits snugly between the feed tank and the condenser, and around the frame of the latter.

Exhaust steam enters the top header of the condenser towards the left-hand side and by the pipe F from the engine. Vertical tubes disposed in rows three deep connect the top and bottom headers and the direction of flow of the steam through them is downwards as it condenses and collects in the bottom header, and is removed by the pump already mentioned and returned to the water tank. The belt-driven fan behind the condenser increases its capacity by inducing a more active flow of air around the tubes, which are not spaced in rows three deep behind each other, but are staggered, and it increases the cooling effect under adverse conditions and when the speeds and directions of car and breeze are similar. Any steam that is not condensed may get away through a lightly loaded escape valve which may be seen at the centre of the top header in Fig. 388, and thence through the downwardly leading escape pipe to atmosphere. The provision of the escape valve opening at from half a pound to a pound pressure adds little to the exhaust back pressure, and it economizes the feed water by not permitting an unrestricted escape of exhaust steam to atmosphere.

The top and bottom covers of the headers may, either or both of them, be removed, leaving the tube ends exposed, and facilitating cleaning or repair and renewal of any of the tubes if they should suffer accidental damage.

The condenser used on the 10-HP. car has approximately the same tubular cooling surface, but the tubes are arranged horizontally and, as has been found in surface-condenser practice, are not so efficient in this position, due to less effective draining of water from the lower internal surface and **Feed-water purification** some reduction of the useful cooling surface. A grease filter is used with the 10-HP. car for the removal of the greater part of the grease and impurities in the condensed steam. The use of the filter is discontinued with the 15-HP. car, and the water of condensation is returned directly from condenser to water tank. At each side of the water tank there are wash-out plugs through which the floating oil may be scummed off periodically.

Fig. 390 illustrates the steam generator as used on the 10-HP. car. **Steam generator** The number and arrangement of the tube sections and the form of the burner and vaporizer, as far as they may be seen in this view, are the same in the 15-HP. generator. In the construction of the generator eleven sections or coils are connected together in a peculiar way to form one continuous length of tube. Each section is flat and is spirally coiled from the end at the centre to the outer end as a clock spring is formed.

The feed-water inlet is at the top of the generator, and the pipe passing through the asbestos-lined sheet-iron casing to the centre of the top coil is lettered G in Fig. 388. Following the course of the spiral from the centre of the top section outwards, it will be seen that the outer end of the top section is turned upwards and is then taken inwards over the top towards the centre of the generator again, and by means of union sockets and a short junction pipe, is connected to the upstanding centre end of the second section. In the same manner the outer end of each coil is led upwards and inwards over the top of the nest of coils to its junction with the inner end of the next lower coil.

457

MOTOR VEHICLES AND MOTORS

Referring to Fig. 392, which gives diagrammatically the parts and connexions required for the supply and control of the fuel to the generator, the outer end of the lowest generator coil is connected at Y Y to the ribbed thermostat tube. This tube passes straight across the burner, immediately below the lowest generator coil, and the steam from the generator flows along it and out to the engine steam pipe at the downwardly pointing limb of the T-piece at X X. The steam pipe to the engine is connected to the vertical limb of this T-piece, and therefore passes through the edge of the burner box or frame. The direction taken by the steam pipe, which is well covered with lagging material, after leaving the generator, may be

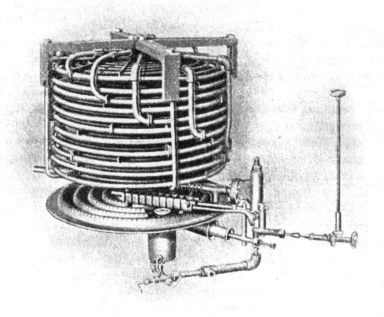

FIG. 390.—WHITE STEAM CAR.—10-HP. STEAM GENERATOR.

seen by reference to the under view of the car Fig. 389, and in Fig. 388 at its point of connexion with the high-pressure steam chest of the engine, it is lettered s. From the arrangement of the generator coils it will be understood that water pumped into them cannot flow by gravity or pass unchecked through the whole length of the eleven sections. Dependent upon the evaporative rate at which the generator is working, there is a variable point in the coils at which the water becomes steam. Before this point is reached, the pipe coils may retain water because of the upward bend at the outer end of each coil, before it is led over the top of the generator to its centre connexion with the next lower coil. The generator differs from those which have come to be described as of the flash type because it has some water content, and generally contains more water than is required at any instant to satisfy the demand for steam.

The tube sections are separated by spacing strips, and are all held together by the four-armed spider frame at the top and the hook bolts which may be seen in Fig. 390. The internal diameter of the steel tubes used in the generator is $\frac{17}{32}$ in. and the external diameter $\frac{21}{32}$ in. A sectional diagrammatic view of the burner as used on the 6-HP. car is given by Fig. 391. It does not show the arrangement of the same parts still used on the recent cars, but it enables a better idea to be gained of their operation than by inspection of external views only, such as are given by Figs. 389, 390 and 392.

The burner is in form a shallow drum, the upper face of which is **Burner** annularly ridged. The ridges G as shown are slit or sawn through at close intervals, and provide openings through which the combustible gas and air mixture combines with additional air flowing upwards through the numerous openings formed by the tubes G^3. The number and arrangement of these tubes, which are in the hollows or valleys between the slit ridges G, may be judged by inspection of the under view of the burner to be seen in Fig. 389.

The position of the vaporizer F^1 (Fig. 391) above the burner plate or drum is shown by Figs. 390 and 392, and in the latter is

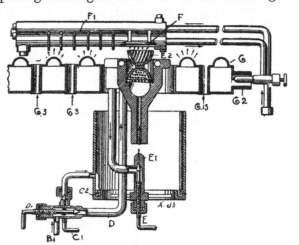

FIG. 391.—WHITE STEAM CAR.—SECTION THROUGH BURNER AND PILOT LIGHT.

Main vaporizer

lettered W. The sectional diagram indicates by arrows the direction of flow of petrol from the supply pipe to the inlet along the length of the vaporizer body and back again and out by the pipe leading to the needle valve. The needle valve through which the vaporized petrol issues at considerable velocity is situated at the mouth of the air inlet opening G^2 to the burner drum, and the mingling of air and vapour commencing inside the drum becomes complete as described on leaving it and mixing with the air flowing up the short tubes G^3. The vaporizer is thus a simple casting with plug holes for cleaning out the flow and return passages; it is transversely ribbed, and where it passes over the pilot burner there is a flat plate cast on to divert the flame and make it wrap the vaporizer and accelerate the preliminary heating up.

The construction of the pilot burner is ingenious, and it may be briefly **Pilot burner** described. Petrol or benzolene, according to which fuel is employed, is taken from the supply tank and by the pipe B^1 to the junction piece in

459

which are the needle valves c^1 and d^1. The valve c^1 is opened and some petrol allowed to flow into the annular groove c^2 inside and at the bottom of the burner box. This is set fire to and kept alight long enough to warm up the pilot vaporizer box and pipes. The valves d and e may then be opened and c^1 shut. The vapour leaving the jet e^1 enters the combining tube, or the vertical passage in the casting immediately above the jet, and burns with a bunsen flame as it leaves the bowl part of the pilot vaporizer. There is, it will be noticed, a double cup piece e^2, having a number of slit and hole passage ways through it, which helps to effect mixture of the gases, and remaining hot serves to re-ignite the vapour supply should the pilot light be blown out at a moment when the main burner has been cut out by the thermostat. The cup piece e^2 is, in Fig. 391, incorrectly shown upside down. The passage cast round the top of the vaporizer bowl and the pipes for the supply of petrol to the vaporizer and of vapour to the jet e^1 from the vaporizer are clearly shown. A diagram view of

Fuel system the parts forming the fuel system is shown by Fig. 392. Of the mountings on the fuel tank a, the internal pipe m leading from the bottom of the tank to the stop valve d supplies the fuel through d and by pipe o to the distributing pipe q, and so to the cocks t, s and r in connexion with the pilot light u. The cocks t, s and r are those lettered d^1, c^1 and e in Fig. 391, and described with reference to the pilot light. Through the right-hand branch of pipe q the fuel flows past union $p p$, hand-valve $f f$, and thermostat valve near z to the outlet pipe $r r$ leading to the vaporizer w. The vaporized fuel returns from the vaporizer as described and enters the burner drum v through the needle valve $o o$ situated at the air inlet opening $n n$. As already stated the outer end of the bottom generator coil is connected at $y y$, and the steam flows along the ribbed pipe and out to the engine steam pipe which is connected at $x x$. The steam in the ribbed pipe

Automatic control of steam temperature will be at its highest temperature, and the duty of the thermostat is to control the maximum steam temperature, and the range of temperature. A brass tube fixed to y, and projecting some distance into the ribbed pipe, contains a steel rod fixed at its outer end to the outer closed end of the brass pipe, and having its other end attached through simple lever or multiplying gear to a valve in the fuel supply passage inside y and at about the level of z. z and n are plug holes. With change in the temperature of the steam, differential expansion and relative movement of the brass and iron elements of the thermostat occurs and results in opening or closing movements of greater amplitude, due to the lever proportions adopted, of the fuel valve at z. It will be understood that the steam flows past the outer surface of the brass tube and that its interior surface and the steel rod it contains are closed from contact with the steam but experience the temperature changes. The thermostat, therefore, depends for its action upon the difference of lineal expansion of steel and brass under changes of temperature, and its duty is the control of the fuel supply and consequently the fire, such control being independent of pressure and dependent entirely on steam temperature. Steel and brass have been mentioned as the metals employed for the elements of the thermostat, but this is not quite correct, special alloys having equally suitable co-efficients of expansion, and capable of withstanding the severe temperature conditions, are used. Means are provided for adjustment of the temperature range through which the thermostat operates, but such adjustment is delicate, is seldom required, and requires to be made by practised hands, and by those familiar with the use of a thermometer or pyrometer.

The hand air pump c is shown with a check valve N in its delivery pipe to the tank, similar to the check valve F in the corresponding pipe from the engine-driven air pump G. A section through G is shown in Fig. 396, and it is a simple single-acting pump with valves in the top cover. The cap over the inlet valve I has the air inlet opening at the top covered with gauze to exclude dust, and both I and the delivery valve cap J are held in place by a centre bolt and a single bridge. The delivery of compressed air is through the pipe K, which is shown led round the back of the cylinder, and the check valve F to the mounting on the fuel tank. A pipe

Engine-worked air pump

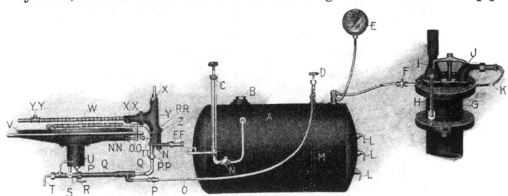

FIG. 392.—WHITE STEAM CAR.—DIAGRAM OF PARTS AND CONNEXIONS FOR FUEL SYSTEM OF 15-HP. CAR.

from this mounting is led to the air-pressure gauge E. The rod H is connected to a short lever pivoted in the inlet valve cap, and this short lever is caused by spring pressure to depress the inlet valve, and by normally keeping it down and open prevents the compression of air by the pump. The plunger is then idly driven, and absorbs little power in friction and air displacement. When the pump is required to raise the pressure in the fuel tank the rod H is pulled through connexion with a foot lever, the short depressing lever is raised from the inlet valve, and the pump may then work. B is the plug at which the tank is filled and L L L are try cocks to determine the quantity of fuel remaining in the tank.

It has been mentioned that the supply of water from the feed pumps to the steam generator is primarily controlled by an automatic regulator. It is of the diaphragm type, and it is shown in section by Fig. 393, and may be seen at F (Fig. 394) and in the side view of the engine (Fig. 397). The diaphragm c is clamped between the flanged end of the case or body A and the end cover o, and is exposed to generator steam pressure on its outer face.

Automatic control of feed water

461

The steam connexion is made at the union R or at Z as lettered in Fig. 394. There is a branch from the forced feed lubricator on the dashboard, and by the use of suitable diverting cocks, oil may be pumped into the space at the front of the diaphragm through S to fill the pipe connexion at R and a pipe connexion not here shown leading to the steam pressure gauge. If these pipes are kept oil charged there will be no difficulty with them, due to freezing of the water they would otherwise contain, during winter weather.

The heavy coil spring G inside the regulator abuts against an adjustable collar M on the right, and opposes the steam pressure on the other side of the diaphragm by thrust against the plate E. The spindle F is screwed into E, adjusted and secured by lock nut H, and at its other end is in contact with the pivoted tongue I. There is a pipe connexion at P with the generator feed-pump delivery pipe and a bye-pass valve at J, which when lifted by

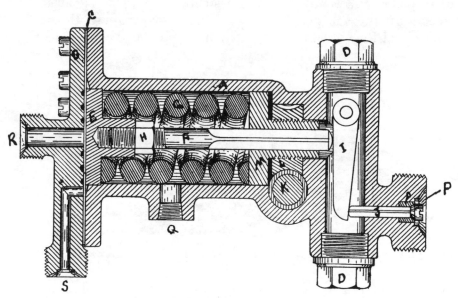

FIG. 393.—WHITE STEAM CAR.—WATER REGULATOR

the tongue I allows the pump-delivered water to flow past it and along the spindle F and out at Q. The end of F, where it passes through and is guided in M, has flats cut on it to provide passages for the flow of the bye-passed water.

When the steam pressure is low, the spring G overcomes the steam pressure on the opposite side of C and pushes the plate E and diaphragm C to the position shown. The bye-pass valve P is then closed and the feed water is pumped into the steam generator. Similarly, when the steam pressure is high, the load on the outer face of C is sufficient to overcome the opposing effort from the spring G, to further compress and move it with E and F and I to the right, and the latter being in contact with the stem of the valve J is caused to lift it from its seat. The method of magnifying or multiplying the movement of J relatively to that of F, by reason of the

long and short arms represented by the distances of the points of contact of
J and F from the pivot pin of I, is direct and simple. According to the
amount of lift given to the bye-pass valve J, the whole, or a part, or none of
the feed water is delivered to the generator, and it will be noted that this
variable delivery is controlled by the steam pressure in the generator and
is not in any way dependent upon the quantity of water in the generator,
except in so far as the steam pressure implies the presence of water which
by variation of quantity with variations in the demand for steam from the
generator results in the steam pressure and temperature fluctuations upon
which the action of the fuel and water regulators depends.

Rotation of the worm nut L by means of the worm screw K results
in movement of M to left or right, M being capable of longitudinal movement,
but a pin and slot way not shown prevents rotation. More or less compression being thus put upon the spring G, the pressure range through which
the bye-pass valve acts may be adjusted.

The method of control of the fuel to the burner and of the water to
the steam generator, dependent, the one upon the temperature of the
steam, and the other upon its pressure, will now be understood. Whether
the demand for steam is great or small the water regulator in obedience to
the ruling pressure allows more or less water to enter the steam generator ;
there its duty ends, and the thermostat then maintains the burner flame in
a greater or lesser degree of activity according to the temperature of the
steam and the predetermined degree of superheat permitted.

The figures reproduced from tests of the 10-HP. type of car on page 473
show what these pressures and temperatures may be when the engine is
running at loads varying from its normal full load to a load equivalent
merely to the horse-power absorbed by the transmission gear.

Fig. 394 is a general diagram of the water pumps and the connexions
between them and the feed tank, steam generator and condenser.

The generator feed pumps E and E¹ both take water from the tank A
through the suction pipe D, which leads directly to the suction valve box
of the upper pump E. A continuation of the pipe D by G and H forms the
suction pipe for the lower pump E¹ The bye-pass discharge from the automatic regulator F, it will be noticed, is connected at the junction of G and
H. The delivery pipe T from the lower pump E¹ is joined by T-piece with
the common delivery pipe I, and the third branch of the T-piece is connected by pipe to U and the bye-pass valve in the water regulator, and in
this short length of pipe there is the junction of the delivery pipe from
the upper pump E. The arrangement of these delivery pipe connexions is
clearly visible in Fig. 397.

On the delivery side of the pump E there is a pipe N leading to the tank
A, and at its junction with the tank at R there is a bye-pass valve which may
be conveniently opened or closed by the driver, by means of the long stem
and hand-wheel V. It has been found in practice that an intelligent driver
can to some extent anticipate the steam requirements and the action of
the automatic fuel and water regulators. It may frequently happen, for

Automatic control of fuel and water

Water system

463

instance, that after the car has descended a hill and the steam demand has been very slight, that an immediate call may come for more steam to ascend **Control of** the following hill. The fuel and water regulators commence to adjust **automatic** themselves to the new requirements as soon as they are felt, but the driver **regulators** possesses and can apply the brains with which the automatic regulators cannot be fitted, and by augmenting the water supply to the steam generator a little before the ascent commences, causes the regulators to respond earlier and allow of the generation of steam and the thorough heating up of the generator in readiness for the period of greater activity during hill climbing.

The lower pump E^1 has sufficient capacity to supply the water to the steam generator required for ordinary steam demands, and usually the bye-pass valve v is open and the upper pump delivers no water to the generator. The delivery from the single-feed pump is then controlled by the water regulator. When the heavy steam demand is about to occur v is closed

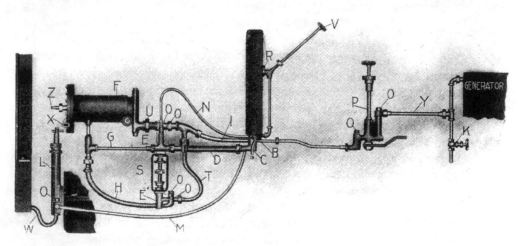

Fig. 394.—White Steam Car.—Diagram of Parts and Connexions for Water System of 15-HP. Car.

and water is delivered from both pumps. A first effect of this increase of water delivery is the reduction of the temperature of the steam leaving the generator, and the second effect, due to the steam temperature reduction and the opening of the fuel valve by the thermostat, is the increased intensity of the burner flame. With the increased water delivery and greater heat from the burner the generator is able to work at a higher evaporative rate and to meet the exceptional and occasional steam demands. Briefly stated, the control which the driver is thus permitted to have over the rate of water supply to the generator enables him to overcome the comparatively sluggish or late action which the automatic regulators must necessarily have, and to hasten the action of the regulators, when change, by increase or decrease in the steam demand, is about to occur. It must be stated, however, that the automatic regulators will take full charge of the water and fuel supplies, but that more regularity in driving and greater

economy of fuel may be obtained by those who choose to avail themselves
of the further control permitted.

With the 10-HP. car a single engine-worked feed pump of greater
relative capacity is used. The activity of the generator can be increased
by augmenting the water supply to it by means of the hand pump, but such
a method is not convenient, and is not so satisfactory as the double pump
system described.

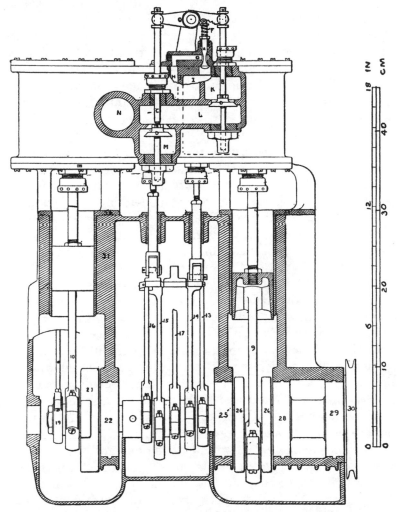

FIG. 395.—WHITE STEAM CAR.—LONGITUDINAL SECTION OF 15-HP. ENGINE.

The hand-worked feed pump is shown at P, and the plunger, like the
auxiliary air-pump plunger, is arranged to screw home on to a valve seat
when the pump is not in use. By this means leakage past the plunger
stem is prevented. The common delivery pipe I from the power-worked
feed pumps passes by B and through the suction and delivery valves of
the pump P to Y, and so to the inlet pipe shown to the generator. The

465   30

valves o o of the pump P thus serve as check valves in the single delivery way to the generator, and when the power-worked pumps are not running and the hand pump is used, then the latter takes its suction through the delivery pipe I and either or both of the pumps E and E¹, and from the common suction pipe D. By thus putting the pump P in the single delivery circuit from power-worked feed pumps to generator, duplication of the suction and delivery pipes is avoided.

Simplification of pipe connexions

The valve K at the bottom of the inlet pipe enables the generator coils to be emptied at the conclusion of a run in frosty weather.

The pump L, as already mentioned, removes the condensed steam from the condenser by the pipe W, and returns it by M to the water tank. The pump L may be seen at the forward or low pressure cylinder end of the engine in Fig. 397. The letter o in Fig. 394 indicates the position of inlet and delivery valves, and at c the water tank may be emptied.

Engine

In addition to the external view of the engine from the water pump side (Fig. 397) a longitudinal sectional view is given by Fig. 395, and a transverse section through the low-pressure cylinder and the air pump by Fig. 396. From these views the arrangement and construction may be studied and reference will now only be made to a few features. The engine is of the completely enclosed double-acting compound high-speed type and has a high-pressure cylinder 3 in. diameter, a low-pressure cylinder 5 in. diameter, and the stroke is 3·5 in.

These cylinder and stroke dimensions are the same in the engines used for the 10 and the 15-HP. cars. The larger car is, however, supplied with a steam generator capable of supplying enough steam to continuously work the engine at the higher load, and the latter has been in some ways strengthened and made capable of working at a higher average load than the engine of the 10-HP. car.

Speed

With the road wheels 30 in. diameter with the 10-HP. car and a bevel-gear reduction at the live axle of 3 to 1, the engine speed when the car is travelling at the rate of 20 miles per hour is 675 revolutions nearly.

At a similar road speed and with the same gear reduction, but with the larger road wheels used with the 15-HP. car, namely 34 in. diameter, the engine speed is about 600 revolutions.

Construction

At the low-pressure cylinder end of the engine the connecting rod 10 works on to an overhung crank-pin in the balanced crank-disc 21, and the high-pressure connecting rod 9 works on to a crank, the webs of which 26 are balanced. At this end of the engine the shaft is supported in two single-row ball-bearing boxes 28 and 29, it projects through the crank chamber as shown, and terminates in the coupling disc 30, which is shown forged solid with the shaft. The edge of 30, it will be noticed, is grooved, and it drives by round belt the fan and lubricator spindle above the engine already described. The fly-wheel, to be seen in Fig. 397, is coupled to 30, and it has a square hole in its boss to receive the squared end of the propeller shaft. There is at this end of the propeller shaft no universal joint of the ordinary form, but the squared faces of the shaft end are slightly rounded, and with the long shaft used with these cars the necessary freedom of movement can be obtained satisfactorily in this way. A universal joint of the usual form is used and may be seen in Fig. 389 at the end of the propeller shaft next the gear box on the back axle. The fly-wheel is also used as a brake drum, and the internally expanding brake ring arranged inside it is operated by a foot lever.

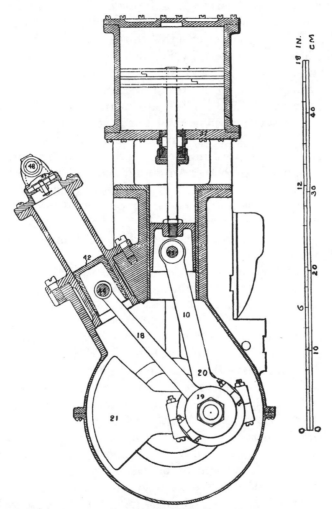

FIG. 396.—WHITE STEAM CAR.—TRANSVERSE SECTION THROUGH LOW-PRESSURE CYLINDER OF 15-HP. ENGINE.

Close up to the crank-webs 21 and 26 the intermediate part of the crankshaft is carried in the ball bearings 22 and 25. This intermediate part of the shaft is in two parts, and the ends are tapered and recessed and cottered together, the same cotter pins being also employed to hold the gang of 5 excentrics, in one piece, in position.

Of the 5 excentrics those connected with the excentric rods, numbered 13 to 16, are obviously in pairs, and through reversing link motion operate

467

the cylinder slide valves. The centre excentric and rod 17 is that which through the rocking lever operates the generator feed pumps already described.

Ball bearings The crankshaft, connecting rod big ends, and the excentric straps all run on ball bearings. The bearings are all of the single-row type and the balls in the shaft bearings are $\frac{5}{8}$ in. diameter, and those in the other bearings $\frac{7}{16}$ in. diameter. Shaft-bearing 29 is formed to take any small amount of end thrust there may be, and 22, 25 and 28 afford support in a vertical plane, but are not designed or required to take any end thrust. The aluminium crank chamber, as shown by Fig. 396, is divided in a horizontal plane at the crankshaft centre line, and the method of holding the bearing boxes in the seats formed for them in the crank chamber may be seen in Fig. 395. The stools upon which the cylinders are carried are each cast in one with the bored cross-head guide liner and the bottom cylinder cover, and are made of phosphor bronze or gunmetal. The cylinders are in one casting with the high- and low-pressure steam chests adjoining, and separated by the partition wall J (Fig. 395). An aluminium jacket in halves is used to cover in the cylinders and to retain the asbestos lagging material with which the cylinders are covered. The one piece or block type of piston and the rings used may be seen by reference to the transverse section. Asbestos

Gland packing packing permeated with graphite is used at both high- and low-pressure piston rod glands.

It will have been noticed from Fig. 397 that the exhaust pipe has two branches at its flanged connexion with the cylinder casting. The larger or left-hand branch communicates with the exhaust port from the low-pressure cylinder N (Fig. 395), and the right-hand branch with the space K. Above the cylinders a transverse shaft is pivoted in bearings,

Bye-pass valves and at one end of it there is the double-armed lever O. When the shaft is partially rotated in one or other direction, the lever O is caused to move the valves and spindles B D and C D in opposite directions. There is also a small valve A normally kept closed by steam pressure and the spring F. There is a short finger or lever on the transverse shaft so arranged that when the shaft is partially rotated it depresses A and allows steam to flow from the high-pressure steam chest I to the low-pressure steam chest H by the passage G. This arrangement of valves constitutes a very complete bye-pass gear. With the valves in their normal positions and as shown by Fig. 395, steam from the high-pressure cylinder exhausts by L and M to the low-pressure steam chest, and the engine runs as a compound engine. When the lever O is moved by the transverse shaft, the valve C D is closed and B D is opened; the depressing finger has also come into contact with A and opened it. The exhaust steam from the high-pressure cylinder then passes from L to K past the valve B D to the exhaust pipe and condenser. The valve C D has closed communication between the high- and low-pressure cylinders, and the latter is supplied with live steam through the valve A.

The capacity of the bye-pass port G is purposely made small, and though large enough to supply enough live steam to the low-pressure cylinder to largely increase the starting effort, it will not supply sufficient to allow the engine to be run at normal speeds and at a heavy overload. The exhaust steam from the low-pressure cylinder passes to the exhaust pipe by the passage N, when the engine is running as a compound or when the bye-pass valve is open. The bye-pass gear is operated by a foot lever which by cranked levers and rod connexions gives the required twisting motion to the transverse shaft. By reference to the plan of the car (Fig. 388) a double-

FIG. 397.—WHITE STEAM CAR.—SIDE ELEVATION OF 15-HP. ENGINE.

ended lever may be seen on the end of the transverse shaft near the letter s. The forward arm of this lever is normally kept pulled downwards by a coil spring connected to its end, and the bye-pass valves are then in the positions shown by Fig. 395, and the other arm of the lever receives the pull from the foot-lever gear when the bye-pass valves are to be opened.

The position of the power-worked air pump 43 may be seen by reference to the transverse engine section (Fig. 396), and the method of carrying it

469

diagonally on a seating formed on the side of the crank chamber. The connecting rod 18, worked from the low-pressure crank pin extension, operates the pump plunger 42, and the latter has no piston packing rings, but is with its cylinder machined and ground to a perfect fit, and has grooves turned in it which retain or become packed with oil. It will have been seen from earlier reference to Fig. 389 that power transmission from the engine is through an exceptionally long propeller shaft and by bevel gear and the differential gear to road wheels. The live rear axle has no special feature other than the arrangement of gear contained in the gear box situated about centrally on the axle.

Fig. 398.—White Steam Car.—Arrangement of Speed Reduction Gearing with 15-HP. Car.

Torque bar

The torque bar T relieves the rear springs and the propeller shaft from the variable stresses imposed by the driving effort, and side distance rods transmit the thrust or drag from the rear axle to the side frame members. A well arranged and correctly adjustable truss bar below the rear axle strengthens it considerably against bending stresses.

Speed-reduction gear

The differential gear is contained in that part of the gear box lettered s, and the arrangement of the speed reduction gear employed with this car is shown by Fig. 398 and by Figs. 398A, 398B,[1] which show the complete transmission and differential gear in vertical and horizontal

[1] From the *Commercial Motor*, 27 July 1905.

section and show the double lever by which the pinion on the main shaft A Fig. 398 and M Fig 398A, and the two wheels on the secondary spindle are simultaneously moved in opposite directions. This gear has been provided for two reasons, primarily to enable the engine to be run without driving the car, and secondly to give a lower gear ratio to enable the car to ascend severe gradients and to travel over bad roads and difficult places. The gear reduction at the bevel drive is 3 to 1, but when the reduction, or emergency gear, as the White Company prefer to call it, is in use there is a total gear reduction of 7 to 1.

The drive from the engine is received by the spindle A Fig. 398 through the universal joint secured upon its squared end. The shaft is supported in bearings at L and J, and near the latter the bevel-driving pinion I fixed to the wheel D may be seen. The wheel D, it will be observed, has internal and external teeth E and F. The pinion C with the collar extension O for the reception of the forked end of a shifting lever, is capable of end-sliding movement upon the squared spindle A. Above the latter is the short spindle B and roller bearing N, upon which the pair of wheels G and H fixed to each other and their shifting collar O may revolve. By means of the rod R (Fig. 389) and the lever arm to which it is connected and the shifting forks inside the gear box, the gears may be caused to occupy three positions. In that shown by Fig. 398 the low ratio gear is in operation and the drive is transmitted from C to G and from H to D. In an intermediate position C is pushed towards D, and simultaneously G and H are moved in

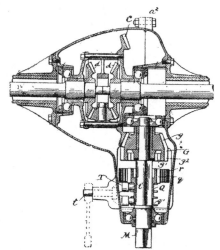

FIG. 398A.—TWO-SPEED GEAR.

the opposite direction and out of engagement with C and D. The spindle A may then, with C, be driven by the engine without transmitting any driving effort to I. For the third position the sliding movement is continued, and C enters the recessed portion of D, the teeth of the former sliding into or engaging with the internal teeth E upon D. I, D, C and A are then locked together, G and H are out of engagement and do not run idle, and the direct drive to I is obtained, the gear reduction then being 3 to 1 or that due to the relative diameters of the bevel pinion I and the bevel wheel upon the differential gear, with which it engages. The ability to run the engine without driving the car enables the power-worked pumps to be used and almost entirely obviates the use of the hand or auxiliary pumps and renders the management of the car still more convenient. The engine may be warmed up and cleared of the initial water of condensation, and it may be here mentioned that no drain cocks or relief valves are fitted to the engine cylinders. The steam generator may also be thoroughly heated up and the capacity of the car to get away quickly, and it may be with an up-hill start, is increased.

In addition to the pedal-operated brake at the engine fly-wheel there are brake drums at the rear road wheels, with internally expanding brake rings operated through side rods and a hand lever to be seen in the **Brakes** various views given of the car. The brakes are all of the internally expanding metal to metal type, and proper provision is made for adjustment after wear.

The White cars have acquitted themselves well in endurance contests in America and in the reliability trials held in this country by the Automobile Club. As a result of the performance of one of the 10-HP. cars, in the 1,000 miles reliability trials of 1903, a special silver medal was awarded for economy in water consumption.

A distinguishing feature in the performance of these cars is their pronounced economy of both fuel and water, resulting from the system of control adopted and from the successful use of superheated steam.

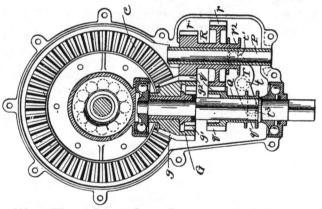

FIG. 398B.—WHITE STEAM CAR.—BEVEL AND TWO-SPEED GEAR.

The 10-HP. cars are provided with fuel and water tanks having capacities of respectively 8 and 12 gallons, and with these supplies the car is said to be able to run an average of 100 miles without any replenishing.

The 15-HP. car has tanks for fuel and water each containing 12 gallons. With the greater condenser capacity and the further economy obtained with the new method of control of the feed pumps described, the car will, it is said, run an average distance of 150 miles without replenishing.

These results have so far not been equalled by any other type of steam-propelled car.

The tabulated results of tests carried out by Professor C. H. Benjamin, here given, are worthy of record, and are of much interest to engineers.

A study of these test results show that the complete performance is remarkable.

From the quantities given it appears that the steam generator efficiency must have been not less than 80 per cent., whilst the engine steam consumption demonstrates, that with superheated steam, and with good internal conditions and construction, it is possible to obtain results with small steam engines directly comparable with those obtained with large engines. This may have been expected by engineers, when the uncertain and variable amount of the condensation, and the variable heat interchange occurring in different sizes of steam engines is eliminated by the use of superheated steam, but there has been hitherto very little direct evidence by careful tests to confirm this opinion :—

Efficiency and consumption test

Fahr. Deg.							Pounds, Avoir.									
Boiler Pressure	Temp. F° Feed	Temp. Steam (Boiler)	Temp. Steam (Engine)	Superheat (Boiler)	Superheat (Engine)	Temp. (Exhaust)	Weight, Gasoline	Weight, Water	E. Evap. from and at 212°	E. Evap. per lb., Gasoline	E. Evap. Per Sq. Ft. H. S. per Hour.	Actual Water per HP. Hr.	Gasoline, per HP. Hr.	Equiv. Steam per HP. Hr. at 212°	Brake, HP.	Cut-off
263	51	763	695	353	285	288	6·6	72	109·3	16·6	5·96	12·6	1·16	16·3	11·4	·75
250	47	779	725	373	319	289	6·5	68	104·7	16·1	5·71	14·9	1·42	19·7	9·15	·75
274	47	786	739	369	325	360	6·0	68·2	105	17·5	5·73	19·7	1·74	26·3	6·92	Open
244	47	772	718	368	314	339	5·7	60·5	93	16·3	5·07	17·5	1·65	23·1	6·92	·75
356	48	782	735	345	298	298	5·3	54·5	83·4	15·7	4·55	15·8	1·53	20·4	6·92	·50
388	46	793	726	347	280	293	4·6	46·5	71·4	15·5	3·89	19·9	1·97	25·9	4·66	·50
483	47	796	711	329	244	344	3·0	29·7	45·4	15·1	2·48	24·8	2·5	31·8	2·4	·50
380	47	737	632	294	189	354	2·0	15·7	23·3	11·6	1·27	..	..	..	0·0	·50
1	2	3	4	5	6	7	8	9	10	11	12	13	14	15	16	17

Further discussion of these test results cannot be now given, but it should be stated that the car tested was one of the 10-HP. type, that the leading engine dimensions are the same as those given for the 15-HP. engine, and that the brake-horse-power given in Column 16 is that measured by the brake or dynamometer employed at the rear road wheels.

The horse-power absorbed in transmission from engine to road wheels when the brake-horse-power was 11·4 was found to be 2 horse-power approximately. The actual figures given are : indicated horse-power 13·47, brake-horse-power 11·4, from which the efficiency is 84·7 per cent.

The weight of the 10-HP. car is 15 cwt., and the wheel base and gauge are respectively 6 ft. 8 in. and 4 ft. 8 in. Correspondingly the 15-HP. car weighs about 18 cwt., has the same wheel gauge or track, and a wheel base of 7 ft. 9 in.

The design of the 18-HP. car for 1906 is in most respects like that of the 15-HP. type. The products of combustion from the burner are discharged downwards through an annular flue surrounding the steam generator, and there is some modification of the burner and of the method of supply of air for combustion. The wheel base is 9 ft. 6 in., and the weight of the car is about 22½ cwt., the increase being in part due to detail changes and strengthening of parts. The leading dimensions of the engine are the same as those of the 15-HP. type here illustrated.

Chapter XXX

THE CLARKSON STEAM AND OTHER CARS

AMONG those firms in this country who have given their whole attention for many years to the construction of steam vehicles, Messrs. Clarkson, Limited, of Chelmsford, take a prominent place.

Subsequent to the period dealt with by vol. i. of this book, some changes have occurred in the types of vehicle made by them and since 1902 efforts have been directed solely to the production of passenger cars and primarily those of the heavier or omnibus type.

In the following description reference will be made to the changes in design of the mechanism which have been found desirable in order that the reader may be acquainted with the arrangement of the car as now built and also with the type that preceded it.[1]

Design of body An external view of the private car or small bus is given by Fig. 399, and it is worth pointing out that this car and type of body with provision for closing or opening side and front windows, and which generally affords the protection now looked upon as desirable, from inclement weather, was introduced by Mr. T. Clarkson in 1902, at a time when the canopy top or closed body was little favoured.

The outline elevation and plan of the car with the body removed (Figs. 401 and 402) shows the general disposition of the parts. The main frame is rectangular in plan, and the channel steel used is 2·5 in. × 1·75 in. × 0·25 in. in section in the smaller vehicles illustrated by Fig. 399, and 2·95 in. × 1·55 in. × 0·35 in. in the single-deck omnibuses like those illustrated by Fig. 408. With the larger omnibuses, seating thirty-four passengers, a bigger section of channel is used, and it is stiffened below by truss bars running fore and aft from a point of attachment below the suspension **Frame flexure** brackets for the differential gear and chain pinion shafts. It was found that the degree of sag or natural flexure of the steel frame, although unimportant in itself, had an injurious effect upon the heavy and inelastic 'bus body mounted upon it, and it was deemed necessary to reduce the frame deflection by the use of truss bars in order that the 'bus body should be freed from any stresses producing deformation.

The boiler indicated in Figs. 401 and 402 at A, is of the multi-tubular

[1] *Automotor Journal*, 8, 15, 22. Nov. 1902; 18 March to 29 April, 1905.

fire-tube type, and the leading dimensions of the shell are given by Fig. 400. It is a large example of the form of boiler which, when it received intelligent treatment, gave very good results in the small steam cars, which have been used in large numbers. Care, however, has to be taken to ensure that the water level is maintained, and to prevent the introduction of oil into the boiler with the feed water when the exhaust steam, or part of it, is condensed and returned to the feed tank. The tubes employed are $\frac{9}{16}$ in. internal diameter, and over 500 are used, giving a heating surface of between 60 and 70 square feet, without taking into account the useful drying effect from the tube surface above water level.

Multi-tubular boiler

FIG. 399.—CLARKSON 12-BHP. STEAM CAR, 1902–3 TYPE.

Messrs. Clarkson no longer use the fire-tube type of boiler, but they have endeavoured to so improve the design and method of construction that it may be as little liable to injury as the flash or semi-flash types of steam generator, now used with the larger vehicles.

The water supply to the boiler was controlled automatically by a thermal expansion device which in its final and simplest form consisted of a steel tube arranged horizontally outside the boiler and at the water level. One end of this steel tube was connected with the steam space of the boiler and the other end to the lower part of the boiler or water space by a syphon tube which served to prevent circulation by convection, and thus allowed the water in the horizontal steel tube to cool more rapidly

Constant level of water in boiler

or to remain cooler. As the water level in the boiler fell, so the horizontal tube became filled with steam from the boiler, and as the water level rose, then the tube again filled with water cooler than that contained in the boiler. The temperature differences caused sufficient movement of the tube by expansion and contraction to open or close a bye-pass valve, taking the form of a circular disc of metal carried by a diaphragm. The bye-pass valve permitted all, or part, or none of the feed water to be delivered to the boiler, according to the amount of its opening, and the bye-passed water was returned to the pump suction pipe.

The burner employed is of the well-known Clarkson form, illustrated in vol. i. p. 531, and paraffin is successfully used as fuel. The burner will not again be described here, though it may be recalled that it takes a distinctive form, and it is retained in use with practically no modification for the new semi-flash type of boiler adopted with the large omnibuses recently constructed.

The position of the burner is indicated at A_2, Figs. 401 and 402, and A_1 is the light sheet-metal fire-box or casing protecting the burner flame from interruption by the wind. The automatic regulator N is shown in position, and a detail of it is shown by Fig. 403. It controls the supply of vaporized oil from the vaporizer to the burner, and therefore the intensity of the burner flame. At the base of the regulator there is a chamber, N_1, containing a plunger packed with a hydraulic cup leather. It is shown in its lowest position and is down on a stop piece at the centre of the cover to the bottom of the chamber N_1. A steam pipe from the boiler is connected to the union stem on the bottom cover and the condensed steam or water in the chamber N_1 is at the same pressure as that obtaining in the boiler. The water pressure on the plunger is opposed by the effort of the spring N_2, under compression between the top and bottom plates. The lower plate has a ball-formed end and rests in a cupped seat provided for it in the upper face of the plunger. At the top of the upper plate there is a pivot piece N_5, forming a fulcrum for the unequally armed lever N_3. The short arm of this lever is connected to the stem N_4, and the long arm, as may be seen from Fig. 401, to the lever operating the

Paraffin fuel

Automatic burner regulator

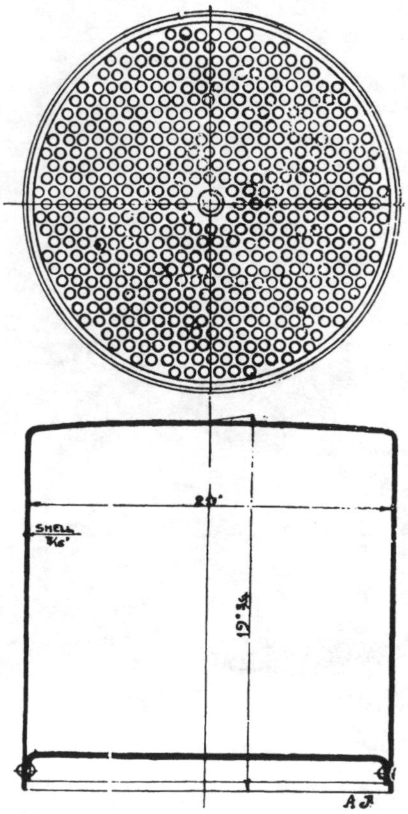

FIG. 400.—CLARKSON STEAM CAR. PLAN AND TRANSVERSE SECTION OF BOILER.

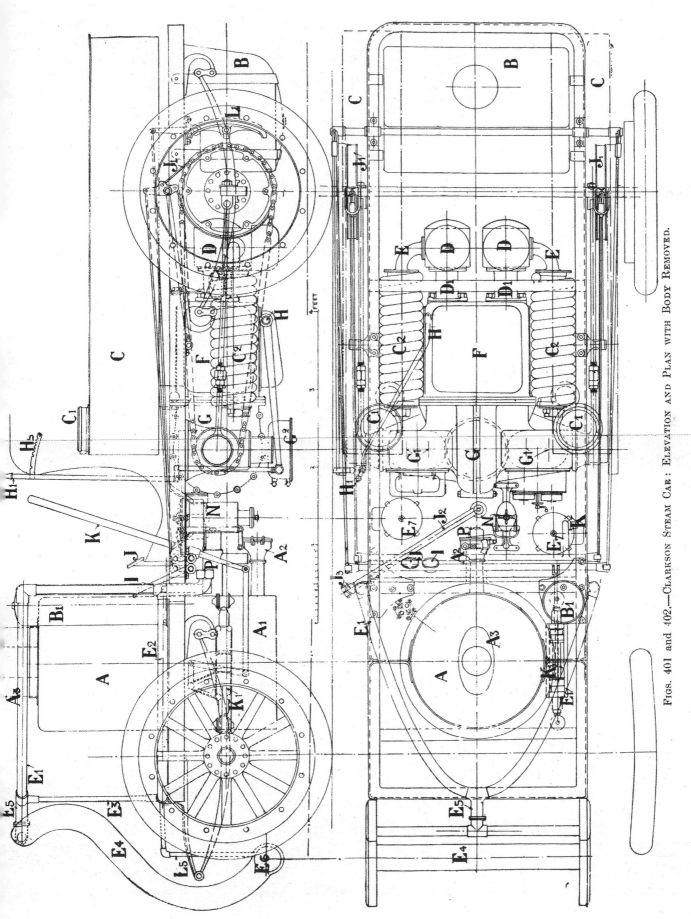

Figs. 401 and 402.—Clarkson Steam Car: Elevation and Plan with Body Removed.

To face p. 476]

The material originally positioned here is too large for reproduction in this reissue. A PDF can be downloaded from the web address given on page iv of this book, by clicking on 'Resources Available'.

gas valve inside A_2. The rise and fall of steam pressure corresponding with fluctuations in the demand for steam will result in slight vertical movement of the stem N_4 and the plunger to which it is attached, either

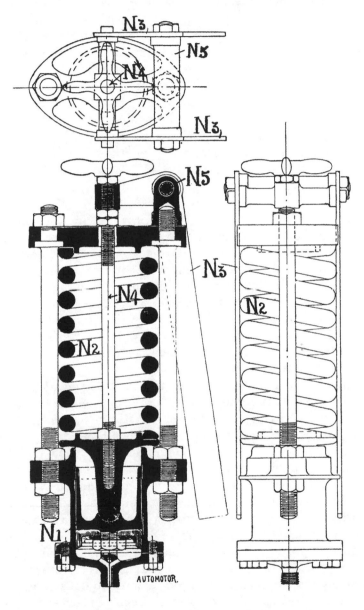

FIG. 403.—CLARKSON STEAM CAR.—ELEVATION PLAN AND SECTION OF AUTOMATIC BURNER REGULATOR.

up or down, in accordance with increase or decrease of steam pressure, and there will be corresponding movements of greater magnitude of the gas valve, giving a variable supply of gas to the burner. The range of pressure through which the regulator operates may be adjusted by means of the double nuts on the lower ends of the side pillars which maintain the spacing of the chamber N_1 and the top plate, and adjustment of the opening of the gas valve may be similarly made by means of the nuts upon the upper end of the stem N_4.

Means of control of fuel and water

With the fire-tube or multitubular boiler the water level is thus automatically maintained by the thermal expansion device, and the fire is controlled by the steam-pressure operated regulator.

With the semi-flash type of boiler or generator, which is now being adopted by Messrs. Clarkson, a regulator of the form described, arranged to open or close a bye-pass valve of slide-valve form, is used to control the delivery of water to the generator, and the supply of vaporized oil to the burner is dependent upon the opening of the engine-starting valve or throttle valve. Thus if the throttle valve is nearly closed, then the burner flame is reduced, or if the throttle valve is opened, then the supply of oil vapour to the burner flame is increased.

An excentric or cam movement is provided in the connecting links between the steam throttle valve and the burner gas or vapour valve, in order that the fire may be active, during the process of getting up steam or for a short period before starting to heat up the generator. During ordinary periods of running, the greater opening of the vapour valve obtainable is not required, but the degree of opening for a given position of steam throttle valve lever may be varied by the driver to meet any special conditions of the route the car is run over, and a pointer and indicator scale enable him to judge the amount of adjustment required or to return to the original setting. A Crompton electric pyrometer arranged to indicate the maximum steam temperature in the lowest generator coil, that exposed to the greatest fire temperature, is in full view of the driver, and its indications may be easily read by him. With steam pressure of from 200

Steam temperature

to 250 lb. per square inch, the mean steam temperature obtained is 800 degrees F.; and if the pyrometer should indicate much variation of this temperature, the necessary adjustment to correct it may be made.

The semi-flash boiler is rectangular in form and horizontal grids or coils of tubing are connected together one above the other. The formation of each grid, or the coiling of the pipe, is like that used in the more recent Serpollet steam-car boilers, and the end of each coil before connexion to a lower one is upturned and at a higher level than the coil. This arrangement is adopted for the same reasons that led to its use with the White boiler, namely, to ensure the retention of water by the upper coils and to prevent the too free passage of water through the generator, with incomplete evaporation. The upper coils, those that to a variable level contain water, and about half of the number, are $\frac{1}{2}$ in. in diameter, the others or the lower steam coils being $\frac{3}{4}$ in. in diameter, and the thickness of metal $\frac{1}{8}$ in. Circular coils are now used instead of rectangular grids.

Capacity of tanks

The paraffin fuel and water tanks are of such sizes that the cars can run an average distance of 100 miles on one filling, and the capacity of the tanks for the omnibuses are 35 and 40 gallons respectively. The cost of the paraffin oil fuel is found generally to be less than one penny per car mile.

With the arrangement of chassis shown by Figs. 401 and 402 there are longitudinal water tanks C beneath the side seats and extending forward to the front seat, below which are the covers C_1 over the filling holes. The paraffin fuel tank B is hung from the frame at the rear of the vehicle and oil is pumped from it to a container B_1 as required by an engine-driven pump to be referred to later. The upper portion of B_1 is used as an air vessel, and now and then requires pumping up by a hand or tyre pump. An air pressure of between 30 and 40 lb. is found to be sufficient to supply the oil to the burner vaporizer, with the method adopted for regulating the flow. A starting lamp P is used to initially heat the vaporizer, and provision is made whereby a definite quantity of oil may be allowed to flow into the starting lamp without necessitating the handling of the oil, or the use of spirit or petrol for the preliminary heating. The oil in the starting lamp having been ignited, a hand-worked fan, not shown in the drawings, is used for a minute or so to produce a hot Bunsen flame from the starting lamp. The process of heating up the vaporizer is conveniently and quickly done. Auxiliary oil and water pumps are required to enable water to be pumped by hand into the boiler, and oil into the container B_1 when the car is standing for any length of time. The pumps K_1 are worked by the hand lever K through the system of links shown ; the oil and water pump barrels are opposite each other and their plungers are connected together and worked from a common rod.

Figs. 404, 405 and 406 show respectively longitudinal sections through the engine, the differential gear and chain-pinion shafts and their transverse casing, and a transverse section of the engine looking towards the crankshaft end, with the parts shown in Fig. 406 removed.

Engine and transmission gear

By reference to Figs. 401 and 402 it will be seen that the engine crank chamber F and the casing G containing the differential gear and chain pinion shafts are connected together, and these parts, containing practically the whole of the active running gear, are completely encased and are protected from mud and dust. The system of lubrication, originated by Mr. T. Clarkson, is very complete, and will be referred to again.

The combined engine and gear casings are hung from the main frame by brackets cast with the transverse casing and near the ends close to the chain pinions, and by hanger straps just behind the rear end of the crank chamber, and attached to the transverse frame member on which are the letters D_1 in the plan, Fig. 402. The two cylinders D are situated at the rear of the crank chamber F and are connected thereto by steel castings D_1. The diameter of the cylinders and the stroke is four inches, and the engine is double-acting with cranks at 90°, and is described as of 12 nominal horsepower. With the steam pressure available, and with the Joy variable expansion and reversing gear, with which the engine is fitted, it could develop the 12-HP. at 600 revolutions per minute, with a mean effective

479

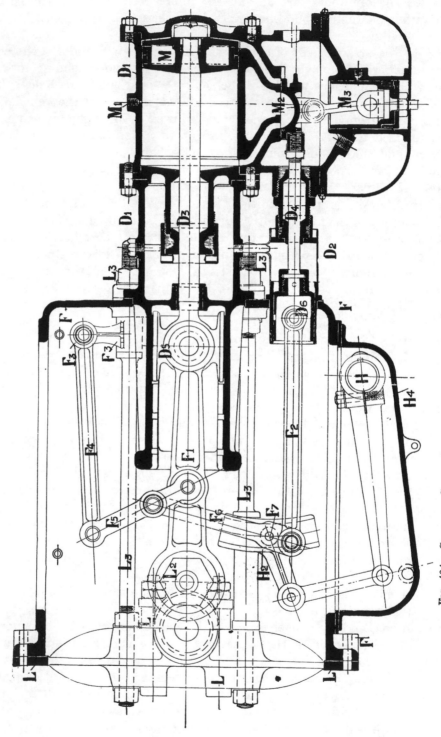

Fig. 404.—Clarkson Steam Car.—Longitudinal Section through One Line of Engine.

cylinder pressure of only about 44 lb. per square inch, and could, therefore, when required, develop higher powers for short periods.

The longitudinal section through one line of the engine (Fig. 404) shows its general design. With the type of piston there shown, a pair of wide adjacent packing rings are used, and as a separate cover plate is employed, the rings do not require to be sprung into position, and may be of stronger form. Stuffing boxes suitable for soft packing material are **Metallic** shown, but a simple form of floating metallic packing is now used for both **packing**

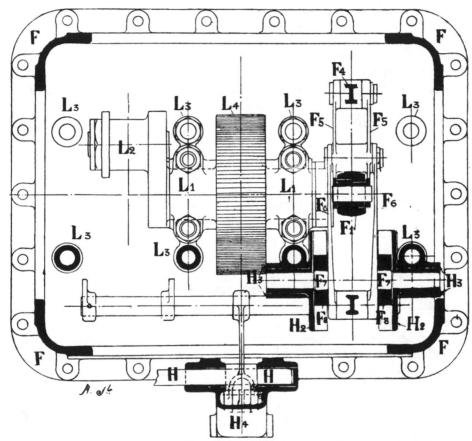

FIG. 405 —CLARKSON STEAM CAR.—TRANSVERSE SECTION ACROSS CRANK CHAMBER LOOKING
TOWARDS CRANKSHAFT.

piston and valve spindle glands. The piston rods and valve spindles, D_3 and D_4, are accurately ground to a precise size, and they work through closely fitting solid packing bushes.

The trunks or stool castings D_1 are cast with the bored cross-head guides, and they and the crank chamber and the bearing frames L are all held together by the frame bolts L_3. There are eight of these bolts and their position may be seen in Fig. 405.

The slide valves M_2, and the driving bridles to which the valve spindles are attached, are circular in plan. This form has been adopted in order to avoid trouble from scoring or grooving of the sliding faces and because of simplicity of manufacture. The valves are said to work round in their bridles and take up fresh working positions. The balancing or pressure relieving piston M_3 was found to be unnecessary, and in recent engines, piston valves have been substituted for slide valves.

Continuous cylinder lubrication

Oil continuously supplied by a mechanically driven lubricator enters the cylinders through bosses M_1 at the top of the barrel, and from this highest point in the engine will distribute itself over the cylinder and slide valve surfaces.

The Joy valve gear has proved very satisfactory, and as in its other applications economizes space and weight ; and it may be seen from Fig. 405 that the engine centres are very close.

Valve gear

The radius bar F_4 is pivoted at F_3, the fulcrum pin being carried in blocks F_3 on the upper bolts L_3, and the outer swinging end of F_4 is connected to a double link F_5, whose opposite ends are driven by the engine connecting rod from the point of attachment near F_1. A pair of links F_6 arranged outside, and at their upper ends connected to F_5, at the position of the boss to be seen in Fig. 405 and about one-third of the length of F_5 from the lower end, have outwardly projecting stud pins F_7 towards their lower ends, and at their lower ends transmit the movement they receive to the valve spindle connecting rod F_2. Fixed to the lower tie bolts L_3 are pivot or fulcrum brackets H_3 in which are journalled pairs of curved guide blocks H_2, their pivot pins being as shown, forged or cast solid with the guide blocks, and lightened at their centres. Sliding pieces F_8, a working fit on the stud pins F_7 are accurately fitted to the curvature of the guide blocks H_2, and from the nature of the movement given to the links it will be understood that when the engine is in motion the pieces F_8 receive reciprocating motion in the blocks H_2.

In Fig. 404 it may be seen how through short lever arms, which are part of the curved guide blocks, angular motion through a limited range may be given to the latter by connexion with the link and lever upon the lay shaft H, which is operated by the reversing lever and its intermediate connexions.

In Figs. 401 and 402 H_1 is the reversing lever, carrying with it the notched quadrant H_5. By varying the angularity of the guide blocks the amount of movement given to F_2, and therefore the slide valve, may be varied to suit different points of cut off for forward or backward running, and through this range the lead, or port opening to steam, at the commencement of the stroke, is constant.

The crankshaft journals L_1 and the crank-pins L_2 are large in diameter, and there is ample bearing surface. The steel driving pinion L_4 is well supported between the bearings L_1, and gears into the phosphor bronze wheel G_4, forming part of the casing around the differential gear, Fig. 406. The latter is of the spur-wheel type and the centre wheels are solid with the divided driving-shaft having the chain pinions G_6 at the outer ends.

The four double-row ball bearings shown have been replaced by plain

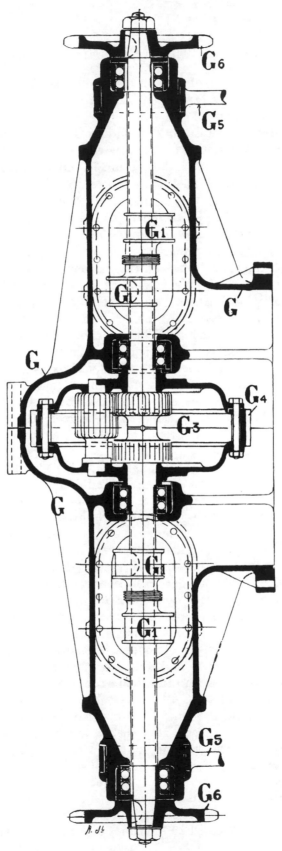

FIG. 406— CLARKSON STEAM CAR.—LONGITUDINAL SECTION THROUGH DIFFERENTIAL GEAR AND CASING.

Bearings

bearings, the makers being of opinion that well made and well proportioned plain bearings, efficiently lubricated, have a longer life and are more reliable than ball bearings.

On the centre-divided driving-shaft there are four excentrics, G_1, which drive four pumps connected to hanging frames bolted to facings on the lower side of the casing G and indicated in Fig. 406 in dotted lines. Covers, G_1, are shown in the plan (Fig. 402) over holes in the casing at the positions of the excentric groups. A section through one of the pumps is shown

Pumps

by Fig. 407, from which the attachment of the pump barrel to the hanging frame will be understood, and the simple method adopted for holding the suction and delivery branch pieces to the centre box. The suction valve seat is cast with the centre box, and the delivery valve seat is a removable piece with a stop on its under side to limit the lift of the inlet valve. The bridle and setscrew by means of which the four pieces are jointed together is shown, but the setscrew has been omitted. The valve faces are flat, and they work metal to metal. This form of valve with a lift of about $\frac{1}{32}$ in. has been adopted as the result of experience with various types, and one form of interchangeable pump and valve box is now used.

Of the four pumps one delivers paraffin oil to the container B_1, from which the burner receives its supply, the second is the boiler feed-water pump, the third removes the condensed steam from the bottom condenser vessel E_6, Fig. 401, and returns it to the feed-water tank, and the fourth is a lubricating oil pump.

The eye-formed ends of the radius rods, G_5, take their bearing on bushes fitted over the machined ends of the aluminium casing G. Their other ends are not concentric with the rear axle, but, as may be seen from Fig. 401, are pivoted in short jaw pieces bolted through it.

The exhaust steam leaves the engine by the bends E, Fig. 402, passes through feed-water heaters, C_2, and thence to the upper tubes, E_1, of the first condenser. In plan this condenser is V-shaped, or pointed, and a large number of vertical tubes, similar to E_3, connect the main tubes E_1 and E_2. The steam condensed in the V-condenser drains by the pipe L_5 to the drum E_6 at the bottom of the S-shaped front condenser E_4. A junction pipe E_5 connects the two condensers at the top and the steam entering the pipes E_1, and not condensed in the first condenser passes through E_5 to the front condenser. The side frames or headers of the front condenser are connected by horizontal tubes, and the steam condensed in them drains to the

Clarkson radiator tubes

common collector vessel E_6. All those tubes which form condensing or radiating surface are wire wound, the wire before application to the tube being spirally wound or coiled like a spring, and then wound or coiled on the condenser tubes. The efficiency of this wire radiating surface is higher than would be expected, and it has been in use for the condenser tubes in the Clarkson cars for some years. The condenser tubes are 0·42 inch internal diameter, and the diameter of the wire coils is 0·32 inch, the wire being about $\frac{1}{32}$ in. diameter, or 22 gauge. With these dimensions the radiating surface is not less than six times the internal surface of the tube.

The external appearance of the condensers may be judged from Fig. 399, whilst the single flat condenser used with the omnibus shown by Fig. 408 may be seen carried at the front of the frame. The tubular surface is about 11·5 square feet.

The condensed steam is removed from E_6 by one of the excentric driven pumps already referred to and is returned to the water tanks through filter boxes E_7, and a third filter or sponge box below C_1 in the

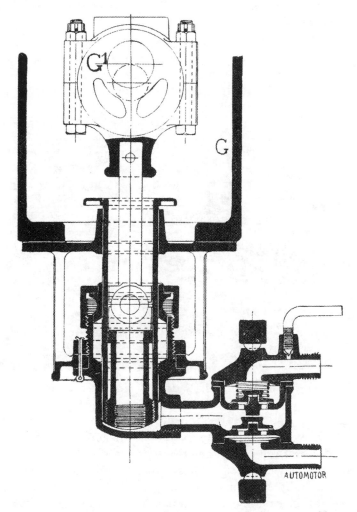

FIG. 407.—CLARKSON STEAM CAR.—SECTION THROUGH PUMP AND VALVE BOXES.

right-hand tank. These filters are only required when the fire-tube type of boiler is fitted to the vehicle, and they have not therefore been used with recent types.

The feed-water heaters C_2 are wire wound in the same way that the condenser tubes are covered, though as indicated with spiral coils of larger diameter. Some of the waste heat in the steam may be utilized, and the external wire-wound surface assists the condensation of the steam, but their use has been discontinued, the reduction of weight effected being of greater importance than the value of the possible gain in fuel economy.

Continuous lubrication system

Referring to Fig. 406, one of the spirals or worms between the pairs of excentrics, G_1, gears with a worm-wheel, operating the forced feed mechanical cylinder lubricator, thus ensuring a constant feed of oil to the cylinders in known quantity. The second worm drives a distributer which controls the supply of oil to all the running parts of the engine and gearing contained in the casing. A copious supply of oil is delivered by one of the pumps worked by the excentrics G_1 to the distributer box, and as the distributer valve is rotated ports are opened, one by one, communicating with pipes through which the stream of oil is directed to each bearing or part of the motion likely to benefit by liberal lubrication. By these means frequent and positive delivery to all the active bearings is ensured as long as the

FIG. 408.—CLARKSON STEAM CAR.—SINGLE DECK OMNIBUS.

engine is running, the necessity for filling lubricators or adjusting sight or drip feeds is avoided, and the system is automatic and thorough, and relieves the driver of the care otherwise demanded.

The omnibus shown by Fig. 408 is one of several that have been running over the route indicated for about a year. They seat sixteen persons, and their weight, in running order, but without fuel and water, is 51 cwt. The fire-tube boiler is used for this type and the engine and transmission gearing are of the form already described. The total gear reduction is $5\frac{3}{4}$ to 1, the ratio of diameters of the side-chain pinions and sprockets being **Brakes** $2\frac{3}{4}$ to 1. Two forms of brakes are shown by Figs. 401 and 402, acting upon the rear road wheels. The inner strap brakes act on drums, J_1, which are part of the chain-wheels or sprocket rings, and brake blocks, one of which is shown near L_1, engage the internal surface of the separate brake rings of large diameter. Both sets of brakes are shown operated through equalizing

486

gear pedals I and J. More recently, and for the heavy omnibuses, duplicate sets of brakes similar to J_1 are used, acting on surfaces at each side of the chain rings. The brakes are all arranged to work directly on the back wheels and the transmission gearing is relieved from braking stresses. The steering lever J_2, connected to the vertical column J_3, provides satisfactory control over the steering gear, but it is not so much used now as the steering-wheel and column visible in Fig. 408. From inspection of this view it will be recognized that the vertical oval funnel lettered A_3 in Figs. 401 and 402 is hardly noticeable, and is not in any way unsightly. The discharge of the products of combustion at the higher level is to be commended, and Messrs. Clarkson have set an example which should be followed by those designing heavy passenger vehicles.

FIG. 409.—LAMPLOUGH ALBANY STEAM CAR.—ELEVATION

The arrangement of the engine, differential gear, chain pinion shafts, and pumps, together with the automatic lubrication system, is excellent. The car entered for the 1,000 miles trials held by the Automobile Club during September 1903, was provided with this design of running gear, and a Gold Medal was awarded for the good features of the design and for the excellence of workmanship.

THE LAMPLOUGH-ALBANY STEAM CAR

The Albany Manufacturing Company are among the few firms who continue to construct steam cars and who have made endeavours to utilize paraffin in preference to the lighter hydrocarbons.

A car of their manufacture, and as introduced by them in 1903, is shown by Fig. 409. In its construction a particular type of steam generator, engine, pressure-controlled feed pump, and other details of Mr. F. Lamplough's design, are embodied, and some of these are illustrated.

The external appearance of the car is distinctive, and its arrangement generally is convenient. There is little to indicate that the car is propelled by steam, and as in others no uptake from the steam generator is used to either prevent down draughts or to ensure that the products of combustion are completely dispersed, and do not become disagreeable to the occupants of the car or other road users. The steam generator is below the rectangular bonnet at the front of the car and an external view of the assembled parts without the protecting sheet-steel casing or the burner in position is given by Fig. 410.

Steam generator

FIG. 410.—LAMPLOUGH-ALBANY STEAM CAR.—ELEVATION OF BOILER, BURNER AND CASING REMOVED.

Referring to the sectional view (Fig. 411), the water inlet is at J, and it is pumped through the pipe D and the passage way shown in the centre block to the spraying plate or cup disc screwed upon and into position with the pipe I. The pipe I, it will be seen, is also screwed into the centre block and communicates by a passage therein with the pipe C. The water entering at the spraying plate becomes vaporized by contact with the interior surface of H, the lower part of H being exposed, like the drying and superheating coils G to the greatest heat from the burner flame. The steam leaving H passes by the pipe I and by C to the junction ring F. By reference to the plan (Fig. 412) the relative positions of the inlet branch J,

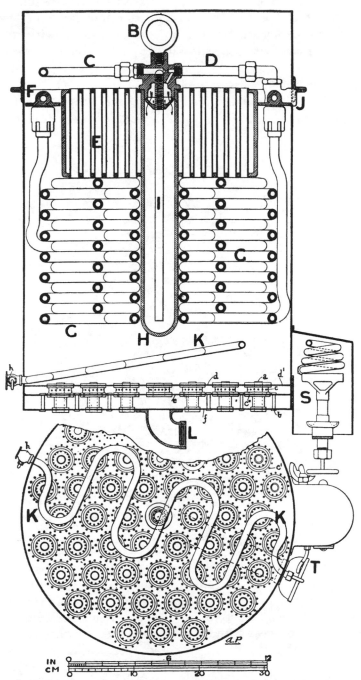

FIG. 411.—LAMPLOUGH-ALBANY STEAM CAR.—SECTIONAL VIEW OF BOILER AND BURNER, AND PLAN OF BURNER.

and of the pipe C, may be noted, and C, after leaving the centre block, is curved round and returned to its junction with the ring F at C.

A separate detail of the junction ring F, and the peculiar form given to the pipe sections, is shown by Fig. 413. From the sectional view, taken across the centre of the ring and through the inlet bend J, it will be plain that the latter is cast with, but has no communication with, F, and that it is attached to or cast with F because it happens to be a convenient point of attachment. The course of the U passage ways, connecting pairs of downwardly pointing union bosses, is indicated by dotted lines. The union bosses are all of them externally threaded, like those shown near F, and the inlet and return ends of the pipe sections are suitably formed for the

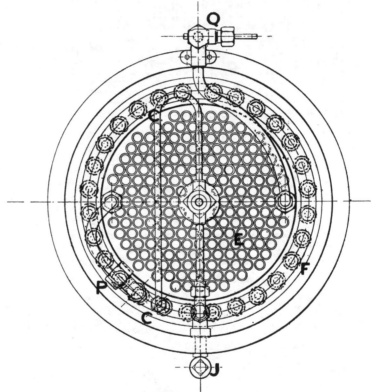

FIG. 412.—LAMPLOUGH-ALBANY STEAM CAR.—PLAN OF BOILER.

metal to metal joints with the union bosses. From the drawing of one of the pipe sections (Fig. 413) and by inspection of the external view (Fig. 410) it will be seen that the inlet and return ends of the pipe sections are adjacent, and are connected by heavy union nuts upon them to the series of U bends cast in F. From this it will be gathered that the coils are all in series or form one continuous length. The inlet to the first or upper section is immediately below C (Fig. 412), whilst the return end of the fourteenth or bottom section is below the connecting bend P, so that the steam has flowed through all the coils and the passage ways in F in a contra clockwise direction, and the inlet and final outlet openings are alongside each other, so that at c and p there are straight through holes, and not u passages, communicating with the starting and finishing ends of the whole pipe length.

Referring to Figs. 411 and 412, steam leaving the superheating coils enters the receiver E by the pipe P, and leaves it by the pipe at the opposite side leading to the stop valve Q, and so to the engine steam pipe.

The heat interchange, during the generation of steam, is a peculiar and indirect method of heat utilization. The steam which leaves the centre vessel H probably carries with it entrained water, due to the contact of the cold feed water, with the highly heated interior surface of H. This wet steam, during its passage through the drying or superheating coils, becomes highly superheated, and the final pipe connecting the lower coil

Heat interchange

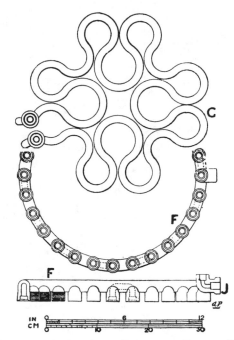

FIG. 413.—LAMPLOUGH-ALBANY STEAM CAR.—DETAILS OF BOILER COIL AND JUNCTION RING.

to the steam receiver E is said by the makers to be at a dull red heat when the steam generator is working properly. The steam, on entering E, may become still further heated by longer contact with the external surface of the fire tubes in E, but as the temperature from the burner would be lower than that experienced by the steam during its passage through the lower superheating coils, the extent of this further heating would be small, so that the purpose of the fire tubes is little more than to provide flue ways for the escape of the hot products of combustion.

The temperature of the steam, as it enters E, is stated to be 1,000° to 1,100° F., and as it leaves the receiver E about 700 to 800° F. The steam, as it passes through the receiver, parts with some of its excessive superheat to the vessel H, upon the interior surface of which the feed water is being

Steam temperature

491

sprayed. The steam, during its process of generation and passage through the generator, is formed as wet or saturated steam in H, becomes highly superheated in the coils G, and has its superheat reduced whilst in the receiver E, so that the steam has a sufficiently low temperature to enable it to be used in the engine without likelihood of damage to the latter by scored surfaces.

Petroleum burners

The forms taken by the petroleum burner are illustrated by Figs. 414 to 417, and a sectional view through the burner drum and a plan view are given by Fig. 411. An external starting burner s is used of the form shown externally by Figs. 414 and 415. This is of the usual construction, and

FIG. 414.—THE ALBANY PETROLEUM BURNER WITH EXTERNAL PILOT LIGHT.

a small quantity of petrol, methylated spirit, or other highly inflammable liquid is used to sufficiently heat the secondary vaporizer coil to enable it to supply vaporized oil to the burner long enough to heat up the main vaporizer coil K, which is above the burner and always in the fire. The oil enters K at *h* and leaves it, vaporized, by the small pipe leading to the air-inlet orifice T. T, it will be noticed from the sectional view, is continued as induction pipe to its connexion with the inlet bend L. The burner drum is double or is divided into two chambers by the partition *e*. The vaporized oil mingles with the air entering with it at L, and occupies the lower chamber between the floors *e* and *f*, passing through each burner block to the upper burner plate by the inner ring of small holes, lettered

492

c_{11} in the sectional view, and visible as the inner concentric ring of holes around the centre air-hole in each burner. Additional air required for combustion may pass through these centre holes, and more finds its way up the small hollow spacing stays b to the upper chamber between the floors d and e, and from this chamber directly to the upper burner plate through the outer concentric ring of small holes around each burner, lettered in the plan view c_1, or by the radial holes c to the centre air passage. The declared object of this design of burner is the imitation, as nearly as possible

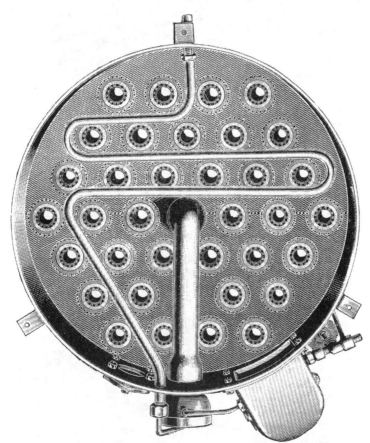

FIG. 415.—THE ALBANY PETROLEUM BURNER WITH EXTERNAL PILOT LIGHT.—PLAN.

with liquid fuel, of the conditions of uniform fire temperature over the grate to be found in coal-fired steam boilers. The arrangement of burner shown by Figs. 414 and 415 is in most respects the same as that already described, but it will be seen that the vaporizer pipe takes a different form to pass by the air and vapour-inlet pipe arranged in this burner above the grate or burner plate.

Figs. 416 and 417 show another modification of the same burner, arranged with an inside starting burner or pilot light. A diaphragm con-

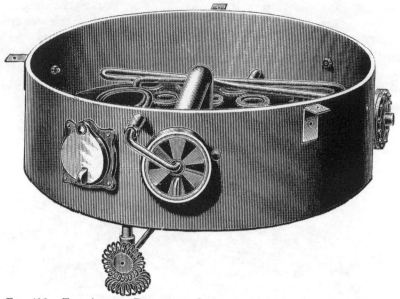

Fig. 416.—The Albany Petroleum Burner with Internal Pilot Light.

trolled valve on the oil inlet end of the main and pilot vaporizer pipes cuts out the supply of oil to the main burner when there is no demand

Fig. 417.—The Albany Petroleum Burner with Internal Pilot Light.—Plan.

for steam from the generator, but leaves the pilot burner unaffected. The air and vapour induction pipe is also above the burner in this design, but the makers prefer to arrange it underneath as shown by Fig. 411, when there is sufficient room. The oil supply to the burner is arranged to be controlled by the steam-pressure diaphragm regulator shown by Fig. 417, or alternatively by a thermostat.

Regulation
of oil
supply

The steam generator feed pump, used by the Albany Company, and constructed by them for various purposes, is illustrated by Figs. 418 and 419. The plunger is actuated by motion received from the engine and through the connecting rod, which has a ball bearing at its big end. At the back of the pump the diaphragm regulator casing is connected by flange with the suction box, and the spindle, which receives motion from the diaphragm or from the heavy coil spring, provides a fulcrum point about which the rocking lever in the suction box may be oscillated. Motion is given to the rocking lever, by connection of the system of links shown, with its upper end. According to the position of the fulcrum point, which is varied by the diaphragm spindle, the lower end of the rocking lever as it oscillates may lift the bye-pass spring-loaded valve, at the lower part of the suction box, to a greater or less extent, or a longer or shorter fraction of the time of a delivery stroke, and so allow the whole, or a part, or none, of the water to be delivered to the steam generator. In Fig. 418 the delivery valve is not visible, but the screw cap above the delivery valve may be seen in Fig. 419, at the same side of the pump barrel and suction box as the auxiliary hand pump barrel shown. The latter is required to introduce some water into the steam generator at the commencement of a run, before steam is up, or to augment the power pump-water supply. The same suction and the same delivery valve is used by the power or the hand-worked pump. There is a lever and an adjustment screw, to be seen in both Figs., for the purpose of adjusting the compression of the coil spring in the water regulator, and the range

Construc-
tion of feed
pump

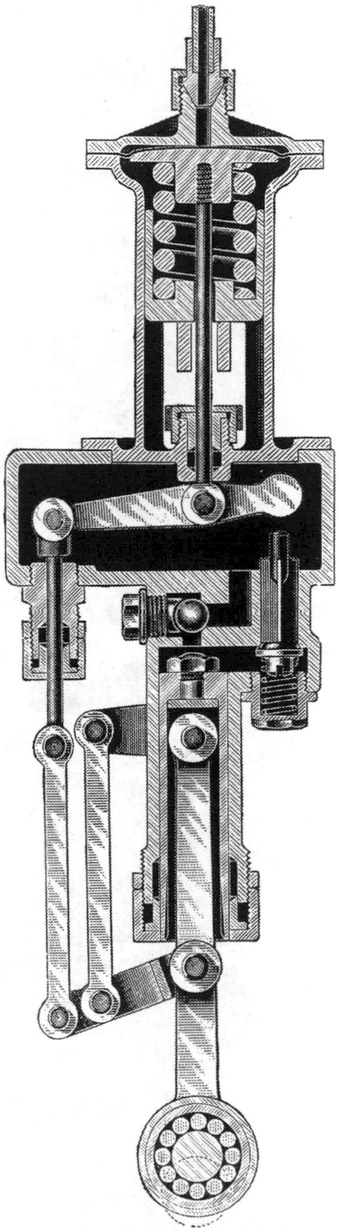

FIG. 418.—THE LAMPLOUGH-
ALBANY PUMP WITH DIAPHRAGM
REGULATOR.
LONGITUDINAL SECTION.

of steam pressures through which there is regulation of water supply. An engine of peculiar construction is employed and it is fitted with the Lamplough valve gear. A feature of this gear in its application to

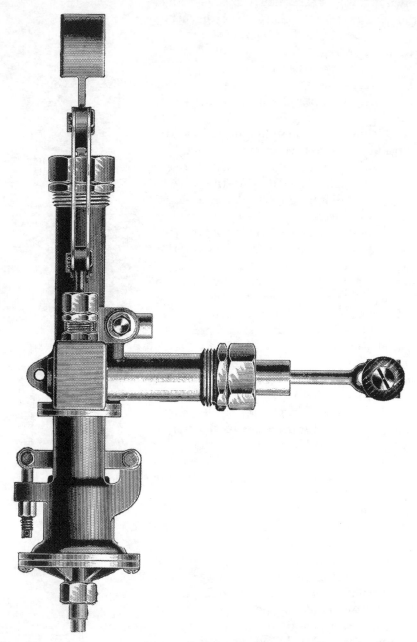

FIG. 419.—THE LAMPLOUGH-ALBANY PUMP, WITH DIAPHRAGM REGULATOR.—PLAN.

engines, having two cranks at 90 degrees, is the derivation of the motion required for lap and lead of each valve from the engine crosshead situated at the same side as the valve or belonging to the same line of the engine, whilst the motion for port opening and cut off is derived from the other engine crosshead.

The drive from engine to live rear axle is by bevel gear and flexible or universally jointed driving shaft. The same form of drive is retained by the Albany Company for use with the petrol cars now constructed by them.

THE MIESSE, CHABOCHE, AND SERPOLLET CARS.

The first of these cars, all of which are constructed to use paraffin as fuel, is of Belgian origin, and is known in this country as the Turner-Miesse. It is shown by Fig. 420 as constructed in 1903, and, with a longer wheel base and minor structural alterations, has a very similar appearance now.

The boiler of the flash type is supported on transverse frame members below the bonnet, and horizontal pipe coils of rectangular gridiron form are employed. The water inlet is at the bottom coil and the steam outlet

FIG. 420.—THE MIESSE STEAM CAR, 1903 TYPE.

at an intermediate point. The water having passed upwards through the lower grids, enters the upper group, and finally returns to the intermediate series of coils, at which stage evaporation is complete, and a variable degree of superheat has been given to the steam. One of the principal requirements in the design of the flash type of steam generator, namely, the provision of an interrupted or indirect course for the water and steam flow, has been obtained in this design in such a way that the coils immediately over the fire receive the cold feed water and the liability to overheat is reduced and the life of the lower coils prolonged.

The solid drawn steel tubing used in the construction of the generator is welded at the joints occurring in the coils, but the connecting pipes for directing the flow from the lower to the upper and to the intermediate groups of coils are not exposed to the fire, and are jointed with heavy unions. The tubing is connected so as to form one continuous length, and the in-

Steam generator

ternal diameter of the first part, that which generally contains water, is $\frac{5}{16}$ in., and that of the second part, forming the steam coils, $\frac{7}{16}$ in. diameter. The normal steam pressure is 150 lb. per square inch. The burner is of

Arrangement of burner

simple construction and consists of vaporizer, vapour jet, air and vapour combining and inlet pipe, receiving chamber, and burner distributing pipes. The vaporizer takes the form of two spirals, the tubing being $\frac{5}{16}$ in. internal diameter and $\frac{9}{16}$ in. external diameter, arranged one at each side of the fire-box. The coils are connected together, and the oil enters at one end and passes out through the vapour jet at the other end after vaporization. The vaporized oil issues from the jet at high velocity and induces a flow of air with it along the induction pipe to the receiving chamber. The distributing pipes, about nine in number, are arranged parallel to each other and in the same horizontal plane, and enter the tube plate of the receiving chamber. At their other end the distributing pipes are closed, and the combustible mixture issues from a large number of holes, about 1200, of small diameter, drilled in double rows at an angle of 45 degrees with each other along the length of the pipes. The burner flame is thus very uniformly divided up over the fire-box area, and the additional air required for complete combustion passes up between the distributing pipes. With the exception of the vaporizer coils, the tubes used for the burner are straight, and the burner may be easily taken down from the boiler for examination. The oil flow to the burner is continuous and under a pressure of 60 lb. per square inch, that being the air pressure automatically maintained in the fuel. tank. Stop valves are provided in the oil supply pipes from the fuel tank to the burner, one controlling the supply to the burner when the car is running and another for limiting the supply when the car is standing and when the small or pilot light only is required. For the initial heating up of the vaporizer a plumber's blow-lamp is employed, and a small tank is carried on the car and contains enough petrol or methylated spirit for starting purposes to last some time.

Three-cylinder engine

A three-cylinder single-acting engine of 10-HP., with cranks at 120 degrees, is used, having mitre or mushroom steam and exhaust valves of the type now almost universally used with petrol engines. The camshafts operating these valves are arranged to slide longitudinally, such movement being required for varying the periods of opening and closing of the valves for variable points of cut off and for reversing. The cam profiles are so formed that longitudinal movement of the spindles brings parts of the cams of different formation below the valve tappets. The arrangement is in principle that devised and used by M. Serpollet for many years in his steam engines, and is illustrated in vol. i., p. 451. A pinion on the extended part of the crankshaft gears into a spur-wheel on the differential, through which the countershaft, or divided shaft, having chain pinions at the outer ends, is driven. The gear reduction from engine to differential spur-wheel is 2 to 1 and the same ratio is adopted for the side-chain pinions, and the sprockets on the rear road wheels.

Pumps

The car is equipped with three pumps, two of which are driven by the engine. One of them, the feed-water pump, is actuated by an excentric on the countershaft and the quantity of water delivered from it to the generator is controlled by a hand-worked bye-pass lever conveniently arranged on the steering-gear column. A pump of similar dimensions and construction is arranged to be worked by a hand lever, in order that water may be pumped into the boiler when starting. The second engine-worked pump is driven from a crank arm fixed on the crankshaft at the end away from the driving gear-wheels. It is an air pump, and its duty is to maintain a pressure of 60 lb. per square inch in the oil tank. A relief or safety valve is arranged to blow at pressures in excess of that required, and the clearance space at the pump chamber is such that the pump will not compress air to a pressure much above 60 lb. per square inch. A hand pump is used to obtain air pressure in the fuel tank after refilling.

Splash lubrication

The engine being single-acting and the pistons of the trunk type, the cylinder surfaces, are lubricated by splash from the running parts in the crank chamber.

No attempt is made to automatically control the fuel and water supplies and the driver depends upon the indications of the pressure gauge. The engine is similar in design to the petrol engine as regards the form of piston and valves employed, and it can doubtless withstand the large variation of steam temperature to which it is exposed. Similarly the steam generator must be frequently exposed to high temperatures, but those coils which experience the maximum burner flame temperature contain the water equivalent to the steam demand from moment to moment, and would, therefore, as already stated, generally be protected from excessive overheating. The conditions under which the car runs may be very variable, and if good or economical running is desired then close attention is required by the driver. The cost of fuel per car mile is said to be about $\frac{1}{2}d$. Although the cost of the fuel is low, economy in its use should, for various reasons, be studied.

The Chaboche steam car is one of the few in which the control of the oil fuel and water supplies was as far as possible automatic and the method of control was similar in principle to that adopted with the White cars. That is to say, the intensity of the burner flame was governed by steam temperature, and the delivery of water to the steam generator was dependent upon steam pressure. The design and arrangement of the automatic regulators and their connexions was very complete, but in some respects unnecessary complication resulted from the use of accumulator vessels for the feed-water and paraffin fuel and of the many connexions to them.

Methods of control

In recent cars the automatic control of the burners has been abandoned, and the supply of oil to the burner vaporizers, six in number, is hand-regulated. A pressure of about 15 lb. per square inch is maintained in the fuel tank by an air pump driven by the engine, and a relief or safety valve prevents any unnecessary increase of pressure. A steam temperature

Steam temperature indicator indicator, under the driver's observation, is arranged so that he can easily watch the rise and fall of steam temperature, and make the necessary readjustment of oil supply in accordance with the temperature indicated and the nature of the road he is travelling over or about to travel over.

The temperature indicator is actuated by expansion of the steam pipe between engine and boiler, just as in earlier cars expansion of the steam pipe actuated the automatic valve in the oil-supply pipe to the burners.

The automatic bye-pass valve for regulation of the feed water to the boiler is retained in conjunction with the accumulator vessel. When a steam pressure of about 350 lb. per square inch is reached the spring-loaded bye-pass valve opens and the feed-pump delivery is returned to the water tank.

The accumulator vessel into which the feed water is pumped contains air, and may be considered the equivalent of an air vessel on the pump delivery of large capacity in relation to the pump displacement. By its use the flow of water to the steam generator is more continuous, and the variation in the rate of delivery is not so abrupt as it would be if the pump delivered directly to the boiler.

With this arrangement of automatic regulation of water supply, combined with hand regulation of the fire, it would be necessary to watch the temperature indicator and to ensure that the steam pressure is maintained as long as the car is running and as long as the stop valve in the feed pipe remains open.

Cylinder lubrication The cylinder lubricating-oil vessel is connected by a pipe with the water accumulator, and the oil it contains is under boiler pressure. As the oil gradually passes to the engine cylinders, its place is taken by water, and the oil is positively fed, since there is always a difference of steam pressure between that in the boiler and that in the engine steam chests. This difference is very variable and may be at times great, and the rate of oil supply to the engine would be equally variable unless the driver observes and adjusts the rate of flow of the drops rising through the glass sight feed tubes on the dashboard. A mechanically driven force-feed lubricator would give a more uniform supply, more in accordance with the requirements of the engine and its rate of running.

Steam generator The boiler or steam generator is of the flash type, and with the exception of the lower section, which is of gridiron form, or like the elements of the recent Serpollet boilers, the coils are spirally wound. The boiler casing is rectangular and oblong in plan, and each of the five horizontal elements, made up of eight spirally wound coils parallel to each other, is arranged with the coils across the width or smaller dimensions of the casing. The union joints between the elements are outside the casing, and away from the fire. The water enters the bottom section and passes from it to the top coil element, and then downwards, leaving by the lowest of the five coil elements. The boiler is fixed transversely above the back of the car frame, and the water tank rests on the frame in front of the boiler.

SERPOLLET STEAM CARS

Six burners are fitted, and each has its vaporizer coil and is in form similar to the Longuemare burner. Arrangements are made to use only three, or the six, burners, according to the nature of the roads.

A double-acting double-cylinder engine of vertical or horizontal form is used to suit particular requirements of body work. It is made in two sizes for light and heavy cars, of 12 and 20-HP. respectively. With earlier cars speed reduction gear in the crank chamber was employed, and the engine could be run free or caused to drive the car through a high or a low speed set of gear wheels. In recent cars a clutch is used, and the engine can be disengaged, but the only speed reduction is through the bevel gear wheels at the rear axle. There is a long, universally jointed propeller shaft between the engine clutch and the bevel and differential gears, and the live axle is of the type in which the driving shafts are surrounded by the fixed tubular housing.

Transmission gear

With the Chaboche steam waggon, designed to carry a load of 2½ tons, the vertical engine used is of 30-HP., and the flash boiler is carried in front and is arranged to be coal or coke fired. Link reversing motion is fitted to the engines, but notches are not provided on the quadrants for running with variable points of cut off, and the throttle valve is used for controlling the speed or power of the engine.

The condenser is made up of gilled tubes like those used for petrol car radiators, and the feed-water tank is at a higher level. Some exhaust steam pressure is necessary to return the condensed steam to the tank, but this back pressure would be small and prevents the too free escape of steam from the condenser. The latter may be of ample capacity to deal with the average quantity of steam entering it, but occasionally there may be such an increase that the whole of it is not condensed. By slightly restricting the escape of this steam, more of it is condensed, and the time during which the exhaust steam is visible is reduced.

Condensed steam returned to tank

The Serpollet steam cars have undergone change in external appearance and in the arrangement and disposition of the mechanism, but as regards the design of the engine and steam generator little modification has been found necessary.

The combined pumps delivering fuel and water to the boiler in constant proportions are retained, but they are now separately driven by a steam cylinder, the arrangement of pumps constituting a horizontal direct acting set with fixed strokes and variable amount of delivery by varying the rate of working. Hitherto the amount of fuel and water pumped has been hand-controlled, but it is now governed automatically by a steam pressure operated regulator connected with a slide valve, the movement of which varies the supply of steam to the pump steam cylinder and so controls the speed at which the pumps work. The exhaust steam from this cylinder is not condensed, but is taken to the boiler flue and is utilized

Donkey pump

501

to assist the draught. Means are provided for operating the automatic regulator by hand, when an increased delivery from the pumps is anticipated to meet a suddenly increased demand for steam for hill climbing. When the increased demand for steam is gradual it would be automatically met, but when, as may often happen, a rapid increase is about to occur, then it is advantageous to prepare for the change. By interference by hand with the automatic regulator the activity of the burner is increased, and the steam pressure raised in readiness for the period of heavy working.

The use of the automatic regulator is thus intended to simplify the control of the car by relieving the driver of the attention which must otherwise be given to regulation of the pumps, and the comparative slowness of reply to the automatic regulator when the occasional sudden changes in steam demand occur may be overcome by use of the hand lever acting upon the regulator or opposing the steam load upon its piston.

In the 1905 type of car, to which the automatic regulator and independently driven pumps are fitted, the engine is fixed transversely to the forward end of the frame beneath the ordinary bonnet, and it drives through a universally jointed propeller shaft and bevel gearing to the rear live axle. The condenser or radiator is of the Loyal type, and is at the front of the car, with the engine bonnet between it and the dashboard arranged in the usual way. With these changes in the form of the transmission gear and in the method of control, and of the arrangement of parts in agreement with what has been found to be most convenient or desirable, much that was characteristic of the Serpollet cars, although largely remaining, is less readily recognized. The flash type of boiler has been adopted by several makers of steam cars, and the single-acting multi-cylinder engine with no packed glands and mushroom valves, a design of engine very suitable for use with superheated steam, has been used by others for both light and heavy steam vehicles.

Chapter XXXI

ENGINE DIMENSIONS, PISTON DISPLACEMENT, AND MEAN PRESSURE

THE growth of the motor vehicle industry has naturally resulted in the use, in very large numbers, of light high-speed engines designed to suit peculiar requirements. An examination of the design of these engines, of their performance, and of the very varied practice obtaining in their arrangement suffices to show that a large amount of experimental work has been done by many unacquainted with the earlier results of efforts directed by engineers towards the solution of similar and frequently precisely the same problems. Sometimes the results obtained have appeared contradictory, and have been such as could not be predicted by those familiar with engineering practice; but more often the success or failure of new adaptations or fresh designs could have been foretold and subsequently proved, had not attention often been diverted from the engine to the more numerous problems connected with transmission and operating gear, and with the many attachments and parts embodied in car construction.

Repetition of experimental work

Table 9 embodies figures and the results of calculations relating to representative types and powers of engines from which some comparative information may be gained, and which serves to illustrate the diversity of practice regarding the number and size of cylinders employed in relation to horse-power developed, the rates of revolution, and from this data, the piston displacement or volume swept by the piston and the resulting mean pressures required on certain assumptions to enable the horse-power developed to be obtained.

In explanation of the table it may be stated that column 3 gives the declared horse-power of the motor, or that actually obtained on test. In most cases the figures relate to engines of 1904; similar figures relating to 1905 engines are given hereafter.

Column 10 gives the displacement per stroke of one piston, and is, therefore, the product of columns 6 and 7.

Column 11 represents the product of columns 9 and 10, multiplied by two, and, therefore, gives the displacement per minute for one cylinder, all strokes being taken into account.

503

Column 12 gives the total displacement of the engine, and is the product of columns 4 and 11.

Column 13 gives the displacements per minute per brake-horse-power, and is obtained by the division of column 12 by column 3.

Column 14 gives the estimated indicated horse-power, a mean pressure P of 70 lb. per sq. in., having been selected, and the expression $\dfrac{P \times L \times A \times N}{33,000}$ represents the horse-power when $N=$ number of working or firing strokes per minute; L and A being taken from Columns 8 and 6.

Column 16 has been obtained from the expression mean pressure $= \dfrac{33,000 \times \text{B.HP.}}{L \times A \times N \times E}$. The mechanical efficiency E has been taken to be 90 per cent. or 0·9, and the same figure has been used in calculating brake-horse-power in column 15 from the estimated indicated horse-power in column 14.

Variation of speed of revolution Referring to columns 7, 8 and 9, it may be seen that the normal speeds of revolution vary from 600 to 1,700 per minute, and that considering speed of revolution and stroke together it sometimes happens that the high-speed engine has a comparatively long stroke, and that its piston speed, or what is of considerable importance, its piston ring travel in feet per minute, is high.

An interesting comparison may for instance be made between the speeds of engines numbers 52, 53 and 54, and engine number 16. The first three mentioned are steam engines, two of them working normally with super-heated steam, and one of them with either dry saturated steam or steam only slightly superheated. With these the rates of revolution are moderate, the piston speed is about 400 ft. per minute, and with the exception of number 53 the mean cylinder wall temperatures would be lower than with the De Dion engine which runs at 1,700 revs. per minute, and has a piston velocity of over 1,200 ft. per minute. The continued satisfactory performance of the De Dion engine indicates, therefore, that if other considerations dictate a limit to speed, difficulty may not be expected to result from the use of piston rings working under severe conditions of temperature and at high rubbing velocity. Reference has been made already to questions affecting the number of cylinders to be used for engines of various powers, and also to the quantities of heat which may be utilized or are lost in the cylinder. The possibility of improving the fuel economy of the engine, although not of primary importance, is a consideration which should be taken into account when determining the design of an engine, in so far as the number of cylinders is concerned.

Hereford trials An interesting result of the Automobile Club Trials of light cars at Hereford in September 1904, was the observed economical running of a number of the competing cars. Most of these cars had single-cylinder engines, and seven or eight, which completed the trials, ran from 30 to nearly 39 miles per gallon of petrol consumed. Some of those which did not run throughout the trials showed marked economy, but Table 8 here given includes the ten cars which completed the trials, and showed the best petrol consumption results. None of these engines had more than two cylinders, and reference to column 16 of Table 9 shows that some of them would require to work at high mean pressures to develop the declared powers, from which it may be gathered that the engines would work at a higher average load and a more economical load than frequently obtains with the engines of cars in which engine power is much greater in relation to car load.

TABLE 8

No. of Car.		Name.	Miles per Gallon.	Total weight with Passengers. lbs.	Declared BHP.	HP. on Hills.	Mechanical Efficiency.
In order.	Official.						
1	14	Swift (1 cylr.) .	38·8	1512	7·0	5·12	73·20
2	18	Siddeley .	38·5	1603	6·5	5·16	79·40
3	30	Croxted (2 cylr.)	33·3	1848	8·0	5·63	62·60
4	22	De Dion .	30·3	1414	6·0	5·19	86·40
5	23	De Dion .	30·3	1414	6·0	4·24	70·75
6	15	Star (2 cylr.) .	29·5	1694	7·0	5·42	77·50
7	10	Wolseley .	28·3	1596	6·5	5·61	86·25
8	38	Star (2 cylr.) .	25·4	1908	7·0	5·74	82·00
9	20	Wolseley .	24·6	1554	6·5	4·75	73·10
10	26	Swift (2 cylr.).	23·6	1610	7·0	5·30	75·75

These tabulated figures are useful as setting forth actual results obtained, but it should not be forgotten that under the conditions of the road tests some variable factors must necessarily be introduced. For instance, in considering the figures relating to cars 14 and 26 it will be seen that whereas the former ran 38·8 miles per gallon of petrol consumed, the latter ran 23·6. Although some gain in economy would be expected with the single-cylinder engine, that actually obtained requires further explanation. The one engine may have been more carefully driven than the other, but it is even more probable that the internal conditions of the single-cylinder engine were more satisfactory than the other as regards the amounts of leakage losses, and the number of places at which some loss by leakage might be expected, while for various reasons the mechanical efficiency would be higher than with the double-cylinder engine. It will further be seen from the Table that the mechanical efficiencies of these two cars, represented by comparison of their brake-horse-powers with the estimated horse-power exerted at the road wheels during the hill-climbing trials, are in agreement within 3 per cent., and that the more economical car was the less efficient. It is, therefore, reasonable in this instance to look to the engines for some explanation of the difference in petrol consumption, and not to debit the driver or the transmission gear with too much of it.

The loss of heat from cylinder to jacket water must necessarily be very variable, but, assuming that two engines of equal horse-power, the one having a single cylinder and the other two cylinders, run at similar speeds, then

Heat loss to jacket

505

with suitably proportioned cooling systems, the heat loss with the single-cylinder engine may be expected to be less.

To give an instance, consider the case of two engines running at, say, 1,500 revs. per minute, the single-cylinder engine having a bore of 3·75 in., and a stroke of 4·5 in., and the two-cylinder engine having cylinders 3 in. diam., and a stroke of 3·5 in.

At the speed stated, and with a mean pressure of 80 lb. per sq. in., the

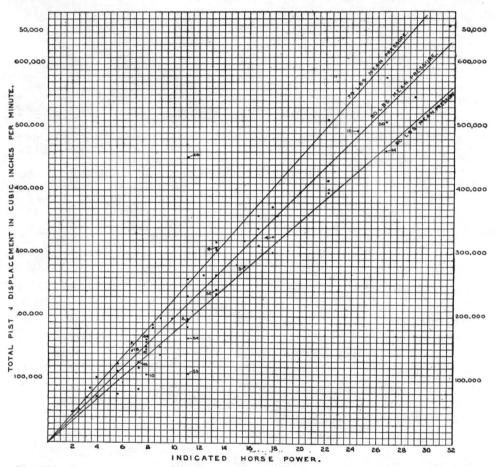

FIG. 421.—DIAGRAM SHOWING RELATION BETWEEN PISTON DISPLACEMENT HORSE-POWER
AND MEAN PRESSURE.

engines will develop respectively 7·53 and 7·46 indicated horse-power, and they are, therefore, of virtually the same capacity.

The areas of the cylinder and combustion chamber internal surfaces exposed to high temperature and water cooled may for the purposes of this comparison be accepted as equal to the piston area or cylinder-head area, with the addition of cylinder-barrel surface equal to the circumference of the bore multiplied by the stroke.

These amounts will be—

PISTON DISPLACEMENT, MEAN PRESSURE, ETC.

For the single-cylinder engine 11·04 +(11·781 ×4·5) —64 sq. in.
„ „ two „ „ [7·06 +(9·424 ×3·5)]2 —80 sq. in.

There is, therefore, 20 per cent. more area exposed for the same time to the same temperature differences with the two-cylinder engine, and it is safe to conclude, the rate of conduction of heat from cylinder to jacket being the same per square inch in each case, that there is a greater loss of heat to the jackets of the two-cylinder engine proportional to these differences of area.

Headless combustion chambers

The results of the recent Tourist Trophy race in the Isle of Man confirm those in this respect of the Hereford trials of last year, and the rather remarkable performance of the Arrol-Johnston winning car fitted with engines having two pistons in each of two cylinders after the manner of the Atkinson differential gas engine of 1885 and the Gobron-Brillie engine illustrated at p. 191 of vol. i., again draw attention to the possible economic value of reduction of the surfaces exposed to the heat of combustion and which must be cooled by conduction to the water jacket. The Tourist Trophy results and cars are referred to in a later chapter.

Fig. 421 shows to what extent the cylinder displacements of those engines given in Table 9 are relatively large or small. There are in addition some air-cooled motor-cycle engines of between two and four indicated horse-power. The values of the indicated horse-power, to which base the mean pressure lines have been plotted, have been obtained from column 3, and by taking an efficiency of 90 per cent. throughout, in order that the comparison may be uniform. The vertical scale represents the total cylinder displacements given in column 12 of Table 9. Since for a given mean effective cylinder pressure the indicated horse-power is known from the formula already given—

$$\frac{P \times L \times A \times N}{33,000},$$

Displacement per HP.

the displacement required may be found from the product of $L \times A \times N$, L being the engine stroke in inches, and N the number of revolutions per minute multiplied by two. It necessarily follows that the displacement per horse-power for any given mean pressure will have a constant value, and is—

For 70 lb. mean pressure 22,630 cubic in. per minute.
„ 80 „ „ „ 19,800 „ „ „ „
„ 90 „ „ „ 17,600 „ „ „ „

Constant mean pressure lines are shown on Fig. 421 for these three pressures, and from the position of the dots representing the horse-powers obtained with the known displacements, it will be seen that a mean pressure of nearly 80 lbs. is, in the absence of specific mean pressure determination, a suitable figure to employ when determining the horse-power to be expected from an engine of given dimensions and which is to run at a predetermined speed.

The figures given in Table 9, plotted as described in Fig. 421, enable an idea to be more readily formed of the extent to which there is agreement or disagreement among engine builders as to the rates at which engines may be expected to work satisfactorily.

The engine numbers used in Table 9 have in some cases been attached to the engine positions plotted in Fig. 421 to facilitate reference.

Argyll For instance the engines used with the Argyll cars, numbers 2, 3 and 4, will be seen to agree comparatively closely, and to require mean effective cylinder pressures of between 86 and 90 lb. per sq. in.

Again, the engine of the Clyde car, number 10, which was a competing car in the trials which took place at Hereford, and which ran very economi-**Clyde** cally, would require, if it developed the horse-power claimed, to have a mean cylinder pressure of 115·5 lb. This figure indicates that the engine must have run during the trials at a higher average load than many of the others, and helps to explain the economical results obtained.

Engine number 26, that of a Lanchester car, and air cooled, is instructive. **Lanchester** It shows that an air-cooled engine must work at a relatively light load to run continuously and without trouble from overheating.

Clement The Clement engine, number 9, requires less than 70 lb. mean pressure, which, considered with the number of impulses obtained from the four cylinders, confirms the observations made as to quiet running.

The steam engines, numbers 52, 53 and 54, are not of course directly comparable with the internal combustion engines; their positions have, however, been plotted so that it may be seen to what extent in practice the greater work obtainable from the steam engine cylinder enables reduction of cylinder volume to be made.

The Gardner-Serpollet engine, number 53, is shown by Table 9 to require a mean pressure of 81·2 lb. to develop its 10 brake-horse-power. During its working strokes it is, therefore, as active as the average petrol **Gardner** engine. It is of the simple single-acting type, so that for similar speeds of **Serpollet** rotation, twice as much work would be obtained from cylinders giving equal displacements. Because twice the number of working strokes are obtained the displacement as shown by Fig. 421 is just half that of a petrol engine working with the same mean pressure.

The Clarkson engine, number 52, is of the simple double-acting type and for similar mean pressures, speeds, and cylinder bore and stroke dimen-**Clarkson** sions, four times the work is obtainable compared with the petrol engine. This engine is, however, of very liberal dimensions; it requires a mean pressure of only 43·9 lb. to develop its normal 12 brake-horse-power, and the displacement is nearly as great as that required by the average petrol engine.

As, however, a working stroke is obtained twice every revolution the product of piston area and stroke may be half that of the single-acting engine, and with similar diameters and strokes only half the number of cylinders would be required, or one-fourth of the number that the petrol engine must have. It is scarcely necessary to point out that although in this case the double-acting engine is shown by Fig. 421 to have a displacement comparable with that of a petrol engine, it still only requires half the

Table 9.—Engine Dimensions, Piston Displacements, and Mean Pressures.

1	2 Name or Maker of Engine	3 Declared Horse-power or Brake-Horse-power	4 Number of Cylinders	5 Diameter, inches	6 Area, inches	7 Stroke, inches	8 Stroke, feet	9 Revs. per minute, normal	10 Piston Displacement per stroke, cubic inches	11 Piston Displacement, cubic inches × stroke × revs. per min. × 2	12 Total Displacement, cubic inches	13 Displacement cubic inches per declared HP, or BHP	14 Estimated IHP, 70 lbs. mean pressure	15 Brake-horse-power, Mechanical efficiency—90 per cent	16 Mean pressure deduced from declared HP, or BHP	17
1	Allday's	7·0	1	4·00	12·58	4·50	0·375	1400	56·6	158,480	158,480	22,640	7·01	6·31	77·7	1
2	Argyll	10·0	2	3·46	9·40	4·72	0·393	1100	44·3	97,460	194,920	19,492	8·61	7·75	90·3	2
3	"	14·0	3	3·54	9·85	4·72	0·393	1000	46·5	93,000	279,000	19,928	12·31	11·08	88·2	3
4	"	16·0	4	3·46	9·40	4·33	0·361	1000	40·7	81,500	326,000	20,374	14·41	12·97	86·5	4
5	Belsize	15·0	2	4·00	12·58	5·00	0·416	1350	62·9	169,830	339,660	22,644	15·00	13·50	77·8	5
6	" Junior	8·5	1	4·52	16·01	4·92	0·410	1000	78·9	157,800	157,800	18,564	6·97	6·27	94·8	6
7	British Motor Fittings Co.	12·0	2	4·12	13·36	5·50	0·457	900	73·4	132,120	264,240	22,020	11·70	10·53	80·2	7
8	Century Engineering Co.	6·5	1	3·30	8·55	3·54	0·295	1400	30·25	84,700	84,700	13,030	3·74	3·37	135·8	8
9	Clement	12·0	4	2·95	6·83	4·72	0·393	1200	32·20	77,280	309,120	25,760	13·66	12·29	68·4	9
10	Clyde	7·0	1	3·35	8·80	4·33	0·361	1400	38·10	106,680	106,680	15,240	4·72	4·25	115·5	10
11	Croxted	10·0	2	3·50	9·62	4·75	0·395	1000	45·60	91,200	182,400	18,240	8·06	7·25	96·7	11
12	Daimler-Coventry	22·0	4	4·13	13·39	5·11	0·426	900	68·40	123,480	493,920	22,450	21·83	19·65	78·6	12
13	Darracq	24·0	4	4·41	15·30	4·72	0·393	1000	72·20	144,432	577,728	24,072	25·56	23·00	73·3	13
14	De Dietrich	24·0	4	4·72	17·52	4·72	0·393	700	82·70	115,780	463,120	19,296	20·46	18·41	91·4	14
15	De Dion	12·0	2	3·94	12·20	4·33	0·361	1500	52·80	158,400	316,800	26,400	14·00	12·60	66·7	15
16	"	6·0	1	3·54	9·85	4·33	0·361	1700	42·60	144,840	144,840	24,140	6·42	5·78	72·9	16
17	Forman Motor Co.	8·0	2	3·54	9·84	3·94	0·328	900	38·75	69,750	139,500	17,436	6·15	5·54	101·0	17
18	"	6·5	2	3·54	9·84	3·94	0·328	750	38·75	58,124	116,248	17,884	5·13	4·62	98·5	18
19	Georges Richard	10·0	2	3·94	12·20	3·94	0·328	1200	48·00	115,200	230,400	23,040	10·15	9·14	76·6	19
20	"	15·0	4	3·74	11·00	5·11	0·426	800	56·20	89,920	359,680	23,978	15·90	14·31	73·4	20
21	Germain	10·0	2	4·13	13·39	5·11	0·426	950	68·40	129,960	259,920	25,992	11·48	10·33	67·8	21
22	Gladiator	12·0	4	2·95	6·83	4·72	0·393	950	32·20	61,180	244,720	20,392	10·84	9·76	86·4	22
23	Humber	28·5	4	4·50	15·90	5·75	0·478	900	91·40	164,520	658,080	23,090	29·10	26·19	76·3	23
24	"	7·5	2	3·50	9·62	4·00	0·333	1200	38·50	92,400	184,800	24,640	8·16	7·34	71·5	24
25	James and Browne	9·0	2	4·00	12·58	6·00	0·500	700	75·40	105,560	211,120	23,456	9·34	8·41	75·0	25
26	Lanchester	10·0	2	5·25	21·64	5·68	0·473	920	122·80	225,952	451,904	45,190	20·00	18·00	38·9	26
27	Mors	18·0	4	4·17	13·60	4·91	0·409	1300	66·60	173,160	692,640	38,480	30·70	27·63	45·7	27
28	Motor Manufacturing Co.	26·0	4	3·94	12·20	5·11	0·426	1100	62·30	137,060	548,240	21,086	24·25	21·81	83·3	28
29	"	8·0	1	4·33	14·75	5·11	0·426	1300	75·40	196,040	196,040	24,504	8·66	7·79	71·8	29
30	New Orleans	12·0	4	3·74	11·00	4·33	0·361	800	47·70	76,320	305,280	25,440	13·48	12·13	69·3	30
31	"	5·0	1	4·50	15·90	6·00	0·500	650	95·50	124,150	124,150	24,830	5·47	4·92	71·0	31
32	Oldsmobile	30·0	4	5·20	21·23	5·60	0·466	750	119·00	178,500	714,000	23,800	31·50	28·35	74·1	32
33	Panhard et Levassor	20·0	4	4·40	15·20	5·60	0·466	750	85·10	127,650	510,600	25,530	22·50	20·25	69·0	33
34	"	16·0	4	3·94	12·19	5·11	0·426	750	62·30	93,450	373,800	23,362	16·55	14·90	75·4	34
35	"	15·0	4	3·54	9·84	5·27	0·438	750	51·70	77,600	310,400	20,692	13·70	12·33	85·1	35
36	"	7·0	2	3·54	9·84	5·11	0·426	750	50·20	75,300	150,600	21,514	6·67	6·00	81·7	36
37	Peugeot-Baby	5·0	1	3·70	10·75	3·94	0·328	900	42·30	76,140	76,140	15,228	3·37	3·03	115·5	37
38	Renault	14·0	4	3·35	8·80	4·13	0·344	1100	36·40	80,080	320,320	22,880	14·15	12·74	77·2	38
39	Rex Motor Co.	5·0	1	3·62	10·31	3·62	0·301	1500	37·30	111,900	111,900	22,380	4·94	4·45	79·0	39
40	Simms-Welbeck	10·0	2	3·74	11·00	4·50	0·375	1000	47·80	95,600	191,200	19,120	8·43	7·59	92·3	40
41	Star	7·0	2	3·50	9·62	4·73	0·394	850	43·20	73,440	146,880	20,982	6·50	5·85	83·8	41
42	Sunbeam	12·0	4	3·15	7·78	4·33	0·361	800	36·80	58,880	235,520	19,627	10·40	9·36	90·0	42
43	"	6·0	1	3·54	9·84	4·33	0·361	1700	42·60	144,840	144,840	24,140	6·40	5·76	72·9	43
44	Swift	7·0	1	3·74	11·00	4·33	0·361	1700	47·50	161,840	161,840	23,120	7·17	6·45	76·0	44
45	Thornycroft	20·0	4	4·00	12·58	4·37	0·364	900	55·00	99,000	396,000	19,800	17·48	15·73	89·0	45
46	White and Poppe	3·5	1	3·15	7·78	3·35	0·279	1400	26·00	72,800	72,800	20,800	3·22	2·90	84·5	46
47	Winton	20·0	1	5·25	21·64	6·00	0·500	800	129·80	207,680	415,360	20,768	18·36	16·52	84·7	47
48	Wolseley	6·5	1	4·50	15·90	6·00	0·500	800	79·50	127,200	127,200	19,570	5·61	5·05	90·0	48
49	"	7·5	2	4·00	12·58	4·00	0·333	900	50·30	90,540	181,080	24,144	7·93	7·14	73·0	49
50	"	24·0	4	4·50	15·90	5·00	0·416	800	79·50	127,200	508,800	21,200	22·44	20·20	83·3	50
51	"	30·0	4	5·00	19·63	5·00	0·416	600	98·20	176,760	707,040	23,568	31·20	28·08	74·7	51
52	Clarkson-Chelmsford	12·0	2	4·00	12·58	4·00	0·333	600	50·30	60,360	120,720	10,060	21·30	19·18	43·9	52
53	Gardner-Serpollet	10·0	4	2·95	6·84	2·83	0·236	700	9·35	27,090	108,360	10,836	9·58	8·62	81·2	53
54	White	10·0	2	3 and 5	19·63	3·50	0·291	600	68·70	82,440	164,880	16,488	14·57	13·10	53·5	54

To face p. 508]

number of cylinders of similar dimensions. In this instance, therefore, only about half of the possible reduction of cylinder dimensions has been made, but compensating advantages have been obtained by using as low a mean pressure as 43·9 lbs. The engine may thus be more economical in steam consumption, and its capacity to work occasionally at an overload is increased.

The conditions obtaining with the White steam engine, number 54, **White** are different from the Serpollet and the Clarkson. The engine is of the compound double-acting two-cylinder type. For the purpose of calculating the horse-power the usual procedure has been adopted of referring the mean pressure developed in the high-pressure cylinder to the low-pressure cylinder. For instance the total mean pressure required is shown by Table 9 to be 53·5, and the diameters of the high and low-pressure cylinders are respectively 3 and 5 in., the stroke in both cases being 3·5 in. The cylinder areas are 7·06 and 19·63 in., and the ratio of areas is, therefore, 2·78 to 1. If the work done in each cylinder is to be equal at, say, full load, then the mean pressure developed in the high-pressure cylinder must be 74·3 lb., and that in the low 26·75 lb. The high-pressure cylinder mean pressure referred to the low-pressure cylinder or divided by 2·78 is then $\frac{74·3}{2·78} = 26·75$, and the addition of the low-pressure cylinder mean pressure of 26·75 lb. makes the required total of 53·5. As the engine is necessarily arranged to reverse, **Division of** there is difficulty in arranging the points of cut off at the valves to give **work** more than approximate equality of division of work between the cylinders at any particular load, and if the reversing gear is utilized to control the degree of expansion and the engine power, then there may be considerable variation in the amounts of work performed in the two cylinders.

The cylinder displacement shown by Fig. 421 is that of the low-pressure cylinder alone. If that of the high pressure is included then the total displacement is increased by about three-eighths of that shown, and it becomes relatively greater than that of the Clarkson engine. For the sake of the economy to be gained in steam consumption by the use of the compound engine less than half the reduction of cylinder dimensions possible, with the steam engine compared with the petrol engine, has been effected. The **Overload** advantage possessed by the Clarkson engine, as regards capacity for working **capacity** at overloads, because of the low normal mean pressure required, is with the White engine obtained to nearly the same extent, partly by the comparatively low normal mean pressure, and by the arrangements made for largely increasing the mean pressure at starting by means of the bye-pass valves provided, and which were referred to in the description of the White car.

Other information or further deductions may be made by a more complete study, but enough has been said to show that an analysis of figures collected and presented in this way provides compensation for the labour involved.

Chapter XXXII

OVERTURNING AND SKIDDING

THE results of calculations of the speeds at which a car, under conditions to be stated, will overturn or skid, under the influence of centrifugal force, when traversing curves, are shown by Fig. 422. The curves there shown enable an estimate to be made of the relation between the overturning and skidding tendencies with racing cars; but as the method of treating the subject is here given modifications of the expressions employed for the calculations may be made to suit other types of car and other conditions.

For the present purpose, the weight of the car with passengers has been taken to be one ton and the height of the centre of gravity from the road 2 ft. 6 in.

A wheel gauge or track of 4 ft. 6 in. has been employed to represent average conditions.

The coefficient of friction or the adhesion of rubber tyres on the road-surface is very variable and is affected in various ways, but a value of 0·66 of the weight on the tyres has been considered suitable.

With this data, the force necessary to overturn the car is—

Overturning force

$$F = \frac{2240 \times 2 \cdot 25}{2 \cdot 5} = 2016 \text{ lb.}$$

2·25 being half the wheel gauge of 4 ft. 6 in., and the value of F. 2016 lb. is the amount of the centrifugal force necessary, with the dimensions given, to commence overturning.

On a curve of given radius the speed at which the force equals this amount may be found from the expression for centrifugal force $\dfrac{Wv^2}{gr}$
Then—

$$\frac{W v^2}{gr} = 2016,$$

from which $v = \sqrt{\dfrac{2016gr}{W}}$;

and when $v =$ speed in miles per hour, then—

510

$$v = \sqrt{\frac{2016 \, gr}{W}} \div 1{\cdot}46$$

The radius of the curve in feet $= r$, gravity or $g = 32{\cdot}2$ and $W = 2240$, the weight of the car with passengers. Thus on a curve of 100 ft. radius, and with no other external or accidental influences considered, the speed at which overturning occurs is—

$$v = \sqrt{\frac{2016 \times 32{\cdot}2 \times 100}{2240}} \div 1{\cdot}46 = 36{\cdot}69 \text{ miles per hour.}$$

and on a curve of 250 ft. radius—

$$v = \sqrt{\frac{2016 \times 32{\cdot}2 \times 250}{2240}} \div 1{\cdot}46 = 58 \text{ miles per hour,}$$

which will be found to be the speeds shown by the overturning curve, Fig. 422.

The speed at which skidding will occur cannot be so readily determined, and the causes of such slipping are variable and sometimes doubtful. It has been here determined in the following way :—

With the value, already given, the centrifugal force must not exceed the adhesion of the tyres on the road, or

$$\frac{W \, v^2}{gr} = 0{\cdot}66 \, W,$$

from which $v = \sqrt{0{\cdot}66 \, gr}$ and if $v =$ speed in miles per hour,
then $v = \sqrt{0{\cdot}66 \, gr} \div 1{\cdot}46$.

In order that the centre of mass of the car may obey the tendency to follow a straight-line path, and to compensate for the departure from that path directed by the steering-wheels, the driving-wheels are called upon to depart from the original path in the opposite direction and to the same extent as the steering-wheels when the weight is equally distributed, and it may be taken that with the centre of mass of the car at the car centre, there is, when the car is travelling around the curve, an equal tendency to skid or slip at the steering and driving wheels.

If skidding or slipping occurs at the front wheels, then the curved path dictated by the steering gear is not followed, and no skidding tendency is set up at the driving-wheels. If, however, the front wheels follow the curved path, then a point in a line situated at or about the centre line of the front axle may be considered as an instantaneous centre, about which the car will slew or endeavour to rotate under the influence of the centrifugal force.

Taking the centre of the car as the point of operation of the centrifugal force, and a point in the centre line of the rear axle as the point at which resistance to slipping is offered, then the amounts of the forces become—

$$\frac{W \, v^2}{gr} = \frac{(W \times 0{\cdot}66) \times 2}{2}.$$

511

The value of the adhesion W × 0·66 is multiplied by two because the point at which resistance is offered to slipping, at the centre line of the rear axle, is at twice the distance that the centre of the car or the centre of mass is from the assumed centre of slewing, namely, at the centre line of the front axle. The same amount is divided by two because in the racing car half the weight is taken to be on the rear wheels and there is therefore only half the adhesion. These considerations do not alter the value of the expression already given.

The extent to which tyre adhesion is affected by tractive effort at the driving-wheels, by irregularity of road surface, by the state or condition of the road surface, and by other special or accidental circumstances,

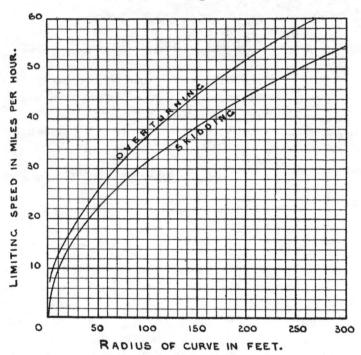

FIG. 422.—DIAGRAM SHOWING CRITICAL SPEEDS ON ROAD CURVES OF DIFFERENT RADII.

may be very variable, and their effect is not here taken into consideration, since the general case only can be presented. The formula employed is, then—

Critical skidding speed

$$v = \sqrt{0\cdot66\, gr} \div 1\cdot46,$$

and on a curve of 100 ft. radius the speed at which slipping or skidding will occur is—

$$v = \sqrt{0\cdot66 \times 32\cdot2 \times 100} \div 1\cdot46 = 31\cdot3 \text{ miles per hour,}$$

or on a curve of 200 ft. radius—

$$v = \sqrt{0\cdot66 \times 32\cdot2 \times 200} \div 1\cdot46 = 44\cdot4 \text{ miles per hour,}$$

both of which speeds are given by the skidding curve, Fig. 422.

512

OVERTURNING AND SKIDDING

From this method of considering the subject and from the curves shown, it is apparent that skidding will occur before the critical speed at which overturning commences.

The nearness of the curves to each other shows that when taking road curves at speeds approaching the critical velocities the lifting or heeling of the car on its springs, accompanied by actual slipping or a tendency to slip or skid, which may be experienced, is a true indication of approaching instability.

The heavier loading of the driving wheels of cars used for ordinary purposes, while rendering the car more secure from slipping on muddy or greasy surfaces, does not favourably affect its behaviour when travelling around curves, and when irregularity of road surface is considered: such heavier loading has an unfavourable effect.

Employing the same form of expression for calculating the critical skidding speeds modified to suit the different distribution of weight and the changed centre of mass of the car, it will be seen that the capacity of the car to resist skidding under the influence of centrifugal force is not affected.

Taking the same total weight of the car and value for adhesion, and assuming that 0·6 of the weight is on the driving wheels, and the centre of mass of the car, therefore, at 0·6 of the distance between front and rear axles and towards the rear axle, then the formula assumes the following form :—

$$\frac{W\,v^2}{gr} \times 0\cdot60 = (0\cdot66\ W) \times 0\cdot60 \times 1.$$

As in the first case, the total adhesion has been multiplied by 0·60, because that is the proportion of the weight on the driving wheels, and it has been again assumed that the steering wheels have not slipped, but have directed the car on to the curved course or changed curvature of the course. The centrifugal force has also been multiplied by 0·60 or the amount of the distance between front and rear axles of the centre of mass from the hypothetical centre of slewing. It is evident that the above expression is the equivalent of that already used, namely : —

$$v = \sqrt{0\cdot66\ gr},$$

when $v =$ speed in feet per second or :—

$$v = \sqrt{0\cdot66\ gr} \div 1\cdot46$$

when $v =$ speed in miles per hour,

and further consideration shows that the question of the distribution of weight, or of the front or rear position of driving wheels, does not in itself affect the resistance offered at the front or back wheels to slipping.

When, therefore, the special or accidental circumstances referred to earlier are not considered, the critical or maximum possible speeds on curves may be arrived at by taking into account only the amounts of the centrifugal or operating force and the power of resistance to slipping represented by the total adhesion of the tyres on the road.

From what has been said it will be evident that when as a result of high speed of travel over road curves having a lumpy or irregular surface, the load on the tyres becomes momentarily reduced by rebounding, then the position of the centre of mass of the car has an important influence on the initial skidding or slewing.

Side slipping

So far the only cause productive of skidding that has been considered is that resulting from traversing curves at high rates of speed. It is, however, common knowledge that on slippery road surfaces skidding occurs when the velocity of the car is low and when it is following, or is desired to follow, a straight course. Those who have considered the subject sometimes advance explanations that appear untenable or which are contradicted by the advice tendered by those who have experienced skidding when driving cars, and by the actual behaviour of the cars themselves. It has been advised, for instance, that the application of the brakes is liable to produce skidding, and again, that when skidding has occurred a momentary and judicious application of the brakes will check it. This advice is sound, though difficult to follow under some circumstances.

[1] It has also been stated that with rear-driven cars the coefficient of adhesion is greatly reduced, and instability produced by locking the rear wheels, and that as a result deviation from the straight course becomes almost inevitable.

The stability or freedom from tortuous running of a front-driven car may undoubtedly be greater than that of a rear-driven car, and the statement that adhesion is considerably reduced when the wheels are locked may be accepted; but under actual circumstances the wheels are not locked, and the forces or efforts productive of the initial slipping are not pointed out.

Darwin shows nothing but that locked wheels, if they slide at all, will slide along the path on which slipping is easiest—secondly, that the wheels which are not locked will roll along a surface and take a path which is not that of locked wheels sliding. If the wheels which are locked are the rear driving wheels and the free rolling wheels are the front steering wheels, then the curves which will be described, if the car is on an inclined plane and moving in a direction not quite in the direction of greatest slope will be a tractrix from commencement of skidding to a position normal to the axis of greatest slope and a similar and symmetric curve for the continued sliding in the same general direction after the car has turned round. But all this teaches the car-maker nothing, and the conclusions arrived at teach nothing, and do not indicate the probable causes productive of the initial slipping. Front drivers are attended with serious practical objections, and the other propositions made condemn themselves. No driver locks his driving wheels intentionally, and what he most wants is an easily locked differential gear and perfectly self-compensating brakes.

Conditions productive of side slip

The conditions productive of skidding on slippery surfaces are very

[1] *Side Slip in Motor Cars*, by Horace Darwin, M.A., F.R.S., and C. V. Barton, D.Sc.

514

variable, and although after the occurrence it may be sometimes a simple matter to trace the causes leading to it, it is not a problem which may be definitely treated mathematically because of those variable and undetermined combinations of conditions.

It must primarily be remembered that the tyre adhesion may be enormously reduced, and that small forces become sufficient to change the position of the car without necessarily affecting the velocity of direction of travel of the centre of mass of the car.

It may, for instance, happen that, with variable road formation or surface, the adhesion of one driving wheel tyre may be, when in any case that adhesion is low, greater than the other, and that as a result of the momentary side effort sufficient slewing from the original course occurs to bring the momentum or energy of the car into play to cause the greater slip or glissade that is known to happen.

The skidding that has been unquestionably caused, according to the evidence of those who experienced it, by the application of the brakes, may be set up by efforts similar in effect to those produced by unequal adhesion. It is quite possible with some of the forms of partially inoperative or useless equalising gear in common use, for the application of the side brakes to be very unequal, and to result in retardation at one rear wheel in excess of that at the other. This, by side or corner retardation of the car, would be as productive of initial skidding or slewing as the side-driving effort resulting from momentary difference of adhesion of the driving tyres. Similar disturbing efforts, though less in amount, would be set up by the use of the foot brake or counter shaft brake, when the differential gear continues to perform its duty of equalising gear, for although the retardation of the driving wheels would be equal, the amount of retardation at the tyres might be unequal by variable adhesion. The use of well-designed equalising or compensating gear is thus of great importance, for apart from other important considerations, it eliminates one common source of side slip. The reasonable lengthening of the wheel base and the greater adhesion made possible by the use of some of the more practical forms of non-skidding devices are well-directed attempts in the right direction, and have already resulted in improvement.

From the foregoing it will be seen that a very little calculation will **Lateral stress on wheels** show the measure of the force which under some circumstances of side slip has torn the spokes and centre clean out of even some very well made wheels. It is not necessary that the wheels of a car should, when side-slip occurs, strike a curb or similar immovable obstacle in order that the side sliding momentum should become evident in either strained or broken wheel, for the same destructive strain may arise from sudden arrest of the side-slipping wheels when sliding from grease on to a dry sandy or rough piece of road.

Thus, for instance, a car which is side-slipping and reaches a sliding velocity of only 5 miles per hour would throw upon the wheels, or upon one of them, a side stress which might reach about $1\frac{1}{2}$ ton with a car weighing a ton. This is assuming that from the moment at which the side-sliding tyre leaves the slippery surface and side-slipping is stopped by entrance upon the rough surface, 3 inches is traversed sideways. This would mean that the rate of retardation would reach 108 ft. per second, and the amount of the effort exerted in the 3 inches of side-sliding and during, say, 0·07 second reaches—

$$\frac{Wv^2}{2gs} = \frac{2240 \times 53 \cdot 8}{64 \times 0 \cdot 25} = 7542 \text{ lb.}$$

The energy stored in the car may affect all four wheels, or it may affect only the back wheels, and in some cases almost the whole of the stress it represents may affect one rear wheel, and smash it, the other back wheel being still on slippery surface.

These stresses should be taken into consideration by the makers and the users of anti-skid apparatus attached to tyres, especially those which would increase the tractive adhesion of the tyre on any kind of road.

Banking Curves on Racing Tracks.

A matter of considerable importance in connexion with the subjects treated is the determination of the angle to which a track must be laid at curves on racing tracks, and indeed, it should be taken into consideration in the formation of ordinary roads. It is not uncommon to find a road in some country districts inclined in a direction opposite to that which is necessary, and the author remembers one place on a road not far from Worthing, which tilts the wrong way and dips toward a pond. The minimum amount of the slope or the banking at the curve must be such that at the limit of speed desired the force productive of side slip shall not exceed the adhesion of the tyres on the track.

The fraction of the centrifugal force to be provided for diminishes as the inclination increases—that is, it varies from the full amount of the force on a level track with no part of it assisting adhesion, until, to take the extreme case of a track at 90 degrees to the horizontal, no part would be productive of slip, and all the force would assist adhesion.

The amount of the component of the force productive of slip on an angle of slope of θ^0 is equal to $\mathrm{Cos}\,\theta\,\dfrac{W\,v^2}{gr}$ and the part of it assisting adhesion is equal to $\mathrm{Sin}\,\theta\,\dfrac{W\,v^2}{gr}$

Similarly the component of gravity assisting adhesion is represented by $\mathrm{Cos}\,\theta\,W$, and that part of it usefully opposing centrifugal force is equal to $\mathrm{Sin}\,\theta\,W$.

Those expressions representing the amounts of the forces assisting adhesion must be multiplied by a coefficient κ equal to the adhesion upon the track. Taking the above quantities, and with selected angles of inclination, the maximum speeds possible on curves of known radius may be calculated from the following equation :—

BANKING CURVES ON RACING TRACKS

$$\text{Cos } \theta \, \frac{W \, v^2}{gr} = K\left(\text{Sin } \theta \, \frac{W \, v^2}{gr} + \text{Cos } \theta \, W\right) + \text{Sin } \theta \, W.$$

In this equation the letters have the same significance as those already used, and in the following table, which is given as an example of its application, various degrees of inclination or banking have been taken, with coefficients of adhesion of 0·66 and 0·40 with a curve radius of 300 feet.

TABLE 10.

Inclination of Track		Maximum Speed in Miles per Hour	
θ	Slope	K=0·66	K=0·40
− 5 deg.	1 in 15·00	47·5	36·8
0 ,,		54·4	42·4
5 ,,	1 in 15·00	59·6	47·5
10 ,,	1 ,, 5·78	65·1	52·6
15 ,,	1 ,, 3·88	71·0	57·9
20 ,,	1 ,, 2·92	77·7	63·3
25 ,,	1 ,, 2·36	85·4	69·0
30 ,,	1 ,, 2·00	94·7	75·5

From this table it is apparent that if an intermediate value for the coefficient of adhesion be taken between 0·40 and 0·66 to represent that likely to obtain under usual conditions on a racing track, then speeds of 80 to 85 miles per hour, on a curve having a radius of 300 feet, would be practicable if the inclination of the track at the curve were 30 degrees equivalent to a slope of 1 in 2.

Calculation has also been made for an adverse slope of 5 degrees, to show how the maximum speed is affected, on crowned roads, if the road curve is followed on the outer side, where the slope would be frequently the wrong way or adverse.

Under these circumstances a part of the centrifugal force equal to $\text{Sin } \theta \, \dfrac{W \, v^2}{gr}$ reduces adhesion, and gravitation equal to $\text{Sin } \theta \, W$ has the same effect by addition to the amount of the centrifugal force tending to produce slipping upon the track.

The equation from which the critical speeds may be calculated then becomes—

$$\text{Cos } \theta \, \frac{W \, v^2}{gr} = K\left(\text{Cos } \theta \, W - \text{Sin } \theta \, \frac{W \, v^2}{gr}\right) - \text{Sin } \theta \, W.$$

Effect of adverse road slope

Inspection of the figures in the table further shows that if the road has a slope of 5 degrees, or is banked, then, as compared with the crowned road, having an adverse slope of 5 degrees, the curve could be followed at a speed of nearly 60 miles per hour instead of 47·5, an increase of 25 per cent. in speed.

With the same amount of useful and adverse slope of road surface and a curve radius of 100 feet, calculation shows that a car, travelling at a maximum possible speed of 34 miles per hour with the road surface inclined 5 degrees in the right direction, could only travel at 28 miles per hour if the road surface sloped the same amount in the wrong direction. Between the lower and the higher speeds there is a difference of over 20 per cent., and with lower speeds and smaller curve radii there would be a corresponding increase of the safe speed possible, with the small amount of slope of the road surface considered.

Chapter XXXIII

CARBURETTORS AND CARBURATION

JUDGING by the way in which an ordinary oil can may be successfully used as a carburettor when others fail, and by the enormous number of different forms of carburettors now on the market, it would seem that almost anything will roughly perform the functions of air carburettor for small engine purposes.

Some of the forms of apparatus used are, however, more satisfactory in their action than others, and the reasons for this may be usefully considered.

The difference in the behaviour of different carburettors is not always due to difference in their construction, but may be due to lifferences in their application or way of using them. Many of those which work badly as used, might work sufficiently well to be tolerated if the distance between them and the inlet valves were lessened, especially the vertical distance. It is generally forgotten that most of the broken-up petrol between jet and inlet valves is merely in mechanical suspension, and that it is only so as long as the air velocity is sufficient to maintain it in suspension and carry it to cylinder entrance. The lower velocity of the air which obtains in the enlarged sectional area of passages and vertical pipes between jet and engine, permits the petrol rain to drop at the slow speeds of starting by hand. It is then only by temporary excess of petrol, or the use of the oil can with petrol, that a start can be made. The disproportion, however, in petrol and air jet passage, and the spaces and pipes beyond these, is not now so often to be noted, but even now it is sufficient to make one sometimes remember that petrol must be wonderfully accommodating stuff, and that its rate of evaporation when sufficiently triturated to present large surface to the air, is fortunately extremely high. **Conditions of working**

The petrol engine is a gas engine or a hot-air engine, in which the air is heated and expanded by the combustion of a gas or vapour charge in the working cylinder ; and the carburettor is, or should be, a simple device for converting liquid petrol into a spray or mist, so that it may be instantaneously combustible throughout the air with which it is intimately mixed, and in this way raise it to a very high temperature and pressure. The really intimate mixture is not attained by all carburettors and modes **Carburettor duty**

519

of air admission. Air thoroughly carburetted or carburised, enters the cylinder with air not so carburised, and it is questionable whether the full effect of the petrol conveyed into the cylinder is as surely attained as when the same quantity of petrol is thoroughly disseminated throughout the same total of air not sò richly carburised. This matter was very well put by a writer in the *Automotor Journal*,[1] who said :—" Carburised air to a certain extent resembles ether-saturated oxygen, and is therefore better fuel than coal gas. This may be the explanation of the fact that more power can usually be got from a petrol engine than from a gas engine of the same size. It, therefore, does not altogether follow that we shall get the same results in a given cylinder if we have the same amount of air and petrol vapour present, irrespective of the conditions under which it got there. For instance, if a certain amount of petrol is necessary for the charge and a certain amount of air, a more effective explosion will be got if all the air has gone through the carburettor and been only partly but not fully carburised, than if the charge consists of two portions, one of fully carburised and completely saturated air, and the other of pure air, more especially if the latter enters through another channel [2] It has, in fact, often been observed that with surface carburettors the best output is obtained from the engine, when nearly all the air goes through the carburettor." To do this, however, the supply of petrol to the carburettor must be variable, according to the engine requirements, hence the spray carburettor, which will uniformly carburise all the air used, provides a more economical method of working, than that which makes it necessary " to mix gulps of pure air and fully carburised air at each intake into the cylinder, varying only their relative proportions." This will be generally conceded. In some cases the spray carburettor is made so that the gulps of pure air and fully carburised air do not thoroughly mix, but in the most recent forms it is probable that the mixture approaches completeness throughout a considerable range of engine speed.

There is, however, something to be said from the stratification point of view, a possible gain, or a plea for no loss from the passage of air and of carburetted air in gulps into the cylinder. This is referred to hereafter. It might appear that the greatest useful effect would be obtainable from **Position** a simple carburettor to each cylinder and as close to the admission valve as practical conditions will permit, so that the fullest charge of air could enter the cylinder with the spray and while the spray is expanding to its full vapour volume, rather than the position which is common, namely, a carburettor down below the car frame with long vertical and branch pipes to the valves.

In this connexion some few figures relating to the specific volumes of petroleum spirit may be referred to. According to some experiments

---

[1] *Automotor Journal*, April 1901, p. 334.
[2] This was practically demonstrated by the use, for a time, of supplementary air valves, not at or near the carburettor, as is now usual, but fixed in the cylinder head or cover.

quoted by Sir Boverton Redwood,[1] dry air will take up of petroleum spirit the following quantities :—

TABLE 11.—EVAPORATION OF PETROLEUM SPIRIT BY DRY AIR.

Temperature of Liquid	Vol. Evaporated by 100 Vols. Dry Air.	
	Specific Gravity 0·679	Specific Gravity 0·70
40 F	0·11	0·095
60	0·175	0·170
80	0·340 ?	0·360
100	0·650	0·480

TABLE 12.—EVAPORATION OF PETROLEUM SPIRIT BY DRY AIR.

Temperature of Liquid	Cubic Inches Absorbed per Cubic Foot of Air.	
	Specific Gravity 0·679	Specific Gravity 0·70
40	1·90 cub. inch	1·64 cub. inch
60	3·10 ,, ,,	2·94 ,, ,,
80	5·87 ,, ,,	6·22 ,, ,,
100	11·20 ,, ,,	8·30 ,, ,,

It must be remembered that the petrol may be broken up into spray or even into a mist and move for some distance from the jet as petrol mist or even vapour not mixed or taken up by the air with which it is for a time carried. In a large straight pipe a few globules thus carried will induce coalescence, and thus make it, not only difficult to start an engine so arranged, but almost impossible to get an automatic carburettor to remain automatic, as it would with properly proportioned, and suitably shaped connecting pipes.

Volume and weight of petrol vapour

It must also be remembered in this connexion that pure petrol vapour, not petrol carburetted air, is not only heavier than air, but very much heavier than air. A quantity of 100 grammes of petroleum spirit, of the composition represented by the formula C_7H_{16}, will yield 22·4 litres of vapour, measured at 0°C. and at 760 mm.; at the same pressure, and at 15° C. = 59° F. the volume will be $(22·4 \times \frac{T}{T_1}) = 23·63$ litres.

Expressing these quantities in the usual weights and measures :—

$$100 \text{ grammes} = \frac{100}{453·6} = 0·2205 \text{ lbs.,}$$

and 23·63 litres $= \frac{23·63}{28·3} = 0·835$ cubic feet. Therefore 1 lb. of this petrol

[1] *Petroleum and its Products*, vol. ii. p. 686.

vapour at 760 mm. and 15°C. will occupy only $\dfrac{0\cdot835}{0\cdot2205} = 3\cdot78$ cubic feet,

and is therefore $\dfrac{13\cdot09}{3\cdot78} = 3\cdot46$ times the weight of air.

Other petroleums and petroleum spirits have different vapour densities, of which the following are examples :—

TABLE 13.—SPECIFIC GRAVITY AND VAPOUR DENSITY OF PETROLEUM AND OTHER SPIRITS.

Name	Formula	Boiling Point		Specific Gravity at 0° C., 32 F.	Vapour Density Air = 1
		°C	°F		
Pentane	$C_5 H_{12}$	33·6	93·0	0·642 to 0·650	2·69
Hexane	$C_6 H_{14}$	64·9	149·0	0·682 to 0·695	2·97
Heptane	$C_7 H_{16}$	94·2	201·5	0·700 to 0·724	3·46
Octane	$C_8 H_{18}$	123·5	254·0	0·744	3·94
Benzine	—	81·0	178·0	0·830	2·69
Benzene	$C_6 H_6$	80·5	177·0	0·889	2·70
Alcohol	$C_2 H_6 O$	80·0	176·0	0·835	1·593

Change of specific gravity with temperature change There are other grades between some of those mentioned in this table, but these are sufficient to show the physical properties with which we are most concerned. The effect of temperature on the specific gravity must, however, be noted. A grade of petrol, for instance, which at the normal temperature of 60 F. has a specific gravity of 0·68 will at 0 degree C., or 32 F. have a specific gravity of 0·694.

The following table gives the specific gravity of such a grade of spirit at a few different temperatures :—

TABLE 14.—SPECIFIC GRAVITY AND TEMPERATURE OF PETROL.

Temperature		Specific Gravity	Temperature		Specific Gravity
F	C		F	C	
32	0·0	0·694	55	12·8	0·682
38	3·5	0·691	60	15·7	0·680
42	5·7	0·689	70	21·3	0·675
48	9·0	0·686	80	26·3	0·670
50	10·0	0·685	90	32·5	0·665

These differences do not, however, within a considerable range of temperature seriously affect the working of any carburetting device, as the air to be carburised is also affected by the change in temperature; but it is necessary to notice that a petrol of specific gravity, say 0·700, at 32° F., is considerably lower than this, or only about 0·670 at 80° F.

CARBURETTORS AND CARBURATION

From the foregoing it will be seen that although petrol vapourises very readily, its great vapour density makes it necessary to design a carburettor and the passages connecting it with the inlet valves, so that the vapour and the air cannot pass in parallel streams, but must be mechanically mixed. If this be completely effected at or near the carburettor, the air will be carburetted to a degree which will remain without much change for a considerable time.

It may be of some interest to inquire into the effect of the vapourisation of the small quantity of petrol which is taken from a carburettor jet at each suction stroke of an engine. A gallon of petrol, 277·5 cubic inches, of specific gravity, say 0·693, will give 40 cubic inches to 1 lb., and approximately 3·2 cubic feet vapour at normal temperature, and 760 mm. $= 29\cdot92$ inches of mercury pressure, or 138 cubic inches of vapour per cubic inch of petrol.

Rate of consumption

A car which runs, say 30 miles per gallon of petrol and makes 20 miles per hour, will use $\dfrac{6\cdot93}{30} = 0\cdot23$ lb. per mile, or $\dfrac{20 \times 0\cdot23}{60} = 0\cdot07$ lb. per minute, or $0\cdot076 \times 40 = 3\cdot07$ cubic inches of petrol per minute. If the engine has a single cylinder and makes 800 revolutions per minute, it will use $\dfrac{0\cdot0767 \times 2}{800} = 0\cdot000192$ lb. or $\dfrac{3\cdot07 \times 2}{800} = 0\cdot0077$ cubic inches per stroke, or 4 cubic eighths of an inch of the liquid at each suction stroke.

The volume of pure vapour corresponding to this will be $\dfrac{0\cdot0767 \times 3\cdot2 \times 1728 \times 2}{800}$, or $\dfrac{3\cdot07 \times 138 \times 2}{800} = 1\cdot06$ cubic inch; and if the cylinder be 4·5 in diameter and the stroke 5 in., the displacement will be $\dfrac{15\cdot9 \times 5 \times 800}{2} = 31800$ cubic in. per minute or 79·5 cubic inches per stroke,

Relative quantities of air and petrol vapour

so that the proportion of pure vapour to air will be about $\dfrac{79\cdot5}{1\cdot06} = 1$ to 75.

The cylinder, however, never receives the whole of the air represented by the piston displacement except when the engine is turned slowly by hand, and then the engine is said to require a richer mixture, which is, of course, not strictly true.[1] More air enters per stroke than at higher speeds, and so more petrol vapour is required to produce an ignitible mixture. This is, of course, apart from the apparently larger quantity of petrol required at very low speeds, when air velocity is insufficient to induce flow.

It requires chemically about 195 cubic feet of air for the complete combustion of 1 lb. of this petrol; or, taking the volume of 1 lb. at normal

---

[1] It may be urged that at working speeds the momentum of the air due to the highest speed of the piston causes an inrush which fills the cylinder, just as the momentum of a long column of gas engine exhaust is operative as a means of scavenging, as used by Atkinson.

temperature as 13·14 cubic feet, it will require $\dfrac{195}{13\cdot14} = 14\cdot85$ lbs. per lb. of petrol for complete combustion under ideal conditions.

The quantity of petrol used by the engine under the above-mentioned practical conditions being 0·000192 lb. per working stroke, the air required for its combustion will be $0\cdot000192 \times 14\cdot85 = 0\cdot00288$ lb. This weight of air $= 0\cdot00288 \times 13\cdot14 \times 1728 = 65$ cubic inches, and there are by piston displacement as above 79·5 cubic inches. There is, therefore, only about 22 per cent. more air entering the cylinder than is required for perfect combustion of the petrol used under ideal conditions, a result which confirms the high efficiency of these engines, and shows that the large excess of air commonly assumed to be necessary for combustion under practical conditions is wrong when applied to the combustion in the cylinder of this class of engine; or it shows that there is a hole somewhere in the foregoing calculations.

To return, then, to the carburettor itself, it will be seen that the very small volume of true petrol vapour which enters the cylinder per stroke makes it highly desirable that means should be adopted which shall cause its intimate mixture with the air, more especially toward the middle and latter part of the suction stroke. This requirement is fortunately met **Complete-ness of mixture** to a great extent by the increase in the velocity of the piston from zero at the commencement of the stroke to a maximum past the centre. At the beginning and early part of the stroke it may be imagined that only air enters, the velocity not being sufficient to induce much petrol flow. Then, toward the middle and three-fourths of the stroke, the velocity being at the highest, the flow is rapid and the richest mixture enters toward the last and is in the end of the cylinder when compression is completed and where it is best placed for ignition, and on combustion, heating the poor mixture. Stratification may, in fact, be appealed to.

All carburettors act by an inspiration or vacuum-produced current, and not by a plenum-produced current. The mention of this fact may, without explanation, seem to be without any particular significance. It may seem to mean only that the difference in pressure which is the cause of the movement of the air, is produced by a lowering of the pressure at the delivery or cylinder end of the air pipes, instead of by raising the pressure at the inlet end of the pipes leading to the carburettor; and it might seem, as has often been urged, that the resulting action would be the same whether the pressure difference be brought about the one way or the other. This, however, is not true, any more than was the theory that the more complete a vacuum the greater the velocity with which air would rush into it; or the statements that air current or transmission effects would be the same with forced current as with induced current.

The passage of air through pipes and passages of various shapes, bends and sizes, like those of and connected with a carburettor, is very different from that of passage through a uniform pipe, and the difference would be intensified if the air were pushed in instead of being induced to flow in by removal of resistance.

This has been overlooked in the design of many carburettors, and it is fortunate for many others that it is a fact, for otherwise they would not be the success they are.

It must be assumed in this connexion that the air passes from inlet to engine as though it were an air rope pulled in by the piston. It therefore takes a path which ignores everything but path and does not impinge

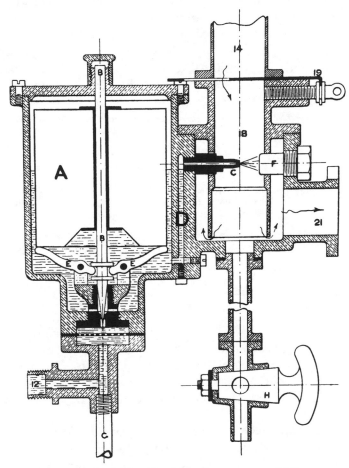

FIG. 423.—PEUGEOT CARBURETTOR, 1899.

upon or stop to examine obstructions, except where they minimise the sectional area of the path. Stepped or roughened cones, for instance, are often put in the air path, but the carburetted air is practically ignorant of their existence as used in some carburettors. Petrol is supposed to impinge on cones and surfaces it never sees, and to be triturated on them. In some places the air rope is caused to contract in area and increase in speed, and then immediately to enter where it may expand, drop in speed and drop much of its petrol.

The Peugeot carburettor (Fig. 423), here reproduced from vol., i., p. 162, is an example of one of the many wrong designs in which one wrong almost corrects another and makes, not a right, but a more or less workable right. Here the air, entering at 14, passed at a high speed through a hole in a throttle valve, then fell in velocity as it passed the jet piece C, and under all circumstances of working had its velocity increased as it passed under the bottom of the mixing chamber tube 18 in its continued passage to the outlet 21 leading to the engine. The jet of petrol was supposed to leap across and be broken up by impinging on the plug end F, but of course it did nothing of the kind. The air current which induced the flow of the petrol, carried it from the jet nipple orifice directly downward in its own direction, and the spray, not completely broken up and evaporated, fell to the bottom, and there, in passing under the edge of the pendant tube, was further triturated and evaporated by the very high speed air. There can be no doubt that this carburettor acted to a considerable extent as a surface evaporator of the shallow pool of petrol which, under some circumstances and medium speeds, would collect there. A cock H was provided for draining off the excess which was likely to collect at starting, and was especially necessary when stale petrol had collected or was delivered from the float vessel.

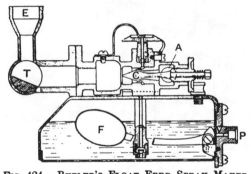

Velocity of air flow

FIG. 424.—BUTLER'S FLOAT FEED SPRAY MAKER CARBURETTOR.

In this carburettor, as in many others, the velocity of the air past the nipple was almost inversely that which is required, as it was slowest when the engine was running slowest, and had not the tube 18 been carried down to the bottom of the mixing chamber, so as to form a spray maker there, the carburettor would not have worked even for the limited range of speed for which it was successful. Like almost all others, it had a fixed area of air passage at the spray jet.

Variable rates of flow of air and petrol

The carburettor made and used by Edward Butler in 1885–87, shown by Fig. 424, went much nearer to a complete fulfilment of all the requirements than many of those made between 1894 and 1903. See Patents, No. 15598 of 1887, 9203 of 1889, and *The Engineer*, July 27, 1890, p. 6. The velocity of the air at the spray jet was variable through a small range automatically, and this range, as well as the quantity of petrol passing to the spraying jet, could be adjusted while the engine was running.

In Fig. 424 the air drawn through the carburettor by the engine connected at E, entered at A, and passed in an annular stream past the valve C, which was surrounded by an annular petrol nozzle, to which the petrol was fed from the tank below, in which the level was maintained by the float F and inlet valve V at P. The petrol was drawn up the pipe seen

CARBURETTORS AND CARBURATION

in the centre of the tank, and at the bottom of which was a non-return valve, into a small space surrounding the end of the adjusting screw valve, with a large head seen above C. From this it passed the valve to the small Butler annulus outside the cone containing the conical valve C, and thence was carried by induction with a small quantity of air entering through the centre of the screw admission valve, and the main air supply past C, triturated and vaporised, or at all events converted into a readily ignitible mist. The spray nozzle valve C was capable of a short inward range of motion according with the engine speed and suction, permitted by a spring at its outer end. The inward movement of C increased the area of the air passage, but, as will be seen, that area could never be greater than the sectional area of the throat minus the sectional area of the stem of C. The range of speed possible with an engine supply by it could therefore not be large, and if the carburettor were of the proper size for and set for working an engine at normal full speed, it could not be suitable for a very low speed. Moreover, the annular form of petrol spray jet would give the spray an irregular character when only a small quantity was passing.

Since the appearance of vol. i., the Longuemare carburettors, shown on Longue-mare p. 323, have received considerable modification. As shown by Fig. 424A, the groove-edged mushroom-shaped jet piece J is still used, but the form of the petticoat tube surrounding it and the means of varying the quantity of air passing under it and inducing the petrol flow have been re-arranged. This tube is carried by the perforated annular plate under R, which is in turn carried by the perforated plate above it, carried by the central stem on which is a controlling lever I. The slotted plate R and the perforated one above it are kept in position by the spiral spring shown, the plate R sitting upon a fixed perforated ring which also carries the petticoat tube. When the engine is running at a slow speed the spaces between the slots in the plate R cover those shown in the plate beneath it, and the whole of the air passing through the carburettor rushes through the small annular space surrounding the jet J, and thus passes through the perforated plate above R; for higher speeds, the arm I is moved so that the holes in the plate R more or less correspond with those beneath it, and air is thus allowed to pass to the engine which does not go through the petticoat tube. The velocity of the air past the jet is thus prevented from increasing with the speed of the engine, and the relation between the air passing the jet and passing direct can thus be regulated by hand to suit the required speed, this being done in conjunction with the use of the throttle valve U moved by the arm I^1 which is fixed upon the upper part of the boss of the internal piece of which the throttle valve U forms a part. The total area of the slots in R equals the area of the annular space round J. The jet chamber is surrounded by a jacket for hot water or exhaust gases entering through a tube C and out through a corresponding tube. More recently the Longuemare carburettor has been modified so as to carry a supplementary air inlet valve, so that it becomes, through a considerable range of engine speed, automatic in its

527

action.[1] The area of the passage past the jet, however, remains fixed as before.

The designers of some of the carburettors which have been made with a view to the supply of a mixture suitable for a large range of engine speeds, **Variable jet opening** have endeavoured to secure this in part by varying the size of the petrol jet by a needle valve moved by a diaphragm or piston affected by the engine suction.

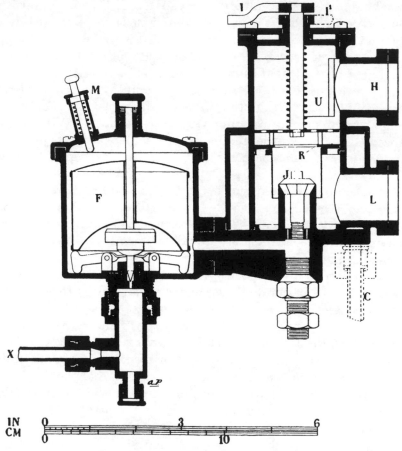

FIG. 424A.—LONGUEMARE CARBURETTOR.

Among this class are the Chenard-Walcker, Brotherhood-Crocker, and earlier in a different form, the Winton. The first of these is shown in section by Fig. 425, in which A is the orifice for pipe connexion to the engine by which air is drawn in at B, through an annular valve seat C, upon which the valve D formed by the bottom of the piston sits. The upper part of the piston is perforated and its periphery grooved to make it, though an easy fit, sufficiently air tight. When the engine moves the piston is sucked upward against the resistance of the spring S and air rushes in through the annular opening B and under the edge D of the valve formed by the piston

[1] *Automotor Journal*, No. 179, p. 707.

bottom. It is deflected to some extent in its inrush toward and round the spray nozzle E, and carries with it an induced spray of petrol, varied in quantity according to the lift of the piston by the taper needle valve G fixed to the upper part of the piston. In this carburettor it will be seen that the air admission passage is of fixed area, both at B and at C, the air inlet being varied by the lift of C above B. The petrol supply would not vary directly with the quantity of air, were it not for the needle valve. The carburettor is simpler than some, and a great deal has been written about it by Mr. Walcker, but except that any petrol escaping the nozzle and falling in a liquid state must be met by the inrushing conical film of air from C, there does not appear any particular reason for expecting specially effective working.

The Brotherhood-Crocker carburettor is a much more elaborate combination of parts, a piston at the top of the carburettor being connected, not only to a needle valve, but also to a conical petticoat pipe surrounding the nozzle and through which all the air to be carburetted must pass. The increase in suction by increase of engine speed lifts the piston and at the same time the needle valve and air cone. The sectional area of the lowest part of the petticoat cone is about two and a half times that of its waist and these two dimensions fix the velocity at lowest speed of engine past the jet and the speed and quantity of air when the engine is at maximum speed; and even then the whole of the air entering the bottom of the air cone must pass through the waist. The needle valve is taper squared, so that as it rises after a certain lift, the petrol orifice increases. When the piston and cone have lifted a short distance, further lift is

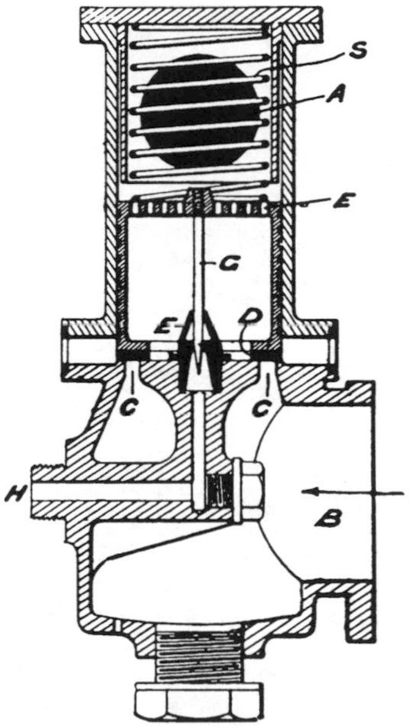

FIG. 425.—THE CHENARD-WALCKER CARBURETTOR, 1904.

attended with a supplementary inlet of air from ports uncovered by a sleeve, so that the highly charged air from the bottom cone surrounding the nozzle is diluted by pure air, giving the same effect as in the Germain, Krebs and other carburettors; but these last-mentioned have no moving needle valve.

The Krebs carburettor illustrated on page 144, originated the recent return to that which was less efficiently done by Maybach and by the Daimler Company, who used a supplementary air valve on their old 6-HP. engines with the cast-iron induction pipe as shown at p. 93, vol. i., but the subject was much better understood when Krebs' combination was designed. In Krebs, like nearly all others, the air inlet at the jet is of fixed sectional area. It is of a sufficient size to produce an adequate inductive effect on the petrol nozzle at the lowest predetermined speed of the engine.

At higher speeds the air inlet would not only be insufficient, or would cause considerable resistance, but too large a quantity of petrol would be drawn in with it. Before this effect arises, the piston valve or supplementary valve opens through the suction action of the engine on the large area of the Krebs piston and diaphragm. This is much larger than that of the previously used snifting or supplementary valve. Hence the valve moved by it is moved with a smaller minus pressure in the engine cylinder than would be required to move an ordinary spring mushroom valve of sufficiently gradual action, and satisfactorily small and gradual admission of air, with considerable piston movement, is secured. The total quantity thus admitted is such as to secure a nearly constant velocity at the nozzle.

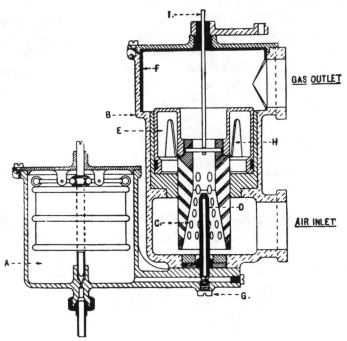

FIG. 426.—THE ROVER CARBURETTOR.

Rover The carburettor of the Rover car designed by Mr. E. W. Lewis is an exception in this respect. It is shown by Figs. 426 and 427. In this the engine suction lifts a piston E and with it a conical perforated petticoat c surrounding the jet D. When in its lowest position and the engine moving at its slowest speed, the jet is at the smallest part of the cone, and hence, with even a small volume of air passing to the engine, the velocity at the jet is sufficient to induce an adequate petrol spray. When the cone is in this position, no air passes through the inclined perforations in the cone, all passes up through the centre to outlet, there being no outlet from the hollow piston E in the position shown. As the engine speed increases, the piston E lifts and the Λ-shaped ports H rise above the level B. There is, then, a clear passage for the air through some of the inclined passages

530

into the hollow piston E and thence through the ports H into the space above, the outlet from which is controlled by the throttle valve F. At full speed the largest diameter or bottom of the air cone is at about the level D, and some air can then pass through the inclined passages into the hollow piston; but even then, most or a good deal of it must pass the top of the nozzle. Some of it being thus free to escape to the inclined passages, the quantity and hence the velocity of that which must pass up the central passage is lessened, and the inductive effect on the petrol is lessened and excess supply prevented.

Thus none of the carburettors at present in use, with this one exception, pass the whole of the air used past the nozzle; all have a supplementary air supply in some form, many of them having merely a lightly spring-loaded inlet valve at or near the carburettor and between it and the engine. All of them, including the Rover, have a permanently fixed minimum clear way for air past the jet piece, determined by the minimum velocity of air necessary for inductive action on the petrol or other fuel. This determines the minimum speed of engine piston at which the air current can have a sufficient velocity to induce a petrol flow sufficient to make the carburettor provide a working mixture.

Passages of fixed area limit speed reduction

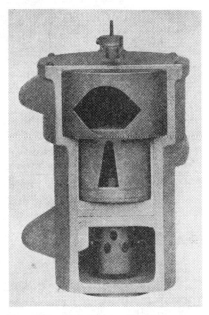

FIG. 427.—THE ROVER CARBURETTOR.

The carburettor illustrated by Fig. 428 has been designed with the object of obtaining automatic and equable carburation for all speeds.

The whole of the air used passes the spray jet and in quantity it may vary from zero to a maximum; it acts as a governor valve without throttling and as a cut off at maximum speeds. There is no fixed minimum air passage which must remain open through a considerable range of low speed, as the air passage varies in area from closed to full open with the jet nozzle always in the centre of air flow. It may be used as a hand-controlled regulator, so that there is no necessity for separate throttle valve. The jet nozzle, indicated in dotted lines, is a sliding fit in the oblong slot in the bottom of the cyclinder C.

Automatic carburettor design

Air enters at A through wire gauze, or at B from a hot-air communication pipe. A cylinder C, with an open end, except for the spider (Fig. 2) carrying the rod D, is bored out in conical form and the end closed afterwards by a screw. At the part F the cylinder is cut away from the cone to the outside by a slot as shown in plan in the sectional plan. This permits the air passing the jet piece to enter the expansion chamber G, which

531

may be jacketed as indicated by dotted lines. In the position shown, the air way is completely or nearly closed. With the movement of the engine piston the internally coned cylinder c immediately moves and permits air to rush through the small passage, until then closed, at the upper part of the jet piece. More rapid movement of the engine pulls the cylinder farther in the direction of the arrow and allows more and more

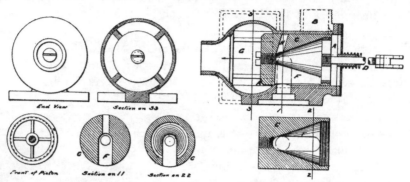

FIG. 428.—AUTOMATIC CARBURETTOR AND GOVERNOR.

air to surround and pass the jet piece, the air passage F through the cone cylinder and opening being proportioned so that the air velocity past the jet piece secures the proper relation between air and petrol output. A hole H or a slot at that side of similar length to that shown in Fig. 3, but narrow, is used to create similar divergence of air current.

Excessive speed of the engine will pull the cone cylinder until its end seats as a valve at J and thus stops the engine; or by making a few grooves in the seat J or on the end K of the cone cylinder a sufficient quantity of carburetted air may be allowed to pass to keep the engine going but prevent its running away. The governor and the hand control may both be connected at M.

With the form shown by Fig. 429 a throttle valve is required. In it a piston A with a conical interior surrounds, and when in its lowest positions as shown, encloses the top of the jet tube B, the air passage being closed. The slightest movement of A will allow air to pass by the top of the jet and through two small holes c c into the enlarged part c of the casing, and thence to the engine at D or d As the engine increases in speed the piston rises and the three or four holes e e rise at different levels so as to deliver to c, these holes being rather larger in diameter than the top holes. Greater speed of the engine is obtained by the further lifting of the piston A by the engine suction or by raising it by hand or pedal connexion to the rod E, to which also the governor

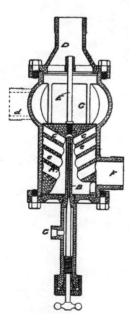

FIG. 429.—AUTOMATIC CARBURETTOR.

may be connected. Air enters at F and petrol at G. For use with the heavier spirits either of these carburettors should be at or above the level of the valve chambers and the distance between carburettors and valves should be short.

Either form may be used horizontally or vertically, and for use with some of the heavier kind of spirits the pipe connexion to the engine cylinders should be short and of the Royle type or a flat tube of section of such proportions as $1\frac{1}{2} \times \frac{3}{16}$ in.

In all this concerning carburettors I have not said anything concerning surface carburettors, as only the Lanchester Company now use them as far as I know.

The spray jet carburettor has displaced nearly all others; even the measured positive feed used by MM. Gobron-Brillie, p. 193, vol. i., has been superseded by a float-feed jet carburettor in which a collar surrounding the jet [1] (which as usual is in an open air way of fixed diameter) can be moved up or down by the driver to restrict or permit full flow of air past the jet to the engine. A jacket is formed round the tube between jet and throttle valve.

The de Dion-Bouton carburettor for two or more cylinders is, like the Brooke and the Argyll, provided with two jets and sets of passages, a duplex carburettor in most respects.

Experience shows that the inductive effort necessary for raising the petrol to the jet orifice should be very small, and for this reason the level of the liquid in the float vessel is kept by the float valve very near the level of the jet nozzle. It is, in fact, kept in unstable equilibrium at this point. It is, therefore, unnecessary to go into the question of the velocities necessary to maintain a column of liquid a given height or the slightly superior velocity necessary to raise the liquid to the overflow point. It is, however, important to note in connexion with the working of a jet carburettor, that a given increase in the velocity of the entering air current is not attended with an arithmetical increase in the depression of the pressure in the cylinder or induction effect at the jet. **Inductive effort at jet**

Taking as example the dimensions and speed of single cylinder previously referred to, namely, cylinder 4·5 in. diam., stroke 5 in., revolutions 800, then, the area of the piston being 15·9 square inches, the velocity of the air entering a tube of 0·5 square inches sectional area or $\frac{13}{16}$ in. diameter, will be
$$\frac{15 \cdot 9 \times 5 \times 800 \times 2}{12 \times 60} = 352 \text{ feet per second, corresponding to a minus pressure}$$ **Velocity of flow**
in the cylinder of a little over 1 lb. per square inch, assuming that the greatest resistance to the air passage is at the jet tube entrance.

The following table gives the velocity of air entering a vessel in which there is a partial vacuum of the minus pressures given in lbs. below atmospheric pressure :—

[1] *Automobile Review*, December 31, 1904, p. 631.

TABLE 15.—VELOCITY OF ENTRY OF AIR INTO PARTIAL VACUUM.

Pressure below Atmosphere.	Velocity in Feet per Second.	Pressure below Atmosphere.	Velocity in Feet per Second.
0·5 lb.	249	4·5 lb.	619
1·0 ,,	337	5·0 ,,	636
1·5 ,,	405	5·5 ,,	650
2·0 ,,	459	6·0 ,,	661
2·5 ,,	503	6·5 ,,	669
3·0 ,,	540	7·0 ,,	673
3·5 ,,	571	7·5 ,,	675
4·0 ,,	597	8·0 ,,	673

From this it will be seen that the greatest velocity of entry of air into a vacuum is at about 7·5 lb. absolute, or half the pressure of a perfect vacuum.

The greater depression sometimes shown by an indicator diagram is thus due, either to resistance, resulting from tortuous and restricted passages between entry and the cylinder, and valve resistance, or to incorrect record by the indicator pencil.

Change of atmospheric pressure The fall of atmospheric pressure and density at high elevations may possibly have some effect on the action of a carburettor, but it is questionable whether in the ordinary course of automobile travelling, its effect will be noticeable in the engine performance in this country. The barometric fluctuations at or near sea level between the conditions conventionally indicated by " stormy " and " very dry " are shown by a change in barometer readings as great as those which, under equal temperature and other conditions, would show a hypsometric difference of over 2,600 feet, and the difference in atmospheric pressure would be 1·48 lb. per square inch. The difference in compression would, therefore, be almost negligible on nearly all the roads of the United Kingdom, and part of this would be neutralized by the accompanying difference in evaporative facility. The difference at or near sea level of a change from say 30 to 28·5 in barometric height is not accompanied by any material effect on the running of a motor, so there is no reason for fearing trouble from a change in altitude of 1,300 or 1,400 feet.

Carburettors and Vaporizers Reference should be made here to some of the vaporizers now used separately or in conjunction with petrol carburettors for working with oils such as those known as paraffin or kerosene—ordinary lamp oils. This subject was treated at some length in vol. i., chapter xvii., but I am afraid that few of those who have since turned their attention to vaporizers as compared with carburettors have learned the lessons they might have done from a study of the descriptions given of the results obtained by oil-engine experimenters. Reference has already been made (page 436) to the combined petrol carburettor and paraffin vaporizer used in some of the Canstatt or Marienfelde Daimler wagons. The most recent forms in use show that

534

the formation of a paraffin mist is, as shown in vol. i., p. 327, the most important function of the apparatus, that a moderate heating, or rather warming, of the chamber wherein the liquid is converted—chiefly mechanically—into a very fine spray or mist, is necessary to keep the mist from condensation ; and, equally important, that it is necessary to get the mist-forming apparatus as close as possible to the admission valves and above rather than even the least below them. One of the best forms of vaporizer is that of the Wolseley Tool and Motor Co. (*Automotor Journal*, 22 April, 1905, p. 506). Others, including that recently tested on a 500-mile trial, under the Automobile Club, on a Darracq car and made under the patent of Bryan and Watling ; and the Sutton (*Automotor Journal*, 4 June, 1904, p. 677) may be mentioned.

Chapter XXXIV

ELECTRIC MOTOR VEHICLES

IN vol. i., chapters xxii. and xxiii., there were illustrated a number of the electrical vehicles then made by the most prominent in this field. No great advances have been made since then, very little invention, only the less attractive but important plodding work of improvement in details.

Although there has been slow but steady improvement in the design and in the results obtained with the secondary batteries or accumulators no radical change has been made, and the combined weight of the cells and their containing boxes still limits the speed at which the vehicles can be run satisfactorily, and the distances they can run on one charge before it becomes necessary to either make a stop of some hours' duration for recharging or a shorter delay for replacing the group or groups of cells with others freshly charged. It is therefore evident that unless special arrangements are made for recharging and for replacing the cells the period during which the vehicle may be kept in service is seriously limited and its convenience and usefulness proportionately reduced.

Prohibitive weight of battery

The natural outcome of these limitations has been the growth of companies owning and maintaining considerable numbers of electric carriages which may be hired by arrangement for long or short periods, those hiring or owning them being relieved of the care of supervision and maintenance, in a condition of readiness for service. The thorough organization of, for instance, the City and Suburban Electric Carriage Co., and the Electromobile Co., of London, has resulted in more satisfactory running, with lower costs for repairs and for renewal of accumulator elements due to the regular and systematic attention given to the batteries. These companies and others who have made similar arrangements for the housing and maintenance of numbers of cars must necessarily select sites for their depôts as near as possible to the districts served, or if possible in the centre of those areas in order that those using the cars may not be inconvenienced by the delay or time occupied in running from and to the depôt, and that undesirable and wasteful expenditure of electric energy may be minimized. Since the success or continued use of the accumulator-equipped electric car is at present largely due to the reduced running and maintenance costs

Central garage and maintenance systems

536

made possible by the universal garage system of the companies referred to, it is not surprising to find that the majority of electric cars are in use for travelling in towns and that fewer attempts are now made to provide what have been called electric touring cars. Although, as already mentioned, it has been found possible to reduce the weight of the accumulators without at the same time reducing their capacity, it does not appear that this reduction has much changed the weight of the cells in use. Now that it is generally realized that long-distance travelling is not yet feasible with the electric car the capacity of the accumulators has been in some instances purposely reduced in order that an appreciable saving in weight might be effected. The weight of the electric broughams **Weight** and landaulets at present in service may be taken to be from 27 to 30 cwt., and they are provided with accumulators having sufficient capacity to run the vehicles from 25 to 40 miles and which weigh between 10 and 14 cwt.

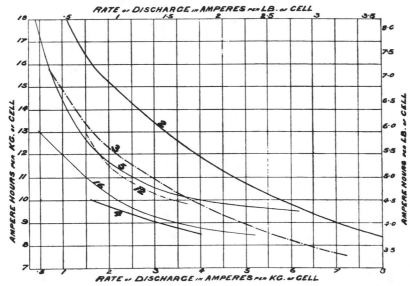

FIG. 430.—CURVES SHOWING THE VARIATION OF THE SPECIFIC CAPACITY WITH THE RATE OF DISCHARGE FOR BATTERIES USED WITH ELECTRIC AUTOMOBILES.

Fig. 430 indicates the performance of a number of the cells given in Table 16, and may be taken to show the results now obtainable with those cells of the lead type which are most largely used and their weight. The curves show the variation of specific capacity with the rate of discharge, the weight of cell considered including the electrolyte, connexions, **Performance** and containing jar or box. Thus curve number 3 relates to the ATI5 type **and weight** of E.P.S. cell, and shows that at a discharge rate of, for instance, 1·5 **of cells** amperes per lb. of cell its capacity was 4·75 amperes hours per lb. of cell. The weight of this cell is 31·7 lb. and is heavier than the E.P.S. cell used by the City and Suburban Company, which is of the ATFII type, and weighs 21·7 lb. The lighter cell at a discharge rate of 1·5 amperes per lb. has a capacity of 4·61 ampère hours per lb. The weight of a cell for a given

Space occupied by cells

capacity in comparison with others of known dimensions provides some index to the space occupied, but Table 16 enables a definite idea of the relative area occupied by batteries of different cells of the same nominal capacity to be formed from the dimensions of length and width of the containing box, and the areas there given. Most of the curves shown by Fig. 430 are the results of tests conducted by M. Henri Joly, and I am also indebted to him for the table of dimensions of cells here given which is taken from a diagram prepared for an article on the " Development of the light accumulator since 1900." In this article[1] illustrations of the accumulator plates are given, and there is much interesting information relating to the construction, dimensions and performance of light auto-mobile accumulators.

TABLE 16.—LENGTH, WIDTH, AND AREA OCCUPIED BY DIFFERENT CELLS OF 150 AMPÈRE HOURS CAPACITY.

	Name of Cell.	Length.		Width.		Area.
		Inches.	Centimetres.	Inches.	Centimetres.	Sq. inches.
1	Berliner	7·40	18·80	4·50	11·4	33·3
2	Contal (Electricia) . .	5·00	12·70	4·40	11·1	22·0
3	Electrical Power Storage	5·50	13·90	4·40	11·1	24·2
4	Exide	6·15	15·60	4·25	10·8	26·1
5	Fulmen	7·00	17·80	4·40	11·1	30·8
6	Gould	6·25	15·85	4·25	10·8	26·5
7	Hart	8·00	20·40	3·60	9·3	28·8
8	Heinz	6·25	15·85	4·40	11·1	27·5
9	Helios-Upton . . .	6·65	16·80	4·20	10·6	27·9
10	Leitner	6·25	15·85	4·40	11·1	27·5
11	Lithanode	6·25	15·85	4·75	12·0	29·7
12	Max	6·25	15·85	4·15	10·5	25·9
13	Phenix	7·30	18·50	3·75	9·5	27·4
14	Porter	5·25	13·30	4·15	10·5	21·7
15	Rosenthal	5·00	12·70	4·40	11·1	2 ·0
16	Progress	—	—	—	—	—

Long distance travelling

From time to time cars have been specially equipped for long-distance runs, and distances of 180 and 190 miles have been travelled without recharging or changing cells.[2] These results are interesting, but they do not possess much practical value, and any comparison of such results with those easily obtainable from petrol or steam propelled cars is very unfavourable to the electric car.

In the existing state of the industry their use is restricted to town

[1] The *Electrochemist and Metallurgist*, Oct. 1903, No. 17. Sherard, Cowper-Coles, Ltd. See also Dec. 1903, No. 19, and succeeding numbers for references to the performance of the Edison and other batteries, in a series of articles by M. Henri L. Joly.

[2] " Electric Automobiles," by H. F. Joel, *Proc. Inst. C.E.* vol. clii.

running and is at present confined to those who are able and prepared to consider them as luxuries.

Heavy commercial accumulator-equipped vehicles have been used in **Heavy vehicles** the United States, but in all cases the heavy running expenses and maintenance costs have rendered them unsuccessful, although in some instances suitability of design and ease of control favoured their use. As with the lighter passenger vehicles, everything pointed to the necessity of their use in sufficient numbers to justify the existence of running depôts with facilities for replacing and recharging and for overhauling or repairing.

The total weight of one of the " Columbia " lorries [1] capable of carrying a 5-ton load is 12,000 lb. or 5 tons 7 cwt., and the cells accounted for 3,500 lb. of this total, equivalent to 29 per cent. of the weight of the vehicle. With this larger and heavier vehicle the relation between the weight of the accumulator and the total weight remains about the same as that already given for the average passenger car. The design of the Columbia lorry is distinctive in that it is equipped with four motors each capable of working at a maximum rate of 4·5 HP. The motors run at 850 revs. per minute and independently drive the four road wheels through single-reduction spur gearing, the second gear reduction being through $1\frac{1}{2}$ in. × $\frac{3}{4}$ in. roller chains. With this system of driving no differential gear is required and the whole weight is available for tractive adhesion. The battery consists of 44 Exide cells (see Fig. 430 and Table 16), having a capacity of 280 ampère hours, and with these it is stated that the lorry can be driven at 6 miles per hour with a 5-ton load. The energy consumption is declared **Energy consumption** to be 35 watts per 1,000 lb. per mile, from which it appears that a distance of about 30 miles could be run on one charge under favourable road conditions.

The total length of the vehicle is 20 feet, the wheel base 11 ft. 6 in. and the wheel gauge 5 ft. 10 in. The wheels are 36 in. diameter and Turner solid rubber tyres 7 in. wide are used. Heavy roller bearings are used for the road wheels. Locking plate or " fifth-wheel " steering is employed and with the front truck road wheels acting as drivers this form of steering is advantageous and manoeuvring in confined spaces is possible that could not otherwise be attempted.

Interesting experiments have from time to time been made with what are known as petrol electric cars, and, as the name implies, a combination **Petrol-electric cars** of petrol engine, dynamo and accumulators is employed to supply current to an electric motor or motors driving the road wheels.

The results obtained have not been sufficiently encouraging to lead to the construction of such cars in numbers. The combined losses in conversion and transmission must necessarily be greater than obtains with the simple petrol engine and transmission gearing alone, although at the heavier rates of working the difference in economy may not be pronounced. An advantage that is sometimes claimed for the system is the ability to utilize the reserve power represented by the energy stored in the accumu-

[1] *Horseless Age*, vol. xi. p. 705.

lator at the same time that the maximum output is being obtained from the dynamo. The dynamo and accumulator are then in parallel and the motors receive the energy from both sources and are able to work at considerably augmented powers for short periods when climbing steep hills. The modern well designed and constructed electric motor can withstand heavy overloading for short periods without injury and the motors need not be heavy or designed for powers much above the normal. If, however, much advantage is to be derived from the use of the accumulators, the addition to the weight of the vehicle becomes a matter of importance, and in any case the combined weight of the dynamo, motor or motors, controller, resistances, switches and leads is greater than the equivalent change-speed and transmission gearing in the simple petrol-propelled vehicle. It is interesting to record that, with the Fischer heavy commercial vehicles working on this system, troubles due to neglected accumulators are very much reduced, due, no doubt, to the cells being kept constantly charged, and because the load factor or the mean rate of working of the cells is low and their useful life in months is consequently prolonged. This, however, means that the load represented by the cells has to be continuously carried and that recognized as a comparatively inactive part of the car it must also be looked upon as weight inefficiently utilized.

Petrol consumption During a six days' test of a 5-ton lorry running in New York city and over some hilly routes in Long Island, 23 loads of from 4 to 5 tons were carried over 139 miles, equivalent to 620 useful ton miles, the consumption of petrol being $84\frac{1}{2}$ gallons. This is equivalent to 7·35 useful ton miles per gallon of petrol or to a consumption of one gallon for every $1\frac{1}{2}$ to $1\frac{3}{4}$ miles run.[1]

This compares very unfavourably with the results obtained with the Milnes-Daimler lorry as given in Table XX., vol i. p. 636, and is in itself a sufficient proof of the inefficiency of the combination.

During the summer of 1902 a private omnibus, seating eighteen passengers, including the driver, was to be seen in London. It was provided with a three-cylinder engine direct coupled to a 5-K.W. multipolar dynamo, and ran at 600 revs. per minute. The cylinders were $4\frac{3}{4}$ in. diameter and the stroke was $5\frac{1}{2}$ in. Two bipolar shunt-wound motors, each rated at 5-HP. at 600 revs. per minute, were arranged to drive the rear road wheels through double-reduction spur gearing. The front wheels were 38 in. diameter, and the rear wheels 46 in. diameter, 4 in. solid rubber tyres being used. A battery of fifty Madden Chloride cells was used, having a capacity of 90 ampère hours. The controller was of the series parallel type with five forward running positions and three for backward running, and speeds up to 10 miles per hour could be maintained. The weight of the omnibus was 4 tons.[2]

Other vehicles of Continental and American origin have been designed in which a petrol engine is directly coupled to a dynamo which supplies

[1] *Horseless Age*, vol. xiii. p. 3.
[2] *Automotor Journal*, June 14, 1902.

current to motors driving the road wheels either directly as in the early Lohne Porsche cars or through spur or chain gearing. These, like the Fischer type, cannot be described as electrically propelled vehicles, but they involve the use of electrical machinery. One of the Dynamobile [1] cars of German make, capable of seating six persons and provided with a two-cylinder 12-HP. Fafnir engine is declared to weigh one ton. The Siemens and Halske [2] road cars are of interest and have been successfully **Siemens and** used in Germany and notably on the country roads in the neighbourhood of **Halske** Dresden. They have also been used in the Bielathal in Saxon Switzerland and in Italy at Turin. The vehicles are of tramcar or omnibus design and take their current supply from overhead wires through trolley poles. Double trolley poles are required and two overhead wires, for the supply and return circuits, since the rails available with the tramcar for the return circuit are not used. The weight of a twenty-two-seated omnibus is a little over 2,000 lb. Some of the vehicles have been equipped with accumulators to enable them to cover intermediate parts of the routes where the overhead wires may not be employed. Their weight is then considerably more and the advantage of the higher voltage of the current supply cannot be so well realized as when the overhead equipment covers the whole route. In those districts where water power is available for the cheap generation of electric energy the Siemens and Halske system of road traction may not only satisfactorily meet the traffic requirements but it may also be commercially successful, since it does not entail the heavy cost of track construction, and interference with the public road and obstruction of traffic, the vehicles being able to deviate from their course either to pass or accommodate other users of the road.

Modern electrically propelled cars have been able to a small extent to contribute to, and to a greater extent to profit by, the changes and improvements that experience has dictated in construction, and the use of materials now available for car construction. In many instances little change in the electrical equipment of these cars has occurred and only structural or mechanical improvement has been effected.

THE ELECTRIC MOTIVE POWER CO.

One of the vehicles constructed by this company to the designs of Mr. Percy W Northey is illustrated by Figs. 431 to 434. It differs from other cars both in the arrangement of the mechanism and in the means provided for its control. An external view with the body in position is not given, but from the elevation and plan (Figs. 431 and 432) it can be seen that either the Brougham, Limousine, or Tonneau type may be carried upon the straight-line frame adopted. An armoured wood frame is shown **General** with five transverse members, and additional longitudinal bars to support **arrange-** the body and to carry the accumulator boxes and the parts of the mechanism **ment**

[1] The *Electrical Engineer*, vol. xxxv. p. 441.
[2] The *Electrical Review*, 5th Sept. 1902; 6th Feb. 1903.

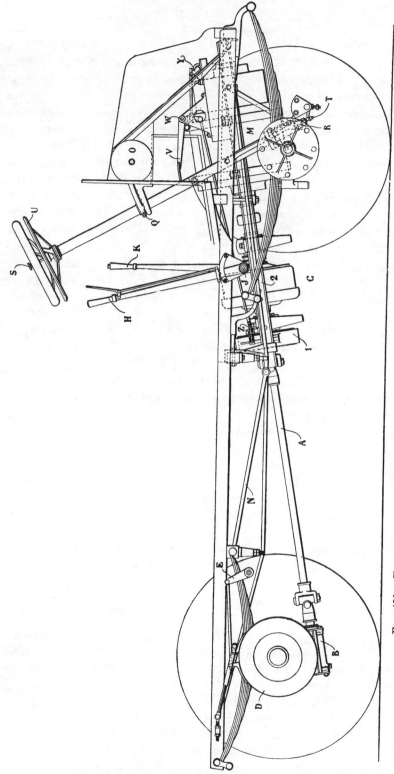

FIG. 431.—ELECTRIC MOTIVE POWER CO.—SIDE ELEVATION WITH BODY AND WHEELS REMOVED.

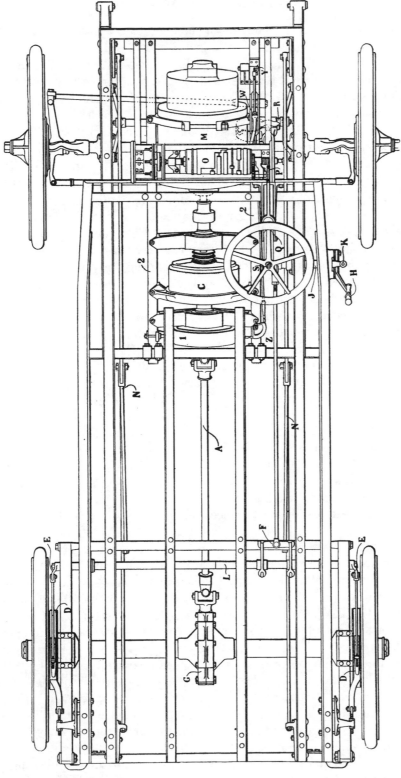

FIG. 432.—ELECTRIC MOTIVE POWER CO.—PLAN OF VEHICLE WITH BODY AND CASINGS REMOVED.

connected to the frame. The motor M, magnetic clutch and gear box C, and brake drum 1 are secured to the inclined frame members 2, and motion is transmitted through the universally jointed shaft A to worm and worm-wheel driving gear inside the casing G, which also contains the differential gear. The frame is stiffened against vertical bending by truss bars N, and by others below N running the whole length of the frame and connected to brackets at the front and back ends visible in Fig. 432.

Two-speed gear and magnetic clutches
The arrangement of the magnetic clutches and two-speed gear is shown by Fig. 433. A^1 is the shaft driven by the motor having a bevel pinion and the annular electromagnet C fixed upon it. The ends of the magnet winding are, as shown, connected to the insulated collecting rings D and E. A second electromagnet C^1 similar to C is fixed through the frame G to the inclined frame bars 2, and does not rotate. Between the electromagnets C and C^1 is a disc armature F which may run free on the transmission shaft A and is capable of the little end play or sliding movement equal to the clearance between the armature and magnet faces. The armature is connected to the box H through driver pins, one of which may be seen, and a bevel pinion loose upon the shaft A is fixed to H, so that these three parts rotate together or when the electromagnet C^1 is excited and attracts and holds F are together prevented from rotating. The transmission shaft A is hollow at its inner end and the shaft A runs within it. On the inner end of A are short spindle or pivot arms J carrying the two bevel pinions always in gear with the pinion fixed on the inner end of A^1, and that loose on the inner end of A.

When the rotating electromagnet C is excited the armature F and consequently the pinion fixed to the box H are obliged to rotate at the same speed as the shaft A^1 and the bevel pinion fixed to it. There can then be no relative motion between the bevel wheels, and the shaft A is driven at the same speed as the driving shaft A^1.

To obtain the low speed the electromagnet C^1 is excited instead of C. The armature F is then held by it and with H and the pinion connected to it is prevented from rotating. The rotation then of A^1 and the pinion fixed to it, causes the pinions on the arms J to roll round the pinion held by H and thereby gives motion at half speed to the shaft A.

To avoid the effect of residual magnetism and to enable the armature F to leave the magnets immediately on opening circuit the faces of the armature are electroplated with copper. When the low speed is in operation the thrust from the bevel wheels is taken up by the ball-bearing shown.

The use of a simple form of change-speed gear with an electrically propelled vehicle is advantageous because it relieves the battery of the occasional very heavy current discharge when the vehicle is being driven uphill with the motor running slowly. If by the use of the speed-reduction gear the vehicle is run at a lower road speed and the motor at more nearly its normal speed and power, the injurious heating effect of the heavy current through the armature and field windings is reduced and damage by burning out is less likely to occur. The two-speed gear used with the White steam cars and illustrated by Figs. 398 is an example of a type of gear that could be usefully employed.

The casing G at the centre of the rear axle containing the driving and differential gear is shown by Fig. 434. The steel worm spindle C, driven by the propeller shaft A (Figs. 431 and 432) has one-half of the rear universal joint cottered upon its tapered end at E, it is carried in ball bearings, and runs drowned in oil in the pocket B below the casing G. The ball bearings

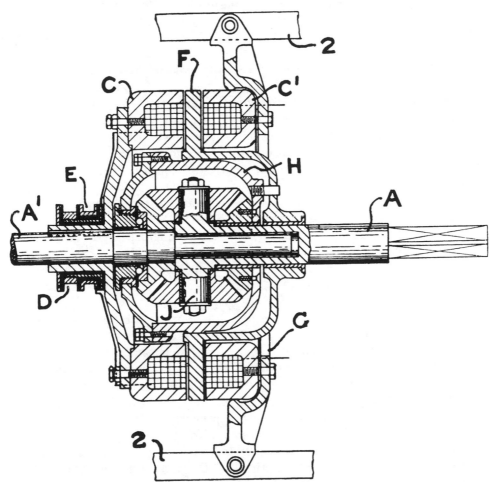

FIG. 433.—ELECTRIC MOTIVE POWER CO.—ARRANGEMENT OF TWO-SPEED GEAR AND MAGNETIC CLUTCHES.

are adjustable by the nut and check nut arrangement shown. The five-threaded steel worm gears with the phosphor bronze worm wheel D secured and spigoted upon the crown wheel or casing F around the differential wheels. The radial arms H, upon which the pinions I take their bearing, are centred through their junction boss upon the inner ends of the hollow driving shafts A and A¹, and at their outer ends are held and driven by the differential casing F. The bevel wheels J, which receive the driving effort are fixed upon the tapered parts of A, A^1, by ring nuts and grub screws. The road wheels are fixed upon the outer ends of the driving shafts which rotate inside the fixed tubular part of the rear axle. Top and bottom screw plugs are provided through which oil may be added or removed and the gear cleaned by flushing through occasionally. The pitch of the worm threads upon the spindle C is necessarily great enough to allow the worm to drive or to be driven by the renewable worm wheel D.

System of control

The control of the car has been simplified as far as possible and side levers and pedals have been arranged to operate switches and brakes so that those accustomed to the control of a petrol driven car can operate similar levers and pedals having equivalent functions.

The controller drum O (Figs. 431 and 432) is arranged transversely in front of the dashboard and below the bonnet. The steering column is pivoted at R, upon a transverse spindle supported by hanger brackets fixed to the frame. It may be canted or rotated through a limited range about the centre R. The casing containing the worm and worm wheel of the steering gear is indicated in dotted lines in Fig. 432, and the universally jointed connecting rod to the near side steering head, the coupling rod behind the front axle, and the arms to which they are connected may also be seen in this view. It will be noticed that the steering wheel pivots are situated inside the dished or recessed wheel hubs and in the plane of the wheels and the steering gear is protected from efforts tending to move it resulting from shocks at the road wheels. There is, it will be noticed, a light wheel U, below the steering hand-wheel, connected to the tube or sleeve terminating at Q.

Operation of controller

At this end of the sleeve the projecting end of a rack is connected. When by slightly gripping or lifting the wheel U and its attached sleeve, the detent Q is disengaged from the notched quadrant plate below it, the steering column may be pushed forward or pulled backward, thus giving movement to the toothed rack and the pinion P (Fig. 432) with which it gears. Rotation of P gives motion to a second pair of wheels visible in Fig. 432, and so revolves the controller drum O. When the desired running position has been reached, the wheel U is allowed to drop, and the detent Q engaging with the nearest notch on the quadrant locks the controller in the set position. Notches are provided for four speeds, rotation of the controller drum to the corresponding positions producing various electrical combinations of the motor windings and series or parallel grouping of the cells. The side lever K is connected to a reversing switch so that the speeds of running permitted by the controller are also available for reversing. A knob S actuates, by a central rod, a two-way switch T at the base of the steering column. When the knob is pulled up to the position shown the electromagnet C (Fig. 433) is excited, and the direct drive obtains. When it is pushed down the electromagnet C^1 is excited and the low-speed gear is in operation. One pedal when depressed operates a controller switch for recuperating, the momentum of the car on down grades driving the motor as a generator. A second pedal V pivoted at W controls a circuit breaker at Y and also by rod connexion with a bell

Reversing and change-speed gear

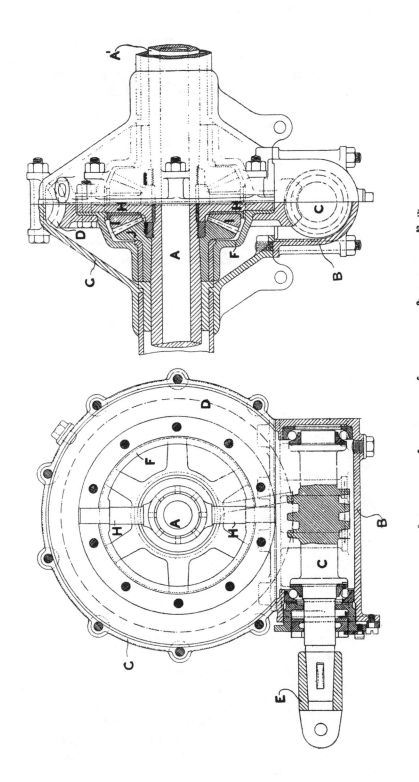

FIG. 434.—ELECTRIC MOTIVE POWER CO.—WORM AND WORM WHEEL DRIVING AND BEVEL DIFFERENTIAL GEARS.

crank z tightens the band I around the brake drum upon the rear end of the transmission shaft. The use of one or both of these pedals is thus equivalent to declutching and using the foot brake on a petrol-propelled car. The side lever κ operates the switch for reversing and by means of **Brakes** the second lever H also pivoted at J the band brakes, acting upon the drums D fixed to the rear road wheels, may be applied through the rod z connected to the equalizing crosshead at F. Connecting links transmit the effort to a sleeve and cross shaft L and at the outer ends of these are arms E connected by tension rods with the brake straps. The vehicle is carried upon semi-elliptic springs, those at the front being longer and made up with more plates than those at the rear axle.

The battery consists of 44 cells, having a capacity of from 160 to 180 ampère hours when discharged in four hours. When the vehicle is arranged as a landaulet the cells are contained in two boxes carried at the front and at the rear of the car.

THE ELECTROMOBILE COMPANY.

The design of the vehicles belonging to this company has been changed from time to time, and the illustrations that are here given do not embody some quite recent improvements that have been effected. The general arrangement of the underframe and mechanism is, however, well shown by Figs. 436 and 437, and the departures from design shown in these views will be referred to. The external appearance of the car with brougham body is shown by Fig. 435, and although it does not differ materially from others of the type, attention has been paid to details, such as the formation of the dash, the curvature of the guards, and to other external features, while the height of the frame from the road and its straight line form, and the lengthened wheel base all contribute to the handsome appearance of the complete vehicle.

FIG. 435.—THE ELECTROMOBILE CO. STANDARD BROUGHAM, 1904.

Frame and springs The frame is built up of channel steel suitably strengthened at the points of connexion of the transverse members by angle or gusset plates.

The top line of the frame is little more than 2 feet from the ground, and in order that there may be room for deflection of the long semi-elliptic springs used at the rear of the vehicle it has been necessary to space them wider than the frame longitudinals and to use the offset dumb irons shown at the back, to which the springs are connected by shackles. At their forward ends the springs take on to stud-pins or pivots fixed to the centre

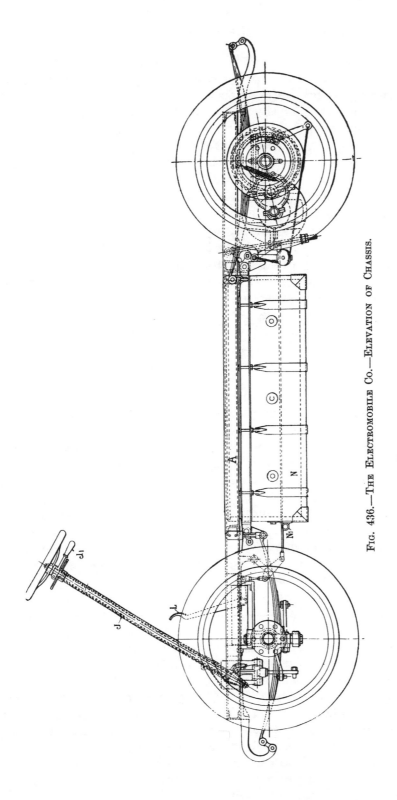

Fig. 436.—The Electromobile Co.—Elevation of Chassis.

or webb of the steel channels. Still easier suspension has been obtained in the recent frames by the use of a rearwardly projecting semi-circular bearer bar resting at its centre upon a transverse spring connected by shackles to the rear ends of the side springs. There is less difficulty in obtaining the requisite clearance for the front springs since the fixed front axle may be set as low as is necessary for this purpose and it will be seen that the front springs are situated in plan directly below the longitudinal frame channels A. Instead of the dumb irons shown the front ends of the springs are now connected to short upper laminated springs fixed to the frame, the rear ends of the main springs being shackle connected.

The steering column J is inclined at 30 degrees with the vertical, and motion is given from the hand-wheel through a concentric inner tube to bevel gear, having a ratio of one to four. The steering gear is not, therefore, of the irreversible type, but this is not necessarily a disadvantage when the speed of the vehicle is moderate.

Battery box The battery box N is carried below the centre part of the frame in such a way that it may be comparatively easily disconnected and taken away for recharging the cells, a new box of cells freshly charged taking their place. When arrangements are made for lifting or taking the weight of the battery and its box and for readily transferring it to the charging room at the garage, the moving of the weight of 11 cwt., which it represents, becomes a simple matter. When, however, the cells are carried above the frame, sometimes in two groups, their withdrawal and replacement cannot be so conveniently effected as when a single box is underslung in the manner adopted by the Electromobile Co. The teak containing box is strengthened at the corners and is slung by four straps to its carrier frame. It is suspended at the four upper corners from brackets fixed to the transverse frame channels A^1. By rotating the right- and left-hand threaded cross bolts G at the ends of the box the latch bolts which support the weight may be caused to enter or withdraw from holes in bosses in the brackets on A^1, when it is required to fix the box in position or to get ready for its removal, the weight being already taken upon a lift table or by other means. At N^1 there is a plug for use when the cells are to be charged in position.

Capacity and type of cells A good opinion may be formed of the volume occupied by the 44 " Contal " cells and their containing box from inspection of Figs. 435 and 436. They have a capacity of 135 ampère hours and provide sufficient power to enable the carriage to travel 25 to 35 miles over ordinary town roads, according to the type and weight of body fitted. There are seventeen plates in each cell of the battery, each plate taking the form of a closely divided lead grid containing the active material in 252 square holes. The plates are 4·72 inches long, 4·05 inches wide and 0·137 inches thick. They are spaced and insulated from each other by perforated ebonite sheets.

Life of cells The performance of these cells in actual service has been very satisfactory. From the running records of the accumulators kept at the garage it is found that the cells of the capacity given will drive one of the Electromobile Co.'s carriages over town roads 2,500 and in some instances 3,000

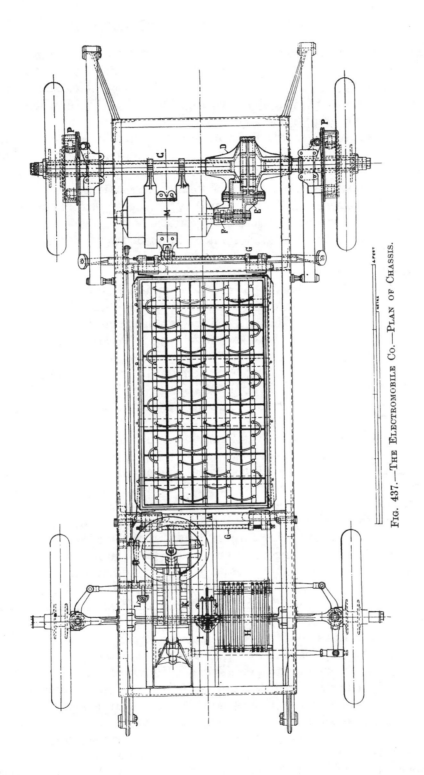

FIG. 437.—THE ELECTROMOBILE CO.—PLAN OF CHASSIS.

551

miles before it is found necessary to wash out the loose and precipitated material. Before renewing the positive plates distances of 3,500 and up to 4,100 miles have been travelled. The relation between capacity and weight and the area occupied is shown by Fig. 430 and Table 16, where the Contal cells bear the number 2.

As shown, the motor M is radially supported at one side by a pair of arms, cast with the steel casing and magnet frame, upon the tubular axle casing C, and at the other side it is spring carried from a suspension block on the transverse frame channel behind the battery box. Completely encased double-helical double-reduction gearing transmits the power from the motor to the bevel differential gear inside the casing D. The outline of the pinion upon the motor spindle F, and of the wheels upon the intermediate spindle E may be seen from the plan (Fig. 436). In recent vehicles precisely the same form of transmission gearing is used, but the casing around it, and a single radial arm at the opposite end of the motor are used to support the weight and to maintain parallelism instead of the pair of arms shown. In addition a pair of parallel angle bars above the motor are fixed longitudinally between the transverse frame channels behind the battery box and at the rear of the vehicle. A pair of suspension bolts one at each side of the motor casing hang from these angle bars and support the motor through springs. A horizontal spring buffered torque bar above the differential gear casing gives spring relief to the gearing under heavy starting or driving efforts.

Gear

The motor is of MM. Contal and Gasnier's design, bipolar, series wound, and with a slotted armature having two windings and two commutators, one at each end. It is described as capable of developing 8 BHP. at 1,500 revs. per minute continuously, or up to 16 BHP. for short periods. The steel motor casing or the magnet frame is horizontally divided through the centre of the pole pieces, and the half frames are separated by a plate of non-magnetic metal. There are no joints in the magnet circuits and by the use of the separating plate stray field losses are minimized, and field distortion due to armature reaction is reduced, and as a result it is claimed that there is no sparking at the carbon brushes on the commutators.

Motor

The conditions of running of these small traction motors are very severe. They must run satisfactorily either as motors or generators at very variable loads and the brushes must necessarily occupy a fixed position and the assistance of positive or negative lead to the brushes for the suppression of sparking is not permissible. The vibration they are subjected to would, in the case of the majority of fixed motors or dynamos, be considered sufficient explanation of failure to run sparklessly and of irregular wear and unsatisfactory condition of the commutators.

The carbon brushes are of very simple and satisfactory construction, a single coil spring connecting the positive and negative brush holders and keeping the carbon blocks up to the commutator with a uniform and lively spring pressure. These carbon brushes are of the type which Messrs. Greenwood and Batley, of Leeds, the manufacturers of the English Electromobile cars, have found to be satisfactory in service with their small electric motors.

Controller

The position of the controller is indicated at K (Fig. 437); it is arranged longitudinally in a horizontal position below the floor line of the car and is protected from dirt and damp by a sheet metal casing below it. It takes the usual form of an elongated drum in this instance of wood with contact segments screwed to it, and the use of a notched cam or star wheel ensures that the running positions are taken up without any dwelling between contacts.

The drum is rotated by means of the short hand lever J^1 below the steering-wheel, connected through a tube inside and concentric with the fixed column J to a bevel wheel gearing with a second wheel on the controller drum spindle. An index plate below J^1 indicates the eight running positions or the stop position when the vehicle is at rest.

It is not necessary to give detail illustrations of the controller or to minutely describe the course taken by the current through the different circuits provided when the controller is in its various positions, because in chapters xxii and xxiii, vol. i., the subject was completely dealt with. Fig. 438 gives the development of the controller drum, diagrams for each of the eight running positions, and the general scheme of connexions.

Diagram of connexions

The forty-four cells are permanently connected in series in order that difference in the condition of the cells due to the unequal discharge which might occur through circuits of slightly different resistance, if parallel grouping was resorted to, may be avoided.

The contact segments of the controller are shown black and the brushes are figured I to XIV, similar figures being used at the points of connexion of leads from the brushes, to facilitate tracing the circuits.

Operation of circuit breaker

A magnetic blow-out circuit breaker with carbon contacts lettered I in Fig. 437 is operated through the pedal L Fig. 436. The first movement of this pedal opens the circuit breaker indicated in the negative circuit in Fig. 438 at II, II, and interrupts the supply of current to the motor. A further movement applies the compensated internally expanding brakes acting upon drums P on the rear road wheels. There is no connexion between the brake pedal and the controller, and if the speed of the car has been much reduced, or if it has been brought to rest, the controller must be returned to the starting position. Failure to do this, or negligent use of the controller, is indicated to the driver of the car and to the passengers by abrupt and jerky restarting, and is not likely to be repeated.

Mechanical and electric brakes

When the speed of the vehicle is insufficiently reduced by operating the circuit breaker and applying the mechanical brakes, through the pedal L, the controller may be set to either of the electric braking positions, when with the motor short circuited and with more or less resistance in circuit, additional braking effort is obtained by causing the car to drive the motor as a generator.

A rheostat H (Fig. 437) with variable resistance is put in circuit for the first speed forward, and for the two electric braking positions EB^1 and EB^2

(Fig. 438). A shutter operated by the controller is so arranged that except when the latter is in the stop position the interrupter plug P (Fig. 438) cannot be removed or put in place. At xv there is a proportional shunt in the return or negative circuit from which the leads are taken to the am-

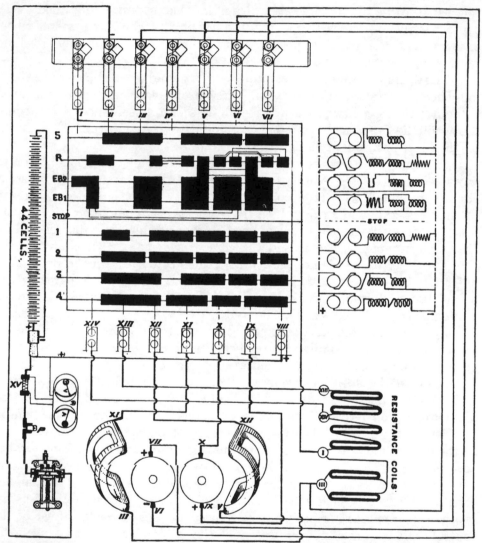

FIG. 438.—THE ELECTROMOBILE CO.—DIAGRAM OF ELECTRICAL CONNEXIONS AND CIRCUITS ESTABLISHED BY CONTROLLER.

pèremeter A. Connexions to the voltmeter v are taken from the positive and negative accumulator leads.

Circuits established by controller When the controller is in the first position for forward running figured 1, the connexions are as indicated in the diagram to the right, the armature windings being in series, and in series with the field windings, and the whole of the resistance similarly arranged.

Similarly for position 2 the armature and field windings are still in series, but the resistance has been cut out.

In position 3 the connexion of the armature windings is unaltered, but they are in series with the field windings connected in parallel.

In position 4 the armature windings are in parallel connected with the field windings in series.

In position 5 for the highest speed forward both armature and field windings are in parallel. Returning to the stop position and thence to the first braking position EB^1, connexion with the battery is interrupted, and the circuit is through the armature windings in parallel to the field coils in parallel, and through the whole of the resistance coils back to the armatures again. The motors are then driven by the car as generators, short circuited, and the braking effect is cumulative and proportional to the speed of the car and the current generated.

In the second braking position EB^2 the connexions are virtually the same, but there is less resistance in circuit and with the greater resulting current there is a proportionately greater braking effect.

In both the braking positions there is, in order that the circuit described may be established, electrical connexion between the controller contact segments VI and X and I and XIV and also between VII and IX and V and XI.

For reversing, the controller is set to the position R, and as shown by the diagram of connexions to the right the armature and field windings and some resistance are all in series, as they are for the first speed forward. The polarity of the motor brushes, and the direction of flow of current through the armature windings has been reversed, and with no change in the course taken by the current through the field coils the change in the direction of running of the motor is effected.

As an example, to illustrate the use of the similar figures given to connected points, the circuit that obtains when the controller is in position R will be described. Current from the accumulator flows through the + lead to the controller brush VIII, and to the contact to the right of the line R, which it is over. This contact segment is as shown cross connected to that which is under brush VI, and the latter will be found to be connected to the negative motor brush — VI, which by the cross connexions at the controller drum becomes the positive brush. The current then flows through the first armature winding to + VII, and thence by the lead shown to controller brush VII, and the contact segment it is over or in line with. By the cross connexion shown the circuit is continued to the contact segment in line with brush X, from which the current is conducted to the motor brush X. The current leaves the second armature winding at the brush + IX, which is now of negative polarity and passes by controller brush IX, and its contact segment through the cross connexion at the controller drum to contact segment and brush V, and so to the end V of the first field coil. Leaving the latter at XII the circuit continues through controller brushes and contacts XII and XI to the positive end XI of the second field coil, leaves the latter at III and passes through the resistance coils from III to XIV. The point XIV in the rheostat is connected through the controller brush XIV with the contact segments XIV and II, and from contact brush II the circuit is completed through the circuit breaker II, II, through P and XV to the negative terminal of the accumulator. In the same way similar figures will be found to be useful for tracing the circuit for the other controller positions. Brush IV, it will be observed, serves no useful purpose for the different circuits as here arranged.

Underslung battery box The outline of one of the Electromobile vehicles with brougham body, shown by Fig. 439, has been superseded by that already described and shown in elevation by Fig. 435. It serves to show, however, that in 1903 the straight frame F had been adopted with the underslung battery box B, and that the inclined steering column and hand-wheel S was in use instead of

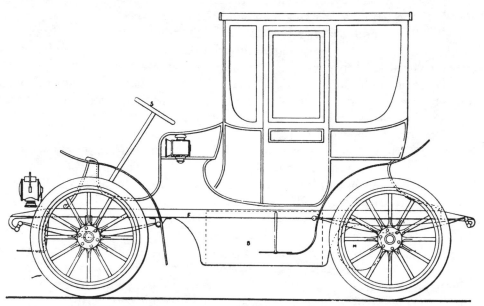

FIG. 439.—THE ELECTROMOBILE CO.—OUTLINE OF BROUGHAM, 1902-3.

the more commonly employed tiller steering lever for these vehicles. The position of the motor driving through single-reduction gearing is indicated at H.

Chambers voiturette The electric voiturette,[1] a rear view of which is shown by Fig. 440, introduced by Mr. T. Chambers, of the Electromobile Co. in 1902, attracted considerable attention. It was equipped with a battery of 42 Leitner cells having a capacity of 135 ampère hours. The total weight of the voiturette was 24½ cwt. and on good roads it could travel 50 miles on a single charge. A four-pole compound-wound motor, nominally of 3 HP. but capable of working up to 6 HP., was partly supported on the live rear axle and radially connected with it, but the weight was mainly taken by the frame through the vertical spring suspension bar visible in Fig. 440. The motor

[1] The *Electrical Times*, Sept. 25, 1902.

is, as shown, hung to one side of the vehicle with the single-reduction gearing arranged close up to the near side driving-wheel. The differential gear was of the spur-wheel type. The controller provided for series parallel

FIG. 440.—THE ELECTROMOBILE CO.—BACK VIEW OF VOITURETTE, 1902.

grouping of the battery cells, for electric braking and recuperation, the latter being of some importance for a vehicle that was intended for use on country roads.

THE CITY AND SUBURBAN ELECTRIC CARRIAGE COMPANY

This company, incorporated in 1903, was the first to adopt the central garage system in London, and by very complete organization to provide for the maintenance of its vehicles in a constant condition of readiness for service. The principal depôt is at " Niagara," Westminster, and as many as 230 carriages of different types are housed, examined and repaired, and their accumulators kept in an efficient condition. Independent electrical generating plant is not installed, but current taken from the Westminster Electricity Supply Company's mains is delivered to motor generators having an efficiency of conversion of from 82 to 85 per cent., the voltage being stepped down from 400 to about 110 volts. With this system of utilization it is found that the actual cost of electric energy per vehicle mile averages 1·25 pence.

Source and cost of energy

Examples of two types of vehicle are here shown in external elevation, Fig. 441 representing the " Alexandra " landaulet and Fig. 442 the " Essex "

Jointed tubular frame

cabriolet or hansom cab. The jointed tubular frame separately shown by Fig. 444 is used, and as it offers no resistance to deformation, only retaining its rectangular form in plan, it will be realized that the body of the carriage receives no support from it and must be capable of resisting the greater tendency to change of form set up by the passage of the carriage over irregular road surfaces. The construction of the body is necessarily heavier and stronger than would be necessary if a frame capable of supporting it and resisting change of form was employed, and it may be questioned whether any advantage is gained or weight reduction effected by the use of this peculiar type of tubular frame. The side members of the frame serve as

FIG. 441.—THE CITY AND SUBURBAN ELECTRIC CARRIAGE CO.—THE "ALEXANDRA" LANDAULET.

reaches and maintain the spacing and parallelism of the front and rear axles, and they are curved to a greater or less degree to suit the particular type of body used. The exaggerated degree of deformation of the frame shown by Fig. 444, suggests that the problem of suitable motor attachment **Motor suspension** or suspension would be a difficult one. The two motors are radially supported at the rear and towards the inner ends of their casings upon the rear axle, and at the front, and at the outer ends of the casings they are spring supported on the side frame tubes or reaches. The points of suspension of each motor are, therefore, at opposite corners of the motor frames or casings, and such straining action as results from movements of the vehicle frame has to be relieved by the freedom of the spring suspension or support.

The motors, each rated at 2·0 BHP. at 1,000 revs. per minute, are of the four-pole series-wound totally enclosed type with carbon brushes in two holders at 90 degrees. On the outer end of each armature spindle there is a steel and fibre pinion gearing into an internally toothed ring fixed to the rear road wheel, and on the inner ends of the armature spindle there are brake drums surrounded by straps brought into action through compensating connexions, and a foot pedal.

The efficiency of these small motors reaches 84 per cent. at full normal **Efficiency** load and falls to 77 per cent. at 100 per cent. overload. The falling off in **of motors**

FIG. 442.—THE CITY AND SUBURBAN ELECTRIC CARRIAGE CO.—THE "ESSEX" CABRIOLET.

efficiency is, however, more pronounced below the normal full load and at about one-third load it drops to 65 per cent.

The use of driving pinions made up of outer steel cheeks and a centre piece of fibre is found to result in more silent running of the gear due to the fibre always remaining a little proud of the steel cheeks, and the wear of the gear teeth is minimized by the slight elastic relief afforded at the contact faces.

By means of a side lever and through compensating gear, brake straps are arranged to act upon the outer surface of the internally toothed gear rings on the rear road wheels.

With the landaulet (Fig. 441) the accumulator consists of 44 E.P.S. **E.P.S. cells** cells of the " C " or ATFII type, each cell weighing, complete with electro-

559

lyte containing box and connexions, 21·7 lb.   Their capacity on a five-hour discharge rate is 125 ampère hours.   The accumulator in its containing box occupies the space below and at the back of the driver's seat.   The latter forms a removable top which when taken away leaves the tops of all the cells open to view   The controller in this carriage is at the base of the steering-wheel column, and it is operated by the hand lever to be seen below the steering-wheel.   The various running speeds are obtained by series or parallel grouping of the cells with the motors connected in series or parallel.   This allows of four combinations of connexions and consequently four speeds, reversing being effected by changing, through the controller, the direction of flow of current through the motor armatures.

No resistance is put in circuit for the purpose of speed control.

Weight   The weight of the " Alexandra " landaulet is about 30 cwt., and of the " Essex " cabriolet 35 cwt.   The latter is of distinctive appearance, and it is provided with the same type of frame and running gear as the landaulet, but the battery is carried in two groups, one below the driver's seat and the other at the back below the cab seat.   The controller is arranged horizontally and is operated through one of the side levers shown.

The line drawings (Fig. 443) show to better advantage the proportions of another design of landaulet in which the two parts of the battery are accommodated at B, B¹.   The tubular frame already referred to may be here seen in plan and the angular position of the jointed ends of the reaches A, A¹ at the four points of connexion P is indicated.

The motors are at M, M, and they drive, as already stated, through pinions at the outer ends of the armature spindles on to internally toothed rings R, R. The arrangement of the springs supporting the body is evident from the four views given.

Diagram of connexions   The general scheme of connexions shown by Fig. 445 possesses some interest, but the complete circuits are not there shown since the system of control already mentioned need not be described again. The power leads or connexions between the motors and accumulators are shown in full lines and those for the lighting, heating and bell circuits in dotted lines. Some of the City and Suburban Company's carriages were provided with two batteries, each composed of 22 cells, and each battery could be connected with the cells all in series or divided into two groups in parallel, as shown by Fig. 445, where the four groups of cells are lettered A, B and A¹, B¹. The motors M and M² have their brushes, the equivalent of the armature windings, and the field coil terminals suitably lettered and the significance of the letters indicating controller contacts will be readily understood.   Thus, for instance, assuming the motors to be in parallel, and considering for the moment motor M², it receives current from the controller brush or finger on 2 AP or number 2 motor armature positive, through the running plug Q and the lead shown connected to AP, representing armature positive or positive brush.   Passing through the armature winding and leaving it at the negative end or brush AN the current returns to the controller contact 2 A N and then goes by contact 2 F P and connexion to the positive end of

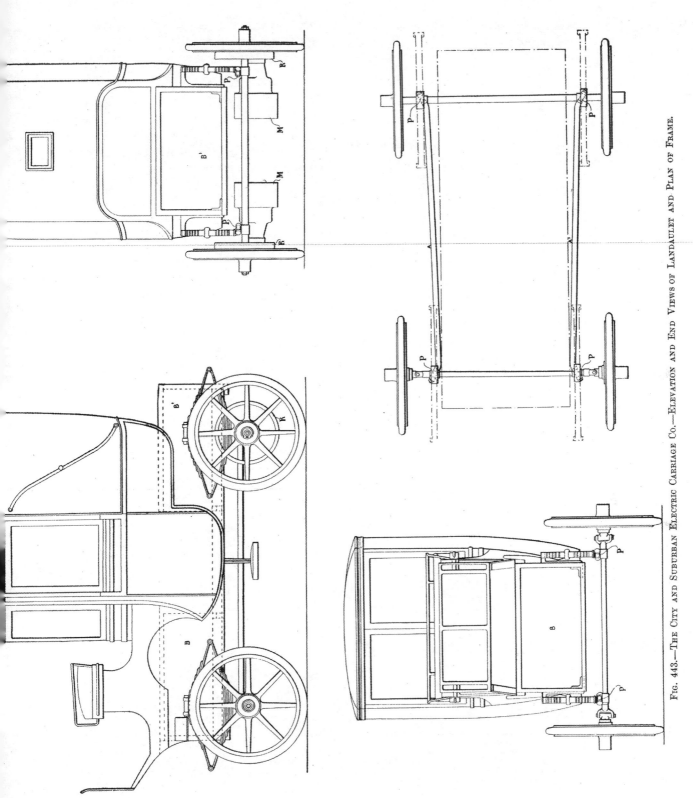

FIG. 443.—THE CITY AND SUBURBAN ELECTRIC CARRIAGE CO.—ELEVATION AND END VIEWS OF LANDAULET AND PLAN OF FRAME.

To face p. 560]

the field winding F P. Leaving the field at the negative terminal F N the circuit continues to the controller by junction of the field lead with B B N contact on the controller or B battery negative. The battery controller contacts are shown lettered B A P and B A N for A battery, positive and negative, and B B P and B B N for B battery, positive and negative connexions. When the change over switch E is in its other position batteries A¹ B¹ are in service. D is a special plug through which the cells may be recharged, P a safety or isolating switch and S P and S N are the terminals of the proportional shunt in the main circuit to which theампèremeter is connected. The voltmeter terminals are indicated at S P and R.

The dotted connexions at the lower part of the diagram are clear and require only passing reference. The bell C is connected across the + and I sides of A battery. When the driver depresses the foot push H¹ he gives

Lighting, heating, and bell circuits

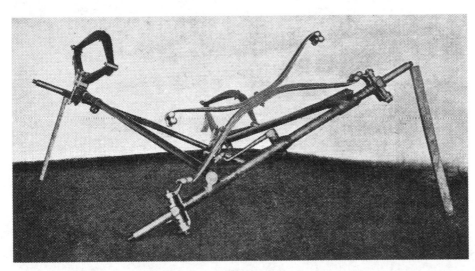

FIG. 444.—THE CITY AND SUBURBAN ELECTRIC CARRIAGE CO.—JOINTED TUBULAR UNDERFRAME.

warning to other road users of his approach, and by means of the push H inside the carriage body the passengers may signal to the driver. The same battery also supplies the carriage side lamp J¹, and the ceiling rose lamp fitting I inside the carriage body, connected with the switch K. Battery B supplies current for the second side lamp J and through the switch O, and wall plug N to a foot warmer or radiator not shown. With the exception of the bell wires, the connexions are all taken to the fuse and switch-board L.

Hitherto it has been found that the average annual cost of maintenance of the batteries for these vehicles is about £53, whilst the cost of upkeep of the running gear and of the solid rubber tyres are respectively £42, and £30. The yearly cost of current at the rate already given of 1·25d. per vehicle mile, averages £32. The whole cost of upkeep of one carriage

Cost of upkeep and rnnning

per year, including all expenses except those relating to directorate and offices, amounts to £288, the number of cars considered as garaged at one depôt being 84. It is estimated that if the number garaged was increased

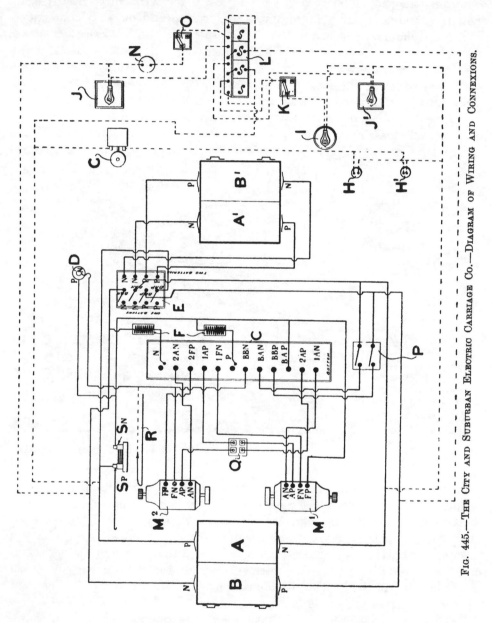

Fig. 445.—The City and Suburban Electric Carriage Co.—Diagram of Wiring and Connexions.

to 120 it would be possible to reduce the maximum costs per annum to £257.

The types of vehicle illustrated are, it will be understood, typical of many designs introduced by this company. At the Automobile Exhibition at the Crystal Palace held in February 1903, a combination petrol electric car [1] was exhibited. It possessed features of design common to the simple accumulator equipped car, the driving gear and suspension of the motors being similar. In addition to the jointed tubular underframe a straight

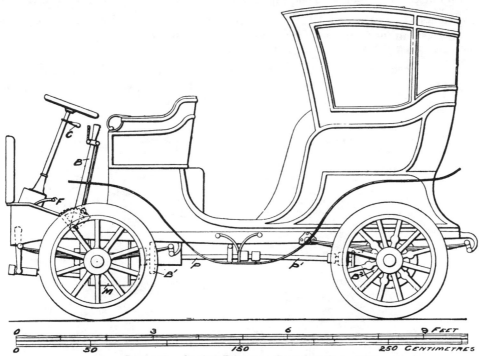

FIG. 446.—GARCIN-RENAULT ELECTRIC CAB, 1905.

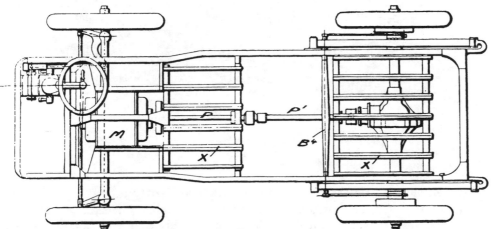

FIG. 447.—GARCIN-RENAULT ELECTRIC CAB.—PLAN WITH BODY REMOVED.

channel steel frame supported on full elliptic springs carried the body, engine and dynamo and accumulators.

The Garcin-Renault car (Figs. 446 and 447) instances the adoption

[1] The *Electrical Review*, Feb. 20, 1903.

563

Garcin-
Renault

for electric automobile purposes of types of frame and transmission gearing successfully experimented with and first used by manufacturers of other classes of motor vehicles. A pressed steel frame with straight top line is employed supported on semi-elliptic springs, those above the live rear axle being spaced wider than the frame channels. Towards the front of the car the frame is reduced in width to give clearance between it and the steering wheels when set to a considerable steering angle.

The series motor M, with two armature windings and commutators is hung from a short central frame longitudinal, bearing upon transverse frame

FIG. 448.—JEANTAUD ELECTRIC COUPÉ.

members at its ends. From observations independently made, with the car running at different speeds up to 18 or 20 miles per hour, and over roads having gradients varying up to 1 in 10, it appears that the motor worked at rates of from 3 HP. on flat roads at 10 miles per hour up to nearly 4·5 HP. at a little over 18 miles per hour, whilst when ascending a gradient of 1 in 10, 10 HP. was exceeded. The current from the battery was then 95 ampères, the voltage being 80. Calculation of the probable brake-horse-power required points either to remarkably high efficiency of transmission or to the possibility of inaccuracy of the electrical measuring instruments. The weight

Power

of the car with passengers is declared to be from 1,200 to 1,500 kilos or approximately 30 cwt., the battery weighing half a ton. An extension of the armature spindle P, supported at its rear end in a bearing attached to the car frame has fixed upon it, near the motor, a brake drum B¹, and the pedal F when depressed tightens the brake strap. The driving effort from the rear end of the extension shaft P is delivered through the universally jointed shaft P¹ to the bevel gear and differential gear inside the casing upon the rear axle connecting the fixed tubular parts, within which are the road-wheel driving shafts. The side lever B is connected to the equalizing bar B⁴ (Fig. 447), and through it operates the internally expanding brakes acting upon drums fixed to the rear road wheels.

Trans-mission and other gear

The controller is arranged at the base of the steering column, and it is operated by the small handle below the steering-wheel moving over an index

FIG. 449. JEANTAUD DUC ELECTRIQUE, 1900.

plate. For speed control, reversing, and electric braking, the combinations of field and armature windings, with the battery in circuit, or with the battery disconnected when the motors are short circuited for electric braking, are effected through the controller in the manner common to other electric vehicles. When the brake pedal F is depressed it also interrupts the current by opening a switch which when the pedal is released remains open. The controller has to be returned to the neutral or starting position before this switch is closed and abrupt or jerky starting of the vehicle as a result of careless driving is obviated.

With the cabriolet or hansom cab shown by Fig. 446 the battery is arranged in two groups of 22 cells, below the driver's seat and the passenger's seat. The capacity of the battery is 175 ampère hours at a 5-hours' discharge rate, or 156 ampère hours at a 3-hours' discharge rate. The cells are described as the Garcin-Renault and the plates are structurally identical with the Contal or Electricia type. The position of the grids which support the cell-containing boxes is indicated at x x (Fig. 447).

Jeantaud   As illustrating an electric brougham or coupé of pleasing external appearance (Fig. 448) is here given; it represents one of the Jeantaud carriages infrequently seen in this country, but considerably used in Paris where they are constructed. The body, it will be noticed, is hung very low and the front springs are below the axle pivots of the steering-wheels. The battery is in front of the carriage and below the driver, the rear axle and driving-wheels only being called upon to carry the weight of the motor

Fig. 450.—Accumulator Industries Ltd.—Electric Coach.

and chain-driving gear and a part of the load of the body. The controller and the rear-wheel hub and tyre brakes are operated by the side lever and pedals visible in the illustration.

The light electric car seating two persons, shown by Fig. 449, is also constructed by Jeantaud et Cie. of Paris. It illustrates the early use of the underslung battery box, and the system of spring suspension at the front axle for which two parallel transverse springs are employed. Much of the delicacy of the decoration given to the body work is imperfectly shown in the photo reproduction here given, but it added very much to the appearance of the car. As regards spring attachments, length of springs, and the general proportions of the car, it would seem that the designer had antici-pated what is now commonly considered satisfactory practice.

Fig. 450 is an external view of an electrically propelled coach con-structed by Accumulator Industries Ltd., of Woking, in 1902. It is prob-ably the largest accumulator-equipped passenger vehicle made and used in this country. The battery consisted of 44 Leitner cells of the pasted type, having a capacity of 200 ampère hours. Each cell weighed nearly 32 lb., and a distance of 80 miles could be run without recharging. It seated twelve persons including the driver. The body of the coach is carried

Electric coach

FIG. 451.—ACCUMULATOR INDUSTRIES LTD.—ELECTRIC STANHOPE.

on semi-elliptic springs at the rear, and on elliptic springs above the front axle. The front wheels run on pivoted axles, the steering being on the Ackerman principle with the gear operated through a vertical spindle and hand-wheel. As shown in the illustration large solid rubber tyres are fitted to all four wheels. Two Lundell series-wound four-pole motors, each rated at 7·0 HP., but capable, in common with other well designed electric motors, of running at double their normal load for short periods, and developing at times as much as 30 HP., drive through single-reduction gearing on to gear rings secured to the rear wheels. The controller operated by the side lever to the right of the driver's seat gives series parallel combinations

Power of motors

567

of the double armature and field windings and reversing is also effected through it.

The hand lever on the right hand side and near the dashboard operates brake blocks acting upon the rear-wheel tyres.

On level roads it is stated that the coach can be driven at 20 miles per hour, a high speed for an electrical vehicle of its size and weighing approximately 3 tons.

A lighter carriage constructed by the same makers is illustrated by Fig. 451. It is of a type that was in favour two years ago when large numbers of the Waverley, Baker, and Columbia electric runabouts were sent over to this country. As with these light American vehicles the body is spring mounted on the tubular underframe upon the road-wheel axles, but unlike them the road wheels were of large diameter and would render travelling easier over irregular road surfaces. They are little used now, not because the carriages were unsatisfactory, but because the accumulators limited their use, and because the work for which they are suitable is more satisfactorily performed by other types of automobiles. Two 2·5 HP Lundell motors of the same type as those used for the coach were fitted and the speed of the carriage was controlled in the same way by series and parallel grouping of the motors and windings. The battery was composed of 44 Leitner cells each weighing 22 lb., and having a capacity of 120 ampère hours. The weight of the carriage was 22 cwt.

Light electric carriages

Chapter XXXV

PETROL MOTOR CYCLES

THE construction of motor cycles of various types has added largely to the number of those actively interested in motor vehicles and has served to stimulate general interest and increase of knowledge of these machines. With the large increase in the number[1] of those owning and using them, there has been a commensurate growth of the number of manufacturers and of the energy and ingenuity expended upon their design and production. As might have been expected a number of the firms previously interested in ordinary cycle construction now produce motor cycles, and in general their machines are more carefully or more completely arranged than are those constructed by firms more experienced perhaps in the motor equipment but less so in the art of cycle construction.

A·large proportion of the engines employed and their auxiliary parts have been of Continental design and manufacture, and many of them have proved to be very excellent examples of small engine work. When the satisfactory motor equipment has been satisfactorily combined with the good cycle work to be obtained in this country, motor cycles have been produced capable of the road performances credited to these machines mostly of the tricycle type, between seven and eight years ago, and chiefly in French road races. That is to say, the capabilities of the recent types of motor cycle are such that without the special attention required and the restricted uses of the early machines used for racing, similar results may be obtained from modern standard machines driven by those not necessarily skilled in their use.

During the past three or four years, the motor tricycle, which had previously been used in considerable numbers in the forms illustrated in vol. i. of this book, has been little seen and more attention has been given to the improvement of the motor bicycle, and to the construction of three-wheeled vehicles, some of which may be included in the motor cycle class, but many of which, by reason of their design and the arrangement of

Margin notes: **Increased use** · **Three-wheeled motor cycles**

---

[1] The *Motor Cycle*, 10 April, 1905. 51,000 motor vehicles were registered during 1904. Of these 26,000 were cars and heavy vehicles, and 25,000 were motor cycles. Another estimate of the number of motor cycles registered is 29,000; this number probably includes three-wheeled motor cycles, weighing more than 3 cwt.

mechanism and on account of their weight could be better described as three-wheeled light cars. These three-wheeled motor cycles, frequently termed tricars are designed to carry two passengers inclusive of the driver, and their weight varies from two to three cwt. for the lighter class and may be as much as five or six cwt. for the heavier class.

Following the production of more substantial and reliable motor bicycles, there arose a demand for means of taking or carrying a second passenger. ·Attempts to meet this demand have been only partly successful and require the use of a chair seat, spring mounted on two wheels, and either used as a trailer connected by a universally jointed draw-bar to the frame of the motor bicycle at about the position of the saddle, or arranged at the front of the bicycle as a forecar. With the latter arrangement the front wheel of the bicycle is removed and steering connexions are made from the front forks or from a steering tube replacing the front forks to the two wheels supporting the forecar and part of the weight of the bicycle. The front wheel spindles or axles are mounted in steering heads and the steering gear is of the Ackermann type.

Another attachment, known as a sidecar, is employed to a more limited extent, and with this arrangement the weight of the chair seat and passenger is partly carried by connexion at one side to the motor bicycle by spacing and drag bars, and at the other or outer side by a single or outboard wheel. The sidecar is connected to the near side of the bicycle and steering is effected in the usual way by means of the front bicycle wheel.

None of these arrangements for taking a second passenger, all of them of the nature of attachments to a motor bicycle, have been completely satisfactory, and it could hardly be expected that they would be. The trailer constitutes the most satisfactory method of carrying additional load by means of a motor bicycle, but when that additional load is a passenger it is not the best method, because the passenger is exposed to the dust and under some circumstances the exhaust from the engine, and he is separated or is at some distance from, and not visible to the driver. The forecar and sidecar attachments were troublesome because of their light construction and liability to suffer damage, and because the method of connexion to the bicycle was in some cases such as would result in the setting up of severe stresses in both the car and bicycle. Poor design or bad balance or distribution of weight sometimes added to the difficulties in driving and steering, and excessive vibration occurred with some types.

Convertible machines unsatisfactory

The result of the experience now gained points to the advisability of not using attached vehicles, but of designing the motor cycle as a tricar and not as a convertible machine. It has also been found necessary to modify the method of transmission of power from the engine to the driving road wheel and to employ a simple form of change-speed gear. The use of water-cooled engines of from four to five brake-horse-power is becoming more usual, and water cooling becomes necessary with such engine powers and the heavier type of tricar that is being evolved Briefly stated the light tricar with the smaller air-cooled engine if considerately driven proved

570

a reasonably satisfactory and simple small car. It was, however, possible to drive it at high speed on level roads or up slight inclines and many drivers would attempt to take the heavier load or two passengers at as high a speed as could be maintained up more severe gradients. The difficulties resulting from vibration at high road speeds and from overheating of the engine by overloading it, has called for the heavier tricar, made heavier partly by the use of the higher-powered water-cooled engine, by the changes in the transmission gear, and by the use of more substantial or stronger construction of the framework and other parts. As now made by several firms this type of heavy tricar can no longer be looked upon as a motor cycle, and it must be described as a light car of a particular class. It remains to be seen whether the cost of upkeep and lower first cost will prove advantages sufficient to create a demand for this form of motor vehicle as great as that now existing for light four-wheeled cars or sufficiently great to render its construction a commercial success.

Although most of the early motor bicycles were of Continental make, it is worthy of note that makers in this country quickly took the lead and were the first to produce machines suitable for ordinary road or touring purposes, makers abroad having for the greater part confined themselves to the construction of high-powered machines, in many instances with dangerously light or poorly designed framework, for high-speed road and track racing. A few makers have devoted their attention to the more useful type of machine, and of these the firms of Peugeot, Werner, Minerva, and the F.N. or Fabrique Nationale d'armes de guerre are the best known, and it is only now that their lead is being generally followed. The engine power has been gradually increased from about 1 HP. to 3·5 HP. for ordinary road machines, and a few makers have exceeded these powers. Varying with the requirements and the weight of the rider powers of 2·75 to 3·5 HP. are now commonly adopted, and the weight of the complete machine varies from about 100 lb. to 180 lb., as constructed in this country. The question of reduction of weight is receiving considerable attention, and machines of about 3 HP. and of good design and construction may now be obtained weighing about 140 lb.

Horse-power and weight

The transmission of power from the engine to the road wheels is generally by **V** belt, this form of belt having superseded the round twisted belt, and the flat form is now very little used. The well known firm of Humber, Ltd., have always consistently adopted chain driving, and others have followed their example. A few machines have been made in which transmission through bevel gearing or through a worm and worm-wheel drive has been employed, but these are in the minority, are more costly to make, but are not necessarily unsatisfactory, and they have good features.

Transmission of power

One of these of particularly ingenious construction, introduced by Starley Bros., and constructed later in essentially the same form by the Swift Cycle Company, employed worm and worm-wheel driving gear, and was provided with a compactly arranged two-speed gear. Among other motor bicycles, which have been provided with two-speed gears, are notably

571

the Clement-Garrard, the Phœnix, and the Minerva. The use of a two-speed gear of good design necessarily considerably adds to the hill-climbing capabilities of the cycle, but for obvious reasons not directly in proportion to the additional mechanical advantage obtained, especially with air cooled engines.

Of those few motor cycles which will be described some are of early design, but as they are pioneer designs, and have as such been profitable examples to later makers, and not generally a source of profit to the originators, they will be referred to here.

THE ORMONDE MOTOR BICYCLE

The appearance of this machine and its arrangment may be judged by reference to Fig. 452, which shows one of the $2\frac{1}{4}$ HP. type.

FIG. 452.—ORMONDE MOTOR BICYCLE, $2\frac{1}{4}$ HP., 1902-3 TYPE.

It was introduced in 1902, was preceded during 1901 by a very similar machine but of $1\frac{3}{4}$ HP., and was made in 1903 with a $2\frac{3}{4}$ HP. engine.

An uninterrupted diamond frame is used and with the proportions adopted, room is made for the engine in the position shown. The latter is clamped to the down tube from saddle to cycle crank bracket, and its weight is taken upon a bearer-pad below the crank chamber. While the position selected for the engine makes it possible to arrange the connexions and attachments very snugly, it is open to the objection that the free circulation of air around the engine cylinder is a little impeded by the shelter afforded by the rider's legs and by the position of the cylinder behind the frame tube and the tank R. The makers of the machine recognized that the air-cooling effect might be slightly reduced, but they expected that any such loss would be more than balanced by the declared freedom of the Kelecom engines, which they use, from overheating.

Kelecom engines

The cylinders of these engines are remarkable for the cleanness of the castings and the radiating ribs are deep and of very thin section. The cylinder and combustion chamber are cast in one, and with the valve chamber at one side, and as a result of the small mass of metal used, particularly at the combustion chamber end of the cylinder, and the large cooling surface of the radiating ribs, there is, without doubt, less probability of overheating than there would be in engines arranged with a separate combustion head.

The internal arrangement follows the general design of the de Dion engines, differing only in small details and in the proportions of some of the parts. The cylinder diameter and the stroke with the 1¾ HP. engine are respectively 66 and 74 mm., with 2¼ HP. engine 70 and 75 mm., and with the 2¾ HP. engine 77 and 83 mm., and the respective powers are claimed to be developed at 1,500 revs. per minute.

The use of large fly-wheels is a good feature in these engines and promotes steady running at normal speeds and the engine will continue to run at the low speeds occasionally required when travelling in traffic. The weight of a pair of fly-wheels for a 2¾ HP. engine is 25 lb., most of the weight being in the rim.

The atmospheric or automatic inlet valve is arranged vertically over the exhaust valve and both are in a chamber cast at one side of the cylinder.

The position of the carburettor is indicated at A (Fig. 452), and the throttle valve A¹ is connected by union immediately above it.

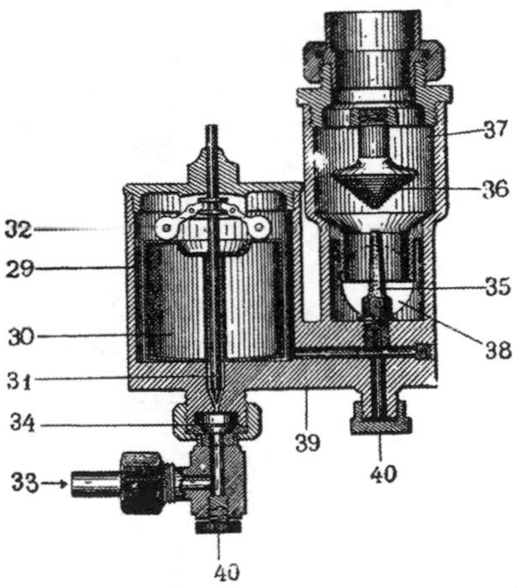

FIG. 453.—ORMONDE MOTOR BICYCLE. CARBURETTOR.

F.N. type of carburettor

Referring to Fig. 453, which is a section through the complete carburettor, it will be seen that the float chamber 29, and mixing chamber 37, on the right are in one casting 39. The arrangement of the little weighted levers 32, and the way in which they close or open the needle valve 31 by the rise or fall of the float 30 is evident from the drawing when it is remembered that the pivot pins of the levers 32 are in the fixed cover above the float chamber. Petrol enters at 33, passes through the filter grid 34, and when the needle valve is open, into the float chamber. The level of the petrol in the float chamber and in the jet 35 is maintained by the float at or near the level of the top of the jet or nipple. Air enters at the opening 38, having previously been warmed by contact with some of the hot surfaces of the engine cylinder or exhaust pipe and cleaned by passing a gauze screen at the mouth of the air inlet pipe. By suction from the engine, the air passes around the jet, scrubs the atomizing cone 36, and continues by

573

way of the throttle valve and induction pipe, to be seen in Fig. 452, to the inlet valve.

The suction or ejector action produced by the rapid flow of air past the jet induces a spray of petrol upwards upon the cone 36, and by the scrubbing action of the air upon the cone it is taken up by, and intimately mixed with, the air. The provision of a plug at 40 below the needle valve, and a cap 40 below the jet nipple render the cleaning or freeing of such small passages, which occasionally become partially obstructed, a simple matter.

Ignition   The form of electrical ignition employed is the accumulator and coil form of the high-tension system and the connexions are shown in Fig. 454.

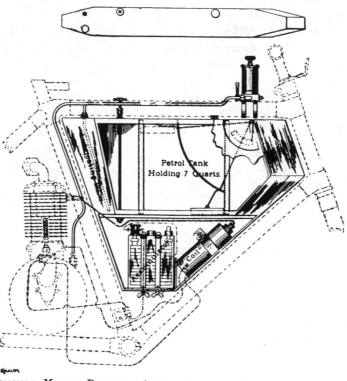

Petrol Tank
Holding 7 Quarts.

FIG. 454.—ORMONDE MOTOR BICYCLE.—ARRANGEMENT OF TANK, LUBRICATION PIPES, AND ELECTRICAL CONNEXIONS.

Starting from the + terminal of the accumulator there is a connexion to one end of the primary winding of the non-trembler coil and a similar connexion from the other end of the primary to the insulated contact screw of the contact breaker, carried on the extension of the half-speed spindle in the usual manner.

From the contact breaker the current goes to the engine frame and through an interrupter switch or contact shown by Fig. 455 and by insulated wire to the plug switch or touche shown in dotted lines at the junction of the frame tube from the saddle with the crank bracket. A short lead from the plug switch to the terminal of the accumulator completes the return circuit.

574

The high-tension circuit is by heavily insulated lead from the end of the secondary winding at the upper end of the coil to the sparking plug and the return circuit by engine frame and connexion to M or the other end of the secondary winding at the lower end of the coil.

The control of the machine is partly effected by means of the combined exhaust valve lifter and circuit interrupter shown by Fig. 455.

Exhaust valve lifter and circuit interrupter

Around the upper end of the exhaust valve tappet guide there is a vulcanite slip 20, carrying the terminal 19b and the contact strip 19a. The terminal 19b, as shown in Fig. 454, is in the low-tension return circuit to the accumulator. A second contact piece 19 formed as part of 18a is normally held in contact with 19a by the spring shown. The piece 18b is fixed by setscrew to the upper end of the tappet guide, and its top face is cam-formed in the same way that the lower face of the loose piece 18a is cut. Except when the valve tappet is raised by its cam it rests, through the collar shown upon its upper end, upon the flat upper face of 18a. By means of a twist handle on the left side of the bicycle handle-bars the Bowden wire shown may be caused to twist 18a or partially rotate it relatively to 18b.

This movement first separates the contact points of 19 and 19a and so stops the engine by interruption of the ignition, and the continued movement by causing the cam face of 18a to ride upon that of 18b, lifts the exhaust valve and so relieves the compression. The same combination in more recent machines is arranged to hold up the exhaust valve before breaking the electrical circuit, but it is questionable whether the practice of running the engine with the exhaust valve slightly off its seat during the firing or working strokes is to be commended, even when only used occasionally. It was not recommended

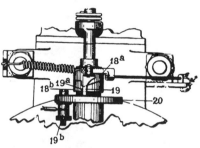

FIG. 455.—ORMONDE MOTOR BICYCLE. EXHAUST VALVE LIFTER AND IGNITION CURRENT INTERRUPTER.

by the makers of the Ormonde machines and the combination described was retained. There are two small levers on the top frame tube of the machine, that lettered B is connected as shown by rod with A^1 and the power developed by the engine and the speed of running of the bicycle is controlled by thus varying the opening of the throttle valve. The second lever C is similarly connected by rod with the contact breaker frame on the opposite side of the engine and by rocking the frame in one or other direction ignition may be advanced or retarded to suit high or low speed running of the engine. The exhaust leaves the engine by the pipe visible in Fig. 452 and passes through the small exhaust box below the engine. The exhaust boxes used with bicycle engines are rarely effective as silencers, but some credit is due to a few makers who have paid special attention to silencing the exhaust without unduly restricting it.

Exhaust silencers

The method of lubrication adopted with the Ormonde machines is good. Referring to Fig. 452 there is, it will be seen, an oil pump, with celluloid barrel, at P, and with a two-way cock at the base. A pipe from the lubricating oil compartment of the tank visible in Fig. 454 is connected to one branch of the two-way cock, and a few strokes of the pump P are sufficient to put the air in the lubricating tank under pressure. The oil will then automatically fill the pump barrel several times when required, it being necessary for the driver merely to turn the cock to the position in which it communicates with the tank. When the cock is turned to its other position to communicate with the engine, oil may flow by the pipe to be seen in Fig. 452 to the non-return valve T. There is an air valve on the engine crank chamber which serves also as a non-return valve. Looking upon the engine piston as a pump plunger, T as suction valve and the air-valve as delivery valve, it will be understood that on the up-stroke of the piston oil will be sucked through the valve T by the small degree of vacuum produced and that on the down-stroke of the piston air in the crank chamber will be expelled through the air valve. As long, therefore, as the engine crank chamber is reasonably tight, and leakage of gases past the piston rings is small, there will be a regular and continuous flow of oil to the crank chamber. This is in principle a better method of lubrication than the periodical supply by hand-operated pump, relied on with most motor bicycles.

Continuous lubrication

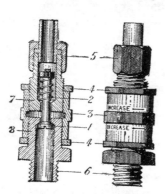

FIG. 456.—ORMONDE MOTOR BICYCLE.—ADJUSTABLE LUBRICATOR CHECK VALVE.

Fig. 456 shows the valve T in section and elevation. The base 6 is screwed into an elbow on the crank chamber and it is connected by a right and left hand threaded collar 3 with the upper piece 7 to which the oil-supply pipe is coupled by the union 5. A valve seat is formed near 8 at the upper end of 6 and the stem of the valve 1 passes through 7. Between the nut on the end of the valve stem and the bottom of the piece 7 there is a spring 2 which keeps the valve 1 up to its seat. By slacking back the lock nuts 4 and rotating the collar 3 in one or other direction, 6 and 7 may be set nearer to each other or separated and thus the compression of the spring increased or decreased. In accordance with the directions indicated by the arrows and the words " Increase " and " Decrease," the rate of suction of the oil past the valve 1 by the engine may be adjusted until judged correct. If necessary an occasional charge of oil may be delivered to the crank chamber by pushing down the plunger of the pump P. With the pump on the top frame tube of the bicycle the rate at which the engine takes the lubricating oil may be readily seen.

Referring to Figs. 452 and 454, it will be seen that although the tank R occupies the whole of the space available in the frame it is thin in outline as shown in plan. It easily contains 7 pints of petrol, or if filled quite full one gallon and not 7 quarts as indicated in Fig. 454. With careful driving **Petrol consumption** on average roads this quantity is sufficient to run the machine 130 to 140 miles, and if means are provided for the supply of more air between the throttle valve and the carburettor considerably more economical results may be obtained. The tank contains a float which operates a pointer moving over a graduated scale outside the tank and the driver may know at any time the quantity of petrol in the tank. This indicator is a convenience, and in recent machines it has been improved and takes a better form.

The shortness of the belt drive and the small arc of contact of the belt with the engine pulley is an undesirable feature, but with the necessary ratio between the diameters of the driving rim and the engine pulley, such as is still commonly used, the arc of contact cannot be much improved, although the length of the belt may be usefully increased. The weight of the complete machine with $2\frac{3}{4}$ H.P. engine and 22 in. frame is about 140 lb.

Short belt drive

The distribution of weight and the long wheel base render these machines comparatively free from the tendency to skid on slippery roads, and they may be frequently seen in use by those indifferent to the many non essential changes in design that have occurred since their introduction.

THE HOLDEN MOTOR BICYCLE

The original bicycle designed by Colonel Holden, F.R.S., was described in vol. i., and some of the many ingenious points embodied in its design were referred to. It was still further simplified and improved in some respects, and the Motor Traction Co., of Kennington, London, were responsible for its construction for public use. Some of the parts of the machine here described and illustrated are of particular interest, because, although they no longer possess the attraction of novelty, they were originated and designed for what was at the time new, and provided examples for the many who can adapt and modify but not create.

An external view of the machine is shown by Fig. 457, and the outline drawing to scale (Fig. 458) enables the proportions of the machine to be judged and the position of the parts illustrated in detail to be recognized.

As in the earlier type, the front wheel, 24 inches diameter with a tyre 2 inches diameter, is the steering-wheel and the four-cylinder engine drives through long tubular connecting rods on to crank arms upon the rear-wheel spindle. The detail given of the driving-wheel I in plan shows that the crank arms T on the squared outer ends of the spindle, are opposite or in line, and are not fixed at an angle to each other, and with the four-cylinder engine the driving-wheel receives four driving efforts or working strokes during each revolution. The triangular openings between the frame tubes are occupied by the petrol tank and surface carburettor A, the cooling water tank D, and the case or box E which contains the accumulators and induction coil required for the electrical ignition system. The form of the tank A and the parts it contains are shown by Fig. 459.

Surface carburettor

Part of the exhaust from the engine passes through the coil 16 and assists carburation of the air entering the tank by warming it and by also warming the petrol in the tank.

The air inlet is at B and the carburetted air before leaving the tank at H[1] passes through the perforated drum shown in dotted lines. By the use of this drum there is little probability that excess of petrol in the form of spray will be carried with the carburetted air up the pipe H[1] to the mixing valve H (Fig. 458). By reference to the same figure the outgoing end of the coil 16 (Fig. 459) may be traced behind the forward end of the engine and its open end points downwards. A separate detail of the funnel or inlet tube at B (Fig. 459) is shown by Fig. 460. The cap to the top piece B forms guide and seat for the light inlet valve O, and the centre stem tube fixed in the lower end of B is a sliding fit in the gland box and guide piece

FIG. 457.—THE HOLDEN MOTOR BICYCLE.—ELEVATION.

fixed in the top of the tank. A double gauze pocket N hangs from the lower end of the stem tube and is immersed or partly immersed in the petrol, and the air entering at O and passing through N bubbles through the petrol in numerous streams and becomes carburetted. Flexible metallic tubing is used for the pipe connexions H[1] and V (Fig. 458) between the tank A and the combined throttle and mixing valve H, and from the latter to the connexion at the front of the engine with the induction pipe beneath the cylinders, from which branches are led to the four inlet valves.

Combined mixing and throttle valve The construction of the mixing and throttle valve H is shown by three views (Fig. 460). The inlet branch is at C, and the outlet at P, and there is an air inlet and valve at K, the valve and its light spring F being of the same form as that used at the top of the carburettor funnel at B. When

578

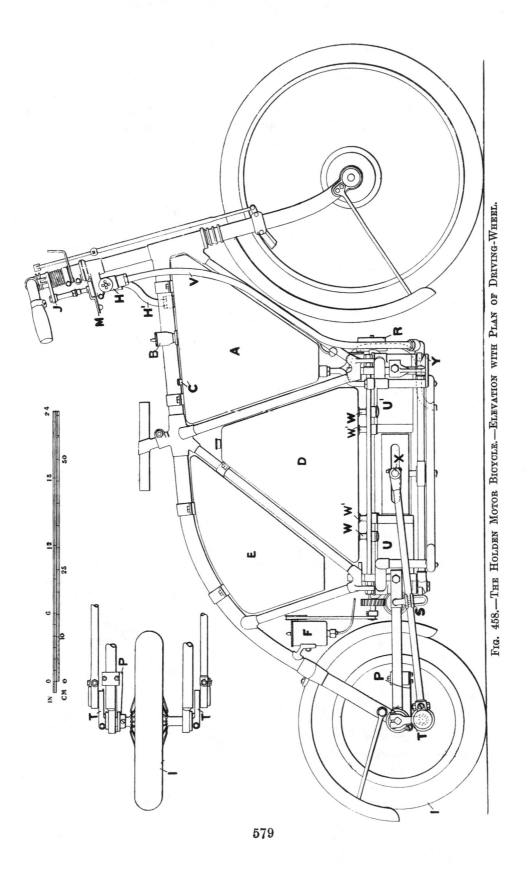

FIG. 458.—THE HOLDEN MOTOR BICYCLE.—ELEVATION WITH PLAN OF DRIVING-WHEEL.

579

the engine is not running these valves are closed and loss of petrol by evaporation is prevented.

The throttle valve may receive both a vertical movement and a limited rotary movement. In the sectional view the valve is shown in the top position and communication between P, C, and K is closed. When the

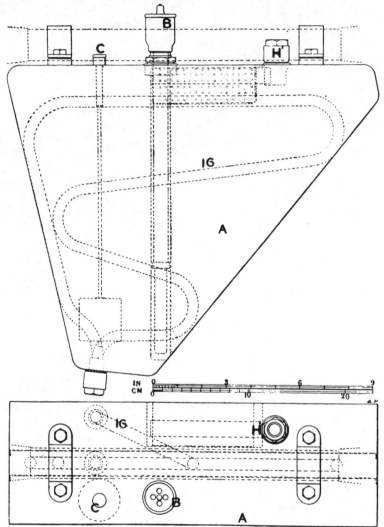

FIG. 459.—THE HOLDEN MOTOR BICYCLE.—ELEVATION AND PLAN OF PETROL TANK AND SURFACE CARBURETTOR.

valve through the spindle E is pushed downwards its upper edge D uncovers the carburetted and fresh-air passages, and the mixture may flow to P and the induction pipe.

The bridge piece just above D, when the valve is in a lower position, partly or completely closes the fresh-air inlet. The amount of opening is regulated by the rotary movement given to the valve by means of the

580

small lever R on the squared part of the valve spindle E, and the driver is enabled to judge, or acquires judgment of the required opening under different conditions, by the position of the pointer end of the lever R over the quadrant scale Q. By inspection of Fig. 458 and the separate detail of the handle-bars (Fig. 261) the method of giving the vertical opening or closing movement to the throttle valve will be understood. By means of the adjustment screw and milled nut at the end of the lever 19, the opening may be conveniently adjusted, and the coiled spring, between the fixed collar and the sliding piece on the handle-bar stem, prevents disturbance of the throttle valve setting by vibration.

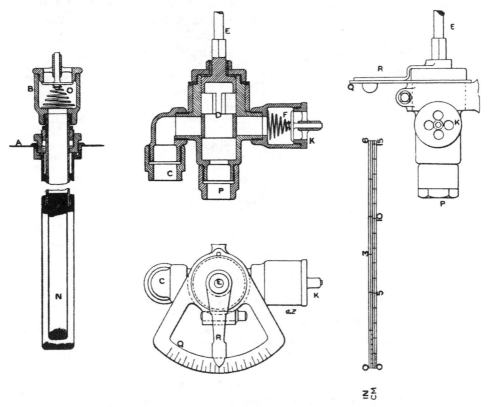

FIG. 460.—THE HOLDEN MOTOR BICYCLE.—DETAILS OF AIR-INLET VALVES AND COMBINED MIXING AND THROTTLE VALVE.

Referring again to Fig. 459, the float there shown serves to indicate the level of petrol in the tank, when the cap C is unscrewed and the float and the vertical rod to which it is attached are allowed to take up their floating position. The tank is filled at B by unscrewing the cap and air-inlet valve there shown.

In the right-hand end of the handle-bars there is a switch by means of which the low tension or accumulator circuit may be closed or interrupted and ignition in the cylinders permitted or prevented as may be desired. A detail of it is given at Fig. 461, and the thumb piece or trigger 18 is shown

Ignition switch

in the open position when the contact buttons x x^1 are separated. The contact strip x is earthed or connected to the frame of the bicycle through the handle-bar tube, and a single insulated wire is connected to the upper contact strip x^1 by the terminal screw to be seen at the right hand end of x^1, and the ebonite insulating slip to which x^1 is attached.

Four-cylinder water-cooled engine

Three views of the four-cylinder 3-HP. engine are shown by Fig. 462, from which its construction may be studied. It is hung from four lugs on the bicycle frame by attachment to the ends of two long tie bolts. These may be seen in the plan view and the frame suspension lugs are there shown sectioned black.

The cylinders, 2·125 inches diameter, are in opposed pairs U U^1, and U^2 and U^3, a length of light steel tube being used for each pair. The engine

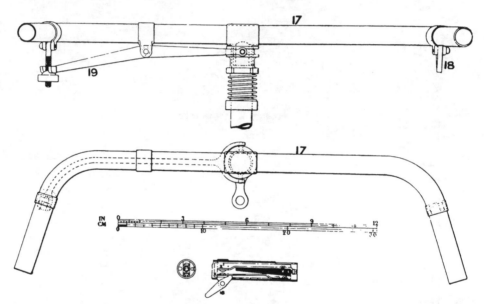

FIG. 461.—THE HOLDEN MOTOR BICYCLE.—DETAIL OF HANDLE-BARS WITH THROTTLE-VALVE CONTROL AND IGNITION SWITCH.

stroke is 4·5 inches. The cylinder ends are thickened up by pieces screwed on and with the additional thickness of the tubular casings forming the water jackets a suitable width of joint face is obtained. Recesses or grooves are turned in the cover blocks to receive the cylinder ends and a very satisfactory form of spigoted joint is obtained, three long tie bolts being employed to hold the two combined valve box and cover blocks up to the cylinders. One of these bolts is figured 3 in the sectional elevation, and the position of the three may be seen in the rear-end view of the engine, the upper two being used for the attachment of the engine to the cycle frame lugs. At the inner ends of the water jackets there are partitions or dividing walls to encourage circulation of the cooling water on the thermo-syphon principle. The inlets are at w^1 and the water flows to the bottom of the cylinders, through the dividing walls, and upwards and around the cylinder barrels to

582

the outlets w. There are eight water legs w and w¹ connecting the water jackets with the tank D immediately above the cylinders.

Pockets are cast vertically opposite in the cylinder heads or cover blocks to receive the removable valve boxes, the automatic inlet valves 13 are at the bottom and the mechanically operated exhaust valves 12 are above them. The method of holding the inlet valve boxes in place by a bridge and centre stud and nut, and the exhaust valve boxes by the double-eared wedge plates may be seen by inspection of the elevation and plan.

The position of the hole in the combustion chamber into which the sparking plug is screwed at the top of the combustion chamber is shown by the cylinder section.

The pistons 5 are cast in pairs and are of trunk form, lightened or cut away at the centre part, enough metal being left to form connecting strips between the gudgeon-pin boss and the piston ends. A narrow bridge strip with an oil groove in the top is screwed to the upper side of the piston body. The oil supplied through holes, one of which is shown, at the top and at the centre of the cylinder barrels fills the longitudinal oil groove and lubricates the cylinders and piston rings, assisted by the Y distributing grooves shown cut in the upper surface of the pistons.

Cylinder lubrication

The tubular connecting rods 4 are cranked or set off to work clear of the projecting parts of the engine and have ball-bearing big ends. The sockets at their other ends are split and fixed upon the outer ends of the single long crosshead pin x by pinching screws which also serve as keys. The pin x works in bearing bosses cast with the pistons and is lubricated from drip pipes which receive their oil supply from the channelled strip already referred to.

Race ways or slots are cut at the centre portion of the cylinder tubes to permit the reciprocating motion of the crosshead pin.

A short transverse lay shaft at the back of the rear pair of cylinders, carried in bearings fixed in lugs cast with the back cover block, is rotated by the connecting rod at the right-hand side of engine through the little slotted crank arm s. In the plan (Fig. 462) and the general elevation (Fig. 458) the driving pin and its sleeve by which it is attached to the connecting rod may be seen. The skew wheel 8 on the transverse shaft gears into and drives the wheel 6 on the camshaft 2, at half the speed of rotation of the crank spindle or back wheel of the bicycle. There are two cams on the shaft 2, each fixed in position as shown by two setscrews and arranged axially at 90 degrees. They give reciprocating angular motion to forked arms, fixed on the inner ends of short spindles which also have fixed upon their outer ends the double-armed valve-operating levers 9. The forked arms are shown in dotted outline in the end view of the engine. Above the camshaft 2 is a light longitudinal shaft 1 having at its forward end a sector lever 7, by means of which and through a Bowden wire attached to it and to a twist handle on the bicycle handle-bars, the shaft 1 may be given a small angular or twisting motion. By inspection of the three views of the motor it will be seen that supplementary bell cranks 10

Arrangement of camshaft

Exhaust valve lifter and relief of compression

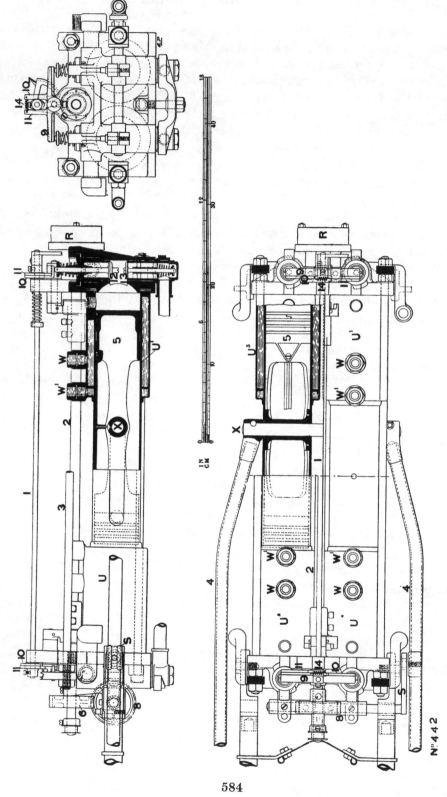

Fig. 462.—Holden Motor Bicycle.—Sectional Elevation and Plan and End Views of Engine.

Nᵒ 442

584

and 11 have their horizontal L section arms interposed between the operating lever 9 and the exhaust valve spindle ends. When the shaft 1 is twisted small expanding cams separate the vertical arms of 10 and 11 and by depressing the horizontal L arms, hold the four exhaust valves open, and so relieve the compression either when starting the engine or when running down hill or enable the driver to control the speed of the bicycle or run at low speed in traffic. The bell cranks 10 and 11 are shown in the normal or running position and not separated by the expanding or separating cam, springs 14 being employed to keep the vertical arms up to the cams and to cause them to follow the movements of the ends of 9 when the compression is not relieved.

The induction pipes connecting the inlet valve boxes are shown broken away, and the pipe connexions from the corner exhaust elbows and the exhaust box or silencer below the engine are not shown in Fig. 462, but are indicated in the general elevation (Fig. 458). The exhaust elbow from the forward right-hand cylinder is shown turned upwards and a flexible metallic pipe, visible in Fig. 457, is led to the inlet end of the warming coil in the petrol tank and surface carburettor (Fig. 459).

On the forward end of the camshaft 2 (Fig. 462) there is a high tension distributer box R, from which the four leads to the sparking plugs are taken. **Ignition system**

The arrangement of the ignition system is substantially the same as that used in connexion with the larger four-cylinder engine described in chapter xxxvii, and the design of the distributer box is there shown. The contact breaker in the low tension circuit is shown at P in the detail of the back wheel (Fig. 458), and the operating cam is shown on the wheel spindle or the equivalent of the crankshaft.

The automatic lubricator F is of the same design as that used in the early machines and is driven by a light belt from a pulley on the rear end of the camshaft. **Continuous automatic lubricator**

A single rim brake is used working on the front-wheel rim. It is shown operated by a lever in Fig. 458, and an improved form is shown in Fig. 457, operated through a Bowden wire and twist handle.

From the same illustration it will be seen that foot rests are provided and that the Crypto gear cranks and pedals used with the earlier machine are not used in the later type, it having been found that the machines were reliable, ran well, and seldom failed to take the rider home.

The machine travelled well on good roads and was a good hill climber, but the small diameter of the driving-wheel was found to be disadvantageous.

The driving-wheel tyre was of a specially strong and heavy type, $3\frac{1}{2}$ inches diameter, and the thickness of rubber on the tread was $\frac{5}{8}$ths of an inch. When much worn at opposite points where the driving effort was felt, the crank arms on the wheel spindle could be easily changed to new positions at 90 degrees to the old positions on the squared spindle ends, and the less worn parts of the tyre tread made use of. **Wear of driving-wheel tyre**

585

MOTOR VEHICLES AND MOTORS

The Humber Motor Cycles

Chain transmission gear

At a time when the use of belts of various forms was looked upon as the only satisfactory method of transmission of power from the engine to the cycle road wheel, Messrs. Humber introduced their machine with the double-reduction form of chain transmission, shown by Fig. 463, which represents one of the 2-HP. bicycles as constructed in 1902. That they retain this method of driving and have extended its use to their two-passenger three-wheeled machines is evidence of its satisfactory performance. By reference to Fig. 463 and Fig. 464, which shows how little it has been found necessary to make changes in the general design of the 1905 3-HP.

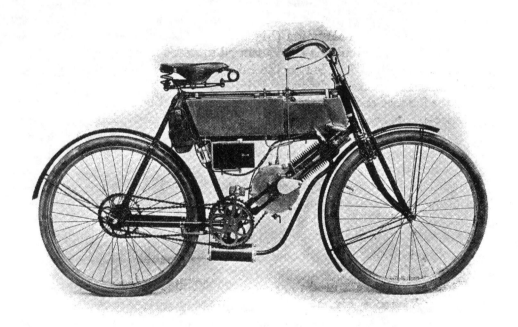

Fig. 463.—Humber 2-HP. Motor Bicycle, 1902.

bicycle, it will be seen that the engine drives, from a pinion on its crankshaft and through a roller chain, a sprocket wheel upon a sleeve, having its bearings in the bottom bracket. Fixed to this sleeve is a second chain wheel, on the inner side and adjacent to that which receives the drive from the engine and from which the driving effort is transmitted by a roller chain to the large sprocket fixed to the road-wheel hub. The pedal crank spindle also has its bearing in the bottom bracket, concentric with the sleeve and chain wheels driven by the engine, and a separate chain and chain wheels are provided on the right-hand side by means of which the bicycle and engine may be started. There are free-wheel clutches at the crank bracket and at the small chain wheel on the road wheel. Through

586

the first of these the engine may be started by pedalling, but it allows the engine to drive the bicycle without driving the pedal gear, and the second allows the pedals to remain at rest after the bicycle has been started.

In order that the irregular driving effort from the engine shall not result in uneven or jerky travelling of the cycle, more particularly at low speeds, a clutch of one or other of two forms on the engine shaft allows a little slip to occur and so relieves the transmission gear from shocks which **Elasticity of drive** it could not withstand and to which it would be subjected with rigid chain driving. With belt-driven motor cycles the natural elasticity of the belt and the slipping which frequently too readily occurs, provide the necessary freedom or relief in transmission. When the belt and its pulley and driving rim are of suitable forms and are maintained in good condition, the elasticity of the belt enables the required relief to be obtained in a more satisfactory manner, in that it need not involve the frictional loss resulting from the slip of the clutch used with the chain-driven machine. It may, however,

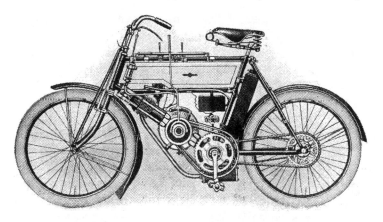

FIG. 464.—HUMBER 3-HP. MOTOR BICYCLE, 1905.

be pointed out that with the forms of transmission gear now employed with most motor vehicles no provision for elasticity or yieldingness is made other than that resulting from slip at the clutch.

The simple form of clutch shown on the engine shaft in Fig. 466 has **Slipping clutch** shown itself to be satisfactory in use with all but the most careless drivers, and is retained as a standard type. The chain pinion H is loosely mounted on a flanged block keyed to the engine shaft and held in position by ring nuts on the reduced end of the shaft. The pinion H is held up against the inner flange of the fixed block by pressure from the spring J upon a washer plate between it and the outer face of the pinion H. The amount of the spring effort and the degree of slip of H between its friction faces may be adjusted by means of the nut on the outer screwed end of the fixed block.

With some of the Humber machines the large chain wheel on the cycle road wheel was mounted upon a free-wheel clutch of large diameter and the machine could then run down hill with the engine and driving chains stationary. This arrangement was satisfactory but necessarily added to the cost of construction and to the number of parts, and on arrival at the bottom of the hill it was necessary to again pedal the machine to restart the engine. This would not be regarded by most as a disadvantage, but the necessity for restarting the engine has been obviated by the use of a leather-faced cone clutch on the engine shaft instead of that just described operated by a hand lever visible in Fig. 464. When descending a hill the clutch may be disengaged, and on arrival at the bottom of the hill the engine may be restarted by gently engaging the clutch.

The simplest arrangement of gearing is that shown by Fig. 463, and it is still retained as a standard type for those machines not fitted with the cone clutch, and hand lever. Only two free-wheel clutches are required in connexion with the pedalling gear, and the engine at all times drives or is driven by the bicycle.

Position and attachment of engine A characteristic feature of the Humber cycles is the inclined position of the engine and its attachment to the frame and utilization as a part of the frame member connecting the head of the machine with the bottom bracket. By reference to Fig. 464 and the separate views of the engine (Fig. 466) the nature of this attachment will be understood. At the lower end of the engine, connexion to the bottom bracket is made through the bossed lug and crown plate L, and at the upper end by a single bolt through the crown plate there shown. The whole arrangement is well designed and executed, and when necessary the engine may be taken down by removing the upper and lower cross bolts. The four long tie bolts which hold the cylinder in position and up to the crank chamber are no longer used, but their equivalent taking the form of stud bolts screwed into bosses cast on the aluminium crank chamber may be seen in Fig. 464.

2- and 3-HP. air-cooled engines The 2-HP. and 3-HP. engines are shown by Figs. 465 and 466, the former having a cylinder with bore and stroke of 2·625 and 3·0 inches respectively, and the latter a bore and stroke of 3·125 and 3·0 inches. The cylinders c are cast with radiating gills on their barrels and around the combustion heads and valve boxes. The pistons P are in each case fitted with three rings, but whereas with the 2-HP. engine taper-pointed setscrews are used to secure the gudgeon pin in position, with the 3-HP. engine the gudgeon pin is in the first instance closely fitted or made an easy driving fit in its bosses, and is there held by the small grub screw at the left-hand end. With the larger engine it has been possible to lighten such parts as the gudgeon pin, the crankshaft M and the fixed spindle upon which the large timing wheel, exhaust-valve cam, and the ignition contact ring E run by forming them hollow and in some cases the hollow formation has been made use of for lubrication purposes by drilling radial oil holes. Suitable channel ways, recesses and oil holes are provided in different places to lead the lubricant to the journals and to assist lubrication of the piston, and at the projecting end of the crankshaft a cap is provided to reduce the leakage of oil and to encourage its return as far as possible through the drain

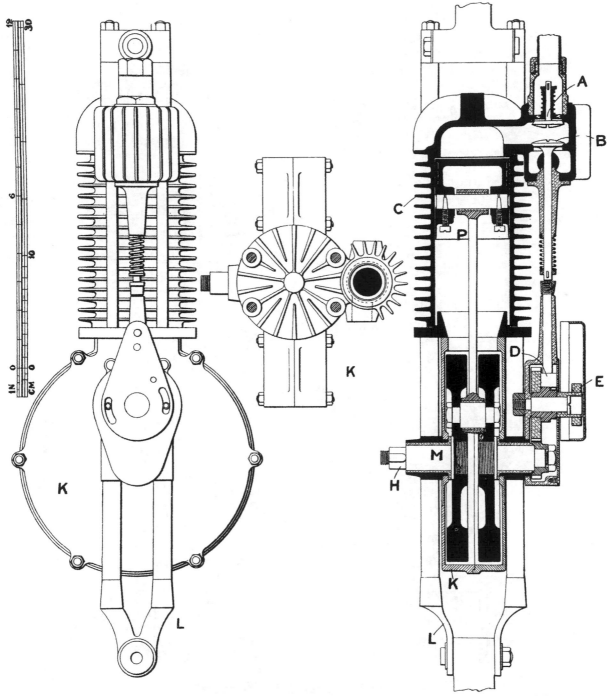

Fig. 465.—The Humber Motor Bicycle.—Side Elevation and Transverse Section of 2-HP. Engine.

589

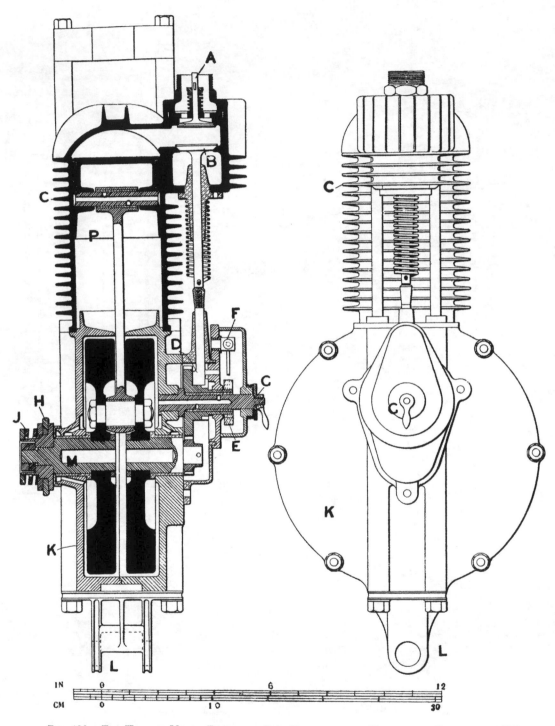

FIG. 466.—THE HUMBER MOTOR BICYCLE.—SIDE ELEVATION AND TRANSVERSE SECTION OF 3-HP. ENGINE.

590

hole shown connecting the crank chamber and the annular recess covered by the leakage cap.

By means of a plunger pump with transparent barrel connected with the engine lubricating oil tank by a three-way cock, a charge of oil may be delivered at intervals through a communicating pipe to the engine crank chamber. **Intermittent lubrication**

The crank chambers K are made of aluminium and occupy little room sideways, their width having been reduced as far as is necessary, bearing in mind the position of the engine, and as far as good design permits. The inlet valves A are atmospherically operated and with their seats are held in position by union bonnets to the upper ends of which the induction pipe is connected. The exhaust valves B are guided in long sleeves separate

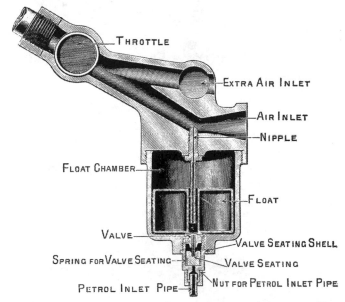

FIG. 467.—THE HUMBER MOTOR BICYCLE.—CARBURETTOR AND AIR AND MIXTURE THROTTLE VALVES.

from but screwed into the under face of the exhaust valve box. On the upper ends of the valve-lifting tappets D, adjustable and renewable pieces are screwed which make it easy to obtain just the requisite clearance between the adjacent ends of the tappet and valve stems, and the full lift and opening of the exhaust valves. Balance weights are cast with the flywheels to partly neutralize vibration from unbalanced rotating and reciprocating parts such as the crank-pin, connecting rod, and piston. In Fig. 465 the chain pinion and clutch parts are removed from the end H of the shaft, and the key or feather which transmits the driving effort from the shaft to the clutch block is visible as well as the reduced threaded end of the shaft upon which are the nut and lock nut which keep the clutch block in position.

The tank occupies the frame area between the top and lower horizontal tubes and is divided into two compartments, a small one forward contain-ing lubricating oil and leaving the greater part of the tank capacity avail-able for containing petrol. The subdivision of the tank was differently arranged in the machine shown by Fig. 463, which was also fitted with a surface carburettor. Carburettors of the float feed spraying type are now alone used of either the Longuemare type or a form introduced by Messrs. Humber.

Humber carburettor
Fig. 467 is a sectional view of the Humber carburettor and in Fig. 468 it is shown in position, the use or duty of each part being clearly

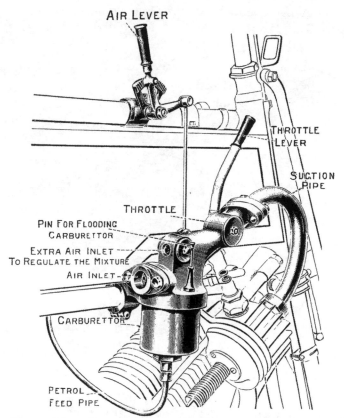

FIG. 468.—THE HUMBER MOTOR BICYCLE.—POSITION OF CARBURETTOR AND CONNEXIONS WITH IT.

indicated. The lower cylindrical vessel which contains the float, and the upper part in which are formed the normal and supplementary air ducts, are made in aluminium. In the upper part are also two throttle valves, both of plug form, one at the entry to the supplementary air passage, which, as shown in Fig. 468, is connected to the air lever mounted on the top frame tube, and the second the main throttle valve close up to the point of connexion of the suction or induction pipe from the car-burettor to the inlet valve bonnet. By reference to Fig. 468 it will be seen that the main throttle valve is opened or closed by a lever directly connected with it. When starting the engine, the air throttle must be kept closed, but thereafter it should be kept open as far as the nature of the road and the speed of travel permit in order that by thus using as little petrol as possible the running temperature of the engine may be lowered, whilst by admitting the greater volume of less highly carburetted mixture to the cylinder a higher compression results with more economical and satisfactory conditions of running.

With the float at the bottom of the chamber as shown the petrol may enter freely and continues to do so until the float has been lifted by the petrol sufficiently to permit the valve seating to be lifted by its spring up against the lower conical-formed end of the valve. The normal level of the petrol in the float chamber is thus an inch or two below the top of the spraying jet or nipple, but very little suction effort is necessary to raise the level and induce the flow of petrol from the jet.

Figs. 469 and 470 show the timing wheels B and the exhaust and half-compression cams C and D in their casing A exposed by breaking

Exhaust valve lifter

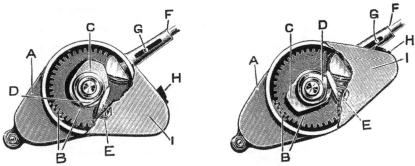

FIGS. 469 and 470.—THE HUMBER MOTOR BICYCLE.—ARRANGEMENT OF TIMING WHEELS, AND EXHAUST-VALVE LIFTING DEVICE.

away that part of the ignition gear base plate I, which is in front of them. The main cam C, which gives the full lift to the tappet F, is behind the little cam D. The latter is of insufficient diameter and eccentricity to give any movement to the tappet stem until the ignition gear base plate is given a partial back rotary movement, in the direction which retards ignition, and so brings the tongue E pivoted to the back of I between the head of the tappet F and the cam D. As shown in Fig. 469, ignition is advanced, the exhaust cam C is just leaving the tappet F and permitting the valve to close at the conclusion of an exhaust stroke. In Fig. 470 ignition has been retarded and the position of the cams indicates that the piston is at the top of its compression stroke. The tongue E is now between the cam D and the tappet F and the exhaust valve is held sufficiently open to reduce compression. Continuation of the backward movement of I in the direction of retardation of the moment of ignition causes the piece H with its inclined face to come into contact with the pin G in the exhaust-valve tappet and so holds the exhaust valve open.

The exhaust silencer is hung below the crank bracket and it is shown in section by Fig. 471. Three tubes differing in diameter are concentrically held between end plates by tie bolts and provide a centre chamber which the exhaust gases first enter, and two annularly shaped chambers between the inner and middle and middle and outer tubes through which the gases progressively pass through a series of small holes. By the increase of volume which the successive chambers give and the final outlet to atmosphere through a number of small holes, the velocity and the noise of exhaust is diminished.

It is unnecessary to completely describe the ignition system, but it may be pointed out that a pair of 10 ampère hour accumulators are contained in a case attached to the back of the tube connecting the saddle and crank bracket, and by means of a change over switch at the top of the case either of the accumulators may be put in service or if the switch is set to mid position the circuit is interrupted, this being the required switch position when the machine is not in use. The trembler induction coil is con-

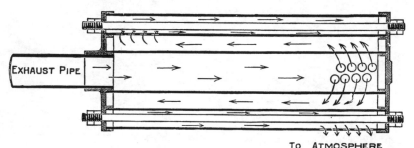

FIG. 471.—THE HUMBER MOTOR BICYCLE.—SECTION THROUGH SILENCER.

tained in the small case hung from the horizontal tube below the petrol tank.

Brake lever switch   The front brake lever is arranged to interrupt the circuit when slightly lifted, so that if the brake is hurriedly applied the engine is simultaneously stopped. It is a convenient method of control when traffic requires momentary stops or reduction of speed. A fully detailed description of these machines cannot be here given, but those who are capable of appreciating excellence of workmanship and finish and the thorough attention given to small details will find by examination of the actual machine much that is of interest.

Weight   Compared with machines of different design the Humber standard type is rather heavy, weighing about 140 lb., but a light machine of 1¾ HP. of somewhat similar design, weighing 72 lb. complete, has been made for those who consider the weight a disadvantage and who are light themselves and do not mind sacrificing to considerations of weight some of the advantages in increased strength, capacity, and durability, possessed by the heavier machines.

Olympia tandem   Earlier in this chapter reference was made to light and heavy tricars or three-wheeled two-passenger motor cycles. An excellent example

594

of one of this type of machine, weighing 3¾ cwt., and therefore of medium weight is shown externally by Fig. 472. The arrangement of the rear part of the machine is similar to that of the bicycle, but the framework is necessarily different, diagonal and longitudinal frame tubes being required for the satisfactory attachment of the fore carriage. A 4½-HP. water-cooled engine with stroke and bore of 3½ and 4 inches respectively, drives through a cone clutch and roller chain to a compactly arranged two-speed gear box attached to the frame at the position of the bottom bracket. By means of the hand lever and quadrant mounted on the top frame tube near the saddle and through the rod shown the direct drive or high gear may be obtained, or an intermediate free position, or the low gear may be put into operation. There are four wheels in the gear box running in pairs always in mesh through which the gear reduction takes place in a manner similar to that common to lathe and other machine-tool back gear.

The road wheels are 26 inches diameter and are fitted with 2½-inch motor-cycle tyres and the total gear reductions for the high and low speeds

FIG. 472.—THE HUMBER OLYMPIA MOTOR TANDEM, 1905.

are 5½ and 12 to 1 respectively. Reversing gear is not provided and is not necessary for a machine of this weight.

The drive to the rear road wheel is through a roller chain at the other side of the machine. The front wheels run on pivoted axles, steering being upon the Ackerman principle and through well designed and arranged connexions to the lower end of the handle-bar steering tube. There are compensated band brakes to the front wheels and a rim brake acting on the driving wheel rim.

A differently arranged machine described as the Humber Olympia tricar is provided with a 5 HP. water-cooled engine, the cooling water being circulated by a gear pump. The change-speed gear box and the arrangement of driving chains is like that adopted in the Olympia tandem, but the engine occupies a vertical position in the horizontal trussed tubular frame. The front seat is hung in the same way and is of the same form as that shown by Fig. 472, but the cycle saddle seat for the driver is replaced by a bucket seat above the engine. There are access doors at each

5-HP. Olympia tricar

595

side of the framework supporting the seat and the latter may be removed to give easier access to the engine and gear. The machine weighs 5¼ cwt., and the driving wheel, 26 inches diameter, is fitted with a motor-car tyre 3 inches diameter. Although the driver may be more comfortably seated and the mechanism is better protected from the weather, being partly housed in, it is open to question whether the work for which this type of motor vehicle is suitable cannot be equally well performed by the lighter machine already referred to. It belongs to an intermediate class, between the two-passenger motor cycle and the light four-wheeled car or voiturette.

Chapter XXXVI

PETROL MOTOR CYCLES (*continued*)

THE WERNER MOTOR CYCLES

SINCE MM. Werner Frères introduced their front-driven type of motor cycle, weighing about 65 lb. and provided with a little 1-HP. engine, described in vol. i., they have maintained their reputation as originators and constructors. Towards the end of the year 1901, they attracted a good

FIG. 473.—WERNER 3¼-HP. TWO-CYLINDER MOTOR CYCLE, 1905 TYPE.

deal of attention by adopting the position for the engine that has since become almost universal, with the engine hung low down in the frame and its crank chamber immediately in front of and supported at the rear by connexion to the frame junction normally occupied by the pedal-crank

597

bracket. The down tube from the head of the machine was forked at its lower end and attached to the front of the engine crank chamber, the latter thus constituting a part of the framework.

In addition to their standard machines equipped with $2\frac{1}{2}$ and 3 HP. engines they have recently introduced an interesting type of machine fitted with a $3\frac{1}{4}$ HP two-cylinder engine, the general arrangement of which may be understood by reference to Fig. 473. The design of the framework is well shown in this perspective view and the method of attachment of the engine is again shown by the separate detail (Fig. 475). With a two-cylinder engine of the design shown with the crankshaft in the usual position, normal to the plane of the cycle frame, it might have been expected that the necessary addition to the width of the crank chamber would have proved undesirable and would have introduced difficulty in obtaining sufficient clearance for the belt drive and for the pedalling gear. The frame tubes have been considerably cranked or set off at the back and chain stays to clear the belt rim, but not much more so than is found to be necessary when the single-cylinder engine is used. In order that the comparatively small additional width taken up by the two-cylinder engine may be realized a front view of it is shown by Fig. 474. The width of the crank chamber from face to face is 5 inches, and the extreme width from the cover of the contact breaker box to the driving pulley nut is 9 inches. The width of the single-cylinder engine crank chamber is $3\frac{3}{4}$ inches and the room occupied by the contact breaker box and driving pulley is in both cases the same. A sectional view of the two-cylinder engine is not given, but from the three external views shown and by reference to its internal arrangement its design will be sufficiently understood. The cylinders with their combustion heads are in one casting secured to the crank chamber through the flange at their lower end by six studs. The bore of each cylinder is 55

Two-cylinder engine

FIG. 474.—WERNER TWO-CYLINDER $3\frac{1}{4}$ HP. MOTOR-CYCLE ENGINE. FRONT VIEW.

mm. and the stroke 76 mm., equivalent to respectively $2\frac{5}{32}$ and 3 inches. The connecting rods work on to one long crank-pin between the fly-wheel discs, and the cylinders fire alternately a working stroke being obtained at every down stroke. The inlet valves are of the atmospheric type, the exhaust valves being operated mechanically through a train of three timing wheels inside the casing at the side of the crank chamber visible in Fig. 475. The gear wheels on the extended end of the crankshaft and on the end of the transverse camshaft have the relative diameters of 1 and 2, the intermediate gear wheel being of larger diameter and serving as a follower wheel to obviate the use of two wheels only of undesirably large diameter, and it does not of course affect the velocity ratio of the train of three wheels. The camshaft takes its bearing in the casing cast in front of the crank chamber, visible in Fig. 474, and it receives ample lubrication by splash from the crank chamber. The inlet valves are situated above the exhaust valves and the valve chambers are in front of the cylinders and in a position to derive the maximum advantage from air cooling. The sparking plugs are screwed into the top of the combustion chamber heads.

The carburettor is of the float-feed spraying type with the jet above and concentric with the float chamber, the tube leading to the jet passing down into the float chamber and forming a centre-guide stem for the float. The arrangement of the lower part of the carburettor may be understood by reference to the sectional view of the Humber carburettor (Fig. 467), which is very similar to the Werner design. The tip of the jet enters the

Carburettor

FIG. 475.—WERNER 3¼-HP. TWO-CYLINDER MOTOR BICYCLE. DETAIL SHOWING ENGINE FROM RIGHT-HAND SIDE.

mixing chamber or back end of the horizontal induction pipe cast with the aluminium cover to the float chamber. A short length of induction pipe is connected by union with the carburettor and at its forward end is attached to the throttle valve, immediately in front of which the induction pipe divides, a branch being taken to each inlet valve. The relative position of these parts may be seen by reference to Fig. 475. The quantity of air entering the induction pipe may be regulated by a supplementary throttle valve, contained in the carburettor cover, operated by the vertically led rod partly visible, and the quantity of mixture passing to the cylinders by the main throttle valve shown operated in Fig. 473 by a lever mounted on the top frame tube and connecting rod. The throttle valve is so arranged that continued motion in the direction of closing it, first completely stops the supply of mixture to the cylinders and then subsequently by a pair of projecting fingers depresses and holds the inlet valves open and uncovers an air port in the throttle-valve casing. The engine may thus be driven by the cycle, when descending hills, with the compression relieved and the internal surfaces of the cylinders become cooled by the air taken in through the port described by the reciprocating pistons. High-tension accumulator and non-trembler induction coil system of ignition is employed, two induction coils are used and double contacts at opposite points in the contact breaker box.

Ignition

The time of ignition is advanced or retarded by the lever shown mounted at the right-hand side of the top frame tube and the ignition circuit may be interrupted for control of the machine by means of a twist handle at the right-hand side of the handle bars, and the usual interrupter plug or touche is provided for insertion or removal when the machine is in or out of service.

The tank, fitted between the top and lower horizontal frame tubes is divided into three compartments. The forward part contains petrol, the middle part accommodates the accumulators and the induction-coil box and the small part at the back contains the lubricating oil supply and the plunger pump. The arrangement of the pump inside the tank is good, leakage of oil is avoided, and when the pump requires examination it may easily be withdrawn. The throttle valve, on recent machines, is controlled by a twist handle on the left-hand side of the handle-bars, an arrangement which is convenient since the control of the machine is chiefly effected by regulation of the throttle valve. The weight of the machine illustrated is 108 lb., which is very moderate for a motor cycle of this power. The Werner machines are now arranged with a well designed spring attachment of the road wheel to the front forks which to some extent relieves the front tyre, and protects the cycle and rider from the vibration which is generally more severely felt at the front than at the back.

Speed control with free engine

A free engine pulley is sometimes fitted to the Werner machines, so arranged that the pulley flanges may be separated or closed allowing the belt to slip or to be driven by the engine. When this is used, in order to prevent excessive racing of the engine, a speed controller is employed which halves the number of working strokes. A supplementary set of small timing wheels is arranged to close an alternative path or shunt circuit in the ignition system, which is the only path open to the current when the engine pulley is opened, at intervals which only permit an ignition to occur once every four revolutions of the engine. The alternate unexploded charges pass away unused to the exhaust box. As a method of governing it is of interest.

The Rex Motor Cycles

The 3¼-HP. Rex motor bicycle illustrated by Fig. 476 is a typical example of a modern belt-driven machine that has been made and is in use in large numbers. Its weight is approximately 125 lb., which may be considered light having regard to the power of the engine. The frame is designed with a long wheel base and both the rider and the engine occupy low positions, and with the centre of gravity near the ground the machine will travel steadily and with less liability to side slip on treacherous road surfaces. The front forks are of the girder type with truss tubes running from the lower end of the forks to a point above the upper end of the frame head tube. The larger tank or case occupying the front area between the frame tubes is divided into three compartments, a small one at the front containing engine lubricating oil, the lower portion accessible by a removable flap on the side visible in Fig. 476 houses the Longuemare carburettor with which all these machines are now fitted, and the largest compartment contains 1¾ gallons of petrol. The smaller case behind the down tube from saddle to crank bracket contains the trembler induction coil and the accumulator and there is sufficient room for a spare accumulator if it is carried

FIG. 476.—THE REX 3¼-HP. MOTOR BICYCLE, 1905.

with the machine. The cradle supporting the engine is a one piece steel casting formed with the lugs for brazing to the frame tubes, from the head and saddle, and to the rear fork tubes or chain stays, and with the crank or pedal bracket barrel, cast as part of it. The engine position is determined by a fitting bolt passing through a boss at the bottom of the crank chamber and the parallel members of the cradle, and by two upper through bolts one of which is visible, and which engage slots or open jaws formed on the exterior of the crank chamber. The neat external appearance of the machine has been carefully studied, and there are few visible connexions or regulating rods or levers. The usual separate silencer box and exhaust pipe have been dispensed with, and their equivalent obtained by casting with the cylinder a chamber communicating with the cavity below the exhaust valve. This chamber may be seen in Fig. 476, projecting forward

601

from the main body of the cylinder, the exhaust gases enter the upper portion separated from the lower by a cast partition or floor, and there is communication between the two by a number of vertical cast pipes perforated with small holes. The exhaust gases enter the lower chamber in a series of streams and pass away to the atmosphere by a second series of small holes. The course of the exhaust is thus rendered sufficiently tortuous without obstructing the flow or creating undue back pressure, whilst the silencing effect is found to be sufficient. No details are given of the engine, the arrangement internally having points of similarity with others and with the 5-HP. engine, a section of which is shown by Fig. 480. The inlet and exhaust valves are, however, both mechanically operated and the smaller of the timing wheels upon the crankshaft gears with a pair of wheels each of twice its diameter, having the inlet and exhaust valve cams formed solid with them. They are contained inside the circular chamber cast with the aluminium crank chamber, and may be got at for examination or removal by taking off the flat cover plate. The spindle of the exhaust gear wheel and cam is prolonged through the cover plate and has fixed upon its end the wiper contact disc for the ignition system. The cylindrical contact breaker box carried on the boss inside which this spindle runs, contains the insulated contact blade, and it may be partially rotated in either direction through a limited range in the usual manner for advancing or retarding the moment of ignition by connexion through a series of levers with the controlling lever mounted on the top tube of the frame. The ignition circuit is so arranged that it may be interrupted by slightly lifting the front brake lever for permitting the cycle to run without being **Ignition cir-** driven by the engine or for automatically stopping the working of the **cuit inter-** engine when the brake is applied, and arrangement is also here made for **rupters** permanently interrupting the ignition circuit when the machine is to be left or is not in use.

There are screw caps above the valves and the sparking plug is screwed into the centre of that above the inlet valve.

A lever pivoted midway between the valve stems has its outer end connected to a Bowden wire leading to a lever on the handle bar by means of which the exhaust valve may be lifted, the pivoted lever at about mid length embracing the end of the exhaust valve stem and lifting the valve by contact with the spring collar near the lower end of its stem. The suction pipe from the carburettor to the engine is short and only that part **Carburettor** of it is visible which projects through the casing containing the carburettor. **housed** The latter has been described in vol. i. and at page 527 in this book and need not be referred to here. Three rods pass from the carburettor up through the tank and project at the top. One, when pushed down, depresses the float and floods the carburettor, the second and third may be rotated or twisted and control the throttle valve and the supplementary or extra air supply. A plunger pump with transparent barrel is arranged at the side of the lubricating oil tank and is connected with it by a three-way cock and with the engine crank chamber by pipe. With the cock in one position the pump may be filled on the suction stroke and with the cock in the second position a charge of oil may be delivered to the crank chamber by pushing home the plunger.

The belt rim is attached to the wheel rim independently of the main spokes by a series of short spokes, and it is formed, for a machine of this power, with a **V** groove to take a ¾-inch or ⅞-inch belt. A brake block acting upon the inside of the belt rim is operated by a conveniently placed pedal on the left-hand side of the machine. A second brake acts upon the front wheel rim and it is applied through the lever at the right-hand side of the handle bars. The road wheels are both 26 inches diameter and they are fitted with tyres 2 inches diameter.

The 2-HP. machine as constructed in 1902 is illustrated by Fig. 477,

" V " belt drive

Fig. 477.—The Rex 2-HP. Motor Bicycle, 1902.

and the changes that have been made in the proportions of the frame and in the general arrangement may be gathered by comparison with the recent machine shown by Fig. 476. The surface type of carburettor was then used and continued to be used with the Rex machines with satisfactory results during 1904. The engine inlet valve was of the atmospheric type and the single control lever shown varied the time of ignition and when retarded lifted the exhaust valve. The tank, hung from the top frame tube, was subdivided to contain the petrol and lubricating oil supplies and to accommodate the accumulator and induction coil.

Surface carburettor

Fig. 478 is an external view of a machine of the heavy tricar type, which has been developed by the Rex Company. A view of the machine with the body work removed to show the frame construction, the position of the engine, and the nature of the transmission gear is given by Fig. 479, and a sectional view of the engine, change-speed gear and the clutch is **Spring-borne** shown by Fig. 480. It is designed to carry two persons and the **load** mechanism is well cased in and protected above by the casing below the driver's seat with sloping extension or beetle back at the rear. The trussed tubular frame is carried at the front and back upon semi-elliptic springs and the rear driving wheel is not called upon to carry any dead weight or load that is not spring carried. An inverted semi-elliptic spring attached at its centre to the fore part of the body and over the front axle, takes bearing at its outer ends near the steering forks, and it is employed to limit the lateral motion or side roll by its damper action. The front wheels run on pivoted axles and wheel steering on the Ackerman principle is employed. The steering column is diagonally braced to the longitudinal

Fig. 478.—The " Rexette," 5-HP., 1905 Type.

frame tubes, and to it are attached the ignition and throttle valve levers and the plunger pump for the supply of lubricating oil to the engine.

Combined Referring to Fig. 480 it will be seen that the engine crank chamber **engine and** and the gear box L are in one piece. This part with the second half of the **gear box** crank chamber and the end cover R to the gear box are cast in aluminium with suitably thickened up places and bosses for reception of the phosphor-bronze bushes in which the engine crankshaft and the gear shafts run. The cylinder diameter and the stroke of the engine are both 3⅝ inches, and by reference to Table 9, p. 508, No. 39, it is evident that the engine can develop the rated 5 BHP. at 1,500 revs. per minute with a mean effective pressure of about 80 lb. per square inch. A Longuemare carburettor is used. The inlet valve, not shown in position, is of the atmospheric type and situated above the exhaust valve, which is mechanically operated as shown through the timing wheels, cam, and tappet, in the usual manner. The projecting end of the crankshaft has a block fixed upon it with cam-formed face for engagement with the starting handle, and the outer end of the short spindle, carrying the larger of the timing wheels with the exhaust cam formed solid with it, has the wiper contact disc fixed upon it. The contact breaker base plate, not shown, is supported on the boss in which the outer end of this spindle runs. The position of the sparking plug between the valves and the boss for the reception of the compression cock are indicated whilst the plug hole at the top of the water jacket facilitates holding the core in position during casting, and its subsequent removal. The cooling water circulates by thermo-syphon action, the water tank being arranged above the level of the cylinder and around the driver's seat. The water is

Thermo-syphon water circulation

FIG. 479.—THE "REXETTE" 5-HP. 1905 TYPE WITH BODY REMOVED.

cooled by the passage of air through horizontal tubes contained in the side portions of the tank, the ears at the front of these parts visible in Fig. 478 encouraging the flow of air through the tubes. Balance weights are, it will be observed, cast in the fly-wheel discs opposite the crank-pin. Attention may also be directed to the liberal dimensions of the crank-pin and the length of the bearings and sufficiency of the support given to the crank-shaft extension T, upon which are keyed the low and high speed pinions A and C, and the upper shaft S. The gear wheels B and D, with teeth one inch wide, always in mesh with the pinions below them are free on the shaft S, but may be driven at the speed of the shaft by engagement of the jaw clutches F E or G H. The centre piece upon the squared part of the shaft may be moved in either direction from the neutral position in which it is shown, by means of a shifting fork working between the collars at J to en-

Operation of change-speed gear

Clutch

gage the wheel B through the clutch teeth F E for the low gear, or the wheel D for the high ratio gear through G and H. The boss of the inner part of the cone clutch M is bushed and upon its extension the chain pinion K is keyed. The outer clutch member N with the part O, through which by the shifting fork it is engaged or disengaged from M, is driven by feathers in the shaft S, upon which it may be moved endwise the small amount necessary. An adjustable ball thrust bearing at P reduces the frictional rub and tendency to drive M when the clutch is disengaged, but there is surface friction or rubbing at the end of the boss of M, and the collar against which it bears. Coil springs on the outer ends of the stud-pins shown normally keep the clutch members in engagement. The spring effort may be adjusted by means of the nuts and washers at the outer ends of these pins. The position of the bosses on the gear case carrying the gear-shifting fork spindle is indicated in dotted lines at L, and one of the bearing lugs for carrying the spindle of the clutch-shifting fork is shown below O. The clutch is 12 inches diameter and 3 inches wide and of ample dimensions for the power to be transmitted. The diameters of the gear and chain wheels are such that total reductions of 10·6 and 5·6 are obtained, and with a driving road wheel

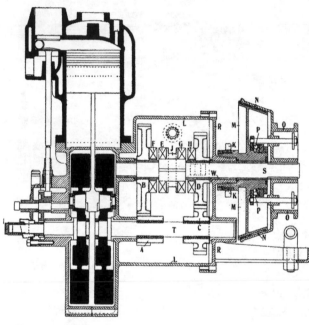

Gear ratio

FIG. 480.—THE "REXETTE" 1905 TYPE 5-HP ENGINE, GEAR BOX AND CLUTCH.

diameter of 28 inches permit the engine to run at its normal speed of 1,500 revs. per minute when the road speeds are about 12 and 23 miles per hour. The final drive is by a ⅝-inch Hans Renold roller chain, cased in and protected from dust and mud. The gear-shifting lever is visible, in Fig. 478, below the driver's seat, and the side lever shown operates compensated band brakes acting upon drums fixed to the front wheels. Of the two pedals one is connected with the clutch and the other with a band brake acting upon a drum fixed to the rear wheel.

The overall length of the machine is 8 ft. 6 in., its width 4 ft. 6 in., and the height to the top of the steering wheel 3 ft. 6 in. It weighs nearly 5 cwt., and is a good example of the modern heavy tricar or light three-wheeled car with well designed spring suspension.

PETROL MOTOR CYCLES

In the new 8-HP. Rexette [1] a twin cylinder V-type engine is used with the two-speed gear box and engine crank-chamber adjacent as in the 5-HP. type. The high and low speed gear wheels are always in mesh with their pinions fixed upon the engine crankshaft extension, but instead of using jaw clutches to bring the high or low ratio gear into operation expanding segmental clutches engaging the internal surface of drums forming a part of each gear wheel are used. The change to this form of clutch renders the use of the cone clutch shown in Fig. 480 unnecessary. A centrifugal governor is fitted to the engine and in the absence of other means of control is required to prevent undesirable racing of the engine when neither of the gears is engaged.

These changes of design simplify the control of the machine and should result in improved performance and quietness of running.

Fig. 481 illustrates a type of motor cycle introduced by the Singer Cycle Co. in 1902. It is there shown as a tricycle with the engine occupying the front wheel, which is also the steering wheel, but it is more commonly employed for motor bicycles, the special motor driving wheel being at the rear. The road wheel with cast aluminium double-spoked centre runs on ball bearings upon trunnions fixed to the front forks and supporting the engine in a fixed position at the centre. The engine crankshaft is not concentric with the road wheel, and a pinion on its projecting end on the opposite side of the engine to that visible in Fig. 481 gears with an internally toothed spring-mounted wheel through which the road wheel is driven.

Singer Motor Wheel

In order to render the engine more accessible and to simplify the connexions for its control the design of the motor wheel has been modified in some respects. The cast aluminium centre leaves the engine exposed, a single row of spokes or spider arms at one side only being used, cranked or off-set sufficiently to provide running clearance. The capacity of the petrol tank need not then be only that afforded by the combined tank and surface carburettor shown, but an additional tank carried on the engine frame may be connected by pipe with it. Instead of the single reduction spur-wheel gearing employed with the earlier design, a double chain form of transmission is possible between the engine and cycle crank bracket and from the cycle crank bracket back to the road wheel, one of the chains, that used for the first stage of the drive being also utilized for the pedalling gear. The machines are now constructed either with a driving wheel 26 in. diam. and a 2-HP. engine, or with a driving wheel 28 in. diam., and a 3-HP. engine. The design is ingenious and the workmanship excellent, but the cost of construction and the appearance of more recent and satisfactory designs will render its continued construction unnecessary. It was introduced in 1902 at a time when the most suitable design of motor cycles had yet to be determined by experiment and actual use, and the

[1] The *Motor Cycle*, 10th July, 1905.

607

pioneer work of the Singer Co. has not resulted in failure and has been of value. The recent types of motor cycles and tricars introduced by them are more in agreement with the present common practice and reference will be made to their characteristic features at a future time.

Figs. 482, 483 and 484 show three types of air-cooled engines that have been largely used for motor cycles.

Clement-Garrard The Clement-Garrard engine illustrated by Fig. 482 is that used by the Garrard Manufacturing Co., of Birmingham, in the construction of their well-known light-weight motor cycles. A distinguishing feature, now common to some other engines, is the use of the external fly-wheel K. This con-

Fig. 481.—Singer Motor Tricycle, 1902.

struction is of importance because it permits the use of a lighter and yet equally effective fly-wheel, the metal being more usefully distributed, and further reduction of weight is effected by the reduced size of crank chamber rendered possible. The crankshaft O with the balanced crank disc and crank-pin are in one piece, and the shaft runs in one long bushed bearing, part of the main piece of the crank chamber 4. To the recessed top of the crank chamber, the cylinder barrel C and the separate cover or head D, are together held by the long side bolts S S, and the cross bar visible in the elevation. The atmospheric inlet valve B and the exhaust valve E are arranged in the cover and open directly upon the combustion chamber or compression space. The exhaust valve is operated through a pivoted lever R and ad-

justable side rod Q, the latter receiving its upward movement from the
tappet actuated by the cam 2. Between the cam 2 and the tappet end
there is a pivoted tongue P, shown in dotted lines, which ensures that a
direct lift is given to the tappet; the movement, too, is smooth and free from

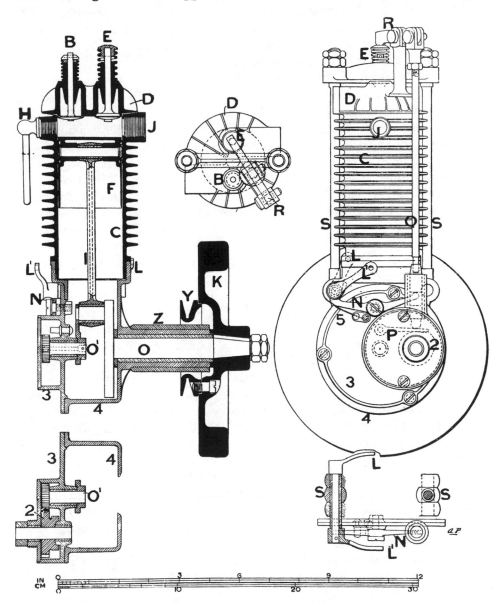

Fig. 482.—Clement-Garrard 1¾-HP. Motor Cycle Engine.—Elevation, Section, and Details.

the side blow which is delivered by impact of the cam against the tappet
at high speed, and quietness of running and freedom from wear of these
parts result. Returning to the sectional view and the detail below it, it
will be seen that a short spindle o^1 formed solid with the smaller of the

timing-wheels has its bearing in a boss in the crank chamber cover 3, and is driven, by the engine crank-pin, through a driver arm fixed upon its inner end.   The larger of the timing-wheels, in one piece with the exhaust cam 2, is fixed to its spindle, which has bearings in the cover 3 and in the outer cover over the small circular chamber containing the timing-wheels.   The spindle on which 2 is fixed has upon its outer end the cam which actuates **Ignition timing and compression relief** the contact breaker blade, and the contact breaker base plate, not shown, is carried on the threaded exterior of the external bearing boss.   A detail below the elevation of the engine renders the arrangement of the timing and exhaust valve-lifting gear clear.   The pin upon which the left-hand bolt s is pivoted to the crank chamber is, as shown, hollow, and through it passes a spindle which may be partially rotated by suitable connexions to the lever L fixed to one end.   At the other end is a lever with a nearly vertical connecting rod, to the contact breaker box which occupies the position described, but is not shown in place.   The boss of this lever, is, as indicated in dotted lines in the elevation, formed as a cam.   Continued movement downwards of the lever L¹ retards ignition by rotating the contact breaker box in a contra clockwise direction, and by bringing the cam in contact with the left-hand end of the pivoted lever N, causes its right-hand end, which enters a slotted way cut in the tappet and its guide boss, to lift the tappet and the rod Q and so hold the exhaust valve open.   A flat spring 5 normally keeps the left-hand end of N up against the boss of L¹, and in this position the right-hand end of N is just out of contact with the top end of the slot in the tappet, and does not, therefore, interfere with the free operation of the exhaust valve.   The points of connexion of the induction and exhaust pipes on the left and right respectively of the cylinder head may be judged from the plan given of the latter and from the engine elevation.   A screw plug H immediately opposite the hole J, into which the sparking plug is screwed, enables the spark to be seen and the correct or incorrect working of the plug to be known without removing it from the cylinder and paraffin or petrol may be injected into the cylinder through the hole occupied by H, when necessary.   It may be pointed out that the plug hole H has in the engine elevation been lettered J.

The piston F is provided with three rings, that at the centre being wider than the others and serving to prevent the gudgeon-pin from working in the bosses, in which it is closely fitted or driven.   The connecting rod 1 is a little steel forging and is of **H** section.

The cylinder diameter is 55 mm. or about $2\frac{3}{16}$ in., and the stroke 60 mm. or nearly $2\frac{3}{8}$ in., and the engine runs up to 2,000 revs. per minute, with perfect regularity.

The performance of this well designed and constructed little engine is very well shown by a series of indicator diagrams and tests taken by Professor H. L. Callender.[1]   One of these diagrams taken at a speed of 2,030 revs. per minute showed that the engine was developing 2·37 IHP., the

[1] Paper read before the Auto-Cycle Club at the Royal College of Science, South Kensington, May 4, 1904. (See *Automotor Journal*, May 14, 1904.)

brake-horse-power measured at the cycle road wheel being 1·18. The mean effective cylinder pressure was 78 lb. per square inch.

The belt pulley Y with **V** groove is, as shown, separate from the fly-wheel but attached to it with three screws. The engine is attached to the cycle frame tube by a suitably formed bracket, the split sleeve z clamped upon the extended crankshaft bearing forming part of it. The weight of the engine is 21 lb.

Fig. 483 is an elevation, partly in section and with the cover over the valve-actuating mechanism removed, of a 2-HP. Minerva engine, one of the first, if not the first, cycle engines to be fitted with a mechanically operated inlet valve. It is there shown with a facing Y on the crank chamber for its attachment to the cycle by clamping to one of the frame tubes, but although it has been largely used in this way it is now more often built into the cycle frame occupying a vertical position. These engines have probably been used in larger numbers than any other, not only with the bicycles constructed by the Minerva Company, but by several English cycle companies, among them the Ariel Company, whose tricycles were described in vol. i., and the Swift Company used these engines of 2¾ HP. for their worm-gear driven two-speed cycles already referred to. The Ariel motor bicycles with the Minerva engine must be specially mentioned as among those which have given the most consistently satisfactory performance, the arrangement of the cycle, the workmanship and the attention paid to detail construction all being good and contributing largely to their success.

The inlet and exhaust valves S and P, with their springs, tappets V and Q, and guide pieces W and R, are of the same size, and one set of these parts forms a complete set of spares. The steel-valve spindle guides, which are of the usual form and each in one piece, are by incorrect sectioning made to appear cut through at the base of the threaded parts where they are screwed into the bottom of the valve chambers. A single cam N upon the spindle of the larger of the timing-wheels, both of which are in the crank chamber, actuates the inlet valve by contact with the end of the tappet V and the exhaust valve through the pivoted bell crank with arms M M¹. The contact breaker cam is carried upon the projecting end of the spindle upon which N is fixed. By means of a finger X, which receives movement through a lever and connexions outside the cover over the valve mechanism, the exhaust valve may be lifted and held open by interference of X with M¹. This exhaust valve lifter is generally interconnected, so that continued movement of the contact breaker in the direction to retard ignition also lifts the exhaust valve. The induction pipe is connected by union with the opening H, and the exhaust pipe is shown at I. There is necessarily a division wall G, cast between the valve chambers and extended up into the combustion chamber. The lower part of the cylinder barrel is left smooth, the radiator ribs being cast on the cylinder head and on the upper part of the barrel and around the valve chambers where they are most wanted. Lubricating oil enters at the upper boss z and the crank chamber may be washed out or excess of oil or dirty oil removed through the lower

Minerva

Valve-operating mechanism

side plug or a cock at z. The cylinder diameters of the 2, $2\frac{3}{4}$ and $3\frac{1}{2}$-HP engines are respectively 69, 76 and 82 mm., or approximately 2·75, 3 and 3·25 in., and the strokes are 70 mm., or 2·75 in. for the 2-HP. engine and the same as the cylinder diameters for the larger engines.

The similarity of the valves and the simplicity of the operating mechanism are good features of these engines.

De Dion-Bouton The small De Dion cycle engine, illustrated by Fig. 484, is an example of compact design. The cylinder barrel E is a plain tube, with no radiator ribs cast upon its external surface, and it is thickened at its top and bottom

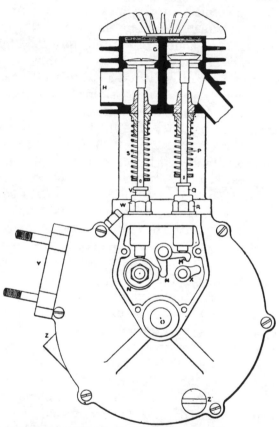

ends where it is jointed to the separate cylinder head D and to the top of the crank chamber. Two long side bolts K, one of which is visible, keep the cylinder head and barrel together and in position on the crank chamber. The atmospheric inlet valve is above the exhaust valve B, and its seating is fixed in place by the flanged inlet elbow A and the gland nut shown. The exhaust valve-actuating gear is novel, and permits the overall width of the engine to be reduced. The half-speed spindle J occupies a position above the engine fly-wheels F and F¹ at right angles to the crankshaft centre line. It is driven by skew gearing, the fly-wheel F with spiral teeth cut upon its periphery serving as one of the gear-wheels and meshing with the pinion

FIG. 483.—2-HP. MINERVA MOTOR CYCLE ENGINE.

G pinned, as shown, to the half-speed spindle. The latter is well supported in bushed bearings and carries upon its squared right-hand end the exhaust cam, and at its other end the contact breaker cam or insulated contact disc. The circular contact box is incorrectly shown as if it was part of the crank chamber. It is necessarily capable of being partially rotated upon the collared oil-retaining bush shown for advancing or retarding the moment of ignition, and lines should be shown indicating its width, which is about 1 inch back from the front cover H. The means provided for lifting and holding the exhaust valve open are not completely shown, but an excentrically pivoted arm, the outer end of which L is visible, may be pushed further in between the exhaust cam and the tappet which it always separates and keeps the tappet sufficiently lifted to prevent the closing of the exhaust valve. A section through the valve chamber taken at right angles to the main sectional view, from which it may be seen that it is cast with the cylinder head, shows the position of the exhaust outlet and the pipe connected with it by union. The sparking plug is screwed in at G. The cylinder diameter is 62 mm. or 2·44 in. and the stroke 70 mm. or 2·75 in. The engine runs at from 1,500 to 2,000 revs. per minute and develops

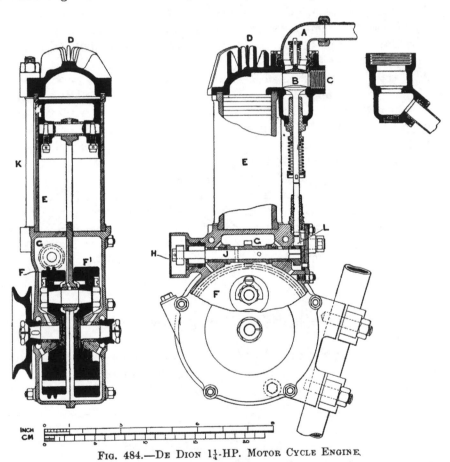

FIG. 484.—DE DION 1¼-HP. MOTOR CYCLE ENGINE.

1¼ BHP., and its weight is 22 lb. Other features of its design are apparent from inspection of the drawing and the dimensions may be found by means of the scale provided. A longer crankshaft bearing at the pulley side of the engine might have been used without adding to the width of the engine. The larger De Dion engines which were illustrated and described in vol. i. have been used for motor cycle purposes in very large numbers, and nearly all the other small petrol engines may be looked upon as of modified De Dion design.

It is impossible in the space of a single chapter to present more than a few typical designs, and in this very active branch of the motor vehicle industry some manufacturers whose machines might with advantage be described cannot be now referred to.

Horse-power It may be of interest to some to give an example showing to what extent the engine horse-power is sufficient for hill-climbing purposes and to call attention to the reduction of power that necessarily occurs with reduction of speed.

Let it be assumed that the engine of a 3-HP. motor cycle develops its rated power at 1,500 revs. per minute.

The machine ready for the road is taken to weigh 140 lb., and its rider 160 lb., the total weight upon the tyres thus being 300 lb.

The diameter of the rear wheel will be supposed to be 26 in. and the transmission gear ratio 5 to 1.

The tractive resistance of well constructed cycles in good order has been found to be about 41 lb. per ton on good roads, and the resistance here taken will be 45 lb. per ton, and it will be further assumed that the efficiency of the transmission gearing between engine and road wheel is 85 per cent. With this data and supposing that a speed of 10 miles per hour is to be maintained up a hill having a gradient of 1 in 8 we may calculate the horse-power required and that available.

The total tractive resistance is equal to the road resistance and that due to gravity, and, neglecting the question of air resistance, which under normal conditions and at this speed with a motor cycle is negligible, it amounts to—

$$\frac{300 \times 45}{2240} = 6 \cdot 0 \text{ lb. road resistance.}$$

and

$$\frac{300 \times 277 \cdot 85}{2240} = 37 \text{ lb. gravity resistance,}$$

or a total tractive resistance of 43 lb. at the road wheel, 277·85 lb. per ton being the amount of gravity resistance on a gradient of 1 in 8.

At a speed of 10 miles per hour and with 43 lb. tractive effort the horse-power exerted at the road wheel must be—

$$\frac{45 \times 5280 \times 10}{33000 \times 60} = 1 \cdot 2.$$

But at a speed of 10 miles per hour the road wheel revolves at 130 revs. per minute and the engine at 650 revs. per minute, and the horse-power it is then capable of developing without overloading is—

$$\frac{3 \times 650}{1500} = 1 \cdot 3$$

or $1 \cdot 35 \times 0 \cdot 85 = 1 \cdot 1$ horse-power at the road wheel.

The rider must then contribute the difference of $\frac{1}{10}$ horse-power between the power available and that required by pedalling or he must overload the engine and run the risk of overheating it. If the hill is long, it is unlikely that a motor cycle engine rated at 3 HP. will be actually able to give off this power or the proportional power at reduced speed for more than a few minutes, without liability of overheating, and the rider would, therefore, under the conditions cited, be obliged to assist the engine by pedalling.

This example, therefore, suffices to show either that 3 HP., or certainly not less than $2\frac{3}{4}$ HP., is desirable for machines that are expected to surmount most hills without pedal assistance, or that a two-speed gear and an engine of about 2 HP. is required.

There are, however, many riders who are content to use a light machine with an engine of little more than 2 HP., and to give pedal assistance for hill climbing. Such a machine is much more economical in petrol consumption and much easier to drive at reasonable speeds on level roads or at slow speed in traffic, and distances of as much as 180 to 190 miles may be travelled per gallon of petrol consumed. It is not proposed to consider the questions of cost of running or the speeds that are possible or have been achieved on special machines, but it may be here stated that there are now two weekly papers devoted to motor cycling matters and that these and the other automobile journals pay attention to the many questions pertaining to light vehicular traffic.

The International Motor Cycle Race—run over the Dourdan course (June 25, 1905) and won for Austria by Wondrick on a Laurin-Klément two-cylinder bicycle—provides a remarkable example of the speeds obtained with racing motor cycles and testifies to the endurance and courage of their riders. Over a distance of nearly 168 miles Wondrick maintained an average speed of 54·3 miles ver hour, driving a machine weighing 101 lb., and having an engine with cylinders 2·84 inches diam. with a stroke of 3·34 inches. A surface carburettor was used, and a flat belt for the transmission of power from the engine pulley to the road wheel.

Inter national motor cycle race

CHAPTER XXXVII

TRANSMISSION EFFICIENCY

THE fundamental question of efficiency of transmission of power, and the efficiency of the engine or motor which delivers power to the transmission mechanism, involves problems of design and selection and utilization of materials, which have at all times engaged the attention of engineers, and more recently of engineers engaged upon the construction of automobiles. Although experimentally, and as the result of observed performance, information has been acquired of the probable degree of excellence now reached, little precise knowledge has been obtained or can be obtained under running conditions of the vehicles.

The stationary trials made by a few prominent manufacturers have furnished interesting figures and are of value although they do not take into consideration certain of the road conditions liable to modify the results obtained, and the experiments carried out under the auspices of the Automobile Club de France did not add much to our knowledge.

Consideration of the design of the vehicle and knowledge of the stresses and movements to which the parts may be subjected, has in the past been instrumental in effecting improvement, and with experience gained by inquiry into, and appreciation of, the causes contributing to failure, will continue to influence future design more than the precise knowledge resulting from tests carried out under conditions which are incomplete or do not completely represent the actual running conditions.

In vol. i., chapter iii. such information as was available relating to the friction of types of bearings and gearing was considered, and it was assumed that with the arrangements of gearing then in general use an efficiency of 70 per cent. approximated closely to the average actual performance. It was, however, in some instances ascertained that such a transmission efficiency was not realized.

Some of the trials of last year and some trials of machinery personally carried out, lead to conclusions as to the possibility, and it may be comparative frequency, of higher efficiencies than have previously been possible. For instance, in the Hereford Light Car Trials held by the Automobile **Deduced** Club, the results of the hill-climbing trials showed that with some of the **efficiency** cars the losses of power in transmission cannot have been more than 10

616

per cent., the maximum power developed by the engines having been previously ascertained. Some doubt may, however, be admitted with regard to the deductions from hill-climbing experiments, and it may be noticed from the table of results given in chapter xxxi. that the least loss in transmission was not always an accompaniment of the simplest arrangements of transmission gear.

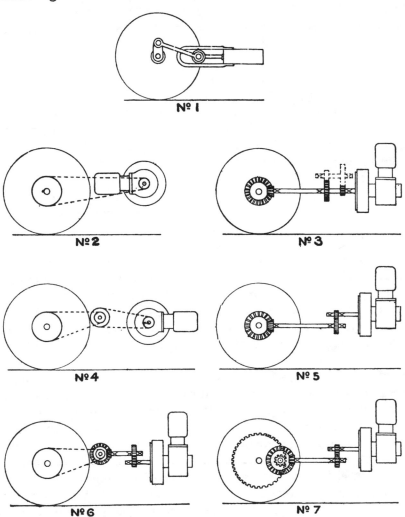

FIG. 485.—DIAGRAMS OF TYPES OF TRANSMISSION GEARING.

The improvements in construction tending to reduced power losses, are those relating not so much to different arrangements of gearing, but to the strengthening of parts, sometimes by increase of dimensions and often by the use of excellent materials which with no change in dimensions give the increased strength or stiffness required. In addition to the elimination of losses brought about by bending or actual deformation resulting from too heavy working stresses, more attention has been given to the

617

support and interconnexion of parts in such a way that comfortable working is ensured, or freedom obtained, when the inevitable relative movement of parts occurs by the passage of the vehicle at high speed over varying road surface. Considerable improvement has also been effected in the direction of enclosing and protecting all parts from dirt, and the methods of lubrication are generally better and more completely carried out.

In Fig. 485 some types of transmission gear now commonly employed are indicated in outline. The first one is the simplest of all and may be said to be reduced to no transmission apparatus unless the peripheries of the road wheels be assumed to be substituted for the free periphery of the engine fly-wheel in other cases. But if this be taken as road resistance, then diagram 1 may be taken as representing maximum transmission efficiency. Diagram 2 represents the simple form of single-chain driving gear, between the engine crankshaft and the live rear axle, employed with a large number of the light cars, made principally in America but also in this country and on the Continent, propelled by petrol and steam engines. During the greater part of the time these cars are running the change-speed gear, in the majority of cases carried upon an extension of the engine crankshaft, but in at least one car upon the road-wheel axle, is inactive, and the only sources of power loss in transmission occur at the bearing supporting the crankshaft extension, in the chain and chain wheels, and in the bearings of the road-wheel axle. When the condition of the parts and the workmanship is good it is probable that the whole loss in transmission does not exceed 10 per cent. of the power delivered from the engine.

When, however, one or other of the epicyclic trains of gearing usually employed is in operation, then an additional loss of, it may be, in some cases approaching 20 per cent. and rarely less than 15 per cent. is incurred.

Diagram 4 is illustrative of two types of transmission gearing which have been shown to give very excellent results. In each case a single chain transmits the power from the engine to a gear box, and from the gear box through spur-wheel gears, the power is finally delivered to the road wheels in the one case by a single chain and wheels to a live rear axle, and in the other by side chains and wheels to the road wheels running upon a fixed rear axle. In either case the same number of bearings may be in active service and although transmission of the same power through two chains may be slightly less efficient than when a single chain is employed there is in the two methods little difference in the amount of loss of power, which probably does not reach or exceed 25 per cent. This is partly due to the fact that the opposite directions of pull on the countershaft in one of the forms as shown puts the shaft approximately in equilibrium and the stress on the bearings is small.

With the largely used arrangement of gearing represented by diagram 6, chain, spur, and bevel-wheel gearing is employed and a larger number of bearings is required than in the other types referred to. It is unlikely that a higher actual efficiency of transmission than 65 per cent. is obtained, but in one prominent type of car 70 per cent. is said to have been reached, with ball bearings used throughout. Diagrams 3 and 5 are representative of the gearing arrangements with cars having live rear axles, and bevel gears driven through universal or Hook jointed shafts. In the first of these it is possible to obtain a direct drive to the bevel gear and the loss of power need not be more than 18 or 20 per cent. When, however, the lower speed or greater reduction gears are active and the conditions are those represented by diagram 5, then with the additional gear and bearings brought into service the transmission loss may be increased to 30 per cent.

Diagram 7 is similar to number 6, the difference consisting in the use of a final spur-wheel drive with internally toothed gear rings on the road wheels. The difference in transmission efficiency would be merely the difference in loss of power in toothed wheels compared with that occurring in chain and chain wheels.

In this brief reference to the diagrams shown it will be understood that consideration has been given to the types and not to methods of carrying them into effect, but it must not be overlooked that constructional excellence or the reverse may have an all important effect on the results obtained and on the working period or life of the mechanism during which the degrees of efficiency here suggested may be realized. For instance, the gear-carrying spindles have in many gear boxes been much too small in diameter with reference to their length, to give necessary stiffness ; hence frictional losses in the bearings.

It is interesting to here show that, if assumed or approximate values be taken to represent the relative amount of the power loss at bearings and in chain and gear transmission, a fairly accurate opinion may be formed and an accurate comparison made of the thorough efficiency of the mechanism. *Estimated power losses*

Let it be taken, for instance, that in every pair of active bearings 5 per cent. of the power transmitted is absorbed, that those bearings which support shafts which transmit only half the power, or constitute guide bearings, result in a loss of $2\frac{1}{2}$ per cent. or an estimated amount in particular cases ; that each chain and its wheels transmitting the whole power be assumed to absorb only 3 per cent. of the power delivered to it, and every pair of gear wheels in service whether they be spur or bevel wheels be considered to result in a loss of 5 per cent.

It may be correctly urged that the loss in bevel gears is greater than in spur wheels, but such loss occurs in the additional bearings and thrust bearings which must be used, and which are included in the number of bearings in service.

In the table here given the types of gearing shown in Fig. 485 are considered in this way.

TABLE 17. TRANSMISSION EFFICIENCY OF DIFFERENT TYPES
OF MECHANISM.

Diagram No.	Source of Loss of Power.	Amount of Loss per cent.	Efficiency per cent.
1	—	—	100
2	When driving direct :—		
	1 chain	3·0	
	1½ pairs of bearings	7·5	89·5
	With epicylic speed gear in operation, add	15·0	74·5
3	When driving direct :—		
	1 set of gear . .	5·0	
	2 pairs of bearings	10·0	
	Partially active bearings	3·0	82·0
	With change-speed reduction gear in operation, add	12·0	70·0
4	1 set of gear	5·0	
	3 pairs of bearings	15·0	
	Equivalent of 2 chains	6·0	74·0
5	2 sets of gear	10·0	
	4 pairs of bearings	20·0	70·0
6	2 sets of gear	10·0	
	5 pairs of bearings	25·0	
	Equivalent of 1 chain	3·0	62·0
7	Equivalent of 3 sets of gear	15·0	
	5 pairs of bearings	25·0	60·0

There are other types of transmission mechanism, particularly some of those in use on heavy vehicles which have not been alluded to. These may be considered in the same way and in all cases regard must be paid to particular conditions of service.

Engine efficiency The question of engine efficiency has not yet been referred to, but the known performances of steam and gas engines of larger powers and sizes than those employed for automobile purposes enables an estimate to be made of the relation between indicated and brake-horse-power in the small high-speed engines. While the power in the large engine may be measured with a close approach to accuracy, little knowledge can be gained, with the indicator instruments at present in use, of the horse-power developed in the cylinders of automobile engines, with the exception of a few which are designed to run at moderate speeds. With high-speed steam engines mechanical efficiencies of 90 per cent. are frequently recorded and are not infrequently a little higher, whilst with gas and oil engines the mechanical efficiency is frequently 85 per cent. and sometimes a little in advance of this. In this volume an efficiency of 85 per cent. has generally been taken for petrol engines, but in one instance 90 per cent. has been chosen as a suitable reference figure for comparison of the maximum brake-horse-power to be expected from steam and petrol engines of stated sizes. There does not appear to be any reason to expect mechanical efficiencies greater than those given for these small engines unless the seldom realized advantage conferred by the absence of rigidity of foundation is greater than has hitherto been imagined. With small engines having many cylinders it is unlikely that a mechanical efficiency of 80 per cent. is reached.

Differences of design and in number and method of working auxiliary parts, including valves, pump, magneto-ignition gear, distributer, mechanical lubricator and fan, and the rigidity of the crankshaft and effectiveness of bearing support, make it necessary to consider each engine separately when estimating the power at the pistons and effective power of an engine the brake-power of which has not been ascertained.

Chapter XXXVIII

SOME ENGINES AND MISCELLANEOUS PARTS

THE engine shown by Figs. 485 to 490, designed by Colonel Holden, F.R.S., is in many respects similar to that of 3 HP. described in chapter xxxv. It will, however, be remembered that with the smaller engine the connecting rods drove on to crank arms upon the rear-wheel spindle, whereas with this larger engine of 10 HP. the connecting rods D D work on to crank-pins, in the balanced fly-wheel discs R R, set at a radius of 2·5 in. for an engine stroke of 5 in. The discs R R are keyed upon a short connecting shaft c running in plain bearings in the projecting bracket arms, which are cast with the pair of cylinders at the fly-wheel end of the engine and rest upon or are supported on the channel section frame indicated in Figs. 486 and 487.

The cylinders A A¹ and B B¹ are 3 in. diameter and are cast in pairs, with their combustion heads and pockets for the reception of the removable valve boxes, and they are bolted together by the transverse facing flanges shown.

The pistons are also cast in opposed pairs, lightened at their centres by leaving only sufficient metal to connect them together and to the centre bosses occupied by the single long crosshead pin M. As with the cycle motor the pin M is lubricated from downwardly pointing drip pipes fixed in the channelled strip N connecting the pistons. The oil filling the channel in the upper side of N finds its way to the piston rings assisted by oil ways cut in the upper face of the piston bodies. A detail is given in section and **Mechanical** elevation by Fig. 490, of the automatic lubricator, used principally with the **lubricator** cycle engine. A light belt, driven from a convenient point on the half-time or engine shaft, rotates the spindle shown with the belt pulley upon its outer end and a collar piece A on its centre part and inside the containing vessel. The oil level must not be higher than the mouth of the drip pipe D, and as long as the engine is running and the lubricator spindle is driven, the endless chain B picks up and delivers oil to the wire stem c resting in a groove between the collars on the piece A. The oil trickles down c, and drips from its end down the delivery pipe D, and so to the distributing pipe at the bottom of the lubricator and the branches to the cylinders. As long as there is oil in the lubricator the engine receives a

622

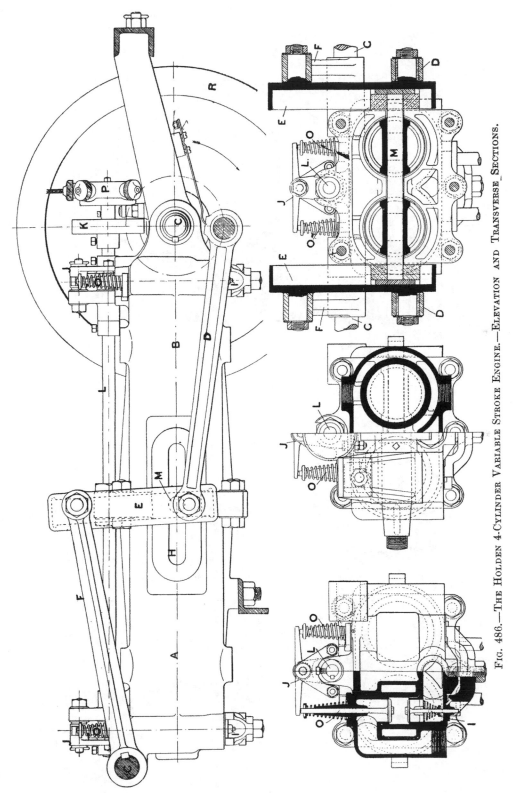

Fig. 486.—The Holden 4-Cylinder Variable Stroke Engine.—Elevation and Transverse Sections.

623

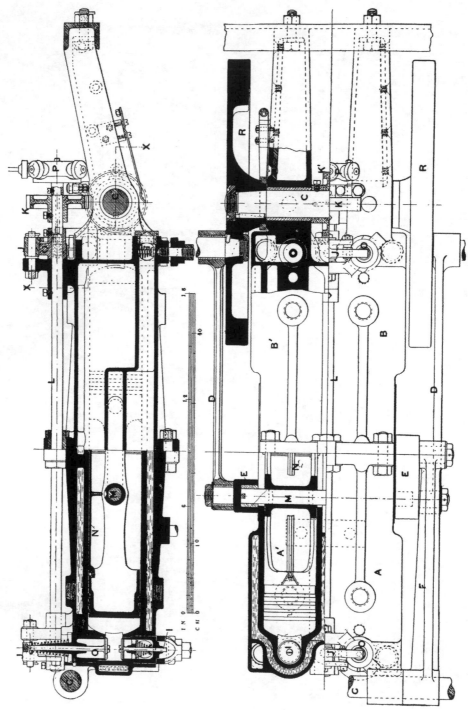

FIGS. 487 AND 488.—THE HOLDEN 4-CYLINDER VARIABLE STROKE ENGINE.—SECTIONAL ELEVATION AND PLAN.

regular supply, which commences as soon as the engine starts and ceases when it stops.

The inner ends of the connecting rods may be fixed, as they were with the 3-HP. engine, upon the ends of the crosshead pin M or they may be journalled upon stud-pins at the lower ends of the channel section links E as here shown. Radius bars F are attached at their outer ends, to stud-pins upon which they work at the upper ends of E, and at their inner ends they are keyed upon a shaft C, which is normally stationary, but may be partially rotated to raise the links E to the top position shown in the side elevation of the engine, or to lower them until the stud-pins to which the arms F are connected are a little above the blocks on the end of M and upon which the links E slide. With the links in the position shown the piston stroke is very nearly as great as that of the connecting rod ends, but in successive lower positions of the links E, although the amount of the connecting rod movement is unaltered, the pistons have a shorter and shorter stroke with lessening degree of compression in the cylinders. When fitted with this arrangement of links the engine thus has a variable piston stroke

Variable piston stroke

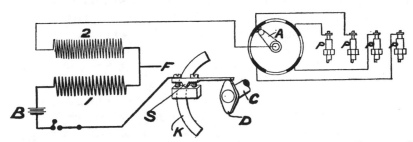

FIG. 489.—THE HOLDEN 4-CYLINDER ENGINE.—DIAGRAM OF ELECTRICAL IGNITION SYSTEM.

and the power developed may be controlled by movement of a lever rotating the shaft C through the necessary angle.

The arrangement of the inlet valves I, and the exhaust valves O, and the operation of the latter by lever arms J and cams upon the half speed shaft L is the same as in the cycle engine already described and to which readers are referred. The shaft L is driven by skew gear wheels K K[1], the latter being fixed as shown in Fig. 487, upon the centre of the crankshaft C.

Of the transverse views, shown by Fig. 486, that on the left is an end view partly in section through the valve chambers, and partly in end elevation. The centre view shows the engine in end elevation from the opposite end, and in section through the cylinder barrel at the position of the inlet and outlet water jacket bosses. The third is an end view at the joint line of the pairs of cylinders, and with one pair removed. The blocks upon the ends of M and the means provided to ensure that they and M move together are clearly shown, the links E being in section.

The electrical ignition system must be referred to as instancing the early, if not the earliest, use of a single high-speed trembler induction coil and a high tension current distributer, with a multicylinder engine.

Electrical ignition system

By reference to the circuit diagram (Fig. 489) it may be understood to what extent the ignition system is simplified without the addition of any delicate electrical parts.

When the cam D upon the crankshaft C closes the low tension circuit by making contact with the insulated spring blade S, current may flow from the battery B through the primary winding I of the induction coil, and to the engine frame at F. From the engine frame the circuit is completed through D, S, and a switch, to the negative battery terminal. The induced high tension current in the secondary winding 2 of the induction coil passes by a highly insulated lead to the centre contact of the distributor A and through the brush at the end of its rotating arm to each of the four segments shown. From the sparking plugs P, connected with the four distributer segments, the high-tension circuit is completed through the engine frame and the connexion at F common to the zero potential ends of the primary and secondary windings of the induction coil. Referring to the sectional elevation and the plan (Figs. 487 and 488) the spring contact blade X may be seen, and its cam with two opposite contact points is shown in dotted lines in the sectional elevation and in plan upon the inner end of the boss of one of the discs R. Contact is thus made twice every revolution, corresponding to a firing or working stroke in alternate engine cylinders at every stroke.

The cycle engine was not intended to run at a speed exceeding the equivalent of a road speed of 25 miles per hour, and no provision was considered necessary for advancing or retarding the moment of ignition. It will be observed too that with the larger engine the contact blade X is shown fixed in Figs. 486 to 488, but it was subsequently arranged as indicated in Fig. 489 and capable of movement about the centre of the shaft C through the requisite arc for timing the ignition. The high-tension

High tension current distributer

distributer is shown at P in the different engine views and its base plate is fixed by two bolts to a bracket on the engine frame, with its centre coincident with that of the half-speed shaft L. A separate view of the distributer is given by Fig. 490 and it consists of an ebonite box of circular form held, with its cover, by two bolts visible in the end elevation to the base plate which, as already stated, is fixed to the engine frame. The half-speed shaft carries upon its extended end E an insulated sleeve carrying in turn the brush F which makes contact as it revolves with the interior surface of the segments H, H, H, H, connected by short highly-insulated lengths of wire with the sparking plugs. The high tension current from the induction coil is taken through J to the light spring blade G and from the centre contact button to the brush F. No movement of the distributor box is necessary for timing the ignition, the segments H being wide enough to ensure that the brush F is in contact with them, when the time of ignition is advanced or retarded at the contact breaker on the engine shaft. The parts comprising the ignition system are, therefore, few and simple in form and therefore not liable to become deranged, and if faults develop they may be readily localized.

SOME ENGINES AND MISCELLANEOUS PARTS

A careful study of the illustrations of this interesting design of engine will reveal much that has not been touched upon or has been only briefly mentioned, and which at the time of its construction and when these drawings were prepared was original, and has since been usefully embodied in other engine designs.

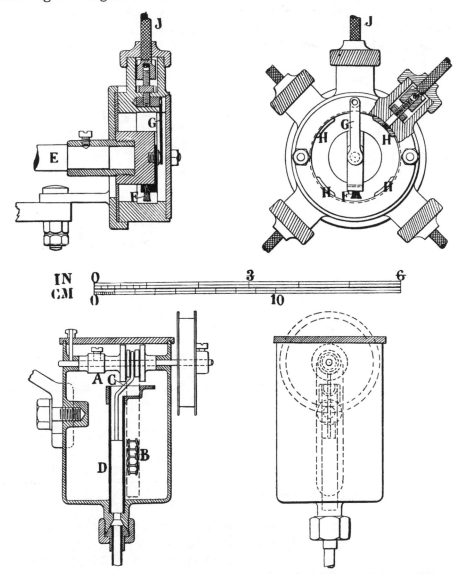

FIGS. 490 AND 491.—THE HOLDEN 4-CYLINDER ENGINE.—HIGH-TENSION CURRENT DISTRIBUTER AND AUTOMATIC LUBRICATOR.

The high-speed Aster engines made originally by the firm having that name and supplied by the Begbie Manufacturing Co., of Willesden, have been used with the cars of a number of manufacturers. That here illustrated by Figs. 492 and 493 of nominally 6 HP. was largely used for voiturette

Aster engines

purposes and an engine having the same cylinder dimensions is now constructed with various detail improvements now generally adopted. From the sectional view (Fig. 493) it will be seen that the engine is of the enclosed fly-wheel type with the timing wheels and exhaust valve cam contained in a circular box at one side of the crank chamber and the contact breaker box 30 mounted upon the extended end of the half-speed spindle 21. The clutch or driving member is fixed upon the coned end of the crankshaft 11 and the opposite reduced end 22 is extended to receive the starting handle

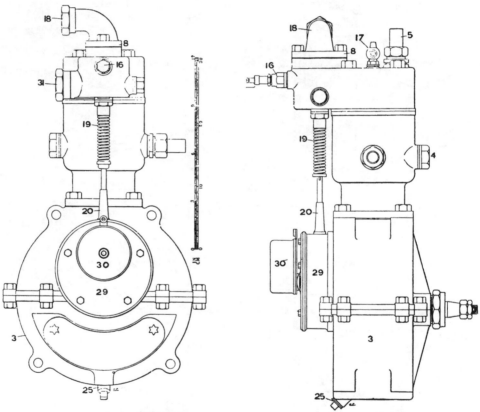

FIG. 492.—ASTER 6-HP. SINGLE-CYLINDER ENGINE, 1902-3.

and to drive the water-circulating pump when not driven by other methods. The inlet valve 14 is directly above the exhaust valve 15, and its seating is fixed in position by the loose flange 8 around the collared end of the inlet elbow 18. A peculiar feature of the design is the long curved port connecting the valve and combustion chambers, with the sparking plug at the extreme end, and in a position where it is unlikely to become sooted up as a result of excessive lubrication. Unlike most single-cylinder engines it will be observed from Fig. 492 that the crank chamber is divided about a horizontal plane at the crankshaft centre line. The use of the separate cylinder head with two water joints has been discontinued, but a large

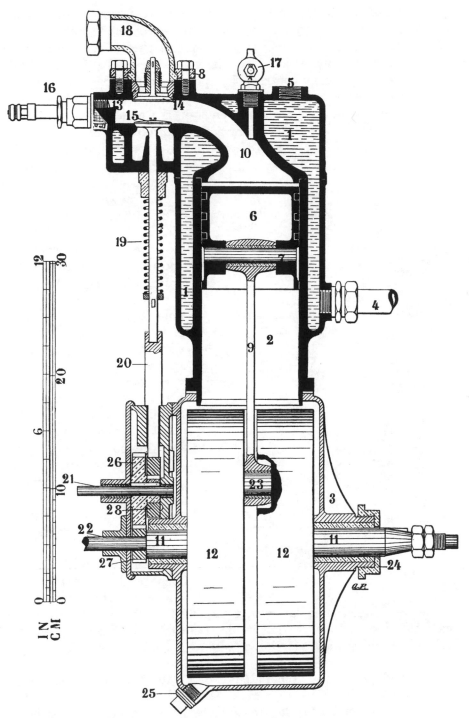

FIG. 493.—ASTER 6-HP. ENGINE, 1902–3,—TRANSVERSE SECTION.

plug hole is cast in the top of the cylinder head to facilitate the removal of the core. In recent designs of four-cylinder Aster engines of powers up to 24 HP. the cylinders are separately cast and the valves are arranged at opposite sides of the cylinder. The inlet valves are mechanically operated and a centrifugal governor controls the speed of the engine by regulation of a throttle valve in the induction pipe. The three timing-wheels, upon the end of the crankshaft, and at the ends of the camshafts, are outside the crank chamber but are protected by a casing around them. The normal engine speeds range from 1,500 revs. per minute for the smaller sizes of 4 HP. and upwards, to 1,000 revs. per minute for the larger powers of 20 to 24 HP. The 24-HP. engine has cylinders 105 mm. diameter and a stroke of 140 mm. equivalent to 4·13 and 5·5 in. respectively.

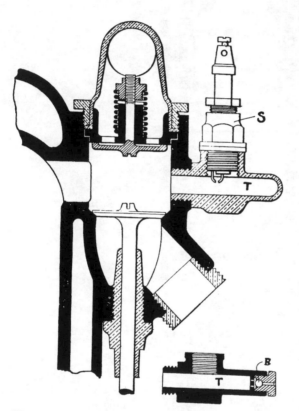

FIG. 494.—THE CHICAGO SPARKING PLUG AND TUBE.

The Chicago sparking plug (Fig. 494) is illustrative of means adopted to reduce the causes of failure of sparking plugs, at one time very common. The plug s is, as shown, screwed into a branch at right angles with the tube fitting T and the sparking points are sheltered. With the alternative form of tube separately shown a snifting valve R is fixed in the outer end through which air is drawn on the suction stroke of the engine to ensure the removal of the burnt products from the tube which on the compression stroke becomes filled or is intended to be filled with the new explosive charge. It is likely, however, that if effective whilst the engine is running, it would be troublesome at starting or when the engine is working at light loads.

Simms-Bosch magneto A longitudinal section from Patent No. 17249 of 1901 (F. R. Simms and R. Bosch) is shown by Fig. 495. It shows the continuously rotating type of Simms-Bosch low-tension magneto generator, patented subsequently to those types referred to in vol. i.

A gear wheel or other form of positive driving gear is fixed upon the coned end of the spindle H[1] and gives the rotary motion it receives to the shield F to which it is fixed. The inner end of H¹ is bored and bushed to form a bearing for the spindle I¹, and the spindle I at the opposite end of the stationary armature D passes through the hollow spindle H, and has fixed upon its outer coned end the arm K, shown separately, with the radial slot K¹ formed at a radius from the spindle axis the same as that of the collar stud J. By loosening the nut at the outer end of J the position of the armature D in relation to the magnet pole pieces may be adjusted through the range permitted by the slotted arm K. This adjustment once correctly made is permanent, and it is provided in order that the armature may be set in the most effective position, and so that when the igniter contact points are separated, the induced electromotive force or armature voltage is at a maximum and when the flow of current through the igniter and the resulting spark upon separation of the contacts is greatest. The

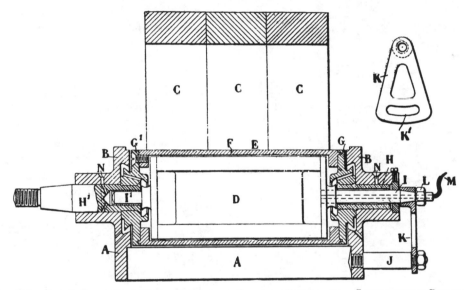

FIG. 495.—SIMMS-BOSCH LOW-TENSION MAGNETO-IGNITION GENERATOR.—LONGITUDINAL SECTION.

armature winding is not shown upon the core D, but one of its ends is earthed or directly connected to the armature core, the other M being taken through a hole at the centre of the spindle I and to the insulated contact of the igniter. Other constructional details may be gathered from the illustration, and those readers who are not familiar with the theory of action of these small dynamo-electric machines may be referred to vol. i, chapter xx. The continuously rotating type of magneto has obvious advantages for application to high-speed engines, but the reciprocating or oscillating type requires simpler driving gear and is largely used for stationary gas and oil engines.

A small magneto of the type illustrated by Fig. 495 is shown in Fig. 496 mounted upon a bracket cast on the crank chamber of a motor cycle engine. It is given a reciprocating, not a rotary, motion by a rod partly visible, from an excentric or excentric pin driven by the half-speed shaft at the other side of the engine. For varying the instant of ignition the lower end of the vertical rod which rests upon the cam at the end of the half-speed shaft is pulled in a plane normal to the axis of the camshaft in one direction for advancing and is allowed to return under spring influence for retarding ignition. The vertical rod receives a gradual upward movement, and an abrupt downward stroke when the stepped part of the ignition cam is below it, and the separation of the igniter contacts is effected quickly. At the present time that which is described as the high-tension magneto system of ignition is in frequent use. It comprises the

FIG. 496.—SIMMS-BOSCH LOW-TENSION MAGNETO APPLIED TO A MOTOR CYCLE ENGINE.

use, in combination, of an induction coil and condenser with a low-tension magneto machine. The latter may be considered the equivalent of the accumulator used with the ordinary high-tension system of ignition, and sparking plugs of the usual form are used. As arranged by some manufacturers the armature core and its winding are utilized as the core and primary winding and the secondary or high tension winding is also wound upon the armature. It remains to be seen whether the use of a high tension rotating winding in a confined space will prove satisfactory. The subject of high tension magneto ignition and several types of apparatus in use were recently very fully described in a series of articles in the *Automotor Journal*.[1]

It may be said that the high-tension accumulator and coil and the high-tension magneto systems of ignition, both require a similar and comparable number of minor parts for control of the ignition current, but

Magneto and accumulator ignition systems

whereas in the former system the minor parts as now generally made and arranged are satisfactory, in the latter system the constructional details have in some instances shown themselves to be unsatisfactory. Extended use and further experience will no doubt lead to the required improvement in this direction. Earlier in this book reference has been made to the igniters or mechanical contact breakers and their operating gear as designed by some manufacturers for use with the low-tension magneto system of

[1] "High tension magneto ignition systems." Reprinted from the *Automotor Journal*.

ignition, and having due regard to recent improvements and simplication effected in connexion with these mechanical parts there will be an extended use of this method of ignition.

Figs. 497 and 498 illustrate a design of water cooler or radiator due to

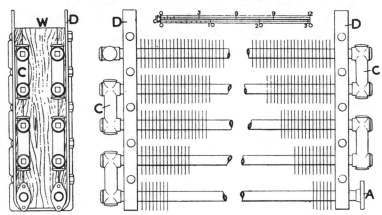

FIG. 497.—CROWDEN'S RADIATOR.—ELEVATION AND END VIEW.

Mr. C. T. Crowden of Leamington, and used successfully on the vehicles constructed by him. A series of horizontal-gilled circular tubes are arranged in an armoured wood frame w, and have their ends connected, to form a continuous length of pipe, by pieces c of the form shown in detail by Fig.

Crowden radiator detail

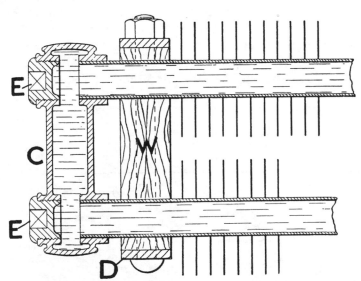

FIG. 498.—CROWDEN'S RADIATOR.—DETAIL OF PIPE CONNEXIONS.

498. Accumulated deposit, or oil in the cooling water, adhering to the tubes, may be removed by provision of the screw plugs E. The tubes may be occasionally cleaned and the efficiency of the radiator maintained without difficulty. The inlet and outlet ends of the coil are visible in the end

view (Fig. 497) at the bottom of the radiator and are indicated at A in the elevation.

Gear repairs A method adopted by Mr. Crowden for the repair of gear wheels, originally made solid with the sliding sleeve, as in the earlier Daimler vehicles, without unnecessarily rejecting the sleeve and it may be some of the wheels upon it in good condition, is here illustrated, as typical of good repair work. In Fig. 499 it is assumed that of the three gear wheels

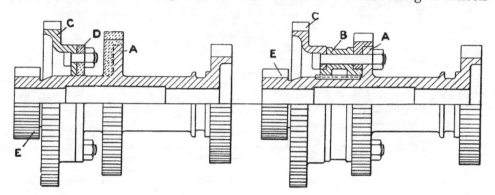

FIG. 499.—CROWDEN'S METHOD OF REPAIR OF DAMAGED GEARS.

formed solid with the sleeve that lettered A has been injured and requires renewal. The left-hand sectional view shows in dotted lines the parts that are to be removed in the lathe, and in the right-hand view the repair is shown completed. A carrier block B, shouldered for the reception of the wheel C and a new gear ring to replace the faulty one at A, is keyed in position, and through bolts hold the three parts together and to the sleeve. A similar sleeve, and wheels upon it of the same diameter, is shown in section by Fig. 500. Here it will be seen a method of renewing the four gear

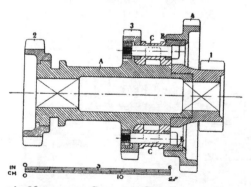

FIG. 500.—CROWDEN'S METHOD OF COMPLETE RENEWAL OF DAMAGED GEAR WHEELS.

wheels is indicated. The new first and second-speed wheels are screwed upon the remaining part of the sleeve A after alteration from the original form in the lathe, and the new third and fourth-speed wheels are held in position by means of a carrier block C, an externally threaded centre piece

B, and through bolts with sunk nuts to hold the parts together and to the sleeve as before. It will be observed that the piece B may be either the centre part of the old wheel 4 if this is in a condition to require renewal with a new toothed ring screwed and fixed upon it, or if the teeth of the wheel 4 are in good condition then the wheel may be altered to the form shown and be screwed and fixed upon a new piece B. It will be observed that some of Mr. Crowden's details are embodied by makers in the construction of their gears.

A compact and simple water-circulating pump made by the Auto Machinery Co. of Coventry is shown by Fig. 501. The spindle has upon its outer end a driving pulley T or a gear wheel, and is of greater diameter at its inner end within the pump chamber. In a slot cut in the enlarged end of the spindle are two vanes or diaphragm pieces R R¹ separated by the small springs shown and kept by them in light rubbing contact with

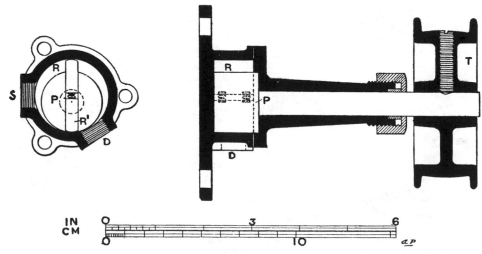

FIG. 501.—THE AUTO MACHINERY COMPANY.—TYPE OF WATER-CIRCULATING PUMP.

the interior cylindrical surface of the pump chamber, which is, as shown in the transverse section, excentric to and not coaxial with the pump spindle. With the vanes in the position shown the volume of the crescent-shaped chamber on each side of them is equal, but continued rotation in a clockwise direction will result in diminution of volume at one side with an increase in volume at the other side of the vanes, S being the suction inlet opening and D the delivery. The sliding vanes require to be closely but freely fitted in order that leakage or slip may be small with as little surface friction as possible. The pump of the size illustrated weighs 2¾ lb. and its capacity at 600 revs. is 3 gallons per minute.

The Bradley-Pidgeon combined clutch and reversing gear is shown by Fig. 502 in transverse section and sectional end elevation. The flywheel B is fixed upon the tapered end A of the engine crankshaft and is

Bradley-Pidgeon disc clutch

635

cast with a centre box provided with a bossed cover. Between this cover and the fly-wheel disc are three stout-shouldered pins on which the sets of wheels J J¹, fixed to each other, may turn. The centre wheel H is fixed to the sleeve and brake drum C and the centre wheel K, through which the driving shaft A¹ is driven, is formed solid with the latter. The reduced end of A¹ enters and is supported in the hollow end of the crankshaft A. The clutch is of the multiple disc type, one series of annular discs being threaded upon a number of carrier-pins D and driven by the fly-wheel, the second series of alternate discs or friction plates being threaded upon other carrier-pins D at a smaller radius than the first set. The second series of friction plates are centred upon the drum shown and by the carrier-pins D which pass through the centre disc of the brake drum C. The series of plates when allowed are kept in engagement by the effort from six springs

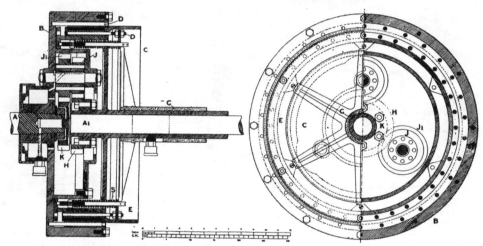

FIG. 502.—THE BRADLEY-PIDGEON CLUTCH AND REVERSING GEAR.—TRANSVERSE SECTION AND SECTIONAL ELEVATION.

S. When the clutch is to be disengaged the spring effort is opposed or balanced by end pressure, through a shifting fork and collar upon the sleeve G not shown, transmitted by the six-armed spider cast with G to the plate against which the springs thrust. The spring effort, and therefore the rate at which the clutch takes up the load, may be adjusted by screwing the spring-pins in or out of the nuts at the inner ends of the springs and by setting the spring plate closer in or further out between the lock nuts on the screwed ends of the carrier-pins D. A later form of the Bradley clutch without the reversing gear has been described in the *Automotor Journal*, March 11, 1905, p. 320.

Reversing gear   When the clutch is disengaged and the vehicle is stationary the gear wheels J J¹ are revolved by the engine fly-wheel around the stationary wheel K, the wheel H with the brake drum C and the second series of friction plates being caused to rotate idly in the same direction as B but at a lower speed, which will be equal to

636

$$R - \left(\frac{K}{J^1} \times \frac{J}{H} \right) R$$

when R =revs. per minute of the engine fly-wheel. With the proportions of wheels shown H will revolve at about three-quarters of the speed of A. If when the clutch is disengaged the band brake is applied to the drum C and H held stationary then the vehicle may be driven backwards by rotation of A^1 in an opposite direction to that of A and at a speed equal to

$$R \left(\frac{H}{J} - \frac{K}{J^1} \right)$$

The rate of rotation of A^1 will then be a little more than one-third of A.

The Bradley-Pidgeon multiple disc clutch, and the modified type recently designed by Professor Hele-Shaw,[1] and other forms employed by the author in various ways for variable speed gearing [2] in 1893, may be looked upon as adaptations of the Weston multiple-disc friction clutch patented in 1863.[3]

Clutches of the Western type

The Weston clutch has been made in a great variety of sizes, from very small to the very large used in heavy plate-rolling mills in combination with reversing gear. One of these very large Weston clutches was illustrated in *The Engineer*, vol. lxxvi. p. 259. It is also being used in the Itala car now well known as made in Turin and sold here, and in some English cars.

As used in the Itala car it is illustrated in its main features by the sectional view, Fig. 503. The clutch box A is fixed to the engine fly-wheel and is centred upon the projecting end of the engine crankshaft or upon a registering boss. The box is closed by the cover B, through the centre boss of which the driven shaft C passes and takes a bearing or is guided. The shaft C is bored out at its forward end, and slides longitudinally upon the reduced end of the crankshaft during engagement or disengagement of the clutch. The external surface of C, within the clutch box, is formed with grooves or featherways milled out to suit the clutch discs upon it, and, as shown in the section and by one of the plates, D, separately drawn to a reduced scale. The interior surface of the cylindrical part of the box A is also grooved to receive the notched plates driven by it and of the form shown at E.

Itala multiple disc clutch

The driving plates E and the driven plates D are arranged alternately and are packed and pressed together for frictional engagement by the effort of the coil spring F, which, it will be seen, abuts upon a ball-thrust bearing seated in the centre boss of the cover B. The spring effort is delivered through the closed end of the annularly recessed part of the shaft C and by the flange H to the stout outer or external plate of the series of

---

[1] *Proc. Inst. Mech. E.*, 1903, part 3, p. 483.
[2] *Variable Speed Gearing.* Patent Nos. 659, 1893, and 4204, 1894.
[3] *Coupling and Brake.* T. A. Weston's Patent No. 263, 1863.

discs.   There is a similar stout plate, at the inner end of the pack up against the distance block J, to receive the frictional rub from the adjacent disc.

A piece K, with collars to receive the end of the clutch-operating pedal lever, is keyed upon the end of the shaft C, and is formed at the outer end as part of a coupling which will transmit the power with freedom for the necessary longitudinal movement which occurs when putting the clutch into or out of action.

With a clutch of this design there are as many rubbing faces to resist abrasive action as there are plates plus one, and the surface available for sharing the work to be done through the clutch may be increased or decreased by using more or less plates.

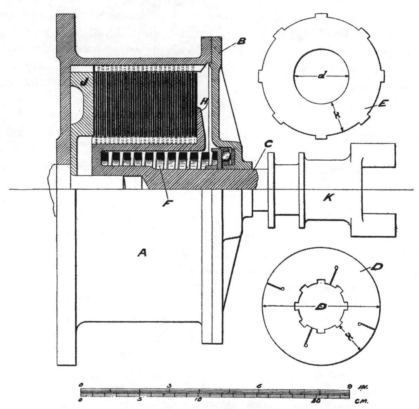

FIG. 503.—THE ITALA MULTIPLE-DISC FRICTION CLUTCH.

When, as in this adaptation of the Weston clutch principle, all the plates are allowed to pack and are not separated or spaced, then the power to be transmitted through the clutch is equally shared by the plates, and the effort required by spring or other means to obtain the total frictional face engagement, equivalent to the driving effort experienced by the clutch, is comparatively light.   An example may be given to illustrate this.

Calculation of power of clutch

Let it be taken that, as in the Itala clutch, there are fifty plates, that the mean radius of effective frictional effort is 3 inches, and that 20 HP. has to be transmitted through the clutch at 1,000 revs. per minute.

638

The effort at a radius of 3 inches is then :—

$$\frac{20 \times 33000 \times 12}{1000 \times 3 \times 2\pi} = 420 \text{ lb.}$$

The clutch box is filled with lubricant, but the plates may be assumed to be in a merely greasy condition, when by the pressure upon them the lubricant has been squeezed from between their surfaces. Under these conditions the frictional coefficient is not likely to be greater than 0·06, and the equivalent load or the combined face pressures must be equal to $420 \div 0·06 = 7000$ lb. In the example chosen there are, however, 51 rubbing faces and the required load upon the pack of discs is $7000 \div 51 = 137$ lb. The distributed pressure over the face or area of the discs is thus at no time heavy, and if the proportions of the lever arms of the clutch-operating pedal are say 3·5 to 1, then an effort on the pedal of only $137 \div 3·5 = 39$ lb. is required to relieve the load on the clutch plates. Although the frictional elements in these clutches are called discs they are in reality flat rings and the frictional surfaces brought into effective use in their employment is, as already stated, one greater than the total number of plates. For purposes of calculation of the transmitting power of these clutches, and for this purpose of the radius of mean effective frictional effort, the width of the ring must be taken as only the width of frictional contact.

For instance, in Fig. 503 the width in this connexion is only x as shown on the separate view of the plates. The radius of mean effective frictional effort between the surfaces of such discs will be $\sqrt{\dfrac{R^2 \times r^2}{2}}$ where R and r are the maximum and minimum useful radii of the plates whose diameters are as shown D and d. In a clutch such as that illustrated with diameters D and $d = 8$ inches and 4 inches respectively, the useful diameter is $\sqrt{\dfrac{64 + 16}{2}} = 6·33$ inches or the radius $\dfrac{6·33}{2} = 3·16$ inches.

An arrangement of cone clutches adapted more particularly for use with many of the American cars which have been fitted with two-speed gears, and a few of similar type in this country, is illustrated in two forms by Figs. 504. The blocks H are keyed upon the shaft A and are connected through toggle links to the centre sliding sleeve with the strap G in a groove at its centre. Through a shifting fork pivoted on G the sliding sleeve may be moved longitudinally on the shaft to engage one or other of the clutches or may be caused to take up a mid position when both clutches are disengaged. In the lower view both clutches are shown disengaged and it will be noticed that the rollers F between parallel pairs of toggle links are up against the inner faces of the male members of the clutches D and have moved them outwards against the resistance of springs E and out of engagement with the female parts of the clutch C and the bushed casings B, connected with them. It may be pointed out that since there is an equal disengaging effort required at each clutch, the number and strength of the clutch springs being similar, the forces are balanced and there is no collar friction at G or end thrust.

In the upper view the sliding sleeve has been moved to the left, disengaging the clutch on that side and permitting the right-hand clutch to engage itself by the action of the springs E. The toggle arms on the left at F, connected with the sliding sleeve, take, as shown, a vertical position,

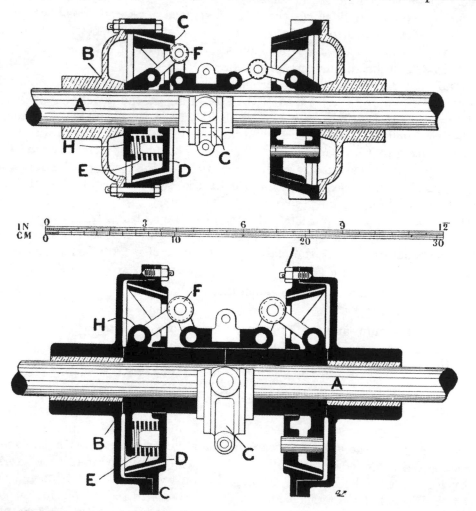

FIG. 504.—THE CHAMPION FRICTION CLUTCH.—LONGITUDINAL SECTIONS OF TWO FORMS.

and the effort of the springs E, in the direction of the axis of A, directed through D against the roller F, is powerless against the toggle action to reengage the clutch. The shaft A experiences no end thrust except during the moment of disengagement of either clutch, and except at such moments there is no collar friction at G. The shaft A may be the driving or driven part and the different ratio gear or chain wheels not shown are fixed to suit requirements upon the external parts of the clutches B.

SOME ENGINES AND MISCELLANEOUS PARTS

VARIABLE SPEED GEARS

Much as a variable speed gear is desirable for petrol-driven vehicles, none of the many forms at present made have been adopted. A few of the Hall's hydraulic transmission gears[1] are in use in 3-ton lorries with chassis made by the German Daimler Co., and a few of the frictional gears of the Nurnberger Motorfahrzeuge-Fabrik Union have been seen running, but generally it may be said that the driving gear made on the silent-feed gear principle, including the Barber[2] and the Newman,[3] have only received experimental use. The Renault-Panhard-Levassor-Mercedes type of sliding gear continues to be the favourite and is more and more recognized as a very satisfactory compromise. Other forms are constantly being designed, one of the most recently published being the Mitchell epicyclic gear,[4] which has as its main features those of my epicycloidal gear of patents Nos. 659 of 1893 and 4204 of 1894.

It does not appear probable that gear with pawls or wedge pieces operated by reciprocating parts at high speed, will give any permanently satisfactory results as driving gear, even if the resulting intermittent motion at the lower speeds be not objectionable.

**Hall
Maurer
Union**

**Newman
Barber**

---

[1] *Automotor Journal*, June 25–July 16, 1904.
[2] Ibid., May 7–21, 1903.
[3] Ibid., September 12–October 3, 1903.
[4] *The Autocar*, October 14, 1905, p. 464.

Chapter XXXIX

ON COST OF MOTOR TRANSPORT

THERE is no question more frequently asked than " What is the cost of running ? " and no question which involves so many others to be put to the questioner. Even if the question be put with regard to one type and **Conditions** one size of vehicle, the answer depends on quantitative replies from the **of use** user. He must be able to say how much he will use the vehicle, what he will carry, and what weight, what sort of country or districts he will generally run in, what speed he will run, what the hour-load factor, and how much skill, care and attention he will give and secure for it.

Of the cost of running light vehicles carrying two or four persons it may be safely said that with care and attention and moderate speeds, the cost is less than for the same mileage of horsed vehicles. The cost per year as found by many users is no guide whatever to the questioner who has in mind the cost as compared with that of keeping a horse and vehicle of the same seat or carrying capacity. The possession of a horse leaves no **Comparison** doubt or option as to the care that must be given it to keep it in working **of costs** order. The same care and solicitude given to the motor car will enable it and the user to do so many more miles per day and year that the cost must always be compared not with that of one horse but of several. No one who possesses a car will keep its mileage down to that of one horse, whatever may have been his original intention. Even if the total mileage per year be only the same as with a horse (a most unlikely case) the miles covered in one day will certainly be more than could be covered by one horse. Assuming, however, such a case, namely, that the work done by the car be kept rigidly to that of one horse, then the answer to the question is that the cost of running will be less than with a horse if the speed be moderate, and proper care taken with the car. The interest on capital will, however, be more, but the time occupied in the journeys will be less, even if the speed be always kept below the legal limit. Such cases can, however, occur only with the private user of a small pleasure car, and are very unlikely.

Bad service The cost of running light vehicles for business purposes is affected by **conditions** many adverse conditions. They have hitherto, when once put into service, been " run off their legs." They are kept at work the greatest number of hours men can be induced to drive them, kept running when they ought to be in the mechanic's hands, driven at high speeds and often by careless or incompetent drivers, and put into the hands of incompetent repairers or into the hands of repairers who are given short time for their work or only night hours.

For some businesses, such as butchers' order delivery cart work, local grocers and fishmongers and greengrocers, the time has not yet come for the profitable use of motor vehicles ; firstly, because the careful, responsible driver is not common enough, secondly because few buyers will now purchase a car geared to only about 15 miles per hour, as they should be, and thirdly, because no makers have yet made the simple car specially suited to this work.

With regard, however, to the trading vehicle for light work, but longer distance running between points of delivery, the suitable vehicles may be now or very soon obtainable, and it will pay to use them when careful drivers can be obtained and when the owner has learned that there is a limit to the mileage without repair and adjustment attention, and when the maximum speed is restricted. It will not, however, pay those to run a motor vehicle who think that because they have discarded the horse they have got rid of everything that requires care and attention. **Light commercial vehicles**

The cost of running the heavier types of vehicles may be incurred by many classes of traders, provided, again, that all proper care be taken by the owners, and, again, provided that moderate speeds be at all times observed. **Heavy commercial vehicles**

There are, however, some traders who could not use heavy vehicles profitably. The building contractor, for instance, who buys a 4-ton lorry and expects it to carry 4 tons out of gravel pits or along the founderous, unmade roads to a building in course of erection, will only court a disappointment properly deserved. Again, a purchaser of motor lorries or wagons for letting out or for ordinary occasional cartage work, varied in character and quantity, is almost sure to lose his money, even if he purchases a number of vehicles. The profitable use of these vehicles depends so much upon continuous employment, both for the vehicles and for an adequate staff and establishment, that very varied and precarious employment so reduces the paying load-hour factor that the margin of profit to pay interest, depreciation and establishment charges disappears.

Motor lorries and wagons may be profitably, and even very profitably, used by those traders who are able to do that which is necessary to the paying expenditure on any machinery, namely, provide continuous employment. Brewers, for instance, who require a number of vehicles, and who have a proper proportion of spare or standing vehicles and workshop appliances, can save considerably on the cost of horse haulage, and even more at busy times by despatch and time convenience, which often has considerable money value. The greater distance the motor can cover with a given load per day is of great importance in comparison with horse **Continuous employment** delivery and opens up fields of business to millers and others that cannot be reached by horses except at a cost that is prohibitive.

Public service vehicles

Of the many forms of motor vehicle concerning which the cost of working is a first consideration, the public service vehicle and omnibus are most likely to run under conditions which make the cost of running or the receipts and expenses comparable with those of transport by horses.

The work is regular and continuous throughout a long day, every day, and the capital invested is thus idle a proportionately short part of the year, and the establishment charges spread over a greater earning period.

The work to be done is of one kind and self-supplying, the prices at which it is done are fixed, and for cash, with no accounts per customer, except tickets and a bell-punch; and this last could be dispensed with if the books of numbered tickets, to be torn out by the conductor, as used in Geneva, were adopted.

As with every other system of transport on a large scale, there is great scope for thoroughly capable management, and without it no fleet of omnibuses will pay for any length of time. Not only must there be good traffic management, general management and technical management, but each of these departments must be controlled by men who are specially qualified for the work, and who act up to the responsibilities of their own spheres, but recognize equally the responsibilities and importance of the work of the other departments. This is as real and as essential to success as with a successful railway, and more than it has been with the great horse-drawn omnibus companies.

Badly managed omnibus companies

There is, however, reason to fear that there will be periods of disappointment with some of the motor-omnibus companies now at work. They are running, or trying to run, motor omnibuses which are each equivalent to the locomotive and the railway carriage, without heed to either a locomotive superintendent's requirements or those of the superintendent of the carriage department. They are all running the omnibuses from 20 to 40 miles per day more than they should; they all leave too little time for overhauling, adjustment, and repairs, and most of them run their omnibuses at too high a speed, considering their weight and the condition of much of the roadway they traverse. They offer as reason for some of these suicidal practices, that they cannot get their omnibuses fast enough, which is no reason at all for spoiling those that they have. The public appreciate and want the motor 'bus, but they will wait until they get them, and they do not force the managers of the existing companies to put omnibuses on the streets or on new routes when they should be in the car-house as standby for those on established routes, or in the repair sheds.

Importance of repairs and examination

The importance of timely repairs is not sufficiently recognized, nor is the imperative necessity for adequate repairing workshops, plant, and mechanics' working convenience. So long as important work has to be done on the machinery and underframe and running gear during the night hours only, often between 1 a.m. and 6 a.m., the work will not be fully and completely done. Hurried night work of the kind cannot be expected, as a regular thing, to be good, especially when the repairs and adjustments have to be made while, or in the place where, the washing is done. Men will stand any discomfort in an emergency case, but, as a regular thing, wetness, darkness, cold and hurry are inconsistent with careful and conscientiously thorough mechanical precision or completeness. Even in the workshops where new chassis are fitted and erected it is necessary, on commercial as well as other grounds, to warm the shops during from four to five months of the year; and it is even more necessary for motor omnibus and road car-houses. The washing, examination and repair sheds and shops must be arranged with a veiw to completeness and expedition of the work to be done; and without this commercial success cannot be sustained.

In chapter xxxvi. p. 573 of vol. i. I gave a number of estimates of the cost of working different types of vehicles. Experience since 1900 has confirmed these, although in some cases the running costs are now somewhat less than then estimated.

The following figures of the cost of running omnibuses when a considerable number are used under one management, are from a report[1] I prepared at the request of the Corporation of Oxford.

Cost of running motor omnibuses

COST OF WORKING MOTOR OMNIBUSES

Running Expenses—90 miles per day (per 'bus mile).

Petrol (6⅝ miles run per gallon at 6d.).	0·90d.
Oil and grease	0·28d.
Tyres	2·00d.
Driver and conductor	1·75d.
Repairs and renewals	0·55d.
Depreciation at 20 per cent. charged to 300 days	1·35d.
Washing, stabling, gas, water, sundries	0·50d.
Total running expenses	**7·33d.**

Traffic Expenses.

Licences, ticket services, superintendence, compensation, lighting, and sundries	0·45d.

General Administration.

Law, rates, taxes, office and building insurances, printing, manager, engineer, etc.	0·35d.
Insurance of 'buses against fire, accident, and third-party risks at £70 per year each	0·56d.
Total working expenses	**8·69d.**

The insurance sum here allowed is very high to allow for speculative third-party cases. This in many places could be reduced, but the cost of repairs and renewals should be raised to 0·75d. in most cases.

The average receipts of the London General and of the Road Car

Traffic receipts and expenses

[1] *The Local Government Journal*, September 28, 1905, pp. 601, 602. *The Contract Journal*, Septemper 13 and 20, 1905. *The Electrical Engineer*, September 9, 1905. *The Commercial Motor*, October 5, 12, 19, 26, 1905.

Companies' 26-passenger horse omnibus is 8·8*d*. per mile. The receipts of the City of Oxford and District Tramways Company's thirteen small tramcars and three smaller omnibuses in the half year ending December, 1904, was 10·23*d*. per mile run and the expenses 8·44*d*., the services being thirty minutes, forty-five minutes, and 1 hour, and the mean speed, including stops at turn-outs, not more than about four miles per hour. I proposed for an entirely new service, the existing old tramway and stock being admittedly of no further value, a service of eighteen motor omnibuses. These would have practically doubled the old accommodation, partly owing to the increased average speed and no stops at turn-outs, at a cost of about £20,000, including standby omnibuses and workshop and car-house improvements. The receipts are estimated at 11*d*. per mile run. This of course would have left a handsome balance on the year's mileage. The old little tramcars and omnibuses ran together 378,457 miles and carried 3,271,930 passengers. Assuming 450,000 miles by the motor omnibuses, and only 2*d*. difference between receipts and inclusive expenses, this would have left £3,750 to pay interest and redemption and a handsome profit. The omnibus would have left the City of Oxford free from the objections to tramways in such a city, and the service would in every way have met the requirements.

In London the receipts per mile run are much higher than above assumed, but this high rate of receipt cannot be expected to continue when the motor omnibus is much more general and the competition is more equal between motors and motors than it now is between motors and horses. There is still every reason to look for a long-continued satisfactory profit, and it will be the omnibus company's fault if profits fall to below a good return on the capital invested.

Horse-drawn omnibus companies' traffic figures The great horse-omnibus companies make their very satisfactory dividend with from 8·4*d*. to 8·9*d*. receipts per mile of 26-passenger omnibuses, and total expenses of 7·75*d*. to 8·35*d*. per mile, leaving an average of from about 0·40*d*. to 0·69*d*. profit per omnibus mile. This seems a very small margin, but it means from 17*s*. to 21*s*. per week profit per 'bus. During the half year ending December 30, 1904, the London General Omnibus Company ran 16,968,618 'bus miles with 1,420 omnibuses running, and carried 108,034,641 passengers. The gross receipts for the half year were £632,075 6*s*. 3*d*., of which £20,201 19*s*. 1*d*. were for advertising on omnibuses and £324 17*s*. 9*d*. for manure. Deducting the two latter sums, we have as omnibus receipts and some private lines, £611,548 9*s*. 5*d*. The average passenger receipts per 'bus per day were £2 7*s*. 1¼*d*. The total expenses in the same time were £582,811 7*s*. 8*d*., leaving £28,737 1*s*. 9*d*. as balance on the half year's working of 1,420 'buses. The average earnings from passengers per omnibus per year were thus about £40 10*s*., a profit which would, of course, be useless to a small proprietor. The revenue from advertisements must, however, be taken as part of the earnings of the omnibuses, and this raises the total earnings by the 'buses to £48,939, or nearly £69 per year.

ON THE COST OF TRANSPORT

The mean passenger receipts per omnibus mile were 8·65*d*. and the expenses 8·25*d*., leaving a balance per 'bus mile from passenger receipts of 0·42*d*. Including advertisement receipts, etc., the receipts were 8·94*d*. per mile, leaving a profit per 'bus mile of 0·69*d*. When one considers that twelve horses are kept to work an omnibus, and that while running, two horses, two men and the 'bus are employed all day for the sake of a net profit of from 2*s*. 6*d*. to 3*s*. 6*d*., it seems extraordinary that even on the large scale of working this can be sufficient to make a dividend, as it does, of about 8 per cent. It also shows why it is that owners of a few omnibuses are not so generous to their horses as the big companies.

Profit per 'bus per day

The advertisement revenue, it will be seen, is a valuable addition to the passenger profits, and is, in fact, 41 per cent. of the whole. A similar valuable aid may help the motor 'bus; but it is to be hoped that they will never be so hideously bedaubed as are most of the horse omnibuses.

For the purposes of showing the probable cost of running different types of vehicles for commercial purposes I recently drew up the series of estimates given in Table 18. The object of those for whom the estimates were made was to ascertain whether these various vehicles could be let on hire or worked under hiring contracts at prices and under arrangements which would give a commercial success. The prime cost of the vehicles given in the table were supplied to me and said to be those at which the vehicles could be obtained if ordered in the proposed numbers. They are, however, distinctly low.

Estimated cost of motor transport

The cost per ton-mile shows how very much the cost of transport depends on the load carried, on the question as to whether a return load is carried or not, and on the total mileage per year. The last three rows of figures given assume that the vehicles are working the number of days and mileage given in the first three rows of figures. From these results the commercial success of a hiring and jobbing project will be seen to depend on the continuous supply of freight and such arrangements as will secure as far as possible the least fetch and light running. Without practically continuous work for all the vehicles owned by any proposed company except those under repair or standby, and extremely good management, such a project does not promise immediate success, and the main thing, namely, continuous work, would be difficult to get. On the other hand, those whose businesses require more than one vehicle, or even one vehicle in some cases, would generally prefer to run them themselves than to pay a profit on the actual cost of maintenance, working and depreciation to a hiring-out corporation.

Referring to Table 18 it must be remarked that the heavy item of the cost of comprehensive insurance might be reduced, but experience shows that speculative third-party actions makes this doubtful. The cost of tyres may perhaps be reduced, depreciation might undoubtedly be taken in some cases as less than 20 per cent., and with care the cost given for oil, grease, wipers, etc., might be reduced to about two-thirds the amount given, but the cost of maintainance and repairs would probably have to be increased in many cases.

Cartage costs with horses

In comparing, however, the cost per load-ton mile it must be noted that a common charge by contracting carters for a one-ton, one-horse, one-man cart is about 15s. per day of ten hours. This cart will convey 1 ton about 12 miles, or it will do 12 load-ton miles, so that the cost to the hirer is about 15d. per load-ton mile. The contractor for this has, of course, to provide two horses each day to the one cart and man and a proportion of standby horse-power. His receipts are £4 10s. per week, and his costs from £3 8s. to £3 12s., made up as follows :—

	£	s.	d.
Horse keep of two horses	1	10	0
One man	1	4	0
Depreciation of horses	0	10	0
Cart depreciation and repairs	0	5	6
Accident insurance	0	0	6
Rent, etc.	0	2	0
Total	£3	12	0

As against this the vehicles of columns 4 and 5, Table, 18 promise a considerable profit, and they would earn that profit with the added value of quicker transport if the load were always provided and always to time. In the possession of a private owner, such as a miller or brewer, this proviso is generally met, and there is a sufficient margin of profit to allow for some loss of time. For comparison with the costs of working vehicles of columns 1, 2 and 3, other bases of comparison are required from horse-vehicle use at higher speeds; but it is known that in many cases these vehicles are run at a profit by trading owners in whose estimates of cost and of value returned, money value is attached to convenience and quickness.

TABLE 18.—COST OF WORKING VARIOUS TYPES OF MOTOR VEHICLES.

	2-Ton Mail Van, Solid Rubber Tyres. Petrol Engines, 4 Wheels.	½-Ton Delivery Van with Solid Tyres on 4 Wheels. Petrol Driven.	1-Ton Delivery Van with Solid Tyres on 4 Wheels. Petrol Driven.	2-Ton Delivery Van with Solid Tyres on Front Wheels and Steel Tyres on Rear Wheels. Petrol Driven.	Steam Vehicle capable of carrying 4½ tons on body and 4½ tons on trailer—9 tons total capacity. Steam Engine.	2 Seated Travellers' Car to take 2 passengers and 2 cwt. of luggage. Pneumatic Tyres. Petrol.	2nd Class 4-seated Passenger Vehicle. Pneumatic Tyres. Petrol Driven.	1st Class 5-seated Passenger Vehicle. Pneumatic Tyres. Petrol Driven.
Prime cost of vehicle, exclusive of tyres	£440	£250	£300	£400	£700	£160	£180	£300
Average miles run per day	60	30	30	35	35	50	75	80
Working days per annum	300	300	300	300	300	300	150	150
Total mileage per annum	18000	9000	9000	10,500	10,500	15,000	11,250	12,000
Prime cost of spare parts	£27	£16	£19	£25	£44	£10	£11	£19
Depreciation on spare parts. Vehicle	£4 0 0	£2 8 0	£2 17 0	£3 15 0	£6 12 0	£1 10 0	£1 13 0	£2 17 0
,, ,,	88 0 0	50 0 0	60 0 0	80 0 0	140 0 0	32 0 0	36 0 0	60 0 0
Maintenance and repairs	45 0 0	29 0 0	35 0 0	58 0 0	75 0 0	25 0 0	30 0 0	35 0 0
,, ,, to tyres	130 0 0	56 0 0	65 0 0	56 0 0	35 0 0	55 0 0	57 0 0	60 0 0
Fuel—petrol at say 7d. per gal.	65 10 0	17 10 0	26 0 0	45 0 0	—	22 0 0	16 10 0	17 0 0
Fuel—steam coke at 15s. per ton	—	—	—	—	120 0 0	—	—	—
Lubricant at 1s. 9d. gal. & grease, etc.	15 0 0	10 0 0	12 0 0	15 0 0	15 0 0	10 10 0	10 0 0	10 0 0
Water	0 5 0	0 4 0	0 4 0	0 5 0	5 0 0	0 3 0	0 4 0	0 4 0
Driver's wages	92 0 0	85 0 0	92 0 0	92 0 0	92 0 0	85 0 0	45 0 0	45 0 0
Boy's wages (steam vehicle only)	—	—	—	—	30 0 0	—	—	—
Insurance	25 0 0	25 0 0	25 0 0	25 0 0	25 0 0	25 0 0	25 0 0	25 0 0
Licences	4 2 0	2 0 0	4 2 0	4 2 0	4 17 0	2 0 0	2 0 0	2 0 0
Sundries, including rents, rates, taxes, stabling, tolls, fees *en route*, superintendence, lighting, towage for breakdowns, washing, printing, offices, etc., etc.	55 0 0	35 0 0	39 0 0	55 0 0	59 0 0	26 0 0	25 0 0	31 0 0
Total cost per annum	£523 17 0	£312 0 0	361 3 0	£434 2 0	£607 9 0	284 13 0	£248 7 0	£288 1 0
Cost per vehicle mile	*d.* 7·00	*d.* 8·34	*d.* 9·65	*d.* 9·92	*d.* 13·88	*d.* 4·55	*d.* 5·29	*d.* 5·76
Cost per load ton mile, loaded one way only and assuming uniform cost per vehicle mile	7·00	33·3	19·30	9·92	3·08	—	—	—
Cost per load ton mile, assuming cost with load 60 per cent. of whole	8·4	40·00	23·20	11·90	3·70	—	—	—

Chapter XL

MISCELLANEOUS

Motor Car Steels

THE demand by motor car manufacturers for metals of very high mechanical properties has met with a most remarkable response by Continental and British makers of high quality steels. Without the aid of these practical steel makers many every-day achievements in the motor world would be an impossibility. They and their practical metallurgists have, as is usual, preceded the so-called science and so-called scientific investigators, some of whom are now engaged in "research" laboratory experiments so as to be able to explain to the makers that they have succeeded. This "help" is very much appreciated by the technical educationist.

It has become quite common now to use for crankshafts, axles and steering arms a nickel steel which will show on test an ultimate tensile strength of 60 tons to the square inch and of the extraordinary elastic limit of 50 tons, on test pieces of 6 in. length and 0·6 in. diameter, the same steel showing elongation of from 14 to 16 per cent., and contraction of area at fracture of as much as 65 per cent. The same steel hardened will have a tensile strength of 103 tons per square inch, an elastic limit of 98 tons and elongation of 6 per cent. and contraction of area at fracture of 40 per cent. Such facts appear to those whose experience is of many years in other branches of engineering to be incredible, but the high-class motor car builder depends on their truth. Other extraordinary steels with properties required for parts which must be of small weight, transmit enormous stresses and take a glass-hard case-hardening, are also provided, and some equally remarkable, which, being of immense strength and toughness, will stand an unbelievable duration of hard wear in the form of spur and bevel gearing without hardening.

The steels chiefly used are nickel steels low in carbon, but chrome-nickel steels, high silica steels and vanadium [1] steels are also largely employed.

Use of steel of great strength

[1] See *Automobile Club Journal*, paper by J. S. Critchley, February 16, 1905, p. 125.

MOTOR VEHICLES AND MOTORS

High cost of special steels.

At the recent meeting of the Iron and Steel Institute in Sheffield, M. Léon Guillet read a valuable paper on motor-car steels, in which he gave some particulars of the steels used by French makers, and a recent visit to Sheffield works shows that quite a number of the steel-makers have given great attention to the subject and are making their own special high-duty steels. The price of some of them is very high. Messrs. Shipman & Co.'s list, for instance, gives for 5 per cent. nickel steel from 37s. to 50s. per cwt. for bars and rods from 4 in. down to $\frac{1}{4}$ inch, while a 30-per-cent. nickel steel for exhaust valves, etc., varies from 1s. 9d. to 2s. per lb., according to size. A 20-to 25-per-cent. nickel by the same makers has a very high ultimate strength and low elastic limit, a steel that will stand great wearing effort, will not easily oxydise, and one which is softened by the ordinary hardening process, varies in price from 1s. 3d. to 1s. 9d. per lb. A case-hardening steel of high strength and toughness, used by MM. De Dion et Bouton has an average price for bars of 70s. per cwt. There is no wonder, therefore, that forgings from the very best of these steels, chosen for their particular mechanical qualities according to the purpose of the part, are very expensive, though it is difficult to understand why they cost so much more if made here than abroad. Herr Fried Krupp of Essen issues by his London representative a catalogue of the steels made at Essen and forged into the various forms required in motor cars, a catalogue which is a credit to the makers and useful to the steel user. It gives not only the mechanical properties of the steels, but illustrations of the parts for which the different steels are best suited, and the guaranteed strengths of the several sorts. This makes it almost unnecessary to say anything about the means of comparing the mechanical properties of the different kinds for the purpose of selecting that one which for instance is best for a crank, or that one for a steering arm, that one which will weld easily, or that one for gear wheels or for springs.[1] The following table gives useful examples in this respect. It must be noted that the test pieces are all of small sectional area compared with dimensions of parts used in cars.

[1] For paper on the Co-efficients Te and Tr, giving an index to the structural value of steels of ascertained mechanical properties, see Paper by the Author of Modern Steel as a Structural Material, *Trans. Soc. Engineers*, 1880.

TABLE 19. HIGH GRADE MOTOR STEELS

Mark.	Dimensions of Test Piece.	Ultimate Tensile Strength.	Elastic Limit.	Elongation per cent.	Contraction of Area.	Tensile Strength Guaranteed	Elongation guaranteed on Test Piece length 2 in., diam. 0·5 in.	Uses.
	Inches.	lbs. per sq. inch.						
A.7.J.	5·91 × 0·59	78,090	44,237	25·8	58·2	72,000	25·0	Axles, etc., moderate strains.
A.12 P	3·94 × ·96 × ·17	92,740	56,896	21·0	—	—	—	Axles, etc., welding steel.
C.46.O	5·91 × 0·59	117,490	88,616	14·0	55·0	100,000	16·0	Axles, pressed frames, nave discs.
C.46.O	3·94 × 1·18 × 0·18	108,814	73,396	15·0	—	100,000	16·0	—
F.86.O	5·91 × 0·59	162,296	111,801	7·3	42·0	—	—	Gear, not hardened.
EF.60.O	5·91 × 0·59	119,908	100,564	16·0	67·0	110,000	16·0	Special nickel cranks, etc.
EF.60.O	,, ,,	232,662	221,285	6·0	40·2	,,	,,	Same hardened.
EF.36.O.	5·91 × 0··59	111,943	86,909	14·5	64·0	100,000	16·0	Cranks, steering gear, etc.
A.4J. .	5·91 × 0·59	annealed 66,853	43,952	28·5	58·0	62,000	30·0	Parts to be case hardened.
A.4J. .	,, ,,	quench'd 75,249	56,612	22·3	71·6	,,	,,	—
A.2.O.	—	57,000	—	32·0	—	—	—	Same, but softer steel for gearing, etc.
E.120.O.	5·91 × 0·59	79,654	62,017	27·4	71·6	75,000	25·0	Mild nickel, for case-hardening guaranteed Ela's limit 55 per cent.
E.112.O	5·91 × 0·59	78,090	53,909	28·3	66·0	65·000	30·0	Same, but softer.
E.112.O	,, ,,	quench'd 101,417	72,542	13·3	62·0	,,	,,	—
S.J.H. .	7·87 × 0·97 × 0·5	118,628	73,965	17·0	37·5	—	—	Springs.
S.J.H. .	,, ,, ,,	199,705	180,076	7·8	41·8	—	—	Do., hardened.

SOME OTHER CARS

There are several cars not herein referred to of which special mention **Rolls-Royce** should be made. One of these is the Rolls-Royce car in its several sizes from 10 to 30 HP., the latter with a six-cylinder engine. These cars are briefly referred to in connexion with their success in the Tourist Trophy Race. There is also the Albion 16-HP. two-cylinder car. **Albion** It is a side-chain driven car in which, among other details peculiar to itself, is a low-tension magneto, with straight bar magnets attached to a carrier fixed on the forward end of the crankshaft. To the ends of these bar magnets pole-pieces are fixed, of segmental form, and these revolve round a nearly circular horse-shoe laminated core carrying a suitable winding and fixed to a stationary part of the engine frame. One stout wire only connects this armature winding with a rigid insulated omnibus bar on the motor, with good strong electrical contact-pieces between it and the insulated stems of the igniters, the trip gear for which is also of a substantial kind.

The governor operates a throttle valve which is formed with the car-

MOTOR VEHICLES AND MOTORS

burettor and gives a constant quality mixture through a considerable range of speed. The Silver Medal of the Automobile Club was awarded for the magneto and ignition apparatus on the occasion of the 1,000 miles reliability trials in September, 1903. This car is made in various sizes as to length of chassis, form of body, etc., particularly adapted as light- or medium-load vans, or for any industrial purposes.

Star   The Star car is a side-chain driven machine with pressed frame and made in various sizes from 7 HP. and two seats upwards to the 70 HP. racing cars entered for the Gordon-Bennett race, two of which ran in the eliminating trials in the Isle of Man last May. The smaller cars are built on Panhard lines, with various modifications in detail.

Siddeley   The Siddeley cars, made by the Wolseley Co., and to a considerable extent on Peugeot lines, with Wolseley details, has been prominently before the motor world through its successful long-distance trials, including that of 4,000 miles, the first of such severe test runs carried out under the auspices of the Automobile Club. The 18-HP. car, which went through a similar trial in France in competition with M. Paul Meyan during the past summer, when the prize was equally divided, took part in the Tourist Trophy race, but owing to a stoppage ascribed to the choking of the radiator, it was withdrawn at the end of the third round.

Vauxhall   The Vauxhall three-cylinder car should also be mentioned, one of them of 12 HP. having been entered for the Tourist Trophy race but was stopped by the damage of a wheel in collision. The engine is controlled by governor and variable lift to the inlet valves. It has cylinders 3·75 diameter and 4·5 inches stroke, runs at 950 revs. per minute and its power is transmitted by side chains and change-speed gear, the latest form of which gives six speeds.

Other cars not hereinbefore mentioned or not described, but which have acquired a deservedly good reputation, are the Peugeot, Gladiator, Fiat, Martini, Rochet Schneider, Florentia, Itala, Spyker, Bollée, and Hotchkiss. Among the more recent British makers of note are the Brotherhood-Crocker, the Crossley, the Thornycroft, the Sunbeam with enclosed side chains, the Dennis with worm and worm-wheel drive, Legros and Knowles, Ryknield, and the Heatley-Gresham Engineering Co. Of small cars not yet mentioned are the useful Alldays & Onions and the Vulcan.

Big racers no longer instructive   Of the racing cars little need now be said, as there can be no doubt that the race competitions under the new Tourist Trophy rules will encourage improvements in all the directions which originally made the Gordon-Bennett Competition so valuable. The Tourist Trophy will, however, encourage development of all parts of a most useful type of car by experiment with that type. In the last two years the building of the big racers had ceased to direct inventive energy and structural skill in directions that gave results applicable generally in the construction of useful cars or vehicles for industrial purposes. In some cases, indeed, it was misleading in its results. A clutch, for instance, that would be desirable

652

in a car for moderate speeds would not meet the requirements of a racer, in which almost instantaneous acceleration after every slow for a curve or for getting away from controls, was necessary.

Spring Wheels

Spring wheels have been occupying the attention of a good many inventors during the past four years, many of whom have industriously repeated the experiments and traversed the forgotten paths of former designers. For heavy vehicles, wheels are much wanted that provide a very small amount of elastic impact absorption, preferably at or near the tread. A wood wheel provides an example of the requirements, but these, when made strong enough for driving wheels, become so massive that the range of elastic movement becomes nil. So far as the effect on the axle and connected parts of a vehicle are concerned, some improvement in this respect could be obtained by construction of an elastic centre of very small range, on some such arrangement as that used by Mr. E. N. Henwood; but for most purposes wheels with a wood block tread or tread of some very slightly compressible or deforming material appear to afford the relief required, not only to axle but to the road surface. No wheel in which there are moving parts, either formed of springs, spring spokes or spring rims, or which are controlled in their movement by springs, have yet been really successful, and everything points to the desirabliity of complete absence of moving parts in a wheel, whether for driving or mere carrying purposes. A wheel may be so made with springs employed in such a way that their material is subject to the least ambiguity of stresses; but, generally speaking, efforts in this direction have led either to increase in weight of the wheel, increase in cost, increase in the necessary attention, or all of these, and most of them have attained no more than could have been attained by better design and attachment of the vehicle springs. Almost the same may be said with regard to a number of the attempts to make a spring wheel which, with a solid rubber tyre, shall enable the user to dispense with pneumatic tyres. In almost all these cases the fact has been overlooked that the high value of the pneumatic tyre is due to its impact absorption at the point of origin of the impact; when this point of absorption is removed to the centre of the wheel, a large part of the whole of the value of a spring structure is lost, and, except for a slight modification of the severity of impact on the axle, the whole advantage sought would be equally well attained by better design of the ordinary vehicle springs and their axle attachment, at comparatively small cost.

Some spring relief desirable

Unsuccessful use of springs

Impact absorption at tread

Motor Vehicle Development a Road Question

That which I wrote in the end of vol. i. on this subject might be repeated here. There are few things so remarkable as the inertness with which carriage owners of all kinds and road users generally endure the

Bad roads and bad road making costly

remediable badness of our road surfaces, especially in and near London They suffer by discomfort and in pocket, and act as though the cause were irremovable. They complain of the rates and complain of the dust, and instead of complaining of the badness of the road, they complain of the badness of the vehicles the roads ought to be made to carry. There are some parts of the country where the roads are well made, a few where they are well maintained, and they cost in the end less than the badly maintained roads ; but almost everywhere road maintenance is either completely neglected or is attempted by means of unskilled and untaught

The " Roads Improvement Association "

labour. More attention is, however, now being drawn to the subject, and the Roads Improvement Association (of which every one in the least interested in roads, roads-maintenance rates and in vehicles of any kind should become a member) is doing a great deal of important work. Its offices are at 16, Down Street, Piccadilly, Mr. W. Rees Jeffreys being secre-

Special motor roads not wanted

tary. The annual subscription is only five shillings. Many proposals are being put forward for special motor roads and overhead motor ways, and various Utopian schemes were recently put forward in the report of the Royal Commission on the traffic of London. Special motor roads are not wanted ; the great want is improvement of the existing roads, their maintenance and the control of all the great trunk and main roads by one central

Too many highway authorities

authority instead of by the hundreds of different and disagreeing authorities in whose hands they are now unsystematically, inefficiently and expensively administered. There are in England and Wales no fewer than 1,890 separate highway authorities, while a tenth of that number would be sufficient ; and

Great trunk roads should be centrally controlled

with a central trunk road authority something like uniformity of methods, material and supervision could be obtained. Hundreds of rural and urban districts could be grouped together for the management of district and ordinary highways. Many new roads are required, which should be planned with reference to all other roads with which they communicate ; but no roads are wanted specially for motor traffic alone. The roads are wanted for traffic of all kinds equally and alike.

Chapter XLI

TOURIST CARS AND THE TOURIST TROPHY

AMONG the useful results of trials carried out by the Automobile Club those obtained by the Tourist Trophy race on September 14 last must be particularly noted. The race was run over the same course, in the Isle of Man, as that selected for the eliminating trials for the Gordon-Bennett race, and the total distance run was 208 miles, the course being completed four times. The conditions and regulations for the race included maximum and minimum weights of chassis and of body, and a limitation of the conditions of running as to speed, etc., by imposing a limit of petrol consumption fixed by the regulations as follows. "The allowance for the course selected shall be equivalent to an allowance of one gallon of petroleum spirit for every 25 miles of dry average road, the term average road signifying a course similar to the road from London to Oxford viâ Uxbridge, High Wycombe and Stokenchurch." Although not expressly stated, this regulation was intended to indicate that if the course were better than or inferior to the average of the course mentioned, then the petrol allowance would be decreased or increased accordingly. The allowance finally made was one gallon for every $22\frac{1}{2}$ miles, this allowance being based on a report made by the Author at the request of the Club, and presented on August 31 last.[1]

Conditions
of the race

One result of the race was to show that if the speed of the tourist car in these races is to be kept down to anything like a reasonable speed, the petrol allowance must be considerably decreased. The fact that about 25 miles per gallon is shown to be an average best consumption on the London and Oxford road on which the quarterly 100-mile trials have been made, shows that the consumption on a road on which there are many traffic interruptions must be considerably higher than on a racing road with no considerations other than those of running to the greatest advantage ; in other words, a course on which the engine may be kept running at its best load.

The Club regulations were intended to secure representative "tourist"

Purpose
of the
regulations

[1] *Automobile Club Journal*, October 12, 1905, p. 377. See also particulars of course and gradients in the *Automotor Journal*, September 9, 1905, p. 1117.

cars. As a tourist car could not be satisfactorily defined or similarity of cars secured by any precise regulation as to engine power or design of engine machinery and body, it was decided that the weight of the chassis should not be less than 1,300 lb., and the load to be carried, including the body, not less than 950 lb., and that the petrol allowance should be limited as above mentioned. A maximum weight of 1,600 lb. for the chassis was also fixed, but this was an instance of the tendency to excess of regulation, and will not be repeated. It will be possible next year to insist on properly made and finished bodies.

Table 20 gives the leading particulars of the eighteen cars which completed the total run[1] of 208·5 miles. Seven cars found the petrol allowance insufficient ; five had damaged or smashed wheels, mostly due to excessive speeds at corners ; three had tyre troubles, and nine failed from various causes.

Arrol-Johnston The Arrol-Johnston cars which took the first and fourth positions were fitted with engines having two cylinders, with two pistons in each, the combustion space being between the two pistons, as in the Linford gas engines of 1879. They were also fitted with hit and miss governing by the exhaust, which has been found most economical in gas engines. The pistons give motion to rocking levers, to the lower ends of which the connecting rods are attached, the big ends of the rods operating a longitudinally placed crankshaft below the cylinders.[2] Each half of the engine is thus theoretically balanced, but the effort which is the most serious cause of vibration, as mentioned in chapter xxxiv. vol. i. p. 565, remains and is aggravated in comparison with some three- and four-cylinder engines, in which the sequence of cycles is better. This question is dealt with in chapter xxiv. p. 358, from which it will be seen that the vibration which is caused by irregularity of impulse will be intermediate between that of the two-cylinder engines with sequence of impulse, as shown by diagrams 2 and 3, p. 358.

Reference to the jacket and heat-loss questions as bearing on this engine will be found in chapter xxxi. p. 505. The carburettor used is properly placed higher up, with reference to the position of the inlet valves, than in most cars, and forced lubrication by a gear-driven pump in the bottom of the crank chamber is provided.

The change-speed gear used in this car is of the Mercedes-Panhard type, driving a live axle with fixed tubular axle on which the wheels run, ball bearings being used for all the main bearings.

---

[1] The similar particulars of the cars which did not complete the course will be found in the *Automotor Journal*, Septemper 23, 1905, p. 1166.

[2] Illustrated in the *Autocar*, September 9, 1905, p. 304, September 23, 1905, p. 364.

TABLE 20. THE TOURIST TROPHY CARS AND RACE.

Place.	Declared HP.	Car.	Official No.	Weight of Chassis.	Engine.				Ignition.[3]
					Cyl's.	Diam.	Stroke.	Revs.	
				lbs.		Ins.	Ins.		
1	18	Arrol-Johnston .	53	1572	2 [1]	4·75	6·5	800	Ml
2	20	Rolls-Royce . .	22	1565	4	3·86	5·0	1000	B
3	14	Vinot-Deguingand	52	1595	4	3·54	5·11	900	BMl
4	18	Arrol-Johnston .	54	1588	2	4·75	6·5	800	Ml
5	16	Rover . . .	49	1360	4	3·74	4·32	1000	BMl
6	16	Swift . . .	25	1595	4	3·47	5·11	1100	B
7	15	Orleans . .	8	1521	4	3·62	4·32	1100	B
8	14	Argyll . . .	18	1499	4	3·31	4·32	1100	B
9	15	Orleans . .	9	1549	4	3·62	4·32	1100	B
10	18	Napier . .	2	1576	4	3·50	4·00	1200	B
11	16	Standard . .	32	1581	4	4·00	4·00	1000	B
12	16	Rover . . .	48	1471	4	3·74	4·32	1000	BMl
13	10	Peugeot . .	46	1564	2	3·94	3·94	1000	Ml
14	15	Ryknield . .	27	1562	3	4·00	4·50	1000	Mh
15	18	Napier . .	3	1573	4	3·50	4·0	1200	B
16	14	Dennis . . .	36	1600	4	3·31	4·32	950	Mh
17	20	Simms-Welbeck .	14	1387	4	3·74	4·32	1000	Ml
18	14	Dennis . .	35	1579	2	4·13	5·51	950	Mh

[1] The Arrol-Johnston engines are horizontal with 2 cylinders and 4 pistons.

TABLE 20—continued.

Place.	Declared HP.	Car.	Official No.	Transmission Gear and Axle.				Car Speeds in m.p.h. on each gear.[2]				Wheel Base.	
				Type.	Clutch	No. of Speeds	Direct Drive on	1st	2nd	3rd	4th		
												Ft.	In.
1	18	Arrol-Johnston .	53	L	Disc	4	4th	9·5	22·0	32·5	45·0	7	8
2	20	Rolls-Royce . .	22	L	Cone	4	3rd	15·0	24·5	37·0	45·5	—	
3	14	Vinot-Deguingand	52	CC	Cone	3	3rd	—	—	—	—	7	6
4	18	Arrol-Johnston .	54	L	Disc	4	4th	9·5	22·0	32·5	45·0	7	8
5	16	Rover . . .	49	L	Cm	3	3rd	12·0	24·0	36·0	—	8	0
6	16	Swift . . .	25	L	Cone	4	4th	8·0	16·0	26·0	36·0	8	3
7	15	Orleans . .	8	L	Cone	3	3rd	13·5	21·1	38·0	—	8	0
8	14	Argyll . . .	18	L	Cone	3	3rd	18·5	27·0	33·0	—	8	0
9	15	Orleans . .	9	L	Cone	3	3rd	13·5	21·0	38·0	—	8	0
10	18	Napier . .	2	CC	Cone	4		8·0	17·25	25·5	34·5	8	0
11	16	Standard . .	32	L	Disc	4	3rd	4·0	10·00	20·0	31·5	9	0
12	16	Rover . . .	48	L	Cm	3	3rd	12·0	24·0	36·0	—	8	0
13	10	Peugeot . .	46	CC	Cone	4	4th	—	—	—	—	7	6
14	15	Ryknield . .	27	L	Cone	3	3rd	10·0	25·0	36·0	—	8	0
15	18	Napier . .	3	CC	Cone	4	—	8·0	17·25	25·5	34·5	8	0
16	14	Dennis . .	36	L	Cone	3	3rd	8·0	15·00	30·0	—	7	7
17	20	Simms-Welbeck .	14	L	Cone	3	3rd	15·0	30·00	45·0	—	8	6
18	14	Dennis . . .	35	L	Cone	3	3rd	8·0	15·00	30·0	—	7	7

[2] At normal speed of engine. 3 Ml=low tension magneto. B=battery. Mh=high tension magneto. Cm=metal cone CC=two side chains. L=live axle.

TABLE 20.—*continued*.

Place.	Declared HP.	Car.	Official No.	Wheel Gauge.	Tyres. Diam. and size in inches.	Time. Total for Circuit (208·5 miles).	Average Speed for the Circuit.	Fastest Lap. Lap.	Fastest Lap. Speed.	Fuel. Surplus.	Fuel. Miles per Gallon.
				Ft. in.		h. m. s.	m. p.h.		m.p.h.	pints.	
1	18	Arrol-Johnston .	53	4 4	32 × 4	6 9 14$\frac{3}{5}$	33·9	2nd	38·1	8·3	25·4
2	20	Rolls-Royce . .	22	—	32 × 3·5	6 11 23	33·6	3rd	342	6·8	24·8
3	14	Vinot-Deguingand	52	4 7	31·5 × 3·35 } 32·0 × 3·50 }	6 14 35$\frac{2}{5}$	33·4	4th	37·9	2·6	23·3
4	18	Arrol-Johnston .	54	4 4	32 × 4	6 36 58$\frac{3}{5}$	31·7	1st	32·4	6·2	24·6
5	16	Rover . . .	49	4 0	32 × 3·5	6 43 53$\frac{3}{5}$	31·0	2nd	31·7	8·5	25·5
6	16	Swift . . .	25	4 0	34 × 3·5	7 1 12$\frac{2}{5}$	29·7	4th	30·4	11·9	26·9
7	15	Orleans . .	8	4 2	32 × 3·5	7 7 42$\frac{1}{5}$	29·2	1st	30·1	11·25	26·6
8	14	Argyll . . .	18	4 0	32 × 3·5	7 10 25	29·0	4th	29·6	6·25	24·6
9	15	Orleans . .	9	4 2	32 × 3·5	7 19 32$\frac{2}{5}$	28·5	1st	30·9	9·45	25·9
10	18	Napier . .	2	4 1·5	32 × 3·5	7 27 44$\frac{3}{5}$	28·0	1st	29·1	0	22·5
11	16	Standard . .	32	4 6	32 × 3·5	7 28 2$\frac{2}{5}$	27·9	1st	28·6	1·3	22·9
12	16	Rover . . .	48	4 0	32 × 3·5	7 41 23$\frac{3}{5}$	27·1	1st	30·8	8·7	25·5
13	10	Peugeot . .	46	4 6	32 × 3·5	7 58 54$\frac{2}{5}$	26·2	4th	26·4	10·0	26·0
14	15	Ryknield . .	27	4 0	34 × 3·5	8 11 44$\frac{1}{5}$	25·4	2nd	26·6	2·6	23·3
15	18	Napier . . .	3	4 1·5	32 × 3·5	8 17 10$\frac{2}{5}$	25·2	2nd	27·8	6·4	24·6
16	14	Dennis . . .	36	4 3	34 × 3·5	8 26 43$\frac{3}{5}$	24·7	1st	26·8	6·1	24·5
17	20	Simms-Welbeck .	14	4 5	34 × 3·5	8 49 8	23·7	1st	29·2	0	22·5
18	14	Dennis . . .	35	4 3	32 × { 3·5 { 4·0	9 5 48$\frac{4}{5}$	23·0	3rd	25·0	12·7	27·2

Petrol Consumption It will be seen from Table 20 that the petrol consumption was at the rate of 25·4 miles per gallon, while that of the similar car, which took fourth position, was 24·6 miles per gallon. On consumption, therefore, both cars were beaten by several others whose speed was somewhat less, including the Orleans, Swift, Peugeot and Dennis.

Rolls-Royce The Rolls-Royce Car, which took the second place, and was only 2 minutes 8 seconds behind the winner, was also of the tubular fixed-axle type, in which the rotating axle is only subject to torsional stress. It is a well-designed axle and combined bevel drive and differential gear. The engine is also of good design throughout and is fitted with automatic variable lift to the inlet valves controlled by the governor, which at the same time, by simple and effective means, varies the period of ignition without shifting the commutator or distributer, and there is thus no need for slack or loose electric wire connexions. It has an automatically variable air-supply carburettor. There is no striking departure from the design of other good cars of the type, but in numerous details there are improvements, and the car is well thought-out throughout.[1]

Vinot-Deguingand The Vinot-Deguingand Car, which took the third position was, it will be seen, a side-chain driven car, and its engine was fitted with the Caron magnetic igniter[2] instead of the ordinary kind of jump-spark plug. It made a non-stop run.

Among the noticeable performances of cars not in the first few are those

[1] Illustrations will be found in the *Automotor Journal* of December 3, 10 and 31, 1904.

[2] *Automotor Journal*, October 14, 1905, p. 1265.

of the two Orleans cars and the Swift, which made a very satisfactory speed, with high economy—and the Rover. The performance of the two 2-cylinder cars, the Peugeot and the Dennis, is also noteworthy, the Peugeot making an average speed of 26·2 miles per hour, remarkably regular running, and 26 miles to the gallon; the Dennis making an average of 23 miles per hour, and covering 27·2 miles per gallon.

In the Tourist Car race, as in all these trials, so much depends on the driver that it is not easy to trace any obvious relations between types or proportions of engines, revolutions, transmission, carburettors, ignition, etc.

Of the fifty-four cars entered, two were steam, and the two large Daimlers did not appear, and of the fifty others twenty-five were fitted with battery-ignition apparatus only, four with battery and high-tension magneto, two with battery and low-tension magneto, seven with low-tension magneto only, and five with high tension magneto only. The battery coil, etc., is thus very much the preferred method. **Battery Ignition most used**

On the whole, the cars with the smaller engines did best, but this cannot be said to be a positive indication. The greater proportion of the cars, however, showed both a higher speed and higher economy than was expected, and no doubt much of the latter is due to the fact that the road was absolutely clear and the engines left alone, except for gradient-regulation and corners, on the best load through the whole run. Judging from the fact that the average speed of the four most economical cars (omitting No. 35), was 28·1 miles per hour on an average consumption of one gallon per 26·2 miles, it is clear that at least 26 miles per gallon may be asked for next year and an ample speed obtained. The Dennis car, which gave **Dennis** 27·2 miles per gallon, made an average speed of 23 miles per hour, quite enough for a touring car.

There is no doubt that the Tourist Trophy Race encourages the production of a car specially suited to the course, a continuous run, very high top-speed gear for the down hills, a third speed suitable for all but the steepest hills, and an engine suited to racing purposes.

If the best objects of the Tourist Trophy are to be realized the Club will probably have to adopt an arbitrary cylinder capacity. This might be done without any serious inaccuracy. Taking 15 HP. as a representative tourist-car power, the total cylinder capacity need not be more than 135 cubic inches if 20 HP. be considered the better representative for touring cars, that is to say 9 cubic inches of cylinder capacity per actual horse-power.

Index

661

INDEX

INDEX

INDEX

664

D

INDEX

INDEX

INDEX

Honeycomb coolers, surface of, 139.
Horse-power, motor cycle, available for hill climbing, 614.
Hospitalier and Carpentier indicator, 354.
Howard steam wagon, 431.
Humber cars, 209–220.
— motor cycles, 586.
Humberette cars, 218.
Hydraulic air regulator, Napier, 250.
— steering gear, Simpson & Bibby, 404.
Hydraulic variable speed gear, Hall's, 641.
Hydraulically operated gear inadvisable, 394.

I

Iden's change speed gear, 26.
Igniter, Brooke low-tension, 157.
Ignition gear, Albion magneto, 651.
— Benz magneto, 110.
— Brooke duplex system of, 157.
— Clement-Garrard, timing of, 610.
— Clement-Talbot low-tension magneto, gear, 188.
— Daimler system, with high tension distributer, 206.
— De Dietrich system of timing magneto, 192.
— De Dion timing, 612.
— distributer, use of high-tension, 243.
— Duryea high-tension magneto, 311.
— Holden system of, 625.
— Magneto or accumulator system of, preference for, 632.
— Minerva, timing, 611.
— Mors low-tension magneto, 160.
— Napier system of, synchronized, 248.
— Oldsmobile, contact breaker, 277.
— Ormonde bicycle system of, 574.
— Richard Brazier low-tension magneto, 181.
— Simms-Bosch low-tension magneto, generator, 630.
— switch, Holden handle bar, 581.
— system, Decauville, 120.
— systems, accumulator and high-tension magneto, 632.
— timing by governor, 290.
— trip gear, Duryea, 311.
— tubes, Humber, 213.
— tubes, Panhard-Levassor, 141.
— Werner speed control by variable frequency of, 600.
— and throttle levers, interconnexion of, 208.
Impact, absorption of, at tread, 653.

Improvements and changes, 432.
Inclined countershaft, Milnes-Daimler, 439.
— steering pivots, 308.
Indicator, steam temperature, 478, 500.
— use of, for petrol motors, 620.
Indicators, float, for petrol tanks, 57, 581.
— mirror reflecting, 354.
Induction coil, use of single high-speed, and distributer, 625.
Inductive effort at carburettor jet, 533.
Industrial vehicle trials, French, 449.
Inertia forces, 354.
— forces, instance of effect of, 176.
— shocks, relief of gear from, 76, 94.
Inlet valve, variable opening of Darracq, 10.
— variable spring pressure on, 26.
Interconnected brake and throttle valve, Wolseley, 73.
Interconnexion of clutch and change-speed gear, 297.
— of clutch and brakes, 244.
Internally expanding brakes, 81.

J

Jackets, Cadillac thin copper, 281.
— Napier, electrolytically formed, 253.
James & Browne cars, 96–104.
— 9-HP. transmission gear, 99.
Jeantaud electric brougham, 566.
Jenatzy, M., and Gordon-Bennett Race, 123.
Jet, variable, opening, 528.
Joel, H. F., and electric automobiles, 538.
Joints, universal, 227, 234.
Joly, Henri L., and accumulators, 538.
Joy valve gear, 482.

K

Kerosene, use of, as fuel, 436.
King Edward VII., Daimler car for, 201.
Knocking in engines, causes of, 356.
Krebs, Commandant, carburettor, 144–146, 529.
Krupp, Fried, motor car steels, 650.

L

Lagging, importance of, 374.
Lamplough-Albany steam car, 487.

INDEX

INDEX

671

INDEX

INDEX

INDEX

INDEX

Valve throttle, hand and foot operated, 193.
— throttle, Oldsmobile carburettor and, 276.
— throttle, separate, to each cylinder, 160.
— throttle, White steam, 454.
— use of piston and slide, 399.
— variable amount and frequency of opening of, 265, 285.
Vanadium steel, 649.
Vapour, volume and weight of petrol, 521.
Vaporizer, Clarkson method of heating the, 479.
— for petrol, White, 459.
— Turner-Miesse, 498.
Variable lift to inlet valves, 137.
— speed gears, 641.
— stroke petrol engine, 622.
Vauxhall cars, 652.
"V" belt transmission for motor cycles, 571, 577.
Velocity of air flow through pipes, 533.
— of entry of air into partial vacuum, 534.
Velox cars, 234.
Vibration, principal cause of, 352.
Vinot-Deguingand car, 658.
Vivinus belt-driven cars, 29, 33.
Voiturette, belt-driven, Richard, 179.

W

Wagons and lorries, 364.
— very heavy, failure of, 375.
Wantage steam lorry, 431.
War Office heavy vehicle trials, 378.
Water and fuel supply, automatic for steam car, 499.
— circulating pump, method of driving, 119.
— cooled foot brake, 74.
— cooler, Crowden's, 633.
— cooler, Duryea jacket, 308.
— cooling system, Holden thermo-syphon, 582.

Water feed, automatic supply of, 500.
— feed, purification, difficulty of, 394.
— filters, limited use of, 485.
— jacket, Napier electrolytically deposited, 253.
— jacket, unnecessary loss of heat in, 316.
— level in boiler, constant, 475.
— required, quantity of cooling, 321.
— spaces, small, objections to, 321.
— supply to boiler, automatic control of, 475.
— supply to bottom coils of boiler, 497, 500.
— system, White, 463.
Weight, conflicting requirements as to, 348.
— of motor cycle engine, 611, 613.
— of electric motor vehicles, 537.
— of Humber cars, 219.
— of Rover cars, 267.
Werner motor cycles, 597.
Westminster Electricity Supply Co., 557.
Weston type clutches, 637.
Wheel, all steel built up, 385.
Wheels, diameter and width of, for heavy vehicles, 365.
— lateral stress on, 515.
White & Poppe engines, 255.
White steam cars, 450–473.
Williams, C. Wye, heat conduction, experiments by, 320.
Wilson & Pilcher cars, 228.
Winton carburettor, 528.
— cars, 301.
Wolseley cars, 62–88.
— heavy petrol vehicles, 446.
Wood frames, satisfactory use of, 253.
World's record, 1902, 160.
Worm and worm-wheel driving gear, electric vehicle, 545.
Wrong, mechanical, A, 187.

Y.

Yorkshire 4-ton steam wagon, 416.

676

Index to Tables

Butler & Tanner, The Selwood Printing Works, Frome, and London

Printed in the United States
By Bookmasters